MAN-MADE CATASTROPHES

REVISED EDITION

MAN-MADE CATASTROPHES

REVISED EDITION

LEE DAVIS

Checkmark Books®
An imprint of Facts On File, Inc.

MAN-MADE CATASTROPHES, REVISED EDITION

Checkmark Books
An imprint of Facts On File, Inc.
132 West 31st Street
New York NY 10001

Library of Congress Cataloging-in-Publication Data

Davis, Lee (Lee Allyn)
Man-Made catastrophes / Lee Davis.—Rev. ed.
p. cm.
Includes bibliographical references and index.
ISBN 0-8160-4418-X (hardcover)—ISBN 0-8160-4419-8 (pbk)
1. Disasters. I. Title.
D24 .D38 2002
904′.7—dc21 2001054324

Checkmark Books are available at special discounts when purchased in bulk quantities for businesses, associations, institutions, or sales promotions. Please call our Special Sales Department in New York at 212/967-8800 or 800/322-8755.

You can find Facts On File on the World Wide Web at http://www.factsonfile.com

Text design adapted by Rachel L. Berlin
Cover design by Cathy Rincon

Printed in the United States of America

VB Hermitage 10 9 8 7 6 5 4 3 2 1

This book is printed on acid-free paper.

CONTENTS

In memoriam
to the thousands whose lives were so abruptly
and savagely stolen from them
in the cataclysm of September 11, 2001,
and in celebration
of the selfless heroism
of those who sacrificed or endangered their
own lives to save the lives of others.

ACKNOWLEDGMENTS

It is the custom, as if it were an awards ceremony, to thank everybody but your dog and your least favorite relative for help in the birthing of a book. That is not going to be the practice on this page. If I were to name all of the people over the passage of three years who, while I was writing about and therefore living through some of the world's worst times, kept me from getting depressed to the point of paralysis, I would compile a cast of—well—hundreds.

So, I won't, but will, instead, express my gratitude to the major players in this drama of disasters:

Particularly to Mary Lou Barber, who forfeited half a summer to help me endlessly and immeasurably in the accumulation of a small mountain range of reference material;

To Diane Johnston, who cut through a continent of red tape at the New York Public Library and made the portion of my life spent there infinitely more productive than it otherwise would have been;

To Jean Kaleda, Edana McCaffery Cichanowicz, Joanne Brooks, Patricia S. Tormey, Robby Walden, Karen Miller, Susan Bergmann and the rest of the research staff at the Riverhead Free Library; to Shirley Van Derof, Phyllis Acard, Karen Hewlett, Susan La Vista, Jane Vail, Elva Stanley, Robert Allard, Jan Camarda and Nancy Foley of the Westhampton Free Library; to Selma Kelson and the research staff of the Patchogue-Medford Library; to the research staff of the library at the Southampton Campus of Long Island University; to the research staff of the print division of the Library of Congress;

To Elizabeth Hooks, of the American Red Cross photo library; to Reynaldo Reyes, of the United Nations photo library; to Pedro Soto, of the CARE photo library; to Michael Benson, Larry Crabil and particularly Bill McGruder of the National Transportation Safety Board; to Tim Cronen of the library of the Smithsonian Institution Air and Space Museum;

To Tom Deja for his picture research; to Fred Robertson for his patient photography of disintegrating copies of old newspapers;

To my agents, Elizabeth and Ed Knappman, for causing this to happen in the first place;

To my editor at Facts On File, during the long course of the birth of the revised edition of this book: Frank K. Darmstadt.

And to all of those unnamed friends and spiritual advisers who kept me sane for three years—my deepest and heartfelt thanks.

Lee Davis
Westhampton, N.Y.

INTRODUCTION

Stupidity.

Neglect.

Avariciousness.

The three weird sisters, the archetypal three of man-made disasters, wend their way through practically every one of the several hundred entries in this volume, often in triplicate and duplicate.

Although "human error" is the euphemism that is used in journalism to describe the reasons for most man-made disasters, it is not altogether accurate. True, the mistakes that a conductor or engineer makes in judging distances when rounding a blind curve are human errors, as are the misreadings of instruments in an industrial plant about to blow apart. Human error is present when an airline pilot, using his best judgment, miscalculates the fuel left in rapidly emptying tanks or the distance to a runway. And human error is present when a navigator of a ship, in a panic situation, steers out of the safety of deep water into the disaster of a reef.

But more often than not, other forces have made that human error easy to commit, and certain to cause a cataclysm. Human sloth and corporate greed often figure in the faulty instrument provided the engineer in the doomed plant, in the failure to provide a proper evacuation plan for a nuclear facility, in the decision of a captain who goes to bed and leaves the bridge to a midshipman in treacherous waters, in the failure of the management of a building or a discotheque to provide the proper fire exits for its patrons, in the neglect of the owners of a shipping line to provide the proper number of lifeboats or the correct filling in life jackets for its passengers.

And in man-made disasters, government often plays an ill-starring role. The cover-ups that are universally present after nuclear disasters have occurred, the misinformation before they happen and the failure to conduct proper inspections of such vital parts as the O-rings in space vehicles have all indicated government culpability.

If, then, there is any constant thread that weaves through the fabric of man-made disasters, it is the presence of those three weird sisters, Stupidity, Neglect and Avariciousness, their pervasiveness before, during and after the disasters and the uncomfortable truth that without them, some of the worst of these disasters never would have occurred.

To carry the Shakespearian analogy still further:

Another basic characteristic of man-made disaster is its inevitability, the inexorable passage of fate once a particular person or group of persons sets that fate in motion. The Shakespearian tragic hero (as distinguished from the Greek tragic hero) has a series of choices. If he makes the correct choice, he wins and faces only complications that result in comedy. If he makes the wrong choice, or a series of wrong choices, he faces his ineluctable doom. For once the first domino has been knocked over, that's it. The rest are bound to follow, with all the inevitability of a law in physics.

And that, too, is another characteristic of many man-made disasters. One error in design is committed; one fatal shortcut is tried by management; one chance too many is taken by a pilot or an engineer; one unwise challenge to fate or inevitability is made by anyone, and the rest is sadness.

At least in the realm of natural disasters, which seem to conform to the Greek theory of tragedy, one can blame it on fate itself, represented by the overwhelming presence of the overpowering forces of nature. People may admittedly be in the wrong place at the wrong time when a tidal wave or a hurricane or an earthquake strikes. But they did not *initiate* the coming of the tidal wave or hurricane or earthquake.

And in the broad spectrum of man-made disasters, it is true that natural forces *have* taken a hand. Ships have gone down in sudden storms and faceless fogs; airliners have been struck by lightning or have been the victims of sudden wind shears; small fires have been fanned into conflagrations by sudden wind gusts. Freezing temperatures certainly played a role in bringing about the *Challenger* space disaster. And it might even be argued that natural, explosive gases in coal mines were placed there, not by man, but by nature.

But except for very few instances, these presences are *secondary,* and it is what occurs *before* or *during* these emergencies that matters in man-made disasters. The judgment of the captain of a ship or an airplane, the decisions made by fire chiefs or rescue squads, the advice given by experts to engineers fighting to bring an industrial plant under control spell the difference between disasters and accidents. And once those Shakespearian dominoes have been set in motion by that act of bad judgment, ignorance, badly placed cowardice or misplaced bravado, the dividing line between trouble and cataclysm is crossed. And there is no going back.

Now, a word about degree:

In his introduction to his play *Death of a Salesman,* Arthur Miller separates the merely pathetic from the truly tragic by using the image of a man being hit by a falling piano.

The situation is this:

A piano is being moved into a fifth-floor apartment via a block and tackle. It hovers outside a window, five stories above a city sidewalk.

An unsuspecting man turns the corner, whistling. He strolls down the sidewalk, and then, just as he gets underneath the piano, a rope breaks. The piano falls, crushing the man.

The next day, an article, headed "Man Hit by Falling Piano," appears in the newspapers. It reports the facts and nothing else.

Is that, asks Miller, pathetic or tragic?

It's pathetic, according to Miller, because you don't know where the man came from or where he was going. If, on the other hand, you knew, for instance, that he had just paid the last installment on his mortgage and was on the way to the jewelry store to pick up the engagement ring to give to the love of his life, it would be tragic. Summing it up, Miller concludes, "You are in the presence of tragedy when you are in the presence of a man who has missed his joy. But the awareness of the joy, and the awareness that it has been missed must be there."

Well, there is both tragedy and patheticness in this revised and expanded edition of *Man-Made Catastrophes.* Because of its encyclopedic nature, there is neither the space nor the information to include on these pages all of the joys that have been missed by the millions who have died. But it is there, by implication, and wherever it has been possible, it has been included.

There is no greater disaster than the one that occurred to you last year, yesterday or in the last instant. And the reason for this is that you knew the joy and you know the vacuum that is left when it has been missed. So, in the catastrophes that consume this volume, there are millions of people to whom that particular disaster was far more than pathetic. It was the supreme tragedy of their existence.

And all of this would be terribly depressing, if that were all. But there is a reason that we revere our tragedies more than our comedies, why we feel, when they are over, the uplift that the Greeks termed catharsis and that Aristotle, in his *Poetics,* decreed must be present in every true tragedy. It's composed of two qualities: bravery and knowledge. Tragic heroes go to their deaths bravely and learn from their errors before they die.

So, while there is terrible, horrible, disgusting cowardice on these pages—particularly, for some odd reason, in the recital of maritime disasters (they seem to have brought out the very worst in us)—there is also noble bravery, too. Just look at the men and women who gave their lives to saving others on September 11, 2001. Time after time, there are vignettes of remarkable, indelible acts of courage that shine like stars in an otherwise dark sky of disaster: The families who went back to their cabins and donned their evening clothes before stoically going down with the *Titanic,* for instance; the heroic conductors and engineers of out-of-control trains who hurtled to their deaths with their hands on brakes that burned out beneath them; the rescuers that risked their lives to go into burning buildings or soon-to-explode mine shafts; the pilots of planes that brought their crippled birds in with minimum loss of life; the flight attendants and crews who faced down terrorists. And particularly, the hundreds of New York City firefighters and policemen who risked—and in many cases, lost—their own lives to save the victims of the terrorist holocaust that rained out of the sky on September 11, 2001, at the World Trade Center in New York City; and the small, ineffably courageous band of passengers who overpowered the suicide hijackers on American Airlines Flight 77 and, by crashing the plane short of its destination, saved hundreds, or possibly thousands, of other lives. The list is long and bright, and proof that there is a goodness and a courage in human beings that no disaster can entirely destroy.

There is also a unique quality in the comparison of categories in this volume. Since the publication of the previous edition nearly a decade ago, I have added more than 70 new incidents and 22 evocative photographs. Some categories have almost disappeared in the lexicon of extreme disasters. Lighter-than-air catastrophes, mine explosions, railroad wrecks and the sinking of transatlantic liners, for instance, are clustered in the past. Others, such as political unrest, fires, airplane crashes and industrial disasters, remain constant. And still others, such as terrorism, nuclear accidents and space disasters, grow in number, frequency and size as time unfolds. And there is the undeniable conclusion that the worst of these has yet to happen.

Bungled investigations and government agencies in competition or at cross-purposes have, in too many circumstances, delayed or derailed the prevention of terrorist threats. The first World Trade Center bombing in 1993 could have been averted, as 20–20 hindsight has revealed, if the FBI had not dismissed as unbelievable the warnings of its own undercover agent. On the other hand, the heeding of admonitions in the months following this incident brought about the thwarting of another plot to destroy a great deal of New York City.

Still, time and custom have led to further relaxation of vigilance and the entrance of other priorities, and so as this is being written, congressional committees are readying inquiries into further security lapses and possible dismissal of evidence that allowed a synchronized and sophisticated suicide mission that on September 11, 2001, killed thousands and ignited a war.

There are old favorites in this volume: the *Titanic,* the *Hindenburg,* the Chicago Fire. And there are new favorites: Bhopal, Chernobyl, the *Challenger,* Three Mile Island and of course the World Trade Center attack. And there are disasters that, for one reason or another, either have not found their way into record books or, because of lack of information or withheld information, remain incomplete stories.

Take, for instance, the worst disaster at sea ever reported. Supposedly, 6,000 Chinese Nationalist soldiers lost their lives in the sinking of a troopship near Manchuria in 1949. But there are no official records, no eyewitness reports, no historians' loggings of this incident that this writer could find after exhaustive research.

Or take the strange case of the *Wilhelm Gustloff.* Its sinking brought about the worst loss of civilian life at sea in all of history. And yet it has scarcely been mentioned in history books of its period, and finding even the few details available took considerable digging. That the *Wilhelm Gustloff* was a German hospital and troopship and that she was sunk by an unidentified Soviet submarine at the very end of World War II undoubtedly accounts for the lack of information. And yet here was a disaster with casualties that were nearly five times those of the *Titanic,* and the incident has remained buried for 45 years in some back room of history.

Finally, take the silence of the Soviet Union after the enormous explosion that shook the Ural Mountains, at a nuclear dump site near the city of Kasli, in 1957. Although the CIA and, presumably, the governments of other Western countries were aware of the explosion, no news of it leaked out until a Soviet scientist, Dr. Zhores Medvedev, emigrated to the West and published a reference to it in a scientific journal. And even then, heads of atomic energy commissions worldwide scoffed at the news. If it had not been for the determination of Medvedev to assert his newfound freedom of expression, this catastrophe might well have remained buried under an international mountain range of official denials.

That these little-known or unknown disasters exist is a final characteristic of man-made disasters. There is no necessity to cover up a natural disaster. But because of the origin of man-made disasters, there has often, unfortunately, been ample—if persuasive—reason to alter or suppress the facts, figures, origins and particularly, in the case of nuclear disasters, the implications of these catastrophes.

Still, the search for these tales with unhappy endings is worth the effort and the collection—to sometimes show human beings at their worst, true, but also to often show them at their most noble and heroic, meeting disaster with courage or simple acceptance and learning from it.

And it is for this last reason, incidentally, that, except for five cases in which helpless civilians were the victims, disasters that took place during a war were omitted. War is, in itself, humankind's very worst self-created disaster. And the fact that humankind has not yet learned that war's endless horror and universal devastation are the most eloquent argument against its recommitment is yet another reason to exclude it from a survey of disasters created by human beings. It is, using Arthur Miller's definition, the most pathetic and least tragic of human enterprises culminating in disasters, one that brings to mind John Wilkes Booth's last words, "Useless, useless, useless . . . "

It is hoped, then, that, utilizing this criteria, this volume of man-made cataclysms does contain some nobility, some surviving dignity and an ultimate sense that, even given catastrophic circumstances, some of us, as King Arthur says at the end of *Camelot,* "do shine."

A word about semantic differences that time and a world economy have brought about: On October 1, 2000, the Library of Congress abandoned the Wade-Giles system for transliteration of Chinese names and places in favor of the internationally recognized pinyin system. This meant that Mao Tse-tung in Wade-Giles became, for many American libraries, Mao Zedong; the Ch'ing dynasty became the Qing dynasty; etc.

The change brought romanization of Chinese into line with U.S. government agencies, the Board of Geographic Names, and many worldwide bodies that had made the slow and piecemeal transition from Wade-Giles to pinyin beginning in 1956 or thereabouts. So, to conform to worldwide standards, where appropriate this revised and expanded edition of *Man-Made Catastrophes* has rendered Chinese names and places in pinyin.

AIR CRASHES

THE WORST RECORDED AIR CRASHES

N.B. For air crashes caused by terrorist bombs or hijacking, see CIVIL UNREST AND TERRORISM.

* Detailed in text

Africa
- Mediterranean coast
  - * French dirigible *Dixmude* (1923)

Antarctica
- * New Zealand DC-10 (1979)

Atlantic Ocean
- French Latecoere 631 Flying Boat (1948)
- * KLM Super Constellation (1958)
- * Eastern DC7-B (1965)
- Air India Boeing 747SR (1985)

Austria
- Innsbruck
  - * British Eagle International Britannia (1964)

Bahrain
- Off Manama
  - * Gulf Air Airbus A-320 (2000)

Belgium
- Berg
  - Sabena Boeing 707 (1961)
- Bali
  - Pan Am Boeing 707 (1974)

Brazil
- Azul
  - Argentine Airlines Boeing 707 (1961)
- Rio de Janeiro
  - U.S. Navy transport; REAL DC-3 (1960)

Cameroon
- Douala
  - * Trans-African British Caledonian DC-7C (1962)

Canada
- Newfoundland
  - Gander
    - * Arrow Air DC-8 (1985)
- Nova Scotia
  - Off Peggy's Cove
    - * Swissair McDonnell Douglas MD-11 (1998)
- Ontario
  - Toronto
    - * Air Canada DC-8 (1970)
- Quebec
  - Issoudon
    - * Maritime Central Airways DC-4 (1957)
  - Montreal
    - * Trans-Canada DC-8 F (1963)

Chile
- Andes Mountains
  - * Chilean Línea Aera Nacional DC-6B (1965)

China
- Guangzhou (Canton)
  - * China Southwest Airlines Boeing B-737 (1990)
- Liutang, Guangxi
  - China Southwest Airlines Boeing 737 (1992)
- Shanghai
  - Three China Air transport planes (1946)
- Xian
  - China Northwest Airlines Tupolev TU-134M (1994)

Colombia
- Bogotá
  - * Colombian military stunt plane (1938)
  - Avianca DC-4 (1947)
- Buga, Valle del Cauca
  - * American Airlines Boeing 757 (1995)
- Medellín
  - * SAM Airlines Boeing 727 (1992)

Comoro Islands
- Moroni (See CIVIL UNREST AND TERRORISM)

Cyprus
- Nicosia
  - Swiss Gobe Britannia (1967)

Czechoslovakia
- Bratislava
  - * TABSO Bulgarian Ilyushin-18 (1966)

Dominican Republic
- Off Puerto Plata
  - * Alas de Transporte Internacional Boeing 757 (1996)
- Santo Domingo
  - Dominican DC-9 (1970)

Egypt
- Aswan
  - United Arab IL-18 (1969)
- Cairo
  - TWA Constellation (1950)
  - * Pakistan International Boeing 707 (1965)

France
- Beauvais
  - * British dirigible R-101 (1930)
- Gonesse
  - * Air France Concorde (2000)
- Grenoble
  - * Canadian Curtis-Reid Air-Tours (1950)
- Nice
  - Air France Caravelle (1968)
- Paris
  - * Air France Boeing 707 (1962)
  - Brazilian Boeing 707 (1973)
  - * Turkish Airlines Douglas DC-10 (1974)
- Pyrenees
  - British Air Ferry Ltd. DC-6 (1967)

Germany
- East Berlin
  - * Interflug Ilyushin-62 (1972)
- Edelweiler
  - Two U.S. Air Force Flying Boxcars (1955)
- Johannisthal
  - * German naval dirigible (1913)
- Munich
  - * U.S. Air Force C-131 Convair (1960)

Great Britain
- England
  - London
    - * BEA Trident 1 (1972)

Wales
Cardiff
* British Avro Tudor V (1950)

Greece
Athens
Olympia Airways DC-6B (1960)

Guam
World Airways DC6-B (1960)
* Korean Airlines Boeing 747 (1997)

India
Bombay
Alitalia DC-8 (1962)
* Air India Boeing 747 (1978)

New Delhi
* Japan Airlines DC-8 (1972)
* Saudi Arabian Airlines Boeing 747 (1996)

Indonesia
Bali
Pan Am Boeing 707 (1974)

Buah Nabar
* Garuda Indonesia Airlines Airbus A300 (1997)

Ireland
Shannon
President Airlines DC-6 (1961)

Italy
Milan
TWA Super Constellation (1959)

Palermo
* Alitalia DC-8 (1972)

Japan
Hokkaido
U.S. Air Force C-46 (1954)

Morioka
* All-Nippon Boeing 727 and Japanese Air Force F-86 (1971)

Mount Fuji
* BOAC Boeing 707 (1966)

Mount Ogura
* Japan Airlines Boeing 747R (1985)

Nagoya
* China Airlines Airbus A300 (1994)

Tokyo
* U.S. Air Force C-124 (1953)
* All-Nippon Airways Boeing 727 (1966)

Kuwait
Kuwait City
Iraqui Airways Ilusia IL76 (1990)

Libya
Ghadames
Air France Starliner (1960)

Tripoli
Libya Arab Airlines (1992)

Mexico
Guadalajara
Lockheed Constellation (1958)

Morocco
Casablanca
Sabena DC-6B (1958)
* Czechoslovak Ilyushin-18 (1961)

Imzizen
Alia Boeing 707 (1975)

Mount Mallaytine
Sabena Caravelle (1973)

Rabat
Air France Caravelle (1961)

Nepal
Katmandu
* Pakistani International Airlines Airbus A300 (1992)

Netherlands
Amsterdam (Duivendrecht)
* El Al Boeing 747 Cargo Jet (1992)

New Guinea
Biak Island
* KLM Super Constellation (1957)

Nigeria
Nigerian Airlines DC-10 (1969)
Nigerian Airlines Boeing 707 (1973)

Norway
Spitsbergen
* Vnokovo Airlines Tupolev 154M (1996)

Pacific Ocean
Flying Tiger Super Constellation (1962)

Pakistan
Karachi
* Canadian Pacific Comet Jet (1953)

Peru
Arequipa
* Faucett Airlines Boeing B-737 (1996)

Cuzco
Peruvian Electra (1970)

Jungle
Peruvian Electra (1971)

Lima
* Varig Airlines Boeing 707 (1962)

Philippines
Samal Island
* Philippines Air Boeing 737 (2000)

Puerto Rico
San Juan
* Pan Am DC-4 (1952)

Russia
Irkutsk
* Balkal Air Tupolev 154 (1994)
* Vladivostokavia Tupolev 154M (2001)

Omsk
Aeroflot Tupolev 154M (1994)

Saudi Arabia
Jidda
* Nationair DC-8 (1991)

Riyadh
* Saudi Airlines Lockheed Tristar (1980)

South Vietnam
U.S. military C-44 (1966)

Saigon
U.S. Air Force Galaxy C-58 (1975)

Spain
Barcelona
* Dan-Air British Comet (1970)

Canary Islands
Santa Cruz de Tenerife
* Spantax Airlines Convair 990-A (1972)
* KLM Boeing 747 and Pan Am Boeing 747 (1977)

Granada
* Union Transports Africain Douglas DC-6 (1964)

Ibiza
* Iberia Caravelle (1972)

Sri Lanka
Colombo
Dutch DC-8 (1974)

Switzerland
Basel
* BEA Vanguard (1973)

Zurich
Swissair Caravelle (1963)

Taiwan
Taipei
* China Airlines Airbus A300 (1998)
* Singapore Airlines Boeing 747 (2000)

Thailand
Suphan Buri
* Lauda Air Boeing 767 (1991)

Turkey
Ankara
Turkish Air Force C-47 (1963)

United States (see also Guam; Puerto Rico)
Alaska
Anchorage
* Northwest DC-7 charter (1963)

Juneau
* Alaska Airlines Boeing 727 (1971)
Arizona
Grand Canyon
* TWA Super Constellation (1956)
United DC-7 (1956)
California
El Toro
U.S. Air Force C-135 (1965)
Lake Tahoe
Paradise Airlines Constellation (1964)
Los Angeles
USAir Boeing 737 (1991)
Simi Mountains
Standard Airlines (1949)
Florida
Everglades
* Valujet McDonnell Douglas DC-9 (1996)
Hawaii
Honolulu
U.S. Navy DC-6 (1955)
Illinois
Chicago
* American DC-10 (1979)
Hinsdale
TWA Constellation (1961)
Indiana
Shelbyville
* Allegheny CD-9 and Piper Cherokee (1969)
Maryland
Elkton
* Pan Am Boeing 707 (1963)
Fort Deposit
Eastern DC-4 (1947)
Massachusetts
Boston
* Eastern Electra (1960)
* Delta DC-9 (1973)
Off Nantucket Island
* EgyptAir Boeing 767 (1999)
Michigan
South Haven
* Northwest DC-4 (1950)
New Jersey
Asbury Park
Venezuelan Airlines Constellation (1956)
Coast
* U.S. dirigible *Akron* (1933)
Elizabeth
Miami Airlines C-46 (1951)
Lakehurst
* German zeppelin *Hindenburg* (1937)
New York
Brooklyn
* United DC-8 (1960)
TWA Super Constellation (1960)
Cove Neck
* Avianca Boeing 707 (1990)
Long Island
* TWA Boeing 747 (1996)
New York City
* Army Air Corps B-25 (1945)
United DC-4 (1947)
American Boeing 707 (1962)
Eastern Boeing 727 (1975)
New York City/ Washington, D.C. (See CIVIL UNREST AND TERRORISM)
Queens
* American Lockheed Electra (1959)
* American Airlines 587 (2001)
North Carolina
Hendersonville
* Piedmont Boeing 727 and Cessna 310 (1967)
Ohio
Ava
* U.S. Army dirigible *Shenandoah* (1925)
Pennsylvania
Aliquippa
* USAir Boeing 737 (1994)
Mt. Carmel
United DC-6 (1948)
Texas
Dawson
* Braniff Lockheed Electra (1968)
Utah
Bryce Canyon
* United DC-6 (1947)
Virginia
Richmond
* Imperial Airlines Lockheed Constellation (1961)
Washington
Moses Lake
* U.S. Air Force C-124 (1952)
Washington, D.C.
Eastern DC-4 (1949)
Bolivian P-38 (1949)
Wyoming
Laramie
United DC-4 (1955)

USSR
Irkutsk
Soviet Aeroflot Tupolev-104 (1971)
Kanash
Soviet Aeroflot Tupolev-104 (1958)
Kharkov
Soviet Aeroflot Tupolev-104 (1972)
Kranaya Polyana
* Soviet Aeroflot Ilyushin-62 (1972)
Leningrad
Soviet Aeroflot Ilyushin-18 (1970)

Venezuela
La Coruba
* Venezuelan DC-8 (1969)

West Indies
Guadeloupe
* Air France Boeing 707 (1962)

Yugoslavia
Ljubljana
* Britannia Airways 102 (1966)

Zaire
Kinshasa
* Africa Air Antonov AN-32B (1996)

CHRONOLOGY

N.B.: For air crashes caused by terrorist bombs or hijacking, see CIVIL UNREST AND TERRORISM.
* Detailed in text

1913
October 17
* Johannisthal, Germany; German naval dirigible LZ-18

1923
December 21
* Mediterranean coast of Africa; French dirigible *Dixmude*

1925
September 3
* Ava, Ohio; U.S. Army dirigible *Shenandoah*

1930
October 5
* Beauvais, France; British dirigible R-101

1933
April 14
* Coast of New Jersey; U.S. dirigible *Akron*

1937
May 6
* Lakehurst, New Jersey; German zeppelin *Hindenburg*

1938
July 24
* Bogotá, Colombia; Colombian military stunt plane

1945
July 28
* New York, New York; Army Air Corps B-25

1946
December 25
Shanghai, China; 3 China Air transport planes

1947
February 15
Bogotá, Colombia; Avianca DC-4
May 29
New York, New York; United DC-4
May 30
Fort Deposit, Maryland; Eastern DC-4
October 24
* Bryce Canyon, Utah; United DC-6

1948
June 17
Mt. Carmel, Pennsylvania; United DC-6
August 1
Atlantic Ocean; French Latecoere 631 Flying Boat

1949
July 12
Simi Mountains, California; Standard Airlines
November 1
Washington, D.C.; Eastern DC-4; Bolivian P-38

1950
March 12
* Cardiff, Wales; British Avro Tudor V
August 31
Cairo, Egypt; TWA Constellation
November 13
* Grenoble, France; Canadian Curtis Reid Air-Tours

1951
December 16
Elizabeth, New Jersey; Miami Airlines C-46

1952
April 11
* San Juan, Puerto Rico; Pan Am DC-4
December 20
* Moses Lake, Washington; U.S. Air Force C-124

1953
March 3
* Karachi, Pakistan; Canadian Pacific Comet Jet
June 18
* Tokyo, Japan; U.S. Air Force C-124

1954
February 1
Hokkaido, Japan; U.S. Air Force C-46

1955
August 11
Edelweiler, Germany; 2 U.S. Air Force Flying Boxcars
October 6
Laramie, Wyoming; United DC-4

1956
June 30
* Grand Canyon, Arizona; TWA Super Constellation; United DC-7

1957
July 16
* New Guinea; KLM Super Constellation
August 11
* Quebec, Canada; Maritime Central Airways DC-4

1958
May 18
Casablanca, Morocco; Sabena DC-6B
June 2
Guadalajara, Mexico; Lockheed Constellation
August 14
* Atlantic Ocean; KLM Super Constellation
October 17
* Kanash, USSR; Soviet Aeroflot Tupolev-104

1959
February 3
* New York, New York; American Lockheed Electra
June 26
* Milan, Italy; TWA Super Constellation

1960
February 25
* Rio de Janeiro, Brazil; U.S. Navy transport; REAL DC-3
September 19
* Guam; World Airways DC6-B
October 4
* Boston, Massachusetts; Eastern Electra
December 16
* Brooklyn, New York; United DC-8; TWA Super Constellation
December 17
* Munich, Germany; U.S. Air Force C-131 Convair

1961
February 15
* Berg, Belgium; Sabena Boeing 707
May 10
* Ghadames, Libya; Air France Starliner
July 12
* Casablanca, Morocco; Czechoslovak Ilyushin-18
July 16
* Azul, Brazil; Argentine Airlines Boeing 707
September 1
* Hinsdale, Illinois; TWA Constellation
September 10
* Shannon, Ireland; President Airlines DC-6
September 12
* Rabat, Morocco; Air France Caravelle
November 8
* Richmond, Virginia; Imperial Airlines Lockheed Constellation

1962
March 1
* New York, New York; American Boeing 707

March 4
* Douala, Cameroon; Trans-African British Caledonian DC-7C

March 16
* Pacific Ocean; Flying Tiger Super Constellation

June 3
* Paris, France; Air France Boeing 707

June 22
* Guadeloupe, West Indies; Air France Boeing 707

July 7
* Bombay, India; Alitalia DC-8

November 27
* Lima, Peru; Varig Airlines Boeing 707

1963

February 4
* Ankara, Turkey; Turkish Air Force C-47

June 3
* Anchorage, Alaska; Northwest DC-7 charter

September 2
* Zurich, Switzerland; Swissaire Caravelle

November 29
* Montreal, Canada; Trans-Canada DC-8F

December 8
* Elkton, Maryland; Pan Am Boeing 707

1964

February 29
* Innsbruck, Austria; British Eagle International Britannia

March 1
* Lake Tahoe, California; Paradise Airlines Constellation

October 2
* Granada, Spain; Union Transports Africain DC-6

1965

February 6
* Andes Mountains, Chile; Chilean Linea Aera Nacionale DC-6B

February 8
* Atlantic Ocean; Eastern DC7-B

May 20
* Cairo, Egypt; Pakistan International Boeing 707

June 25
* El Toro, California; U.S. Air Force C-135

1966

February 4
* Tokyo, Japan; All Nippon Airways Boeing 727

March 5
* Mount Fuji, Japan; BOAC Boeing 707

September 1
* Ljubljana, Yugoslavia; Britannia Airways 102

November 24
* Bratislava, Czechoslovakia; TABSO Bulgarian Ilyushin-18

December 24
* South Vietnam; U.S. military C-44

1967

April 20
* Nicosia, Cyprus; Swiss Gobe Britannia

June 3
* Pyrenees, France; British Air Ferry Ltd. DC-6

July 19
* Hendersonville, North Carolina; Piedmont Boeing 727; Cessna 310

1968

May 3
* Dawson, Texas; Braniff Lockheed Electra

September 11
* Nice, France; Air France Caravelle

1969

March 16
* La Coruba, Venezuela; Venezuelan DC-8

March 20
* Aswan, Egypt; United Arab IL-18

August 9
* Shelbyville, Indiana; Allegheny DC-9; Piper Cherokee

November 20
* Nigeria; Nigerian Airlines DC-10

December 8
Athens, Greece; Olympia Airways DC-6B

1970

February 15
* Santo Domingo, Dominican Republic; Dominican DC-9

July 3
* Barcelona, Spain; Dan-Air British Comet

July 5
* Toronto, Canada; Air Canada DC-8

August 9
* Cuzco, Peru; Peruvian Electra

December 31
Leningrad, USSR; Soviet Aeroflot Ilyushin-18

1971

July 30
* Morioka, Japan; All-Nippon Boeing 727; Japanese Air Force F-86

September 4
* Juneau, Alaska; Alaska Airlines Boeing 727

December 24
* Jungle, Peru; Peruvian Electra

1972

January 7
* Ibiza, Spain; Iberia Caravelle

March 18
Kharkov, USSR; Soviet Aeroflot Tupolev-104

May 5
* Palermo, Italy; Alitalia DC-8

June 14
* New Delhi, India; Japan Airlines DC-8

June 18
* London, Great Britain; BEA Trident 1

August 14
* East Berlin, Germany; Interflug Ilyushin-62

October 14
* Kranaya Polyana, USSR; Aeroflot Ilyushin-62

December 3
* Santa Cruz de Tenerife, Canary Islands; Spantax Airlines Convair 990-A

1973

January 22
* Nigeria; Nigerian Airlines Boeing 70

April 10
* Basel, Switzerland; BEA Vanguard

July 11
* Paris, France; Brazilian Boeing 707

July 31
* Boston, Massachusetts; Delta DC-9

December 23
* Mt. Mallaytine, Morocco; Sabena Caravelle

1974
March 3
* Paris, France; Turkish Airlines Douglas DC-10
April 27
* Bali, Indonesia; Pan Am Boeing 707
December 4
* Colombo, Sri Lanka; Dutch DC-8

1975
April 4
* Saigon, South Vietnam; U.S. Air Force Galaxy C-58
June 24
* New York, New York; Eastern Boeing 727
August 3
* Imzizen, Morocco; Alia Boeing 707

1977
March 27
* Santa Cruz de Tenerife, Canary Islands; KLM Boeing 747; Pan Am Boeing 747

1979
May 25
* Chicago, Illinois; American DC-10
November 8
* Antarctica; New Zealand DC-10

1980
August 19
* Riyadh, Saudi Arabia; Saudi Airlines Lockheed Tristar

1985
June 23
* Atlantic Ocean; Air India Boeing 747SR
August 12
* Mount Ogura, Japan; Japan Airlines Boeing 747SR
December 12
* Gander, Newfoundland, Canada; Arrow Air DC-8

1990
January 25
* Cove Neck, New York; Avianca Boeing 707

1990
October 2
* Guanghzhou, China; China Southwest Airlines Boeing 737
October 2
* Kuwait City, Kuwait; Iraqi Airways Ilusia IL76

1991
May 16
* Suphan Buri, Thailand; Lauda Air Boeing 767
July 11
* Jidda, Saudi Arabia; Nationair DC-8
December 1
* Los Angeles, California; USAir Boeing 737

1992
May 19
* Medellín, Colombia; SAM Airlines Boeing 727
September 28
* Katmandu, Nepal; Pakistani International Airlines Airbus A300
October 4
* Amsterdam, Netherlands; El Al Boeing 747 Cargo Jet
November 24
* Liutang, Guangxi, China; China Southwest Airlines Boeing 737
December 22
Tripoli, Libya; Libya Arab Airlines

1994
January 3
* Irkutsk, Russia; Balkal Air Tupolev 154
April 26
* Nagoya, Japan; China Airlines Airbus A300
June 6
* Xian, China; China Northwest Airlines Tupolev TU-154M
September 8
* Aliquippa, Pennsylvania; USAir Boeing 737
October 11
* Omsk, Russia; Aeroflot Tupolev 154M

1995
December 20
* Buga, Valle del Cauca, Colombia; American Airlines Boeing 757

1996
January 8
* Kinshasa, Zaire; Africa Air Antonov AN-32B
February 6
* Off Puerto Plata, Dominican Republic; Alas de Transporte Internacional Boeing 757
February 29
* Arequipa, Peru Faucett Airlines Boeing 737
May 11
* Everglades, Florida; Valujet McDonnell Douglas DC-9
July 17
* Off Long Island, New York; TWA Boeing 747
August 29
* Spitsbergen, Norway; Vnokovo Airlines Tupolev 154M
November 12
* New Delhi, India; Saudi Arabian Airlines Boeing 747

1997
August 6
* Agana, Guam; Korean Airlines Boeing 747
September 26
* Buah Nabar, Indonesia; Garuda Indonesia Airlines Airbus A300

1998
February 16
* Taipei, Taiwan; China Airlines Airbus A300
September 2
* Off Peggy's Cove, Nova Scotia; Swissair McDonnell Douglas MD-11

1999
October 31
* Off Nantucket Island, Massachusetts; EgyptAir Boeing 767

2000
July 25
* Gonesse, France; Air France Concorde
August 23
* Off Manama, Bahrain; Gulf Air Airbus A-320
October 31
* Taipei, Taiwan; Singapore Airlines Boeing 747

2001
July 4
* Irkutsk, Russia; Vladivostokavia Tupolev 154M
November 12
* Queens, New York; American Airlines 587

AIR CRASHES

Commercial air travel has come a long way from the days of the old Ford Tri-Star. In 1928, a trip on TWA (Transcontinental and Western Airlines then) was a true adventure. According to engineer Arthur Raymond, quoted in Paul Eddy, Elaine Potter and Bruce Page's book *Destination Disaster,* "They gave us cotton wool to stuff in our ears, the 'Tin Goose' was so noisy. The thing vibrated so much it shook the eyeglasses right off your nose. In order to talk to the guy across the aisle you had to shout at the top of your lungs. The higher we went, to get over the mountains, the colder it got in the cabin. My feet nearly froze . . . When the plane landed on a puddle-splotched runway, a spray of mud, sucked in by the cabin air vents, splattered everybody."

Improvements were clearly necessary; Raymond and his cohorts at the Douglas Company set about "building comfort and putting wings on it," as Raymond phrased it. By 1933, the commercial airliner as we know it today had been built, at least in an elementary outline. Its configuration—a pilot's cabin up front, engines encased in cowlings on the wings, a row of portholes through which passengers peered (apprehensively or otherwise) at the clouds, the sky and the arrival or departure of the ground—was established early in the DC-1 (D for Douglas, C for Commercial, 1 for first).

The craft was a model of noisy luxury. Each of its upholstered seats had a reading lamp and a footrest. The galley at the rear was equipped with electric hot plates and a lavatory. The cabin was heated and contained a ventilation system designed to let in more air than noise, and the "stewardess" was, before her transformation into "flight attendant," more than a waitress and psychiatrist. She was also a registered nurse.

Fortunately, very few passengers can recall the days when airline travel was a luxury or an adventure and were treated as such by both the airlines and their passengers. For a long time, seats were large and comfortable and wide apart. Amenities multiplied as the number of airlines blossomed. Free food and copious drinks became the norm. The extra experiments included cocktail lounges with piano bars and, most intriguing, observation lounges in the tail section in imitation of the observation cars at the rear of the crack trains that still carried most American travelers across and throughout the country.

But competition between multiple airlines also brought about a reduction in fares, and as the fares dropped, so did many of the amenities. Airliners, which had begun as one-class travel, evolved into three-class travel, and while the cost and amenities in first class remained fairly solid, those in business and particularly tourist class became fluid. Gradually, the food declined from fine to bearable to inedible, then disappeared almost entirely. The days and nights of free drinks became history, as did the piano bars and observation lounges. And as for the comfort of the seating, only children under four and a half feet tall or severely undernourished adults traveling in tourist class can, as this is being written, honestly say that they are traveling in comfort. In an effort to offer competitive fares, airlines internationally have packed more and more seats into areas limited only by the dimensions of the airplane.

Yet, even with all of these inconveniences, air travel is the travel of choice for just about everyone making long-distance trips and for an increasing number of people who are traveling only short distances. Air travel has become as customary to most of the population as catching a crosstown bus, and for good reason. Competition has made air travel affordable and ubiquitous.

And for all the loss in creature comfort, the safety record of airlines is exemplary.

Considering the burgeoning number of flights, the diverse distances and destinations and the number of passengers carried each day in absolute safety, it is impossible not to say that flying is a safe method of transportation.

And yet, large segments of the public are still deathly afraid of flying. Some refuse to fly. Some fly with their hearts in their throats and their fingers digging into armrests or their companion's arms. Others board flights with grim resignation. And their resignation has a basis: Once aboard an airplane, the passenger is, admittedly, helpless. There's no stopping the plane in midflight to debark, no way of seeing ahead or behind, no escape hatch in the sky. Strapped into his or her seat, the airline passenger is at the total mercy of the crew in the cockpit and the fates surrounding the plane.

And, unfortunately, those crews have not always been skillful, nor has fate been kind. As in any man-made disaster, human failure has played a feature role. And there are a huge number of people who can fail in the chain of command behind the orderly flying of a commercial airliner. There are the engineers who design the craft, the mechanics who service it, the flight personnel who fly it and the controllers who guide it from airport to airport. There are the groups who make the rules regulating commercial flying, and the politicians behind the regulatory bodies.

Who, for instance, can argue that for a while, at least, the skies became dramatically more dangerous when, in 1981, President Reagan fired most of the country's experienced air controllers? And as the skies become more and more crowded, as they inevitably will, air control, pilot training and the technology to prevent turning the more than 2,000 near misses that occur in the sky yearly into fatal collisions will become more and more of an issue and a burning necessity, overseen by government agencies and government policy-makers.

Having said all of this, the one human failure that no amount of technology or skill can cure is that of bad judgment under pressure. As highly trained, as experienced and as cool under grueling circumstances as airline pilots are today, they are nevertheless human and can make mistakes. And, in the case of commercial aviation, as the following section will show, these mistakes can—and often do—result in terrible tragedy.

There are other factors besides human failures: Faulty design and the metal fatigue that comes from long use have caused crashes. Weather has been a constant hazard. And terrorism, whether individual or state sponsored, has picked the airlines as its prime target.

So, the fear of flying, which is really a fear of crashing, has its realistic roots. The statistics are grim, and although air crashes occur infrequently now, they receive enormous media coverage because of the sheer magnitude of the loss of life.

New technology—transponders, three-dimensional radar, wind-shear protection, etc.—is being developed in the interest of safer flying. Efforts to build better, safer airports will certainly diminish the number of crashes, for the statistics reveal that most troubles with aircraft and consequent crashes occur during landing or takeoff. And that has held true, whether the aircraft is a lighter-than-air blimp or a jetliner.

Therefore, improvements in safety have concentrated on takeoff and landing procedures, and in the agonizingly slow rebuilding of the air control system to the level of safety it possessed before it was decimated in 1981.

There is no doubt that, overall, air travel is safe. And in a grim and possibly inverted way, the events of September 11, 2001, have made it safer. Though air travel, with its endless lines and time-consuming security checks, may now be less convenient, that inconvenience stops once the passenger has boarded the plane. The long lines created by increased surveillance and baggage checks at the airport have been designed to provide, among many advantages, greater peace of mind for the traveling public, and the traveling public has accepted it.

In addition, this new emphasis on surveillance has created a greater sense of security for pilots and flight attendants. Federal armed marshalls travel aboard many long-distance flights, and their presence, as of this writing, is increasing. Vigilance has escalated not only on the part of airline personnel but on the part of passengers, too. Possibly taking heart from the courage of the passengers on United Airlines flight 93 on September 11, 2001, who subdued three hijackers and saved untold lives when they forced their flight to crash into a Pennsylvania field instead of a target in Washington, D.C., a flight attendant and another group of passengers aboard American Airlines flight 63 from Paris to Miami, on December 22, 2001, thwarted a possible tragedy by subduing accused terrorist Richard Reid, who was attempting to ignite explosives in his shoes.

Airplane crashes—tragic and terrible ones—will still occur, but at least one cause of them will, for the foreseeable future, be severely curtailed.

The criteria for inclusion of crashes in this section centered upon the number of fatalities and the severity of the crash. But statistics alone were not the constant arbiters, and for a logical reason: A 1990 casualty figure could not realistically be applied to the 1930s, when planes were smaller and fewer people were flying. Thus the casualty figures were matched with the time in which they occurred. Other, lesser crashes were included if they were of a particularly unusual nature or the only crash of a particular type.

Airplane disasters involving terrorism can be found in the section CIVIL UNREST AND TERRORISM.

GLOSSARY

Airline lingo, technical terms and the alphabet soup of regulatory agencies have become part of the national vocabulary. Thus, the following glossary explains terms used in this section.

ADF Automatic Direction Finder
ALPA Airline Pilots Association
Alpha-Numeric System A method of radar identification of individual aircraft by letter and numbers
Altimeter A device displaying the height of a plane from the ground or sea level
AOPA Aircraft Owners and Pilots Association
ARTC Air Route Traffic Control
ARTS Automatic Radar Traffic Control System
ASDE Air Surface Detection Radar
ATC Air Traffic Control
Attitude The horizontal direction of an aircraft
AWLS All Weather Landing System
CAB Civil Aeronautics Board
CAS Collision Avoidance System
FAA Federal Aviation Administration
Flaps A series of hinged wing extensions whose angle can be changed to provide greater lift for landings and takeoffs
Handoff The transfer of control of an in-flight aircraft from one radar center to another
HICAT High Altitude Clear Air Turbulence

IFR Instrument Flight Rules
LOCAT Low Altitude Clear Air Turbulence
Mayday The international SOS, from the French *m'aidez*
Overrun The failure of an aircraft to halt its forward progress before reaching the end of the runway
Propjet An aircraft whose motion is caused by a propeller turned by a jet engine; also known as turboprop
Rotate The change of an aircraft's attitude by bringing the nose up, as in takeoff or landing
SST Supersonic transport
Transponder A device that can be actuated to make a specific plane more readily identified on a controller's radarscope
Turbojet An aircraft driven by a jet engine using a turbine-driven air compressor
Undershoot A landing attempt in which the aircraft hits the ground before reaching the runway
VASI Visual Approach Slope Instruments
VFR Visual Flight Rules
VTOL Vertical Takeoff and Landing

AFRICA
MEDITERRANEAN COAST
December 21, 1923

The state-of-the-art airship Dixmude, *one of France's proudest lighter-than-air ships, exploded when it was struck by lightning over the Mediterranean coast of Africa, on December 21, 1923. All 52 crew members aboard died.*

The golden age of dirigibles, from World War I until the *Hindenburg* disaster (see p. 56), was characterized by opulence, experimentation and isolated horrors. When an early dirigible caught fire, there was no hope for survivors, since the hydrogen that kept these liners of the sky aloft before 1923 was enormously combustible, turning metal white hot in moments.

There was a certain sad irony about the loss of the French dirigible *Dixmude* on December 21, 1923, over the Mediterranean coast of Africa: It was the French who had built the first successful power-driven airship, *La France,* in 1884. And in 1923, the very year the *Dixmude* apparently exploded when its hydrogen-filled skin was struck by lightning, noncombustible helium was successfully introduced in the United States-built *Shenandoah.*

The *Dixmude* had actually been constructed by the Germans as a fighting ship during World War I. It was a state-of-the-art airship, fast, sleek and marvelously outfitted. Turned over to the victorious French in 1920 and christened the *Dixmude,* it went about establishing records. On August 10 of that year, it made the trip from Maubeuge to Cuers in a record 24 hours and 25 minutes. It established a world endurance record when it flew from France to Tunisia in 118 hours and 41 minutes on September 25, 1923.

Three months later, on December 21, the *Dixmude,* now a survey craft for the French Navy, set out for Algeria from Cuers-Pierrefeu, on a course that was designed to locate freshwater sources in the Sahara. Fifty miles south of Biskra, the ship's commander, Lieutenant de Vaisseau du Plessis de Grenadan, was informed that he was headed for heavy weather. His reply to the radioed report was a change of course.

But apparently the storm system was all-encompassing, and the *Dixmude* soon found itself in the midst of a raging, towering thunderstorm. An hour and a half later, the airship succumbed to the gale and began to fall toward the Mediterranean. A distraught Commander de Grenadan radioed that he was making an emergency landing in the water.

This was the last message received from the *Dixmude.* Planes and ships sent to retrieve the ship and its survivors in the area from which the commander had radioed found no traces of it, nor did those who scoured the Sahara nearby.

Villagers along the coast reported seeing a huge fireball in the sky at about the time of the last radio message. Experts concluded that the ship had been struck by lightning and had exploded, flinging fragments of itself and its crew to the elements.

Of the 52 crew members aboard, only the body of Commander de Grenadan was recovered—by two Sicilian fisherman, hauling in their nets two days later.

ANTARCTICA
November 28, 1979

An Air New Zealand DC-10 on a sightseeing flight, hampered by low visibility, unpredictable wind currents and a possibly malfunctioning navigation system, slammed into the side of the 12,400-foot-high volcano Mount Erebus on November 28, 1979. Two hundred fifty-seven died in the crash.

Air New Zealand began a series of sightseeing flights over Antarctica in 1978. Popular with tourists, they were looked on askance by representatives of the National Science Foundation, the coordinators of an extensive scientific program in Antarctica. The planes disrupted both wildlife and the atmosphere, as they dipped and circled over the scientific encampments at the end of the earth.

On November 28, 1979, the fourth flight of the season took off from Auckland, New Zealand, at 8:21 A.M. It was a nonstop flight that was scheduled to circle over the South Pole and then land at Christchurch at about 5 P.M. There was a low cloud cover near Mount Erebus, a 12,400-foot active volcano located on Ross Island, off the Antarctica coast about 30 miles north of the U.S. military and scientific station at McCurdo Sound. This was a disappointment; a circle of the volcano was always a high point of the tour.

The temperature in the area was approximately 15 degrees Fahrenheit, and, at that time of year, there was daylight almost around the clock.

At 1 P.M., the pilot of the DC-10, Captain Tim Collins, radioed Auckland that he was descending from his 10,000-foot altitude to 2,000 feet, apparently in an attempt to pierce the cloud cover and give the sightseers a closer, unimpeded view of the volcano.

It was the last anyone would hear from the Air New Zealand flight. It slammed into the side of the mountain about 1,500 feet from its base, exploded and caught fire. No one would survive.

It would be seven hours before rescuers in a U.S. Navy Hercules C-130 would sight the wreckage, pinpointing it for ground rescue crew to arrive by sled. That night, blinding snowstorms roared into the area, keeping the rescue teams from getting to the site and helicopters from landing at it. It would be two days before the search could be resumed and a helicopter pad built.

Only 90 of the 257 bodies would ever be found. The rest, rescuers reasoned, had slipped into crevices or were buried too deeply under the snow to be found.

The "black box" flight recorder was retrieved, and its story was one of instantaneous disaster with almost no warning. Seconds before impact, an alarm sound reverberated in the cockpit from the ground proximity warning system, telling the crew that the plane had descended too low.

Captain Collins was an experienced pilot. A later inquiry determined that there must have been a malfunction in the navigation system of the airplane. This malfunction, coupled with the low visibility and compounded by the hazardous mountain wind currents, was thought to be the cause of the crash.

ATLANTIC OCEAN
August 14, 1958

KLM Flight 607-E, a Super Constellation, bound from Brussels to New York exploded over the North Atlantic 130 miles off the Irish coast at 11:35 P.M. on August 14, 1958. There were no survivors.

At 11:05 P.M., on August 14, 1958, at the height of the tourist season, KLM flight 607-E left Shannon, Ireland, bound for New York, with 99 passengers and crew aboard. The weather was calm in Ireland; there were no obstacles to a safe and smooth flight to the United States.

A half hour later, the pilot radioed his position to Shannon; he was 130 miles west of the Irish coast, and all was proceeding normally.

It would be the last word to be uttered from Flight 607-E. Somewhere west of the last report, the flight disappeared from radar screens and radio frequencies. Rescue ships were dispatched, and by daylight the next day, bodies and pieces of wreckage were fished from the Atlantic. There had apparently been a midair explosion that had blown apart the airliner without warning. This was before the era of terrorist bombings, so only two theories—weather or a mechanical malfunction—were given credence for the origin of the explosion. Neither would ever be proven conclusively. Whether it was from lightning in a sudden electrical storm (not an uncommon occurrence over the North Atlantic) or mechanical failure no one was able to ascertain from the wreckage.

And there were no survivors.

ATLANTIC OCEAN
February 8, 1965

Sabotage was suspected in the crash, shortly after takeoff from Kennedy Airport, of Eastern Airlines flight 663, a DC7-B, off the coast of Long Island. All 84 aboard died.

Eastern Airlines flight 663 to Richmond, Virginia, took off in normal fashion from New York's Kennedy Airport at 6:20 P.M. on February 8, 1965, looped out over the ocean and banked over Jones Beach, Long Island, preparatory to gaining cruising altitude.

Captain Stephen Marshall, the pilot of a flight coming in from Puerto Rico, later testified that he saw the aircraft take "an exceptionally deep turn," hang in the air for a moment and then plunge into the Atlantic Ocean, where it exploded upon impact.

At that same moment, Kennedy radar lost the flight from its screens and radioed an emergency alert. A seaman stationed at the top of a watchtower at Short Beach Station, Long Island, reported seeing a "red ball of fire about ten feet high above the water" and hearing something that sounded like a small firecracker exploding. Other witnesses in nearby Lido Beach reported seeing and hearing the same thing, which led investigators later to suspect sabotage as the cause of the crash.

Coast Guard ships discovered the oil slick that identified the crash site eight miles south of Jones Beach, Long Island. There were no survivors. All 84 people aboard perished.

AUSTRIA
INNSBRUCK
February 29, 1964

A British Eagle International Airlines Bristol Britannia failed to make an instrument landing at Innsbruck Airport on the night of February 29, 1964, and smashed into Mount Glungezer. All 88 aboard were killed.

There are scary descents into certain airports in the world. Several Caribbean islands are noted for "white knuckle" landing sites, including the St. Thomas airport in the U.S. Virgin Islands. But at least these airports are situated in areas in which the weather is usually clear and placid.

Not so the airport at Innsbruck, Austria, in the heart of the Austrian Alps and the midst of some of the best ski country on earth. Ringed by 8,000-foot-high mountains, it lies in a bowl that requires a steep and rapid descent from 10,000 feet. Few unseasoned travelers who have ever landed there forget the experience or wish to repeat it.

The night of February 29, 1964, was foggy and still. British Eagle International Airlines, Britain's largest independent airline in 1964, flew regularly from London to Innsbruck, and one of its Bristol Britannias, piloted by Captain E. Williams, attempted an instrument landing in the midst of the fog. The plane never made it to the runway.

Some 12 miles from the airport, flying 100 feet below the necessary altitude to clear the surrounding peaks, the plane smashed into Mount Glungezer, cracked apart and slid into a gorge. All 88 people aboard were killed.

BAHRAIN
OFF MANAMA
August 23, 2000

On a flight from Cairo to Manama, Gulf Air flight 072, an Airbus A-320, approached Bahrain International Airport in Manama, overshot the runway and, as it circled to try again, suddenly fell into the Persian Gulf, killing all 143 aboard.

A Gulf Air Airbus A-320, on a flight from Cairo to Manama, Bahrain, was obviously going too fast and too high on its first approach to Bahrain International Airport in Manama. And so, realizing that he was going to overshoot the runway and advised so by the tower, the pilot aborted his first attempt at landing.

The controller gave him clearance to make a 180 degree turn and to climb to 2,500 feet. The pilot acknowledged the directions, increased his air speed and began to climb.

But at 1,000 feet the plane suddenly changed direction and headed directly for the Persian Gulf. In a moment, it had plowed into the shallow waters of the gulf at 310 miles per hour.

Ahmed Hassan, an eyewitness, told the BBC that the jet veered to avoid buildings before plunging into the sea. "It U-turned and tried to land, then in 15 seconds it went sharply down, like an arrow, into the sea and there was a huge fire," he said.

Later examination revealed that the pilot—at first described by the *Gulf Daily News* as a devout Muslim and a dedicated flier who had worked his way up through the ranks of Gulf Air over 21 years but who was later discovered to have just been appointed a captain—did not navigate any avoidance maneuvers but seemed to be having difficulty controlling the aircraft.

Rescuers and onlookers immediately flocked to the scene, which was only one mile from the airport. That same day, mangled refuse, the two black boxes and the bodies of some of the victims of the crash were recovered and lined up on the shore. None aboard had survived. All 143 had died in the crash.

Two days later, the Bahrain prime minister, Sheikh Khalifa bin Salman Al Khalifa, and a cordon of government officials attended a service of funeral prayers for the plane crash victims in Manama's Grand Mosque. Three bodies, wrapped in cloth, were laid before the 2,000-plus worshippers in a symbolic tribute to the 107 adults and 36 children who lost their lives. One body was the size of a small child.

Within days, a representative from the plane's maker, Airbus Industrie, arrived on the scene accompanied by French accident investigators. A day later, NTSB investigators from Washington landed.

A month of analysis revealed that there were absolutely no mechanical or electronic problems with the aircraft. Thus, the conclusion was that the crew, who did not signal any sort of emergency, acted incompetently from the very beginning of the first, aborted approach.

CAMEROON
DOUALA
March 4, 1962

There has never been an explanation for the crash on takeoff of a Trans-African British Caledonian DC-7C at Douala Airport, in Cameroon, on March 4, 1962. All 111 aboard died in the crash.

The mystery of the sudden crash, on takeoff, of a Caledonian DC-7C from Douala Airport in Cameroon on March 4, 1962, has never been solved. A charter owned by the Trans-African Coach Company and loaded with 111 persons, it had originated in Mozambique and had made an eventless stop at Lisbon. Its destination was Luxembourg.

All seemed to be in order as the DC-7C left the runway on the last leg of its journey. But a mere two minutes after it became airborne, the plane plunged to the ground, bursting into flames and killing all 111 aboard. No incendiary device was discovered in the wreckage, no distress call was radioed to the airport and no mechanical failure was found in the smoldering wreckage of the plane.

CANADA
NEWFOUNDLAND
GANDER
December 12, 1985

A combination of human error and malfeasance on the part of Arrow Air Charter combined to send an Arrow Air DC-8 plunging to earth after takeoff from Newfoundland's

Gander Airport on December 12, 1985. All 258 American servicemen aboard died.

In 1982, a multinational peacekeeping force was deployed in the Sinai Peninsula after Egypt made peace with Israel. With troops from 11 countries, this 2,500-member force was given the job of patrolling borders and generally supervising the security arrangements of the peace treaty, which had been signed in 1979.

In 1985, the American contingent consisted of men in the Third Battalion, 502d Infantry of the 101st Airborne Division. By December of that year, they had put in their requisite six months of duty, and by December 11, one-third of the battalion had been rotated to its home base at Fort Campbell, Kentucky. At Cairo Airport that day, the second shipment of men boarded a DC-8, chartered by the military from Arrow Airlines of Miami. They were on their way home for Christmas.

By the time they left Cologne, Germany, their next stop, there were 250 soldiers and eight crew members aboard the jet. They flew overnight to Newfoundland, their last refueling stop before proceeding on to Kentucky, where Welcome Home banners, a brass band and eager relatives would be waiting to greet them.

The DC-8 touched down at Gander at 4:08 A.M. on December 12, 1985. Six minutes later, it was "on the blocks," refueling. The weather was miserable but not unusual: There was a light drizzle formed from fog, snow and freezing rain. The temperature was 25 degrees Fahrenheit. Visibility was 12 miles, with broken clouds at 1,200 feet.

It was a quick refueling, and there was no attempt to apply de-icing material to the plane's wings.

At 5:10 A.M. the plane left the maintenance area, and at 5:15 it took off, with 258 people and 101,000 pounds of fuel. It had scarcely cleared the runway when it veered suddenly to the right, shuddered for a moment and then plunged into a spruce and birch forest, skidding through it, leveling trees and then exploding in a huge burst of sound and fire.

"I saw this explosion," Bob Cole, a Gander truck driver, later told reporters. "I never saw nothing to match it in my life."

"It was just like a sunset," added Lucy Parsons, an airport worker.

Within eight minutes, volunteer fire fighters, Royal Canadian Mounties and military personnel from the Gander base had reached the scene. The explosion had ignited brushfires over a wide area. "There were bits of fuselage everywhere, no big pieces," Keith Head, a Gander fireman, said. Charred bodies were strewn over the landscape. There were no survivors. All 258 were killed in the gigantic explosion caused by a combination of the impact of the crash and the full load of fuel.

There were many unanswered questions, and boards of inquiry were convened in both Canada and the United States. Two pieces of information emerged soon after the crash: The right outboard engine was in a reverse mode, and this could have accounted for the veering to the right of the plane just before it went down. But investigators were quick to point out that the impact of the crash might have thrown the controls into this mode; that had happened in other crashes.

Second, it was disclosed that the crew had underestimated the weight of passengers and fuel by at least six tons.

The last disclosure led Congress to investigate Arrow Air, and the disclosures that were aired in this investigation were frightening. Former pilots for the line came forth and told Congress that the airline often pushed pilots until they would fall asleep in the cockpit. And it performed absolutely minimal maintenance on its airplanes.

Further examination of the craft that crashed revealed other disquieting information, stories of a pocketknife having to be used to open the forward cargo door, and windows that were taped shut. The DC-8 had had mechanical difficulties earlier that year and had been forced to abort two takeoffs in the past six months.

In the wake of all of this, Arrow Air declared bankruptcy and went into Chapter 11 on January 21, 1986.

The final conclusion regarding the cause of the crash settled on the right outboard engine. It was delivering less power than the other three engines at the moment of impact. Whether its reversal on takeoff came from an inattentive crew or yet another mechanical malfunction of that particular DC-8 no one will ever know. That information died with the crew and passengers of the Arrow Air DC-8.

CANADA
NOVA SCOTIA
OFF PEGGY'S COVE
September 2, 1998

While it was attempting an emergency landing at Halifax, Swissair flight 111, a McDonnell Douglas MD-11 en route from New York's Kennedy Airport to Geneva, plunged into the Atlantic Ocean on the night of September 2, 1998, seven miles off the coast of Nova Scotia. All 229 aboard the jet were killed.

Swissair has been the choice of transatlantic passengers for years because of its exemplary safety record. The paragon of Swiss efficiency, the airline had had only one accident since its founding in 1931: on October 7, 1979, when a DC-8 overshot a runway in Athens, Greece, while attempting to land. In that crash, 14 people were killed.

But on the cold, rainy, windy night of September 2, 1998, that record would be deeply modified. Swissair Flight 111, the nightly nonstop from New York's JFK Airport to Geneva, left, routinely, at 8:19 P.M. An hour and three-quarters later, it was out over the Atlantic Ocean in Canadian airspace, cruising at 33,000 feet, roughly 320 miles east-northeast of Boston.

The pilot, Urs Zimmerman, radioed air controllers at Moncton, New Brunswick, wishing them good evening and giving his altitude. The Moncton controller replied and informed him that there were "reports of occasional light turbulence at all levels"—an aftermath of Hurricane Bonnie.

Sixteen minutes later, the conversation became less pleasant:

> 10:14:18 P.M. Atlantic Time: Pilot: Swissair one-eleven . . . Pan Pan Pan [a term for an urgent message, short of a distress call] We have, uh, smoke in the cockpit. Uh, request immediate return, uh, to a convenient place, I guess, uh, Boston.
>
> 10:14:33: Tower: Swissair one-eleven, roger . . . turn right proceed . . . uh . . . you say to Boston you want to go?
>
> 10:14:34: Pilot: I guess Boston . . . we need first the weather so, uh, we start a right turn here.
>
> 10:14:50: Tower: Swissair one-eleven, roger, and a descent to flight level three-one-zero [31,000 feet] is that OK?
>
> 10:15:08: Tower: Uh, would you prefer to go into Halifax?
>
> 10:15:11: Pilot: Uh, standby.
>
> 110:15:38: Pilot: Affirmative for Swissair one-eleven . . . We prefer Halifax from our position.
>
> 10:15:43: Tower: Swissair one-eleven, roger. Proceed direct to Halifax. Descend now to flight level two-niner-zero [29,000 feet].

The pilot announced that they were donning oxygen masks in the cockpit, and the Halifax tower took over and directed them to the radio frequency for its runway. Then came a chilling communication from Halifax:

> 10:21:23 P.M.: Tower: Swissair one-eleven, when you have time could I have the number of souls on board and your fuel onboard please for emergency services.
>
> 10:21:30: Pilot: Roger. At the time, uh, fuel on board is, uh, two-three-zero-tons [this was the gross weight of the plane plus its fuel].
>
> 10:22:04: Tower: Swissair one-eleven, uh roger, uh turn to the ah, left, heading of, ah two-zero-zero degrees and advise time when you are ready to dump. It will be about 10 miles before you are off the coast. You are still within about 25 miles of the airport.
>
> 10:22:20: Roger, we are turning left and, ah, in that case, we're descending at the time only to ten thousand feet to dump the fuel.
>
> 10:22:29: Tower: OK, maintain one-zero-thousand. I'll advise you when you are over the water and it will be very shortly.

At this point, the pilot lapses into German to ask the co-pilot if he is checking the emergency list to deal with smoke by turning the air flow on the air conditioner to maximum. If that does not help, the procedure is to turn off one air conditioning pack and run the other, then to reverse the procedure. Suddenly this was interrupted.

> 10:24:45 P.M.: Pilot: Swissair one-eleven . . . is declaring emergency.
>
> 10:24:56: Eleven . . . we starting dump now, we have to land immediate.

This was the last communication from the jet, and at 10:30, Swissair 111 plunged, with a force that registered on instruments that detect earthquakes, into the Atlantic Ocean seven miles off Peggy's Cove, a small, picturesque fishing village known for its wild and magnificent sea views.

"I heard an aircraft," Mieke Martin, the proprietor of the Century House Bed and Breakfast in Blandford, the nearest large town, told reporters afterward. "I was lying in bed with my husband when the hotel began to rumble. [The aircraft] was very low and it had irregular sounds coming from it. It didn't sound right, like a helicopter sound, or like it was tearing the sky apart. Seconds later it crashed. There was no explosion. It hit like a thud. It went into the water. I started to think, 'Oh my God. Those people.' I knew it went in. I whispered a prayer, I think."

Dwellers in Peggy's Cove, nearer the crash site, felt that there was an explosion. "We knew it was an explosion," said Darrell Frallick. "There was a silence for two seconds after the explosion, then my ears popped."

"The motors were still going [when the plane flew over], but it was the worst-sounding deep groan that I've ever heard," said Claudia Zinck-Gilroy, who witnessed the crash from Peggy's Cove.

Immediately, dozens of fishing boats and coast guard ships headed out in driving rain to the crash site. It took them nearly an hour to find it, and when they did, they discovered debris spread over six square miles. The smell of fuel was overwhelming, but it was not on fire, suggesting that there had not been an explosion.

Searchlights from coast guard cutters, fishing boats, helicopters and planes illuminated the area for most of the night in drenching rain, which eased at dawn.

"It's real ugly," said Craig Sanford, the operator of a whale-watching boat that was one of the first vessels on the scene. "You see Styrofoam floating, chunks of wood, panels, the odd body here and there. It's not a nice scene."

Debris was scattered as far as Clam Island and some of the other islets between Peggy's Cove and Blandford. Fuel oil had been dumped into St. Margaret's Bay, inland and north of Blandford.

By morning, 36 bodies had been found, some of them wearing life vests, which indicated that they had been planning for an emergency landing. Dozens of ambulances were rushed to the scene, but their cargo would be only dead bodies. No one survived the crash; all 229 aboard flight 111 died.

By late morning the next day, 1,500 search and rescue personnel were on the scene. A makeshift morgue was set up aboard the Canadian supply ship *Preserver*, which had decks for helicopters and a crew of 250.

As the day progressed, the search grid was widened to more than 10 square miles, and as more bodies were recovered some were flown to a large morgue set up at Shearwater, a Canadian Forces base at Dartmouth, across Halifax harbor from the provincial capital.

No large pieces of wreckage were found. The largest was roughly the size of the roof of a car, and none of the pieces showed any fire damage. It was nearly impossible to identify the bodies.

As the two black boxes were recovered, it became evident that a series of vital operating systems on the plane had been gradually breaking down as the crew fought to remain airborne. The unavoidable conclusion from the flight data recorder was that the pilots were receiving false information from the instruments and computers just before the jet slammed into the Atlantic. One of these sensors, a sensing relay, had malfunctioned before the jet left Kennedy, but it had been replaced—first, improperly, which caused a short circuit, then properly, which allowed it to pass tests given it before takeoff. Furthermore, both black boxes ceased to operate at exactly the same time—six minutes before the crash—which indicated that the jet suffered a devastating electrical breakdown. Parts of the cockpit showing signs of heat damage were recovered, too, indicating a frightening situation there.

As the days passed, a U.S. Navy salvage ship, the *Grapple,* anchored above the main area of the debris. A remotely operated submarine began mapping the wreckage. Divers continued to descend the 180 feet to the wreckage to extract the last of the bodies before *Grapple* began its work.

As the investigation continued, questions arose. In its modernization, the McDonnell Douglas MD-11, a reliable airplane for many years, had replaced a navigator with a computer. If another human being had been in the cockpit, could adjustments have been made that would have prevented the crash? And what of the decisions of the pilot, first to head to Boston, then to circle out over water to drop excess fuel? The plane had crashed with a mere seven to 10 minutes flying time to Halifax airport. Could he have made it if he had not decided to deviate for Boston or open water?

The answers would never be known.

CANADA
ONTARIO
TORONTO
July 5, 1970

Incorrectly deployed wing spoilers caused the crash of an Air Canada DC-8 on landing at Toronto Airport on July 5, 1970. All 108 aboard were killed.

The last moments of pilot Peter Hamilton and co-pilot Donald Rowland, commanding an Air Canada DC-8 over Toronto Airport on the morning of July 5, 1970, were heard and recorded by ground controllers. The succession of events was terrifying, and yet these two went about their work with a calmness that was astonishing. Rowland even apologized at one point to Hamilton for prematurely engaging a braking device.

The flight was a routine one for the veteran team, who had flown together many times, though not always peacefully. Captain Hamilton was a strong believer in challenging his airline's approved procedures for arming jet spoilers (wing slats to brake jets after they land) early, as high as 2,000 feet. Rowland believed in following company orders, and they had reportedly gotten into a heated argument in Vienna earlier that year over Hamilton's ignoring of the rule. As a result, a lopsided compromise was reached between the two: Hamilton agreed to arm the spoilers at 60 feet.

The danger in even arming these spoilers is one of mistakes under stress. If the spoiler is merely armed, it is set to automatically engage when the landing gear touches the runway. To do this, the pilot or copilot lifts a lever.

If the spoiler is deployed, it "spoils" the airflow over the top of the wing and reduces lift, thus braking the plane. To deploy the spoiler, the pilot or copilot pulls the same lever that arms the spoiler.

The cockpit recorder, recovered after the crash, told the tale: Rowland, the copilot, inadvertently *deployed* the spoilers just before touchdown. "No-no-no," the captain shouted.

"Sorry—Oh! Sorry, Pete," answered Rowland, and simultaneously the right wing hit the ground.

The outboard engine fell off. The plane, suddenly lighter by 4,000 pounds, was catapulted back into the air.

Captain Hamilton regained control and announced to the control tower that he would circle once again and attempt to land. But before he could complete the circle, the inboard engine ripped away from the plane, taking part of the right wing with it.

The plane began to break up. Hamilton and Rowland fought to regain control, adjusting, trying to stabilize the wildly pitching, disintegrating plane. Finally, the forces of balance and gravity won out, and the plane heeled over and dove to earth, breaking apart as it fell and crashing into a populated area alongside the airport. All 108 aboard were killed.

CANADA
QUEBEC
ISSOUDUN
August 11, 1957

An overloaded Maritime Central Charter DC-4, piloted by a man with a record of intentionally causing accidents, was flung to the ground in a fierce thunderstorm over Issoudun, near Quebec, on the morning of August 11, 1957. There was one survivor—a baby. Seventy-seven passengers and crew died.

Bizarre as it may seem, Norman Ramsay, the pilot of the ill-fated Maritime Central Charter DC-4 that plunged to earth near Quebec on the morning of August 11, 1957, had been discharged three years previously by Trans-Canada Airlines for intentionally plunging his aircraft into the ground. Since that time he had been under psychiatric care.

That incident had involved a Super Constellation flight from Tampa, Florida, to Toronto on December 17, 1954. Coming in for a landing 400 feet below the minimum required altitude, he had simply flown the plane into the turf near the runway. Miraculously, none of the 44 passengers or crew was seriously injured.

Three years later, after his psychiatrist had pronounced him unfit to fly (all of this was revealed in the later board of inquiry findings), Captain Ramsay was piloting an over-loaded charter, packed with veterans of the Canadian contingent of the British Expeditionary Force in France, who had gone back to view the battlefields on which they had fought in World War I. The DC-4 had been designed to carry 49 passengers. On this trip, from London to Toronto, it would be carrying 73.

The plane took off from Heathrow Airport in London at 5:55 P.M. on the night of August 10. The overloaded plane had difficulty clearing the runway but finally managed it, and the weather report signaled clear weather all the way to Montreal. Refueling stops were planned in Reykjavik, Iceland, and Montreal, unless headwinds forced an earlier stop in Quebec.

By the time the plane left Iceland, at 1:12 A.M., the crew had been on duty for 20 hours and had flown almost constantly for 15 or 16 hours. It would be daylight before they would reach Seven Island and Mont Joli, near Quebec. When they did, they had a mere 122 gallons of fuel left. Most pilots would have elected to land and refuel Ramsay elected to continue on. He radioed his position.

It would be the last communication received from the plane.

West of Quebec, an enormous thunderstorm loomed up directly in the flight's path. Had they had extra fuel, the flight crew could have climbed above it or skirted around it. Ramsay had only one choice: He entered the storm, and that proved to be a fatal decision. Somewhere over Issoudun, near Quebec, the forces of the thunderstorm flung the DC-4 into the ground with such force that some of its fuselage and all of its motors dug in to depths of up to 84 feet. There was only one survivor—an infant child, whom rescuers, arriving a disgraceful five and a half hours later, discovered sitting in the unoccupied pilot's seat, a considerable distance from the main body of the wreck.

Only 20 bodies could be identified. The others were damaged beyond recognition. Local authorities were given the responsibility of protecting the site until Department of Transport crash investigators could arrive the next day. But these local authorities apparently ignored their mandate. Hundreds of curious souvenir hunters from Quebec and elsewhere roamed the wreckage, carrying away personal possessions of the dead passengers and removing thousands of small parts of the aircraft that might have pieced together the last moments of the doomed flight.

CANADA
QUEBEC
MONTREAL
November 29, 1963

Trans-Canada flight 831, a DC-8F, encountered severe turbulence after taking off from Montreal Airport on November 29, 1963, and plunged to the ground. All 118 aboard were killed.

Instrument failure was blamed for the crash, on takeoff, of a Trans-Canada DC-8F at Montreal Airport on November 29, 1963.

Flight 831 was an enormously popular one, for its route and for its timing. "A Friday night businessman's milk-run" is the way it was described, as it made its slow way from Montreal to Toronto and points west every night after work.

At 6:10 P.M. on November 29, 1963, the departure time of flight 831, a torrential rainstorm whipped across the runways of Montreal Airport. Roadways from downtown Montreal to the airport were jammed with stalled traffic, and by 6:22, when the gates closed, eight of the scheduled 118 passengers had still not reached the terminal. The plane nevertheless filled to capacity. Flight 831 always had a full complement of standbys waiting to board.

The flight deck was manned by two veterans, Captain John Snider and co-pilot Harry Dyck. Their weather report read overcast with light rain and fog, visibility four miles, surface wind 12 mph from the northeast. The takeoff, at 6:20, was without event.

The ship climbed to 3,000 feet, and the captain requested permission to make a left-hand turn northwestward, over Rivière-des-Milles-Iles and the small village of Ste. Thérèse, in the foothills of the Laurentian Mountains.

Four miles northwest of this village, something went terribly wrong. The jet dove to earth, striking the ground with such force that the shock was recorded on the seismograph at the Collège Brébeuf in Montreal. The plane dug a crater several times its size in depth and circumference, and its 51,000 pounds of JP4 fuel ignited, sending a pyre of flame and smoke 500 feet in the air. The heat was so intense that rescuers found little left to identify. All 118 passengers and crew apparently died instantly.

Two reasons were later given for the crash: First, there is a particular type of near-ground turbulence called geostrophic convection, which usually happens at night and in rain showers, often to the rear of cold fronts. Simply explained, it is a wind that travels over the ground and is wafted upward because of trees, buildings, hills and other irregularities of the landscape. According to Fred McClement, in his book *It Doesn't Matter Where You Sit,* "[These winds] become dangerous to jets during the takeoff and approach procedures when speeds are lower, turns are being made, and when no trouble is expected."

There was no doubt in the minds of investigators of the crash that the plane encountered some turbulence as it was banking into its northwest turn, and that would mean that the pilot and co-pilot would be relying on their instruments more than usual. The one instrument on which they would be relying most would be the plane's artificial horizon—a circular instrument with a floating ball dial that indicates whether the aircraft is in level flight.

Activated by a vertical gyroscope, this artificial horizon has a built-in danger signal to warn of malfunctioning. The worst-case scenario is one in which the gyro fails, the artificial horizon follows the failed gyro, and the pilot, following the misinformation of the instrument, goes into a dive. At 6,000 feet, the height at which flight 831 was flying when its fatal trouble occurred, it would be impossible to bring any jet out of a dive.

This was the informed speculation that led the board of inquiry to its conclusions about the possible reason for the crash: a failed gyroscope. As a result of the tragedy, the installation of flight recorders in all turbine-powered aircraft in Canada was required, as well as an improved vertical gyroscope system.

CHILE
ANDES MOUNTAINS
February 6, 1965

An ill-maintained Aera Nacional DC-6B with 87 tourists aboard lost radio contact with Santiago, Chile, on February 6, 1965, entered a cloud bank and crashed into the side of San José Mountain. All aboard were killed.

It may be fortunate that the jets of the major airlines fly at altitudes too high to view the majestic but dangerous peaks of the Andes. Most adventurous tourists view them close at hand, at ground level. But others, particularly those who want to get an intimate glimpse of the famous statue of Christ of the Andes, erected in Uspallata Pass on the Argentine-Chilean border, must take smaller, propeller-driven planes, which are subjected to the unpredictable air currents and weather of the high Andes.

Such was the case when a soccer team, sightseers and tourists climbed aboard an Aera Nacional DC-6B at the Santiago airport on February 6, 1965. It was the height of the summer tourist season, and the plane was packed with 87 passengers eager to catch a glimpse of the fabled statue and the peaks that surround it.

The plane was old and, some said afterward, ill maintained. No reason is known for its loss of radio contact 20 minutes after takeoff. A heavy cloud cover shrouded the Andes near Santiago and completely obscured San José Mountain and its ancient gorge, El Volcán Pass. The DC-6B disappeared into the cloud cover.

A waterworks engineer spotted it flying overhead, in the direction of the pass. A loud explosion followed, and the engineer immediately reported his sighting to the Santiago airport. Some time later, rescuers discovered the wreckage of the aircraft strewn along the slope of the mountain. There were no survivors.

CHINA
GUANGZHOU (CANTON)
October 2, 1990

At approximately 9:03 A.M. on October 2, 1990, a hijacked China Southwest Airlines Boeing B-737 crash-landed at the airport in Guangzhou and collided with another plane. One hundred twenty-eight in the two planes were killed.

Until 1980, China did not disclose air crashes that occurred within its borders. But in the fall of 1990, China was seeking favorable world opinion, and the Asian games, held under tight security in Beijing, were its chance to win back international respect after the army crackdown against democratic demonstrators in June of 1989.

The security, apparently, was not enough, for on the morning of October 2, a China Southwest Airlines Boeing 737, bound for Shanghai, took off from the eastern port of Xiamen and was immediately commandeered by a hijacker, who demanded to be taken to Taiwan or Hong Kong.

The captain of the plane circled the airport for 30 minutes while he tried to talk the hijacker out of his determination. Finally, the pilot decided upon subterfuge. He headed to Guangzhou, circled to land and apparently convinced the hijacker that it was a foreign city.

But as they descended, it dawned upon the now distraught man with a gun that they were landing in, not out of, China. He began to struggle with the cockpit crew, and the struggle took its toll on the landing. The plane hit the runway, sideswiped an empty Boeing 707 then slammed into the wing area of a parked Boeing 757 filled with passengers waiting to take off for Shanghai. The hijacked plane flipped over and struck the tail area of the Shanghai-bound plane, which immediately exploded into flames.

"The plane [that fell] was snapped in half like a match stick," one witness told reporters later. "All that was left of the fuselage was charred metal. It looked like a crematorium."

"It landed virtually on top of us, near the wing," said Nils Aliasson, a senior Swedish diplomat who was in the Shanghai plane and who slid down an escape chute to safety after the crash. Only those sitting in the first 14 rows of the plane escaped. The rest of the passengers and crew, along with many on the hijacked B-737, perished. One hundred twenty-eight of the 226 aboard both planes died. Eighty-two aboard the hijacked jet were killed, and 46 aboard the B-757 perished. It would be the worst air crash since the relaxation of reporting of air accidents in China.

COLOMBIA
BOGOTÁ
July 24, 1938

A military stunt plane flying too close to the packed stands at a military air show at the opening of the Campo de Marte in Bogotá, Colombia, on July 24, 1938, broke up and slashed through the crowd of 50,000. Fifty-three spectators were killed by falling debris.

The 155th anniversary of the birth of Simón Bolívar, the patriot-dictator who liberated South America from Spain, and the opening of the Campo de Marte, a military exercise field, were celebrated in Bogotá, Colombia, with a military air show. Fifty thousand spectators packed several metal-roofed stands that sunny afternoon to applaud the daredevil acrobatics of a hand-picked group of military fliers.

The low-flying antics of these pilots did not please Dr. Eduardo Santos, president of Colombia, and he complained more than once to War Minister Alberto Pumarejo about the reckless disregard of safety unfolding before them.

The president's evaluation of the situation turned out to be correct. Within minutes after his last complaint to the war minister, a plane piloted by a Lieutenant Abadia clipped the end of one stand with its wing tip, rolled sideways, sheered away a set of steps leading to the stand and burst into flame. The propeller, separated from its engine, preceded the flaming wreck as it plowed into the stands, incinerating and slashing hundreds of screaming, terrified and trapped spectators.

Some, their clothes on fire, bolted for the open fields; others were pinned to their seats by pieces of wreckage. Fifty-three persons perished in this needless disaster.

COLOMBIA
BUGA, VALLE DEL CAUCA
December 20, 1995

American Airlines flight 965, a Boeing B-757-223 en route from Miami to Cali, Colombia, crashed at 9:38 P.M. on December 20, 1995, in the mountains near Buga, 50 miles north of Cali. One hundred and sixty of its 164 passengers and crew died in the crash.

The holiday season had begun, and on December 20, 1995, the major portion of the 164 passengers and crew boarding American Airlines flight 965 in Miami were Colombian nationals on their way home for the holidays. When the captain, Nicholas Tafuri, radioed to the Cali tower to ask for permission to land, he appended a cheery "Merry Christmas!" to the request. Moments later, the cheeriness led to panic, which in turn led to several fatal errors in the cockpit of the Boeing 757-223.

The approach, in clear weather shortly before 9:00 P.M., was normal. The plane was over Buga, a town roughly 40 miles north of Cali and in a heavily mountainous region of Colombia, when radio contact with the flight suddenly ceased.

But previous to that, a tug-of-war between the ground and the plane had begun. Colombia's flight management system naming convention is different from that published in international navigational charts, and that difference set in motion a chain of tragic events.

In addition, air traffic control in Colombia differs dramatically from that of the United States and nearby offshore areas, which consist of a blanket radar coverage that immediately indicates when a plane strays from its assigned course and allows traffic controllers to warn the crew of the plane about it. In Colombia, by contrast, jets navigate by using radio beacons on the ground that tell instruments in the plane which beacon is sending the signal and from which direction on the compass the plane is approaching the beacon.

The Cali tower requested flight 965 to report over Tulua radio beacon, but the flight had already passed this beacon, so following the directions was impossible. Instead, the crew decided to use the Rozo beacon and entered the abbreviation "R" in the controls. This steered them not toward Rozo, but toward Bogotá.

The plane made a gentle turn toward the Colombian capital, and it took the crew slightly more than a minute to realize that they were heading in the wrong direction. They disconnected the autopilot and began to steer the plane back in the direction of Cali. But this would take them directly into mountainous territory, which they again failed to recognize.

Finally, at 9:35 P.M., the ground proximity warning system went off, telling them of an approaching mountain. The captain pulled up the nose, gave the jet full throttle, and nearly cleared Mt. El Deluvio, but not quite. A mere 200 yards from the mountain's summit, the plane plowed directly into it.

As the flight's 9:45 arrival hour came and went, and no further word was heard from it, it became obvious that the plane had crashed, but no rescue parties went out. General Joseph Serrano, the national police chief, promised to send helicopters at dawn, but not before then, because in addition to darkness, the area into which the plane had apparently crashed was a territory controlled by leftist guerillas protecting the enormous drug cartel based in Cali.

At dawn, helicopters carried rescuers to the region, and the crash site was discovered quickly. The decimated aircraft and its contents were strewn over a huge area on the face of the mountain, and it seemed at first that there were no survivors. But a closer look and reports from the ground soon revealed eight survivors of the horrendous crash.

One of the survivors was spotted with his brother. Juan Carlos Reyes, who was in a helicopter surveying the area, suddenly saw his brother, Mauricio, who was a student at the University of Michigan in Dearborn. Another survivor was a brown dog found alive in a carrier in the cargo hold. The dog earned the name Milagro, Spanish for "miracle."

Peasants from the area told conflicting stories. Some said the plane had exploded; some said it had not. The conclusion of the rescuers was that it had not. "We think that one of the reasons why there are survivors," Deputy Police Chief Fenergal Francisco Montenegro said, "is because the plane hit a soft, mossy area instead of a rocky one."

Nevertheless, wreckage was strewn over a radius of a mile. The plane had obviously been dashed to pieces; bits of clothing and pieces of the fuselage were found hanging from trees.

By the next day, more helicopters from the United States Air Force base in Panama as well as 500 Colombian police began to comb the area. Fog settled in early and hampered efforts, but there were no clashes between guerrillas and workers.

The NTSB and the FBI were soon on the scene investigating the crash site. At first, suspicion of a bomb set by

Colombian rebels surfaced. *El Universal,* a Caracas newspaper, received a letter postmarked from Miami expressing hatred of immigrants and threatening to bomb a plane bound to Miami from Colombia or Venezuela, but the threat was ultimately dismissed as a crank letter.

Discovery of the black boxes confirmed that crew error had led to the fatal crash—the cause of 80 percent of airline accidents—and American Airlines issued an apology. "We've always taken great pride in the vigor of our training and the excellence of our flight disciplines, and we are saddened that human error on the part of our people may have contributed to the accident," said American Airlines' chief pilot Captain C. D. Ewell to the press. "The accident reminds us that aviation, while not inherently dangerous, is terribly unforgiving of any inattention to detail."

The fatal misreading of radio signals might have been avoided had the pilot and co-pilot conducted a mandatory "approach briefing," which reviews the speed, the route to the landing, the positioning and configuration of the plane, what time to begin communication with the air control tower, what to do when the runway is visible and what to do if any problems arise. The cockpit black box revealed that this had not been done.

In addition, preliminary reports from Colombia hinted that alcohol had been found in the body of the pilot, Captain Nicholas Tafuri. American Airlines quickly disputed these findings, citing Captain Tafuri's long and immaculate record. The findings were later corrected. The alcohol found was a result of the partial decomposition of the body before testing.

Of the survivors, all but four died later in hospitals. The death toll was 160 of the 164 aboard American Airlines Flight 965, just before Christmas of 1995.

COLOMBIA
MEDELLÍN
May 19, 1992

A SAM Airlines 727 crashed into a mountaintop while attempting to land blind at the Medellín International Airport in northwestern Colombia on May 19, 1992. All 132 aboard were killed as the plane exploded upon impact.

The primitive and unique radar system at Medellín International Airport in Colombia was damaged and made inoperable by guerrillas in May of 1991. In the following year, it was neither replaced nor repaired, but the radio search system, which allows air traffic controllers to give an airplane its exact bearing before landing, was in order and it allowed planes to land in inclement weather with relative safety.

But on the afternoon of May 19, 1992, the radio search system, too, was out of order and not functioning. Thus, planes flying into Medellín were literally flying blind, their only contact with the airport through voice contact with the controllers.

SAM Airlines flight 501, a Boeing 727, left Panama City at 2:18 P.M. and approached the Medellín airport in steadily worsening weather later that afternoon. There was no sign of trouble in the conversations between the pilots and the ground controller, but the absence of guiding mechanisms brought about tragedy. The aircraft was not at all on course, and it suddenly disappeared from the radar screens at the Bogotá airport. Ground officials declared an emergency, but without an indication of where the plane was and with foul weather increasing and darkness falling, search parties were forced to wait for daylight of the next day.

A radio news helicopter was the first to spot the wreckage of the plane, on a mountain top near the town of Urrau, in the northwestern province of Antioquia. Police, army and Red Cross officials immediately set out from Medellín and arrived shortly to find the remnants of an exploded airliner littering the countryside for a radius of two and a half miles. "There was no fragment of the plane more than two yards long," Gonzalo Bernal, who was in charge of the search, told reporters. "It is going to be very difficult to recognize any corpses," he added.

Bits and pieces of upholstery and luggage and fragments of the fuselage were driven into trees, hung from others and littered the landscape for as far as it was possible to see. It was evident that none of the 132 passengers and crew members had survived the explosive impact of the plane as it hit the mountaintop.

Workers nevertheless dug through bits and pieces of the airliner for as long as they could, which was only to midafternoon. By then, the weather had become foul beyond description, and the teams of rescuers were exhausted from the lack of oxygen at the 11,000-foot height.

The reason for the crash became obvious to all of them: The jet was at least 24 miles off course. With neither radar nor a radio search system, the pilot had no idea if he was headed for the airport. Tragically, he was considerably off course—a disastrous distinction in bad weather in Colombia's mountainous northwestern provinces.

CZECHOSLOVAKIA
BRATISLAVA
November 24, 1966

An Ilyushin-18 owned by TABSO Airlines disappeared into a snowstorm on the night of November 24, 1966, and plunged into a peak of the Carpathian Mountains near Bratislava, Czechoslovakia. All 82 aboard died.

The TABSO Airlines Soviet-built Ilyushin-18 was not scheduled to stop at Bratislava on the night of November 24, 1966. Its normal run took it from Sofia to Budapest and Prague, but for some reason, it made an unscheduled stop at Bratislava.

The stop could have been caused by the weather, which was bad. There were high winds, and just as the liner began to taxi onto the runway, preparatory to takeoff, the skies seemed to open, and it began to snow heavily. Apparently disregarding the weather, the airport gave the aircraft clearance for takeoff.

The four-engine turboprop roared forward and within minutes was obscured from view by the snowstorm. Minutes later, it slammed into the side of a peak in the Carpathian Mountains, which surround Bratislava like a dangerous fence. All 82 people aboard were killed, including Bulgarian opera star Katya Popova.

DOMINICAN REPUBLIC
OFF PUERTO PLATA
February 6, 1996

One hundred and eighty-nine persons lost their lives when a charter Boeing B-757-225 fell into the Atlantic Ocean on takeoff from Puerto Plata in the Dominican Republic. The plane was returning vacationers from Puerto Plata to Berlin and Frankfurt.

Once again, the failure of a crew to take heed to warning signs of disaster and a failure to act promptly to avert that disaster brought an airliner down and killed 189 people.

On the night of February 6, 1996, a group of German tourists boarded a charter plane to take them home to Frankfurt and Berlin from a vacation in Puerto Plata, on the northern, Atlantic coast of the Dominican Republic. There was a slight delay; a Boeing 757 was brought in to replace the original 767 that had been scheduled by Oger Tours of Hamburg but that had developed mechanical trouble.

The 11-year-old Boeing B-757-225 had Byzantine credentials. It was originally built for Eastern Air Lines, was owned by a Turkish Company called Birgin Air and was under lease, with 11 Turkish and two Dominican crew members, to a Dominican airline, Alas de Transporte Internacional, for a single flight.

Some of the passengers were nervous about flying from the Dominican Republic. The crash of an American Airlines flight in Colombia little more than a month before (see p. 19) was on their minds. "I do have mixed feelings," an elderly man said before boarding, "but there isn't going to be another crash right away." Nobody cancelled, despite the change in planes and the steady rain that was falling.

The takeoff at approximately 11:30 P.M. was normal. The jet climbed to 7,000 feet, began to make a right turn, then suddenly disappeared from the radar screens at Puerto Plata. A disconcertingly nervous "Stand by" was the last radio message the airport tower received from the plane. And then it became clear: The flight had plunged from 7,000 feet into the Atlantic Ocean, 13 miles offshore from Puerto Plata.

It would be morning before rescue workers reached the crash site, in which debris was spread on the water in an area two miles long and 150 yards wide. Dominican civil defense personnel and three American Coast Guard cutters arrived and began to scoop debris and bodies from the water. There was jet fuel on the surface, but it had not burned, indicating that the plane had not broken up before it hit the water.

Germany sent two civil aviation officials and three officials from the federal police in Wiesbaden to help identify bodies. The FAA and NTSB dispatched their investigators, but an immediate problem developed. The water in the Atlantic off Puerto Plata is extraordinarily deep. Only some wreckage and some bodies had been salvaged from the surface. The main fuselage of the airplane, most of the bodies and the all-important black boxes lay beneath 4,900 to 6,560 feet of water.

The cost of raising the wreckage would, it was estimated, cost hundreds of thousands of dollars. Although the plane was American built, neither it nor the airline nor the flight was under American jurisdiction and no Americans were on board. Nevertheless, an agreement was reached among the participating countries, and deep water salvage began.

It revealed a grisly and disturbing conclusion. A clogged control tube had produced incorrect airspeed indications for the crew in the ascending jet. The readings were greater than the actual airspeed, and so the crew reduced power to the engines. This led to a stall, which was signaled in the cockpit, but none in the crew apparently took notice of it. The stall in itself was not fatal, but the crew failed to carry out the procedures that would recover the aircraft from the stall and restore its climbing lift. And so the jet slipped backward, 7,000 feet into the Atlantic Ocean, killing all 189 persons on board.

EGYPT
CAIRO
May 20, 1965

Hampered by dangerous landing conditions at Egypt's Cairo Airport, a Pakistan International Airways Boeing 707 crashed on landing on the night of May 20, 1965. All 124 aboard were killed.

Cairo Airport in the early 1960s was a nightmare. For years it had been the talk of the industry, and finally, by 1965, members of the International Pilots' Association refused to make night landings there.

There were numerous hazards. First, there was inadequate lighting, making it difficult to define the parameters of each runway. In addition, some of the runways were eccentrically configured. One in particular was heavily sloped and approachable only by an abrupt dropping of a plane for a distance of almost 1,000 feet. Finally, there was a shortage of rescue equipment.

Heeding the boycott by the IPA, authorities had begun to make improvements. But their priorities and the way they deployed them were mysterious. Their first purchase was up-to-date rescue equipment, but this was kept locked away.

The price for this laxity was paid by 124 people who died in the wreckage of a Pakistan International Airways Boeing 707, which, approaching the airport at far too steep a decline, crash-landed on the night of May 20, 1965.

The tragedy finally brought about the improvements that could have prevented it.

FRANCE
BEAUVAIS
October 5, 1930

A combination of bad weather, bad judgment and faulty design sent the R-101, Britain's mammoth dirigible, to the ground on October 5, 1930. Forty-nine died.

The R-101, Britain's entry in the speed and passenger sweepstakes of giant dirigibles in the 1920s and 1930s, was buffeted by an enormous storm on the night of October 5, 1930. Its passengers and crew were flung from one side of the gondola to the other. Yet Lieutenant H. C. Irwin, the commander of the dirigible, radioed to his home base in Cardington, England: "After an excellent supper our distinguished passengers smoked a final cigar and, having sighted the French coast, have now gone to rest after the excitement of the leave-taking."

A sister ship to the R-100, the R-101 was an imposing construction. Seven hundred seventy-seven feet in length, it had a hydrogen capacity of five million cubic feet. Powered by six Rolls-Royce Condor engines, it was capable of lifting 150 tons and had accommodations for 100 passengers.

But the R-101 had been in trouble from the first moment out of the factory. On its maiden voyage, on October 14, 1929, its engines had malfunctioned. The fabric of its skin had seemed faulty.

The British government had been aware of these problems, and yet, in an international dirigible competition with Germany and France, it had still pushed for a long trip.

Thus, on October 5, 1930, 52 crew members and four distinguished observers, including Lord Thomson, who had argued most strongly in Parliament for the trip, boarded the R-101 for a trip to the Orient.

There were two portents of disaster. Four tons of water used as ballast were accidentally thrown overboard while the dirigible was still tied to its mooring. Second, a storm was brewing on the English Channel, and that was exactly where the R-101 was heading.

Nevertheless, at 6:40 P.M., the R-101 left its mooring mast and headed directly into the storm. Cruising altitude was supposed to be 1,000 feet, but it could not maintain that altitude, dropping abruptly to 700, then shooting up to 1,100. One of the engines stopped partway across the Channel. Still, Lieutenant Irwin pressed cheerily on.

The storm did not abate when they passed over the French coastline. Thirty-five-knot winds buffeted the R-101, driving it lower and lower. By 1:00 A.M., at a village called Poix, the ship's altitude was a mere 250 feet and descending.

An hour later, the R-101 had reached Beauvais. It was now skimming trees and rooftops, headed for what seemed like a gentle and safe landing in a large field outside town. The field, however, was not altogether without obstacles. At one end was a ridge of hills, and the R-101 headed straight for it, crashing into the ridge and igniting its load of hydrogen. The entire ship was consumed in roaring flames in an instant, and 48 crew members and Lord Thomson were burned to death.

FRANCE
GONESSE
July 25, 2000

An Air France Concorde chartered for a group of German tourists attempted to take off from Charles de Gaulle Airport in Paris for a trip to New York on July 25, 2000. The supersonic jet crashed on takeoff into a nearby hotel in Gonesse, killing all 109 on board and 4 on the ground.

The Concorde, the rapid and sophisticated way for the affluent to travel to and from New York to Paris or London, had an enviable safety record. From 1962, when it was developed by a British and French partnership, to its launching of flights in the late 1970s, to July of 2000, there had not been one crash of a Concorde. Thirty years of safety is an unusually long stretch for any type of commercial aircraft.

This dimension of safety, added to the absence of jet lag because of the 1,350-mph swiftness of the flight, encouraged a steady stream of passengers to pay up to $11,000 each to make the trip between Paris and New York or London and New York.

But on the afternoon of July 25, 2000, all of this would abruptly and tragically change. Air France Concorde flight 4590 was chartered by a German tour company for the flight from Paris's Charles De Gaulle Airport to New York's Kennedy Airport. An eager group of German tourists, on their way to a cruise to Ecuador, accounted for 96 of the Concorde's 100 passengers and nine crew members.

The flight was delayed for approximately one hour while a thrust reverser was replaced in the number-two engine. The plane could have flown easily with only three of its four thrust reversers in operation, but the pilot was taking no chances.

And so the passengers finally boarded, and, as usual, they were treated to champagne and luxury and settled back for the thrill of a supersonic flight across the Atlantic. The plane took off into a clear and sunny sky at approximately 4:40 P.M.

But the graceful, needle-nosed aircraft had been in trouble even before it left the ground. Seconds after it began its high-speed takeoff run, black smoke began to pour from its number two left engine. "Failure of engine number two!" one of the pilots radioed to the control tower, which radioed back that the engine was on fire. A 75-foot line of flames was spouting from the engine, and the plane, only feet above the ground, was wavering. It was traveling too fast to remain earthbound, and the pilot was obviously struggling to get it airborne and head it for Le Bourget airfield, where there would be less busy runways.

Two minutes later, the plane was 200 feet in the air and obviously in mortal trouble. The number four engine lost power, then stopped completely. The number one engine began to lose power. The landing gear would not retract.

The plane settled into a left turn, headed for Le Bourget and suddenly went totally out of control. It pointed its nose skyward, stalled and crashed into the Hotel Hotellisimo in Gonesse. Both the hotel and the Concorde burst into flames.

Sid Hare, a Federal Express pilot who was at a hotel near the airport and witnessed the crash, told reporters afterward, "It was a sickening sight, a huge fireball. The airplane was struggling to climb and obviously couldn't get altitude. There was smoke trailing from one of the plane's engines, and it [the plane] started rolling over and back-sliding down to the ground. The pilot kept raising the nose . . . and the airplane stalled, the nose went straight up into the air and the airplane actually rolled over to the left and almost inverted when it went down in a huge fireball."

"It was like some bomb taking off," said John Palmerston, an Australian who saw the plane leave the runway. "The whole back end of the plane was on fire. I thought it was going to crash right here at the airport."

In the tiny town of Gonesse, four miles southwest of the airport, where the Hotel Hotelissimo was located, other witnesses registered similar shock. "First there was a large explosion," an office worker said, "then barely a second later, another large explosion. We went to the window, and the flames were higher than the trees. When we went outside, pieces were falling down on us."

"It was truly hell," another told French television. "There were flames everywhere. There were pieces of the fuselage over more than 200 meters, pieces of plastic, of cloth, bits of metal, pieces of the wing, rivets."

More than 200 fire fighters, 110 police, 13 ambulances, three helicopters, the prefect and the prime minister, Lionel Jospin, converged upon the scene. The president of Air France witnessed the entire drama through a window in Charles de Gaulle Airport.

Firemen placed traffic cones across the blackened field surrounding the hotel, apparently marking the places where charred bodies were found. "Once you get used to looking through the smoke, the horror hits you," one emergency worker told *Agence France-Presse*. "By the dozens, bodies or pieces of bodies, horribly mutilated. Here and there, shoes, books, parts of suitcases that are wasted away."

The hotel was no more. The impact had completely destroyed it. One survivor had escaped through a window as

A firefighter attempts to control the fire from the crash of the Air France Concorde in Gonesse, outside of Paris on July 25, 2000. (AP/Wide World Photos)

Hotel Hotelissimo is reduced to flaming rubble after the out-of-control Air France Concorde crashed into it, killing four in the hotel and all 109 aboard the airplane. (AP/Wide World Photos)

the plane hurtled toward the hotel. Some onlookers smashed a window and tried to get into the wreckage of the building, but the victims were already dead. The toll was total: All 109 aboard the plane and four on the ground were killed.

The entire fleet of Concordes was immediately grounded as an intense investigation of the crash was launched. Supersonic flight at an altitude of 11 miles places different stresses on the Concorde than those experienced by subsonic aircraft. The rush of passing air heats the skin above the boiling point, which can weaken the airplane's construction, and this was the first concentration of the inspectors.

In addition, the Concorde, like other supersonic planes, had a history of wheel and tire problems that could lead to fires. One of these had to do with the fact that the Concorde's V shape requires it to fly about 50 percent faster on landing and takeoff than is typical for subsonic jets. This heats up the tires, and because they retract quickly into a wheel well crammed with hydraulic lines, ignition of the hydraulic fluid could cause a fire. However, this Concorde's landing gear never retracted.

And yet, as it turned out, it was a tire that caused the first fatal Concorde crash: As it sped along the runway at Charles de Gaulle Airport, one of the plane's wheels struck a 16-inch piece of metal. It would not have been possible for the pilot to see it, but this small shard, later identified as a thrust reverser part dropped earlier from a Continental Airlines DC-10 that had departed from Paris to Newark, was sharp enough to shred the wheel's tire.

The whirling pieces of rubber pounded the nearby left wing, sending reverberations through the fuel tanks powerful enough to punch a hole in one of them. And through this hole, a deadly stream of ignited aviation kerosene started the fire that caused the crash. The cause of the fire was assumed to be the huge amount of heat generated, but this conclusion came only from a process of elimination. No trace of any other ignition was found.

During the year-long grounding of Concordes, safety assurances were added to combat the possibility of the recurrence of the freak conditions that caused flight 4590's fiery end. To lessen the effect of an assault on the fuel tanks, six of the 12 were fitted with flexible Kevlar liners—tests showed the other tanks were not at risk of debris strikes. Kevlar is a synthetic, nylonlike material that had not been available when the Concorde was designed and built in the 1960s. Pliable and nearly impenetrable, it is used in bulletproof vests. As a precaution against the ignition of leaking fuel, electrical cabling in the Concorde's wheel recesses was armor-plated so that flying debris could not sever the wires, which would cause a spark. New, reinforced tires, in development before the crash, were exchanged for the older tires. And finally, the interiors of the Concorde were redesigned with lower-weight fittings.

Air France recommenced its Concorde service between New York and Paris in the fall of 2001.

FRANCE
GRENOBLE
November 13, 1950

Bad weather forced a Canadian Curtis-Reid Air-Tours charter flight to ram into Mont Obiou, near Grenoble, France, on November 13, 1950. There were no survivors.

A charter flight loaded with pilgrims returning from a Holy Year pilgrimage to the Vatican encountered severe weather conditions over the French Alps on November 13, 1950. A mere 10 days before, an Air India plane, flying the same route, had flown into the face of Mont Blanc, killing 48 passengers.

The Canadian Curtis-Reid Air-Tours charter flight, carrying 58 passengers who had, that very morning, had an audience with Pope Pius XII, was lost in heavy fog and rain. Some time in the early evening of November 13, it rammed into the side of Mont Obiou, an 8,500-foot peak near Grenoble.

Rescuers would find no survivors among the pilgrims, many of whom, in death, clutched pictures of the pope.

FRANCE
PARIS
June 3, 1962

Pilot error in failing to abort the takeoff of an overloaded Air France Boeing 707 from Paris's Orly Airport on June 3, 1962, was blamed for its fiery crash. One hundred thirty passengers and crew members were killed; two flight attendants survived.

An Air France Boeing 707 carrying 132 passengers—most of them the leading figures in the cultural life of Atlanta, Georgia—gained eight feet of altitude, fell to the runway at Paris's Orly Airport, skidded forward, hit a row of approach lights at the end of the runway and exploded.

The group, all members of the Atlanta Art Association, had been touring some of Europe's leading museums and repositories of art—the Louvre, the Tate, the Uffizi Gallery and St. Mark's. It was a full flight but not overcrowded. However, for some reason never explained, the pilot seemed to have difficulty raising the jet from the runway. A board of inquiry later speculated that he was experiencing heavy loads on his elevators and tried to abort the takeoff before he reached the end of the runway. Apparently, he made his decision too late to arrest the forward ground speed of the aircraft. The brakes were applied, and all of the engines were thrown into reverse, but the plane, loaded to capacity with jet fuel, was blown to bits when it struck the approach lights and some cherry trees beyond.

One hundred thirty people were killed. The only two survivors were two flight attendants, Jacqueline Gillet and Francoise Authie, who were strapped into seats in the tail section, which broke loose just before the explosion and threw them clear.

Horribly, Mr. and Mrs. Milton Bevington of Atlanta carried through with a plan they always practiced when they traveled but never believed would achieve its intended result: They always flew in different airplanes, reasoning that if one crashed, the other would survive to take care of their children. Mrs. Bevington boarded the plane, and her husband watched the entire catastrophe from the terminal.

FRANCE
PARIS
March 3, 1974

In the worst crash in the history of aviation to that date, Turkish Airlines flight 981, a Douglas DC-10, suddenly lost compression and control when a faulty cargo door blew off, forcing the craft to crash in a turnip field near Paris. All 346 people aboard perished.

The worst crash in the history of aviation to that date was caused by negligence, ignorance, tampering and a poorly designed hatch door. And what is worse, a similar accident that occurred two years earlier should have alerted the experts and those in charge to the problems that were repeated in this tragedy.

On June 6, 1972, an American Airlines DC-10 developed trouble 11,000 feet over Windsor, Ontario. Its rear hatch door blew off, tearing a hole in the fuselage, decompressing the interior of the plane, collapsing part of the passenger cabin floor, sending huge clouds of fog and iron filings through the plane and severing some of the control cables.

For a few feverish minutes, the stunned pilot, Captain Bryce McCormick, and his first officer, Peter Paige Whitney, were certain that they were about to dive to their deaths. The throttles had been slammed backward into idling speed, rendering the plane's engines almost useless. But Captain McCormick gave the plane full throttle, thus preventing its nose from dropping.

Fortunately, his hydraulic lines had not been severed, and he was able to control the plane well enough to return to Detroit—the flight's origination point—and land safely, with no casualties.

The board of inquiry investigating this accident determined that the latching mechanism of the cargo door was faulty and weak enough to give way under pressure. Thus, recommendations were made immediately to McDonnell

Douglas to make the following three modifications in all DC-10 cargo doors:

1. The locking pins of the cargo door were to be reconfigured, so that in the "closed" position they would overlap the latches by one-quarter of an inch, instead of 1/32 of an inch.
2. A support plate was required to prevent the upper torque tube from bending under pressure as it did in the Windsor incident.
3. The tube carrying the locking pins was to be modified to prevent a false signal through the microswitch system.

Apparently, the modifications were not strictly enforced. Not only were they not made on some DC-10s already flying, but two were actually manufactured in the following two years with the same defects in the cargo door. One of these, plane number 29, was sold to Hava Yollari of Turkish Airlines.

On Sunday, March 3, 1974, this DC-10, as flight 981, took off from Orly Airport in Paris, headed for London. It was loaded with 346 persons, an unusually high passenger-load caused by an engineer's strike at London's Heathrow Airport. Because of the strike, all British European Airways flights were grounded and their passengers transferred to other flights. THY flight 981 usually flew half full, but this time it turned away passengers, including the British Bury St. Edmunds Rugby team.

Flight 981 had originated in Istanbul at 9 A.M. that morning. Along with all other airlines flying DC-10s, THY had been warned, because of the Windsor incident, to be vigilant when closing the cargo door and to check its locking mechanism carefully.

However, even with careful maintenance, the door's locking mechanism was missing the support plate that was to have been installed after the Windsor incident. Thus, for the 63 weeks it had been operating, ship 29 was in essentially the same condition that the American Airlines flight was in when it had lost its cargo door at 11,000 feet. In addition, a later investigation revealed that the locking pins on this rear cargo door were also misrigged, so that even a slight vibration could dislodge them. Someone had turned the rod on ship 29's rear door the wrong way.

At Orly, THY's maintenance was managed by an outside contractor, SAMOR. The last person to close the rear cargo door was an Algerian expatriate, Mohammed Mahmoudi. He followed his instructions exactly, but not being a trained mechanic, he did not notice that the door closed more easily than it should. Furthermore, he was not assigned to make the final visual inspection of the door through a small peephole. If that inspection had been made, it would have revealed that the locking pins were in the "unsafe" position.

This visual inspection was to be done by either the THY station mechanic or the flight engineer. But on March 3, 1974, Osman Zeytin, the THY station mechanic at Orly, was on vacation in Istanbul. And none of the airport workers saw a flight engineer check the peephole.

The third safety backup was a light in the cockpit that would have revealed to the crew that the door was not securely locked, but it did not light.

Thus, Flight 981 was doomed from the moment it took off, at 12:30 P.M. The captain, Mejak Berkoz, and his co-pilot, Oral Ulusman, both seasoned fliers, climbed to 11,500 feet and headed east over the village of Coulommiers, where they were to change course north, to London.

At 11,500 feet, there were approximately five tons of air pressure pushing against the mislatched cargo door. The door blew open. The bolts gave way, the latch talons were pulled from the spools and the door flew backward, ripped from its hinges by the aircraft's slip stream.

As in the Windsor incident, the DC-10 rapidly decompressed. The loss of pressure buckled part of the passenger cabin, and the last two rows of seats in the left-hand aisle, with the six passengers strapped into them, were sucked into the cavity and ejected through the hole in the plane's fuselage. They would fall two and a half miles and land in a turnip field near the village of St. Pathus.

The other 340 people aboard Flight 981 only had a minute and a half to live. This time, the buckling of the floor apparently severed all of the control cables, as the dialogue in the cockpit, recorded by the cockpit voice recorder, later revealed:

Voices: "Oops. Aw, aw."
A klaxon sounds.
Berkoz: What happened?
Ulusman: The cabin blew out.
Berkoz: Are you sure? Bring it up, pull her nose up.
Ulusman: I can't bring it up—she doesn't respond.
Ozer: Nothing is left.
Ulusman: Seven thousand feet.
A klaxon sounds, warning that the plane has gone over the maximum speed.
Berkoz: Hydraulics?
Ulusman: We have lost it . . . oops, oops.
Berkoz: It looks like we are going to hit the ground. Speed. Oops.
The sound of the initial impact.

Flight 981 hit the ground at 497 miles per hour. None of the passengers and crew survived. The violence of the impact shredded the plane and the people on board. Only 40 bodies, including the six who were sucked out of the plane when the door blew off, were found more or less intact and thus were easily identifiable. The rest were disintegrated into approximately 18,000 fragments, which medical personnel and police methodically and scrupulously gathered over the next days.

The FAA board of inquiry discovered the next day that the hatch door was missing its support plate, and there was evidence it had never been fitted. "Human failure" was the final conclusion of the investigation, and according to Paul Eddy, Elaine Potter and Bruce Page, in their examination of negligence in flight, *Destination Disaster,* the blame was placed, at a later annual meeting of McDonnell Douglas, "specifically [on] the failure of an 'illiterate' baggage handler (Mohammed Mahmoudi) to close the cargo door properly."

Hearing of this, the union to which Mahmoudi belonged threatened to boycott DC-10s at all French airports, and an emissary was dispatched from the American Embassy in Paris to union headquarters, where he apologized on behalf of the U.S. government.

Senate hearings were also held, and inspection and improvements on the latching mechanisms of the rear cargo doors on all DC-10s were made mandatory. If they had been made so after the Windsor incident, 346 people would not have perished in this, aviation's worst disaster to date.

GERMANY
EAST BERLIN
August 14, 1972

One hundred fifty-six persons died in the midair explosion of an Interflug Ilyushin-62 flying from East Berlin to Burgas, Bulgaria. No official explanation for the tragedy was released by East German authorities.

Confusion and mystery have muddled the facts about the midair explosion of Interflug's Ilyushin-62 on a flight from East Berlin to Burgas, Bulgaria. The aircraft apparently took off without incident from Schonefeld Airport in East Berlin on August 14, 1972, climbed to 100 feet, exploded and hurtled to earth, killing 148 passengers and eight crew members.

Western journalists heard conflicting eyewitness reports—one eyewitness saw flames, the other did not—and the East German authorities refused to release the findings of a board of inquiry.

GERMANY
JOHANNISTHAL
October 17, 1913

A ripped hull, overheated gas and faulty design were responsible for the crash of the German naval dirigible LZ-18 on October 17, 1913, over Johannisthal, Germany. Twenty-eight died in the fiery crash.

Felix Prietzker was a prize student of Count Ferdinand von Zeppelin, the designer and constructor of Germany's first dirigible, Luftschiff Zeppelin Number 1. That ship was the prototype for all of the airships that would ferry military personnel and passengers across the continent and across the Atlantic for the next 40 years.

But Prietzker was no slave to a master designer. His design for the L.2, which became known as the LZ-18, was a marked modification of Count von Zeppelin's original. First, the gondolas, which carried the crew, passengers and control instruments for the ship, were fitted consider-

The German naval dirigible LZ-18 plunged to earth in a fiery crash on October 17, 1913, killing 28—the highest death toll in an air crash to that date. (Library of Congress)

ably closer to the ship than in the original design. Then a series of windscreens, designed for greater maneuverability, were attached to its bow. According to other engineers, these windscreens created a dangerous situation because they were of metal and were positioned very close to the skin of the ship's hydrogen-filled hull. As air pressure fluctuated, the hull changed in shape and size, and the abrasion of these windscreens against it was a matter of extreme concern to some aircraft engineers.

Prietzker, however, was unconcerned about what he considered to be a fancied danger, and so he pressed ahead with the construction of this airborne leviathan, which was 518 feet, 2 inches in length and was powered by four 165-horsepower Maybach engines mounted on the ship's two gondolas.

The ship, commissioned by and delivered to the German Navy in the autumn of 1913, made its maiden voyage under ironic circumstances. The flight, on September 9, 1913, was an uneventful one. But it occurred at almost the same time that its prototype and sister ship, the L.1, encountered a fierce storm over Helgoland and plummeted into the sea, killing 14 of its crew.

The LZ-18 had, however, passed its first test, and a more elaborate trip was planned for October 17, 1913. This one would take the designer Prietzker and 27 others, including some zeppelin consultants and admiralty officers, on a tour of the German countryside from its berth in Johannisthal.

The airship launched smoothly, floated serenely upward from its post and reached its cruising level of 1,500 feet. The warmth of the morning expanded the hydrogen in the hull, lightening the ship, and it rose easily. At 1,500 feet, the engines were engaged, preparatory to directing the ship westward.

But no sooner had the engines started than a tongue of flame shot outward from the forward engine gondola. Within seconds, the flame had doubled back on itself—blocked, fatally, by the windscreen—and both gondolas and hull were consumed by a roaring, cataclysmic fire. The ship crumpled, its white-hot ribbing bright against the blueness of the morning sky, and as crew members and passengers frantically broke gondola windows to escape, the great ship nosed earthward.

The gondolas hit the ground first, but before the men inside could escape, the white-hot hull collapsed around them. Only three men survived the crash, but they were too severely burned to survive the day. All in all, 28 died—the highest death toll in an air crash to that date.

A board of inquiry named the cause of the crash: a rent in the hull, which allowed the overheated gas to escape and ignite the ship's skin. Prietzker's windscreens, which had directed the burning hydrogen back to the engines and the gondolas, both focused and accelerated the fatal process.

GERMANY
MUNICH
December 17, 1960

A U.S. Air Force C-131 Convair went out of control over Munich, Germany, on December 17, 1960, and plunged into the crowded city. Fifty-three people died; 20 were on the plane, 33 on the ground.

Airlines are particularly careful to give thickly populated urban areas wide berths. The chances of an airplane falling into a crowded city street are thus remote; yet, bizarre accidents have been known to happen. Just such an incident took place at the height of the Christmas shopping season in Munich on December 17, 1960.

That city's Bayerstrasse was thronged with holiday shoppers in the late afternoon of December 17. The thoroughfare is flanked by rows of shops. The public transportation system of the city serves this wide boulevard well; a major trolley line runs down the middle of it.

Outside the city, the Munich-Riem Airport was also crowded. At the section that serviced U.S. military aircraft, a Convair C-131 loaded 18 military service dependents for the brief ride to London. A thick fog had settled over the city and the airport, but the Convair took off on schedule.

Minutes after it became airborne, the C-131 developed engine trouble. The pilot radioed his plight to the Munich-Riem Airport, which in turn contacted the fire brigades in Munich. The plane was out of control and over the city. There was a distinct danger that it could lose power and plunge into Munich itself.

Tragically, that is exactly what happened. The pilot fought to control his aircraft, banking sharply to avoid several tall buildings, then slicing the steeple from Saint Paul's Church and finally, with a roar, diving into the middle of the holiday crowds on the Bayerstrasse.

The plane collided squarely with a crowded trolley, exploding upon impact, setting fire to the car, the street and some buildings bordering the crash site. Passengers in the trolley who might have escaped the fire were trapped inside behind closed and now useless electric doors. Fifty-three people—20 on the plane and 33 on the ground—died.

GREAT BRITAIN
ENGLAND
LONDON
June 18, 1972

Overloading was the apparent cause of the crash of a BEA Trident 1 shortly after takeoff from Heathrow Airport in London on June 18, 1972. All 118 aboard died, and no official reason for the crash was ever given.

A worldwide pilots' strike had been threatened for the summer of 1972 in protest over the lackadaisical attitude individual Western governments had taken toward the steadily increasing incidents of air terrorism and skyjacking. As a result, flights in Europe in June 1972 were full and occasionally overfull.

This was apparently the case with a British European Airways Trident 1, bound for Brussels, at London's Heathrow Airport on June 18, 1972. The weather was clear; flying conditions were ideal. The plane had been inspected and appeared to be taking off normally, if sluggishly. It was overloaded with 118 passengers and crew.

The plane cleared the runway and retracted its landing gear, preparatory to climbing to cruising altitude. And then, like a wounded bird, it fell out of the sky into a field beyond the runway it had just cleared. The aircraft split upon impact, scattering bits and pieces of its fuselage, wings and engines all over the field.

Rescue crews rushed to the scene immediately, dousing the huge fires fed by the large amount of jet fuel aboard. Two passengers survived the impact but died later. A young girl was pulled from the wreckage by police, but she expired before an ambulance could arrive for her. An older man, found in the field, was taken to a London hospital, but he too died before the end of the day. The precise cause of the crash was never determined.

GREAT BRITAIN
WALES
CARDIFF
March 12, 1950

No reason has been given for the crash on landing of a charter Avro Tudor V at Cardiff, Wales, on March 12,

1950. Eighty soccer fans and crew members died; three men survived.

The legion of unexplained crashes under ideal conditions seems to be endless. One such incident is the odd crash of a chartered Avro Tudor V prop plane that carried 78 Welsh soccer fans returning from a triumphant match in Dublin to their deaths.

It was a one-hour, commuter-length flight from Dublin. The weather was nonthreatening; Llandow Airport at Cardiff was clear and ready to accept the landing of the aircraft.

The Avro circled and approached—far too low. Just short of the runway, the plane's forward section hit the ground, and the plane flipped over.

There was no fire, just an eerie, other-worldly silence surrounding this bizarre but fatal accident. Evan Thomas, a farmer who was one of the first rescuers on the scene, reported to the newspapers, "The smoke from the engines was curling from the wreckage. Through it walked two men. They were the only things that moved."

Three men survived the crash. Eighty other soccer fans and crew members were crushed to death.

GUAM
AGANA
August 6, 1997

A Korean Airlines Boeing B-747 crashed in heavy rain as it approached A. B. Won Guam International Airport in the early morning of August 6, 1997. Two hundred twenty-nine of the 254 aboard were killed.

There was plenty of blame to go around after the crash, in a blinding rainstorm, of Korean Airlines flight 801 at 1:42 A.M. on August 6, 1997. The plane was at the end of a long flight from Seoul, South Korea. The captain had completed several shifts in a row and was noticeably fatigued. Evidence from the black boxes after the crash showed that he was landing erratically.

But all of the fault did not lie with him. Korean Air was discovered to routinely conduct faulty and inadequate training for its crews, and this could have been a contributing factor in the first officer and flight engineer's failure to monitor and cross-check the captain's execution of the approach to the airport.

And this was not all. There are three vital aids that pilots depend on for instrument landings in foul weather. First, there is the glide slope transmitter, which shows a pilot how far he is from the ground. Then, there is the localizer, which shows how the plane is lining up with the runway. And, finally, there is the minimum safe altitude warning system (MSAW).

Two of the three aids were not operating that night. The airport's glide slope equipment had been down for maintenance since July 7—a not uncommon occurrence, and experienced pilots then usually go to a stairstep approach using localizer readings. If they err, the MSAW system is a check on them. But at Guam that night, the MSAW system was also down because of a software problem. Even if it had been operating, there is no guarantee that it would have been of much help. The FAA itself had ordered an "intentional inhibition" of the MSAW system at Guam, and the FAA had over a long period failed to adequately manage the system.

And finally, the reason all the systems at the Guam tower should have been monitored was that it was one of the few towers in the United States defined as a "contract tower," that is, a tower not manned by government-approved controllers. The use of contract towers had begun after President Reagan fired all the union controllers in 1982, and the use of privately contracted controllers in so-called low-activity airports expanded after that time as a government cost-cutting move.

The fatigued captain was fighting all of these odds as he attempted a "nonprecision" landing at the obscured airport at Agana, Guam, that rainy night. He and his plane missed the runway by 3 miles. That distance short of Runway 06R, the jet slammed into Nimitz Hill and exploded, setting fire to the jungle around it.

Miraculously, 22 passengers and three crew members survived the crash, but 229 were killed.

INDIA
BOMBAY
January 1, 1978

No official explanation was given for the midair explosion of an Air India Boeing 747 over Bombay on January 1, 1978. All 213 aboard were killed.

A mysterious and unexplained explosion aboard an Air India jumbo jet bound from Bombay to the Persian Gulf emirate of Dubai blew the plane apart and killed 213 persons on January 1, 1978.

Residents of the western Bombay suburb of Bandra said they saw the Boeing 747 break in two after the explosion and plummet into the Arabian Sea two miles offshore. It had only been airborne for four and half minutes and had begun to bank to the left to set itself on its course when it suddenly blew apart and plunged into the sea.

Search parties and rescue crews failed to turn up any survivors. All 213 aboard perished in this bizarre tragedy.

INDIA
NEW DELHI
June 14, 1972

Terrorism was suspected but never proved in the midair explosion of a Japan Airlines DC-8 over New Delhi on June 14, 1972. Eighty-seven died; six survived.

A suspicion of terrorist sabotage surrounded the mysterious crash of a Japan Airlines DC-8 jet at New Delhi's Palam International Airport on June 14, 1972. Only two weeks before, Japanese terrorists had massacred 25 innocent people in the International Airport in Tel Aviv, and threats of retaliation had been made. But investigators would never adequately prove this theory.

The plane had departed from Tokyo and was proceeding with a normal, clear-weather landing when it suddenly burst into flames, heeled over and crashed into the farmland that rimmed the airport, near the Jamuna River. Authorities reasoned that there was an explosion, either before or on impact, since pieces of the plane and its passengers were strewn over a two-mile radius. Eighty-seven died in the wreck and six survived, but their testimony cast no light upon the events that led to the crash.

INDIA
NEW DELHI
November 12, 1996

Three hundred and forty-nine people were killed when a Saudi Arabian Airlines Boeing B-747 taking off from Indira Gandhi Airport in New Delhi and a Kazakhstan Airlines Ilyushin IL-76TD landing at the airport collided in midair on November 12, 1996.

Midair collisions between airliners are very rare. In the United States, radar information to control towers includes airspeed and altitude, and since the tragic collision on December 16, 1960, of a TWA and United Airlines plane all U.S. carriers have had collision avoidance systems aboard.

But neither the Saudi Arabian Airlines Boeing B-747, taking off from New Delhi for Dharan, nor the Kazakhstan Airlines Ilyushin IL-76TD, landing after flying to New Delhi from Chimkent on the early evening of November 12, 1996, had collision avoidance systems, nor did the New Delhi airport have a radar system that revealed altitudes or airspeeds. In fact, its radar system was severely outdated, ("antiquated" according to pilots who flew in and out of it) and was due for improvement that was only partially completed.

Besides this, the Indira Gandhi Airport at New Delhi had only one runway available for civilian flights. Because it was used for both landings and takeoffs, it had, over the years, contributed vastly to India's reputation of having one of the world's worst air safety records.

And so, as the Saudi plane took off with 312 passengers—mostly Indians on their way to take jobs in Saudi Arabia and Muslims on their way to Mecca—at 6:33 P.M., New Delhi air controllers were simultaneously guiding the Kazakhstan Airlines jet, a cargo plane modified to carry 37 Kazakhstanis on their way to shop in the bazaars of New Delhi, through the same flight corridor in a gradual descent for landing.

The controllers asked the Saudi aircraft to climb to 14,000 feet and hold at that altitude, and the Kazakhstan aircraft to descend to 15,000 feet. According to recordings, they did not add the crucial words "and hold" to that request, but most crews would have assumed that this was their order.

Not the crew of the Kazakhstan plane, which was found, as its cockpit recorder later revealed, to be in an overly relaxed mood. Furthermore, the crew knew only minimal English, which was the language the controllers were speaking. So, instead of holding at 15,000 feet, which would have meant that the two planes would have cleared each other by 1,000 feet, the Kazakhstan Airlines plane continued to descend.

At exactly 6:40, it collided with the Saudi jet in a terrifying fireball and explosion. "We noticed out of our right hand a large cloud lit up with an orange glow from within the clouds," the pilot of a U.S. Air Force C-141 transport reported later. "Two fireballs emerged from the cloud and exploded into a field of fire when they hit the ground about a minute later," he concluded.

The planes, afire and in pieces, plummeted to earth immediately, the Saudi plane crashing into a mustard field, the Kazakhstan plane falling into a rice field in the heavily populated farming region near the village of Charkhi Dadri, 50 miles west of New Delhi in the state of Haryana. Miraculously, no one died on the ground, but wreckage was strewn over an area six miles wide.

The planes were nothing but charred skeletons when rescue parties reached the devastated fields. But if there was panic and horror in the air, there was an equal amount on the ground.

Looters, the grindingly poor people of the area, had reached the crash site before officials, and bodies were dragged from the wreckage and stripped of watches, jewelry and clothing, even socks.

The arrival of police and army contingents did little to better the scene. Ignoring the bodies strewn about, they set about building a helipad for the visit of Prime Minister H. D. Deve Gowda.

Ordinarily, salvage teams cordon off a site and attempt to post identifying flags by bodies that can be correlated with the plane's seating chart. This was not done, and instead, villagers and family members dragged bodies and tried to line them up so that trucks could take them to the hospitals, where they were laid on blocks of ice in the corridors.

Once there, the families were forced to negotiate with touts offering to expedite the embalming of the bodies, which was required for transport home by train or plane, or to obtain some of the scarce plywood coffins available from these touts. Some of the families, whose annual incomes were as little as $250 a year, were forced to pay $110 each for the embalming and the coffin.

Meanwhile, Hindu, Muslim and Christian groups battled over the proper way to dispose of the bodies. Approximately 150 that were beyond identification were buried or cremated at Charkhi Dadri, but it took another five hours of talks to reach a compromise to divide the remaining bodies among the three religious groups according to the percentage of passengers of each faith on the plane.

To add to this, the governments of India and Kazakhstan traded accusations of blame for the catastrophe. Officials of Kazakhstan Airlines refused any suggestion of fault, including failure to understand English by the Kazakhstani crew. Indian government officials denied any responsibility by Indian controllers, despite criticism of the incompetence of the Indian government by Indian newspaper editorials. In an offset to the turmoil on the ground, villagers gathered in the thousands to fetch water for relatives, to cook meals over wood fires and to serve the customarily sweet Indian tea to the distraught families.

Ultimately, dispassionate investigation placed the blame on the crew of the Kazakhstan Airlines plane, whose knowledge of English proved to be inadequate, whose casual behavior in the cockpit prevented them from seeing the possibility of danger and whose ignoring of the order to hold at 15,000 feet led to the inevitable collision.

Three hundred and forty-nine persons died in the tragic crash—312 passengers and 37 crew members. The accident killed more people than any other midtair collision, and it was the fourth-worst crash in aviation history up to that time.

INDONESIA
BUAH NABAR
September 26, 1997

On September 26, 1997, Garuda Indonesia Airlines flight 152, an Airbus A300 from Jakarta, crashed into mountainous terrain as it was attempting to make an instrument landing at Polonia International Airport in Medan because of extreme smoke conditions caused by nearby forest fires. All 234 aboard were killed.

To pilots, air traffic control (abbreviated ATC on official reports) is, in less than ideal landing conditions, like a mother to an infant or a master to a pet. The captain of a fully loaded airliner is effectively and entirely in the controller's hands, dependent upon him for his—and his passengers'—continued existence.

Such was the situation on the early afternoon of September 26, 1997, when Garuda Indonesia Airlines flight 152, an Airbus A300 with 234 aboard, was concluding a flight from Jakarta to Medan, on the northeastern coast of the Indonesian island of Sumatra. For years, irresponsibly set and uncontrollable forest fires had raged in Indonesia. Started by farmers, miners and road builders, they were eliminating much of the remaining rainforests. On September 26, 1997, the fires were continuing, unabated, and the haze from them hung heavily over much of Sumatra, and particularly over the city of Medan.

Earlier in the day, two cargo ships, navigating blind in thick smoke from the fires, had collided in the Strait of Malacca, between Indonesia and Malaysia. Twenty-eight crew members had drowned in the collision.

And so, the visibility for the pilot of Garuda Airlines flight 152 was anything but ideal. In fact, the ground was almost totally obscured. He radioed ahead to Polonia International Airport in Medan to guide him in on an instrument landing. The airport responded. The visibility at the airport was 600 yards; they were equipped with modern guidance systems and would be able to get him down safely.

The crew and passengers aboard the liner prepared for landing. They began their descent, looking forward to an early arrival in Medan. Then, suddenly, all schedules became irrelevant. Where the airport should have been was, instead, a series of tree-covered mountains that abruptly appeared out of the enshrouding smoke. The jet slammed into the ground in remote, hilly terrain 30 miles short of the airport, bursting apart and strewing wreckage and bodies over a wide area. Rescuers who hacked their way to the site found some intact bodies hanging from trees. Others were buried in mud and hardly visible. Others came apart as the plane had. There were no survivors. All 234 aboard the liner had perished in the cataclysmic crash.

What had happened? Investigators could not fathom why the plane was where it was, or why it was at such a low altitude. Normally, a plane still 20 miles from an airport would be thousands of feet in the air.

Attention first focused on Garuda, Indonesia's flagship airline. It had been criticized in the past for poor pilot training and poor performance, racking up a total of six major crashes since 1982.

Then, the two black boxes were retrieved, and a chillingly different scenario revealed itself. The pilot was relying on the air control system and controllers to guide him through the haze into the airport, but the controllers on the ground at Medan were giving him wrong information. They sent him a series of erroneous commands, which led him not to the airport, but into the mountainous terrain in which he crashed. It was not pilot, but air controller, error that brought down Garuda Indonesia Airlines flight 152 in that smoke saturated afternoon of September 26, 1997.

ITALY
PALERMO
May 5, 1972

Pilot error caused the crash of an Alitalia DC-8 into Montagna Lunga, near Palermo, on May 5, 1972. All 115 aboard were killed.

Palermo's Punta Raisi Airport is a treacherous trap. Ringed by high mountains, it has claimed even the most careful of pilots. Night landings are exercises in care, skill and prayer, and most airlines try not to land or take off after dark at this, one of Europe's most dangerous landing fields.

On the night of May 5, 1972, an Alitalia DC-8 jet from Rome attempted to land at Punta Raisi. The pilot, a veteran of the run, apparently misjudged his approach and came in too low to clear the mountain range. Minutes before he should have safely touched down on the runway, he slammed into the side of Montagna Lunga, a 12,250-foot peak near the airport. The plane exploded on impact, setting forest fires that would burn for days. Every one of the 115 crew members and passengers aboard perished in the crash.

JAPAN
MORIOKA
July 30, 1971

A Japanese Air Force F-86 Sabre jet without radar collided in midair with an All-Nippon Boeing 727 on July 30, 1971. Pilot error was blamed in this crash that killed all 162 passengers aboard the jet. The military jet pilot parachuted to safety.

The year 1971 was particularly bad for near collisions in the sky. There were 600 of these reported in the United States that year, and 200 in Japan. There were just too many civilian and military aircraft flying in close proximity to one another in the two countries, and experts felt that a tragedy was bound to occur. It did, on July 30, over Morioka, in the so-called Japanese Alps.

Flight 58, an All-Nippon Boeing 727, was loaded largely with members of a Japanese society dedicated to the memory of war dead. They had just been to Hokkaido on a pilgrimage and were heading back to Tokyo. The takeoff from Chitose Airport was uneventful; it was a clear day, and visibility was good. They reached their cruising level of 28,000 feet easily and without incident.

In the same area, Sergeant Yoshimi Ichikawa, a student pilot with 21 hours in the air, was at the controls of a Japanese Air Force F-86 Sabre jet. Neither he nor his instructor, Captain Tamotsu Kuma, who was flying in another Sabre jet, had the benefit of radar.

The airliner did, and thus it is somewhat puzzling that its commander did not see the military aircraft until it was too late. The planes collided in midair. The airliner exploded upon impact, and pieces of it and its passengers were scattered over mountaintops. All 162 jet passengers died in this tragedy, but Ichikawa, the student pilot of the jet, parachuted, to safety.

He was immediately arrested and charged with involuntary homicide, but he was acquitted.

JAPAN
MOUNT FUJI
March 5, 1966

Pilot error and wind conditions claimed a BOAC Boeing 707 on March 5, 1966. The aircraft collided with Mount Fuji, killing 124 people.

A terrible crash had taken place a month before at Tokyo International Airport (see p. 35), and as BOAC flight 911 taxied out for takeoff, its passengers could plainly see the charred remnants of that catastrophe. None of them could have known that within a few minutes, a second tragedy, almost as grim as the first, would snuff out their lives.

The Boeing 707 was carrying a full passenger load the afternoon of March 5, 1966, but its takeoff was without effort. It reached 6,000 feet and circled close to Mount Fuji, which was clearly visible in the sharp afternoon sunlight.

What was not visible, to either passengers or flight crew, were the killer air currents that circle Mount Fuji. Of the force and nature of tornadoes, they are a pilot's anathema, and captains of aircraft give them wide berth. But for some reason that will never be explained, the pilot of BOAC flight 911 did not avoid the mountain and its lethal winds by a wide enough margin.

Caught in the winds, the airliner blew apart in midair, flinging wreckage and bodies against the slopes of Mount Fuji. One hundred twenty-four people—all those aboard—died, raising the total number of crash victims at or near Tokyo International Airport in one month to 257.

JAPAN
MOUNT OGURA
August 12, 1985

Improper repairs of a Japan Airlines Boeing 747SR led to the loss of part of the airplane's tail section and loss of control of the aircraft, which crashed into the side of Mount Ogura, Japan. The worst air disaster in the world involving a single passenger plane, it claimed 520 lives. Only four passengers survived.

The worst air disaster in the world involving a single passenger plane occurred on a mountainside in central Japan on the evening of Monday, August 12, 1985. A Japan Airlines jumbo jet veered eratically in an uncontrolled path and slammed into the side of Moung Ogura, one of Japan's highest peaks. There were 524 people aboard; only four of them survived.

Japan Airlines flight 123 left Tokyo's Haneda Airport at 6:15 P.M. for its 250-mile flight to Osaka. Travel had been especially heavy in the preceding few days. Thousands of Japanese were returning to their hometowns for a traditional midsummer festival known as Obon, in which the souls of ancestors are honored.

The plane was a Boeing 747SR, especially configured for Japan, where flights with 500 passengers were not uncommon. This one was completely filled.

The aircraft had seen a lot of service and, seven years before, had made a hard landing at Osaka. The impact had damaged the lower rear fuselage, and Boeing maintenance people had supposedly repaired it.

Thirteen minutes after takeoff, flight 123 was clearly in trouble. The pilot, Masami Takahama, radioed that the right rear passenger door had "broken." One of the survivors, Yumi Ochia, an off-duty assistant purser for Japan Airlines, said later that there was a loud noise that originated in the rear of the plane and over her head. Following it, there was an instant "whiteout" as a thick fog filled the cabin—a sure indication that the cabin pressure had dropped and water vapor had condensed. The plane had reached its cruising altitude of 24,000 feet by now, and the crew immediately radioed Tokyo for permission to drop to 20,000 feet.

But a far greater problem than reduced cabin pressure had been created by whatever caused the loud noise heard by the off-duty purser. The impact of it had been great enough to force the plane's nose up. And worse, it had apparently done something major to the craft's control system. Among the 43 sensors placed throughout the airliner, the one in the rear, near the horizontal stabilizer, had registered the greatest shock. It had broken off. Captain Takahama was unable to control the plane.

"Immediate trouble, request turn back to Haneda," a crew member radioed to Tokyo at 6:25 P.M., seconds after the first transmission. Four times over the next 30 minutes, the captain radioed that he had no control of the plane. He was attempting to steer it by alternately increasing and throttling down engine power.

By 6:45, the plane had lost an appreciable amount of altitude. It was flying at only 9,300 feet, and the captain had all but lost complete control. "We may be finished," he said to his co-pilot. At the same time, the black box recorder picked up a voice in the background announcing emergency-landing instructions in the cabin.

Ten minutes later, Tokyo radioed flight 123 that it could land either at Haneda or at a U.S. air base in Yokota, northwest of Tokyo.

But by then, Flight 123 had stopped answering. One minute after the clearance had been given for an emergency landing, it had slammed into the side of Mount Ogura, in Gumma prefecture, 70 miles northwest of Tokyo. It exploded on impact, setting fire to acres of forest on the side of the mountain.

An eyewitness later told the Japan Broadcasting Corporation that she had watched the plane crash in "a big flame" after it flew over her house. The explosion was "followed

Rescuers comb the wreckage of Japan Airlines flight 123 after it crashed into Mount Ogura, north of Tokyo, on August 12, 1985. Five hundred twenty died in the crash of the Boeing 747 en route from Tokyo to Osaka. (National Transportation Safety Board)

by a white smoke which turned into a black mushroom-like cloud," she said.

It would be 14 hours before the first rescue teams, composed of three dozen airborne troops, could be lowered onto the mountain by helicopters, joined by local police who hiked through a steady rain to the crash site.

At first, sabotage was thought to be the culprit. But a Japanese Navy destroyer on routine patrol reported finding a section of the plane's tail in Sagami Bay, just off the Miura Peninsula, southwest of Tokyo and 80 miles southeast of the crash site. The tail had apparently fallen off.

Now the history of the aircraft began to help solve the mystery of the crash and the 32 terrible minutes that preceded it. One of the first questions to be solved was what had caused the rudder and other hydraulic systems to stop functioning.

The answer was in the repairs made seven years previously to a ruptured bulkhead. They had been improperly made, even though they had been done by a team flown by Boeing to Japan. Shaped somewhat like an umbrella canopy, the bulkhead was at the very back of the passenger cabin and separated the highly pressurized cabin from the unpressurized tail section, which was immediately behind it.

A single line of rivets was used for part of the repair, instead of the double line of rivets called for in the manual. With the expected metal fatigue brought on by seven years of constant use, the improper repair eventually gave way, and the results were monumentally tragic.

The blame apparently was not entirely Boeing's, at least in the minds of some officials and employees of Japan Airlines. Shortly after the findings of the investigative panel were released, there was a major shake-up in the executive structure of Japan Airlines. The president and two executives were replaced.

But this was not the end of the aftermath of this record catastrope. On September 21, police in Yokohama were called to the home of Hiro Tominaga, a Japan Airlines maintenance official who had been negotiating compensation payments with families of the crash victims. Tominaga was found on the floor of his home, dead of knife wounds inflicted to his neck and chest. A four-inch-long knife was found near his body. A note nearby read, "I am atoning with my death."

JAPAN
NAGOYA
April 26, 1994

A China Airlines Airbus A300, enroute from Taiwan to Nagoya, Japan, crashed on landing at the Nagoya airport on the night of April 16, 1994, killing 264 of the 271 aboard.

It was a clear and calm night on April 16, 1994, when China Airlines flight 140, an Airbus A300, arrived on a flight from Taiwan to Nagoya in central Japan. Given clearance for a landing at 8:14, both the crew in the Airbus and controllers on the ground relaxed for an orthodox landing. At 8:15 the plane was almost at the airport's single runway, flying on autopilot at a height of 1,000 feet.

And then, something went terribly wrong in the Airbus's cockpit. The first officer inadvertently triggered the TOGA (take-off-go-around) lever; "CAL 140 going around," he radioed to the tower. In the cockpit, the crew frantically tried to override the situation by turning off the autothrottle and reducing their air speed. But it was the wrong decision. The tailplane setting had moved automatically to a maximum nose-up position, and the jet abruptly began to climb.

Even in those seconds, if the crew had disengaged the autopilot, they might have been able to save the situation, but they did not, and the plane shot up steeply, stalled and plunged backward, tail first toward the runway. It hit on its tail, dug into the tarmac and burst immediately into flames.

"Suddenly, when the plane was just at the edge of the runway, it began to move strangely and wobble," Masumi Inazumi, who had just arrived on another flight and was in a bus to the terminal, told television reporters.

Another witness, Minoru Furuta, told another reporter that he saw "flashes from the two wings. The plane rolled to the right . . . then I saw flames and heard explosions."

Firemen and rescue workers rushed to the scene. The flames were furious, trapping survivors in the plane. Firemen were able to extract nine, but the rest were killed by the horrendous heat of the fire. "I could see many dead people lying on the ground," fire chief Jinji Magamori said later, noting that rescuers carried out people who seemed as if they had a chance of surviving. Others, he said, were burned beyond recognition.

Of the nine survivors taken to local hospitals, two died later that night and two during the following week, raising the death toll to 264, which made this Japan's worst plane disaster since August 12, 1985, when a Japan Air Lines Boeing 747 crashed into a mountain northeast of Tokyo, killing 520 of the 524 people aboard (see p. 32). And it would surpass by three deaths the Nigerian DC-8 crash in Jidda, Saudi Arabia (see p. 40), making it the 10th-worst air crash on record up to that time. And it all happened because members of the crew, in the crucial seconds before the crash, made tragically, terminally wrong decisions.

JAPAN
TOKYO
June 18, 1953

Unspecified engine trouble caused the crash of a U.S. Air Force C-124 Globemaster shortly after takeoff from Tachikawa Air Base, near Tokyo, on June 18, 1953. All 129 aboard perished.

A U.S. Air Force C-124 Globemaster, a double-deck transport, was loaded with 129 military passengers and crew when it took off from Tachikawa Air Base, near Tokyo, on June 18, 1953.

The flight was routine during the Korean War. Relief personnel were being ferried to Korea; the wounded and those in need of R and R would board the return trip to Japan. The weather was clear; the takeoff was clean.

But four minutes after takeoff, the flight developed engine trouble. One engine quit completely, and the pilot radioed that he would be returning to the field. He never did. Less than five minutes later, the plane dove into a rice paddy, exploding and bursting into flames. None of the 129 aboard survived, and no cause for the crash was ever established.

JAPAN
TOKYO
February 4, 1966

No official reason was ever given for the crash of an All-Nippon Airways Boeing 727 landing at Tokyo International Airport on February 4, 1966. All 133 persons aboard died.

The year 1966 was a terrible one for air crashes at Tokyo International Airport. Two took place within one month of each other, on February 4 and March 5 (see p. 32) The precise reasons for both crashes would remain forever unexplained.

At a little before 7 P.M. on February 4, 1966, an All-Nippon Airways Boeing 727, arriving from Sapporo and piloted by veteran captain Masaki Takahashi, radioed for clearance to land. The weather was cloudless and calm. The controllers in the tower cleared the flight for landing, and the 727 began its long circle of approach over Tokyo Bay.

At exactly 6:59 P.M., it disappeared from the controllers' radar screens. A few minutes later, when it should have been coming in for its final approach, it did not appear. And then the tragic news arrived by telephone: Fishermen had seen a "pillar of fire" soar overhead and dive into Tokyo Bay.

Rescue efforts went on through the night, but it would be dawn before bodies and the remains of the plane could be brought to the surface. No one survived. One hundred thirty-three people died.

MOROCCO
CASABLANCA
July 12, 1961

A fear of exposing Soviet military secrets to the West caused the pilot of a Czechoslovakian Airlines Ilyushin-18 to ignore instructions to divert from fogbound Casablanca Airport to a nearby U.S. air base. The plane crashed, killing all 72 aboard.

International politics seem to have figured prominently in the terrible tragedy that befell the passengers in a Ceskoslovenske Aerolinie (Czechoslovakian Airlines) Ilyushin-18 flight from Zurich to Prague, via Morocco.

The flight was scheduled to stop at Casablanca on the evening of July 12, 1961. But Casablanca was completely blanketed with a heavy fog, and the airport had closed down. Josef Mikus, the veteran pilot captaining the flight, was denied permission to land and told to divert to the U.S. air base at Nouasseur, a mere 15 minutes away. It was still open and relatively fog free.

The plane had plenty of fuel. It could easily have made the distance to the U.S. air base. However, Captain Mikus knew that there were several Soviet flying instructors aboard, on their way to train revolutionaries in Ghana and Guinea. Furthermore, there was a huge cargo load of films and pamphlets also destined for these two insurrection-torn African republics. To Mikus, landing at an American military base was tantamount to landing in the arms of the enemy.

At least that was the explanation given by authorities afterward, after Mikus ignored the instructions of the control tower at Casablanca and nosed his turboprop in for an instrument landing. But without help from the controllers, Mikus and his plane had no chance of success. Midway to its landing, the aircraft hit a power line. The heavy cable ripped two engines from the plane's wings, hurling them into midair, where they exploded. Two towers collapsed inward, toward the plane, which was flung heavily to earth.

It would take 25 minutes for rescuers to reach the scene of the crash, and even then, according to the *I.C.A.O.* (International Civil Aviation Organization) *Accident Digest,* Number 13, "When the police arrived at the scene calls for help were heard coming from the wreckage, and an attempt was made to rescue the passengers, but a fire started, and it was impossible to continue operations."

All 72 persons aboard died.

NEPAL
KATMANDU
September 28, 1992

A Pakistani International Airlines Airbus A300 struck one of the mountain peaks surrounding Katmandu airport on September 28, 1992, killing all 167 aboard.

The weather was clear and calm on September 28, 1992, when a Pakistani International Airlines Airbus A300 requested permission to land at Katmandu's airport. The flight, containing, among its 167 passengers and crew, 89 Europeans, 10 Nepalese, two Bangladeshis and 12 Pakistanis, was flying from Karachi to Katmandu, a city ringed by mountains. A high percentage of the Europeans aboard were headed for mountain climbing vacations in the Himalayas.

A mere two and a half months earlier, on July 31, a Thai Airways Airbus A310-300 had overshot the runway at Katmandu and smashed into a snowcapped mountain, killing 113. The cause had been faulty wing flaps that the pilot had struggled with moments before the overshoot and crash.

Now, it appeared that the Pakistani flight was in trouble. Its approach should have been at 9,000 feet, but its altitude was only 7,500 feet, and the control tower advised the pilot of this. Still, the plane continued on its dangerously low landing glide. Within moments, it smashed into a mountain top 14 miles south of the airport.

It would take helicopters to get to the site of the crash, and when the first one arrived and lowered the first rescuer to the scene by rope, the report was grim: There were no survivors within the burning wreckage of the plane. Investigation discovered no mechanical or electronic failure aboard the plane, and pilot error became the assumed cause of the tragedy.

NETHERLANDS
AMSTERDAM (DUIVENDRECHT)
October 4, 1992

An El Al Boeing 747 cargo jet with a crew of three, one passenger and 114 tons of cargo, bound from Amsterdam's Schiphol airport to Tel Aviv, crashed shortly after takeoff on October 4, 1992, into an apartment complex in Bijimermeer, just outside Amsterdam. In the worst toll of victims on the ground from an air crash, 100 died, including the three crew members and the passenger.

It was dinner time, 6:30 in the evening, and families who lived in the two 11-story apartment buildings of the Bijimermeer section of Amsterdam were at home after Sunday outings. The apartments housed a great many of the poorest residents of Amsterdam, many of them immigrants from Suriname and Aruba. It was a chilly autumn evening, with a strong wind blowing.

An El Al Boeing 747-2000 cargo jet, with a crew of three, one passenger and 114 tons of computers, machinery, textiles and perfume, took off from Amsterdam's Schiphol airport at 6:27 P.M. bound for Tel Aviv.

Nine minutes into the flight, the pilot, Isaac Fuchs, radioed a Mayday distress signal to the tower, adding that his number 3 engine was on fire, then that his number 4 engine was aflame. First one, then both right engines now dropped from the plane and spiraled earthward. The jet tipped crazily up on one wing as the pilot wrestled with the controls and shouted a request to the airport for permission to first dump fuel into a nearby lake and then return to the airport.

But he never made it. Six miles short of the airport, at 6:33 P.M., he radioed the tower, "Going down," and lost control of the jet. At 6,500 feet it went into a wingtip stall, keeled over and plummeted into the apartment complex in Bijimermeer. Eighty apartments crumbled as if they were mounds of sugar, and an immense fireball raced through the complex, igniting every square foot. Screams pierced the night air as flames engulfed the buildings, and some residents of the apartments leaped from windows to escape the fire. Gusts of white smoke broke free from the smoldering mass and blanketed the area as firemen poured water on the ruins and Dutch Navy helicopters circled above, beaming floodlights on the scene. Pandemonium broke out as relatives and friends of inhabitants of the complex rushed to the spectacle, making the firemen's jobs more difficult.

It would be days before the final count of dead could be assessed, because some of the residents in the apartments were undocumented immigrants. The final toll hovered slightly more or less than 100, including the three crew members and the passenger on the plane. One engine of the aircraft was immediately retrieved from a lake into which it had fallen.

Israeli officials at first suspected and suggested sabotage, but their theories were soon abandoned in favor of a structural weakness in the Boeing 747-2000. The previous December, a China Airlines 747-2000 had also lost its two right engines as it was taking off in Taiwan. The investigation there focused on the fuse pins that connect engine struts to the wing.

At first, Boeing refused to connect the two crashes or admit their likenesses. But by the middle of October, the company conceded that there were at least 15 instances of fatigue-related cracking in the pins in the preceding seven years.

Further examination revealed that a grim domino effect had occurred when engine number 3's fuse pin, because of corrosion, had separated from its pylon and damaged the leading edge of the right wing. Its separation caused the number 4 engine and pylon to come loose and fall from the wing, taking with it several cabling systems that in turn made the plane uncontrollable. No matter how the crew tried at this point, they could neither keep the El Al jet in the air nor land it, and it rolled over and crashed.

In total, it was the worst air disaster in Dutch history and the worst toll ever of victims on the ground in an air crash.

NEW GUINEA
BIAK ISLAND
July 16, 1957

An unexplained engine explosion hurtled a KLM Super Constellation into the sea shortly after takeoff from Biak Island, New Guinea, on July 16, 1957. Fifty-seven died.

A KLM Super Constellation took off in pleasant weather on the afternoon of July 16, 1957, bound from Biak Island in Dutch Guinea to Amsterdam. Within moments an engine exploded. The pilot, Captain Bob de Roos, fought for control of the plane while his copilot went about the methodical closing down of the engine.

But no amount of closing down could reverse the forces the explosion had started. Within minutes, the remaining engines were afire and useless. Without power, and with most of the wing controls also useless, the aircraft nosed over and plunged into the sea, a mere five miles from its takeoff point.

The forward part of the plane disappeared instantly beneath the surface of the water. The extreme tail section broke off and floated. Papuan natives, seeing the plane hurtling over their heads and into the sea, set off immediately in canoes. Their speed, skill and calmness saved 11 terrified passengers huddled in the detached but still floating tail section. Fifty-seven other passengers, trapped in the rest of the fuselage, drowned or were burned to death.

NORWAY
SPITSBERGEN
August 29, 1996

A Tupolev TU-154M, operated by Vnokovo Airlines, on a flight from Moscow to Spitsbergen, Norway, on August 29, 1996, struck a mountain seven miles from the airport. The crash killed 141.

The Arctic island of Spitsbergen is the largest island in the Svalbard archipelago, roughly 350 miles north of the Norwegian mainland. A 1920 treaty gave Norway ownership of the archipelago, but the Soviet Union acquired mining rights in 1935, and, in fact, of Svalbard's total population of 2,500, almost two-thirds are citizens of former Soviet republics. These are working families, and the men work in two mines—Pyramiden and Barentsburg—extracting some 600,000 tons of coal a year, which is approximately twice the output of Norwegian mines.

On the night of August 19, 1996, a Tupolev TU-154M, operated by Vnokovo Airlines, one of the so-called babyflots formed when Aeroflot was broken up in 1991, made an uneventful flight from Moscow to the vicinity of Spitsbergen. It was loaded with a crew of 12 and 129 passengers, most of them coal miners and their families. The workers were a relief force employed by the Trust Arktik Ugol mining company. They were scheduled to replace an equal number of miners and family members who were waiting at the airport to take the same jet back to Moscow.

Flight 2801 approached the airport at Longyearbyen, Spitsbergen's main settlement, at 10:15 A.M. There were, as there often are in this part of the world, banks of low clouds that made visibility difficult. The normal approach to Svalbard Airport is a wide turn over the ocean, but for some reason the pilot of flight 2801 chose an entirely different approach, through a narrow valley.

The mistake could have been caused by unfamiliarity with the airspace—the crew had never made the flight before. Still, when uncertainty in bad weather occurs, the ordinary procedure is to abort the landing temporarily and ask for instructions from the airport tower. None of this happened. At 10:15 all radio communication between the ground and the plane ceased. At 10:22 the plane smashed into the 3,000-foot peak of Operafjellet mountain, seven miles short of the airport runway.

A Norwegian search party set out in two helicopters and a light plane from Svalbard Airport immediately. At 12:06 P.M. they spotted the wreckage, most of it spread crazily about a snowy plateau. The rear section of the jet had tumbled down a cliff, causing a small avalanche.

Rescue workers were lowered to the scene, but they found no survivors. All 141 aboard the plane had perished. Still, the rescuers hunted for evidence of the cause of the accident and extracted bodies and portions of bodies from the decimated plane. By 4 P.M. a heavy fog forced them to temporarily suspend their efforts.

Later, as evening began to fall and the fog dissipated, another Norwegian helicopter crew lifted two prefabricated cabins onto the plateau near the wreckage. Three sentries were dropped by rope to the site to ensure it was not disturbed by any of the 2,000 polar bears in the area.

Vnokova Airlines had suffered from lack of cash and poor maintenance in the months preceding the crash. Many of the aircraft in all of the Russian fleets were long past retirement age, but in this case it was not the plane but the crew that brought about the tragedy.

PAKISTAN
KARACHI
March 3, 1953

Pilot error was blamed for the first fatal jetliner crash in history, on takeoff from Karachi Airport on March 3, 1953. On a test run, the Canadian Pacific Comet 1A turbojet burst into flames on impact, killing all 11 men aboard.

The first fatal jetliner crash in airline history took place while deHaviland was in the process of refining the design of its Comet 1A turbojet. The Comet was not a trouble-free airplane; only six months before this catastrophe, one had crashed near Rome, but none of the technical and flying crew aboard were injured.

On March 3, 1953, the Comet that had been sold to Canadian Pacific Airlines was being tested by 11 technicians and crew members. It had been loaded with 5,000 extra gallons of fuel before takeoff—an enormous addition of weight and enough to tip its nose dangerously up as it struggled to become airborne.

Had he seen that this would put the plane in danger of crashing, the pilot could have aborted the takeoff. The pilot apparently did not, and thus the board of inquiry later attributed what happened next to "pilot error."

Unable to lift off, the plane overshot the runway, climbed a few feet in the air, wavered and struck a brick culvert with one wing. This spun it off course, and it immediately slammed into and then through a wall. The fuel ignited; the plane burst into consuming flames, and all 11 men aboard were incinerated.

PERU
AREQUIPA
February 29, 1996

A Faucett Airlines Boeing B-737-222, landing at Arequipa, Peru, on a flight from Lima on the night of February 29, 1996, slammed into a hillside five miles from the airport, killing all 123 aboard.

Before it left Lima for Arequipa, the Boeing B-737-222 owned and operated by Compañía de Aviación Faucett SA, a Peruvian carrier, was cleaned and polished. According to the manufacturer's instructions, the maintenance crew had, while they cleaned, taped moisture-resistant paper over the plane's six static ports—three on each side of the fuselage. These openings in the fuselage of airplanes measure air pressure, and the information from the instruments contained in them is used in calculating altitude and air speed. If they are disabled, tragedy can readily occur. On the afternoon of February 20, 1996, the Lima maintenance crew failed or forgot to remove some of the masking tape they had used to cover these ports.

The jet loaded its 123 passengers and crew in Lima and conducted an eventless flight until it neared its destination in Arequipa, in southern Peru. At approximately 8:10 P.M. the pilot radioed the airport in Arequipa that his instruments had "gone haywire" and he did not know where he was. The airport tried to reach him, but his radio had gone dead.

One can only imagine what was taking place in the cockpit at that time. The instruments in the covered static ports were sending wrong information to the crew. Most important, they were sending wrong altitude readings. The pilot, now hopelessly lost, thought he was flying at 9,500 feet when he was actually flying at 8,640 feet. The difference was enough to slam the jet into a hillside five miles from Arequipa, killing all 123 aboard. The rocky site was only about two miles from a highway, and so relatives and friends of those who had perished walked to the site and watched that night and the next day as the bodies of the passengers, many burned beyond recognition, were carried on stretchers by firefighters and lined up in the bottom of a ravine.

It was the deadliest crash in Peru's history, but not the last that was attributed to the same cause. In October, a 757 dove into the Pacific Ocean off Peru, killing all 70 aboard. In that crash, too, three of the static ports on the left-hand side of the 757's fuselage were covered with masking tape that the maintenance crew had applied during the polishing process. In that crash one automatic system indicated that the plane was flying too slowly and would fall out of the sky, while another sounded an alarm that it was flying too fast.

"At impact into the Pacific Ocean," the NTSB report announced, "the captain's flight instruments were reading approximately 9,500 feet."

Finally, after this crash the FAA, prompted by the NTSB, directed that static ports on all aircraft be covered not with masking tape but with the brightly colored covers with warning flags attached that McDonnell Douglas and Airbus Industries were already prescribing for maintenance of their planes.

It was a valid directive but one that came too late for the two planes and their passengers that crashed in Peru in 1996.

PERU
LIMA
November 27, 1962

No explanation was given for the crash, in clear weather in the Andes near Lima, of a Varig Airlines Boeing 707 on November 27, 1962. All 97 aboard perished.

The Andes have claimed numerous victims in a number of air crashes. Some passengers have survived to write books and magazine articles about their collisions with these needle-like peaks. However, there would be no survivors of the crash of a Varig Airlines Boeing 707 in the late afternoon of November 27, 1962, and no record of the reasons for the mysterious crash in clear weather.

The flight from Rio de Janeiro to Los Angeles encountered trouble just before it was to land in Lima, Peru. At 5:30 P.M. the pilot radioed the airport in Lima stating that there was trouble with the aircraft. Before he could explain further, transmission abruptly ended.

It would be 3:00 A.M. before a search party would reach the demolished plane, strewn over the face of a peak in the Andes. All 97 aboard were dead.

PHILIPPINES
SAMAL ISLAND
April 19, 2000

A Philippines Air Boeing 737, on its second try at landing at the Davao Airport in the Philippines on the morning of April 19, 2000, crashed into a coconut plantation on a mountain on Samal Island. All 131 aboard were killed, making this the Philippines' worst air crash.

Air Philippines Flight 541, a Boeing 737 traveling from Manila to Davao, arrived over the Davao Airport on time, shortly after 6:30 A.M., on April 19, 2000. The weather was good except for some low-lying clouds, particularly over the mountainous region of Samal Island.

The crew radioed in and began their final approach, but just before they were to touch down they were forced to abort the landing. Another airplane was on the runway.

They ascended normally, banked gently and began another circle of the airport, this time attempting the landing from the opposite direction, but now the low clouds were causing visibility problems. The pilot noted them and told the airport that the flight was seven miles away and on its final approach.

And that was the last message from the plane. A few seconds after it was sent, the jet smashed into Mt. Kalamgen, on Samal Island, a mountainous island near the airport but far enough away not to normally cause problems for ascending or descending aircraft.

The plane exploded and caught fire upon impact, and its pieces rained down on a coconut plantation on the side of the mountain. Rescuers who rushed to the scene found a smoldering, fragmented liner with only a small part of its tail section intact. Bodies, pieces of seats and pieces of the fuselage were strewn over the entire plantation. There were no survivors; all 131 aboard the jet were killed in the crash.

The cause, as of this writing, had still not been firmly established. There seemed to be nothing mechanically or electronically wrong with the airplane, although, ironically, two days earlier an FAA panel had recommended that Boeing redesign the rudder assembly on all models of the 737, which is still the most widely used passenger jet in the world.

The fact that the point of impact of the crash was at 500 feet and the necessary altitude to maintain flight over the area was 1,500 feet led to the supposition that the pilot may have misread his gauges. Whatever the cause, it was the worst air crash in the history of the Philippines.

PUERTO RICO
SAN JUAN
April 11, 1952

The failure of airline personnel to follow basic safety procedures during a ditch at sea near San Juan killed 52 people aboard a Pan American DC-4 on April 11, 1952.

A failure to follow basic procedure by the cabin and cockpit crew aboard a Pan American DC-4 on April 11, 1952, accounted for 52 needless fatalities.

The flight took off uneventfully from San Juan, Puerto Rico, bound for New York. But nine minutes into the flight, the plane developed engine trouble. With one of its motors inoperable, it began to descend. The pilot had time to decide between ditching the plane in the sea, four miles from Puerto Rico, or taking his chances over land. He opted to go down at sea and informed the San Juan tower of his intent.

Once this course had been determined, it was up to the crew to prepare the passengers for the ditching effort by informing them of the location of life jackets and rubber rafts—both of which were aboard in sufficient quantity to save everyone. But this crucial and elementary drill was never carried out.

As a result, when the aircraft struck the water and floated, passengers panicked. To compound the terror, the crew opened the wrong doors to debark the passengers and let enormous amounts of water into the cabin, thus hastening the sinking of the craft. Terrified, confused and without life jackets, the passengers huddled on the plane's wing. Only one life raft was inflated, and this floated off with only seven survivors aboard.

Three minutes later the aircraft sank beneath 15-foot waves, carrying 52 passengers to a watery death. A handful managed to stay afloat until ships could rescue them.

RUSSIA
IRKUTSK, SIBERIA
January 3, 1994

A Balkal Air Tupolev-154 bound for Moscow made a crash landing in a field just beyond Irkutsk, Siberia, on January 3, 1994. The plane destroyed 3 farmhouses before exploding, killing 125 people.

Distances between cities in Russia are great, and so the flights that many of its airliners make are long. The Balkal Air Tupolev-154 that left the ground in Irkutsk, Siberia, on January 3, 1994, had a 2,500-mile flight to Moscow ahead of it, but this particular journey would be considerably shorter than that.

As the jet was partway to its cruising speed, a sudden lurch informed the 120 passengers and crew aboard that something was amiss. Within moments one of the three jets on the plane began to spout flames. The pilot immediately radioed the Irkutsk tower to request permission to return.

He turned the plane, which was loaded with jet fuel, around and headed back to the airport. And then the unthinkable happened. Both of the other jets shut down. Powerless and on fire, the plane began to glide back to the airport, but it never made it. Devoid of power in two engines, the jetliner went into a flat spin and slammed into a field, skidded into three farm houses, shattering them, killing five people who were in them and finally, as the load of jet fuel ignited, exploding in a huge fireball.

Rescue workers were on the scene in minutes, but there was nothing that could be done for either the passengers and crew of the jetliner or the inhabitants of what had once been farmhouses strong enough to withstand a Siberian winter.

All perished. The plane had been in the air for only 12 minutes.

RUSSIA
IRKUTSK, SIBERIA
July 4, 2001

A Vladivostokavia Tupolev 154M, on a flight from Yekaterinburg to Vladivostok, crashed in a Siberian meadow while attempting to land on an intermediate stop at Irkutsk on July 4, 2001. The death toll was 145, making this Russia's worst commercial air disaster.

At a little before 2 A.M. on July 4, 2001, a Vladivostokavia Tupolev 154M, flying from Yekaterinburg in the Ural Mountains to Vladivostok in the far east prepared to land on an intermediate stop in the Siberian city of Irkutsk.

Distances between stops are long on this continent-wide run, and the first officer was flying the jet. Whether he was fatigued or inexperienced no one will ever know, but he missed his first approach, circled and began again. This time, his approach was at too wide an angle, which set off an alarm at 2,500 feet.

Now, consternation rose in the cockpit of the jet as it entered a flat spin. The captain took over the controls and ordered the throttle to be advanced. There was a 22-second fight for control of the plane, which was not only spinning but descending, and at 2:10 the plane slammed belly first into a Siberian meadow near the rural community of Burdakovka, approximately 20 miles east of Irkurkst and on the southern shore of Lake Baikal.

Residents of the village were shaken from sleep by the sound of a huge explosion as the Tupolev hit the ground. The reports were so dramatic, in fact, that president Vladimir V. Putin met with his top aides and instructed Prime Minister Mikhail Kasyanov to form a commission to investigate the crash. There was, in many minds, a suspicion that the airliner had been brought down by a terrorist bomb.

No evidence was discovered to support the supposition, and pilot error seemed to be the cause of Russia's worst air crash. This catastrophe claimed 145 lives, exceeding the death toll of 143 established by the Tupolev 154 crash on the Norwegian island of Spitsbergen in August of 1996 (see p. 37).

SAUDI ARABIA
JIDDA
July 11, 1991

All 261 passengers and crew members aboard a Canadian charter DC-8 were killed when it crashed on landing at King Abdel-Aziz Airport in Jidda, Saudi Arabia, on July 11, 1991.

The *hajj,* an annual Muslim pilgrimage to the holy cities of Mecca and Medina, has frequently been marred by tragedy. In 1990 1,426 pilgrims died in a stampede in a tunnel (see CIVIL UNREST, p. 103). Before that, terrorist attacks had disrupted the pilgrimage, but the 1991 *hajj* was a relatively peaceful one, free from unrest and calamity—until after it concluded.

On July 11, 250 Nigerian Muslims boarded their charter jet at King Abdel-Aziz airport in Jidda, Saudi Arabia, for the return trip to Sokoto, Nigeria. The DC-8 was a Canadian charter, owned by the Montreal-based carrier Nationair and leased by a Nigerian company, Holdtrade. It was a prestigious arrangement; Holdtrade was set up exclusively to carry Muslim pilgrims by Ibrahim Dasuki, the son of the spiritual leader of Nigeria's Muslims, the Sultan of Sokoto.

The weather was clear and hot; the takeoff was uneventful. But shortly after the plane cleared the sands of the Saudi desert, the control tower at King Abdel-Aziz Airport began to receive rapidly delivered distress signals from the Canadian pilot of the jet. There was a fire in the plane's landing gear.

Officials at the airport advised the pilot to dump his fuel and return to the airport, but within instants of this command the plane nosed over and dove for the tarmac. With a bone rattling, thunderous explosion, the jet erupted on impact, flinging pieces of the airplane and the bodies of its passengers and crew far into the desert.

It was a grisly sight. Dismembered bodies and white robes blackened to torn tatters littered the grounds of the airport and the nearby sands. All 250 passengers and 11 crew members died, making this the tenth-worst crash in commercial airline history up to that time.

Its cause proved to be unique: The fire that developed in the landing gear well came from an underinflated tire, which overheated on takeoff. As the plane was struggling to land, the fire spread to the hydraulics and electrical systems, eventually disabling them. Without control, the pilot could do nothing to prevent the crash.

SAUDI ARABIA
RIYADH
August 19, 1980

A combination of the failure of airline personnel to follow evacuation procedures and the Saudi Airlines practice of cooking over open butane stoves aboard its aircraft resulted in death by fire aboard a Saudi Airlines Lockheed L-1011 TriStar on the ground at Riyadh, Saudi Arabia, on August 19, 1980. All 310 aboard died.

Three hundred ten passengers and crew were aboard a Saudi Airlines Lockheed L-1011 TriStar on the afternoon of August 19, 1980. The flight had originated in Karachi, Pakistan, and was on its way to Jidda, after a short stopover in Riyadh. Among the passengers were some devout Muslims, who were allowed to bring two butane cooking stoves aboard so that they could observe their dietary laws in flight. U.S. carriers would blanch at this practice, but it was routine in the Middle East.

The flight departed from Riyadh without incident. But shortly after takeoff, when the plane was approximately 80 miles from the airport, the pilot radioed that there was a fire aboard and he was returning. He was given clearance and landed safely. The plane taxied to the end of the runway and onto a side area.

Rescue teams rushed to it, expecting the passengers to pour from the escape hatches. But before firefighters could reach the plane, it burst thunderously into flames. Heavy smoke billowed from every opening, and the heat was intense. No one could approach the aircraft, not even in a protective asbestos suit. Firetrucks sprayed the consumed airliner with foam. It was all that could be done for the moment.

When teams were finally able to cut their way into the wreckage, they found an appalling sight. Passengers were piled up in mounds of fused and charred flesh near the escape hatches. There had obviously been a stampede when the plane landed. But none of the hatches were open or even unlocked. The crew's escape hatch in the cockpit

The death of all 310 passengers aboard this Saudi Airlines Lockheed L-1011 at Riyadh was a result of the Saudi custom of allowing passengers to cook on board and a failure of airline personnel to follow proper evacuation procedures. (National Transportation Safety Board)

was also, strangely, firmly latched. Every single person on the aircraft had died in the fire.

There was a fire extinguisher alongside one of the butane cooking stoves, and this must have been where the fire began. But why hadn't the crew opened the escape hatches?

Investigators posited three possibilities:

1. From the position of the bodies, it was obvious that there had been mass panic aboard the craft. Those in the back had rushed to the front, and this mayhem might have blocked the efforts of the crew to open the exits.
2. It was entirely possible that by the time the plane had taxied to a stop at Riyadh airport, all aboard had been asphyxiated.
3. The escape system on the plane was designed so that the hatches could not be opened from the inside unless the cabin had been depressurized. For some reason, the cabin pressure had not been released by the crew, and thus the escape hatches could not be opened by anyone.

A series of lawsuits by relatives of some of the 310 who perished aboard the plane attempted to prove negligence in the use of flammable materials within the aircraft, and that certainly contributed to the intensity of the blaze. But the reason that the plane remained sealed will never be known.

SPAIN
BARCELONA
July 3, 1970

No official reason was given for the disappearance of a Dan-Air Airlines Comet over the Mediterranean near Barcelona on July 3, 1970. All 112 persons aboard died.

The Comet turbojet probably should have been retired from service soon after it was first built. In its 25 years of existence, it had many fatal accidents, and one of the worst of these occurred over the Mediterranean, near Barcelona, on the night of July 3, 1970.

A group of tourists, weary of the dampness of England, boarded a chartered, British-built Dan-Air Airlines Comet in Manchester, England, at midday on July 3. All were headed to Barcelona, the Costa Brava and vacations in the sun.

The flight was uneventful until the aircraft was within 12 miles of its destination. The pilot radioed to the Barcelona tower at 7:00 P.M. that he was 12 miles away, flying at an altitude of 6,000 feet. This was the last anyone heard from the ill-fated flight. Observers along the coast reported seeing a large plane disappear into the sea approximately 20 miles from Mataro, up the coast from Barcelona.

Search boats were sent out, but no wreckage or bodies nor a trace of the plane or the 112 persons aboard was ever found.

SPAIN
CANARY ISLANDS
SANTA CRUZ DE TENERIFE
December 3, 1972

In the fourth-worst air crash up to that time, a Spantax Airlines charter Convair 990-A Coronados lost an engine over Santa Cruz de Tenerife, in the Canary Islands, and plunged to earth. All 155 aboard perished.

The Spanish-owned Canary Islands, off the coast of Africa, are a paradise and a winter refuge for thousands of Scandinavians and Germans. The local joke is that there is more German, Danish and Swedish spoken on the islands from September through May than there is Spanish.

A majority of these northern European tourists reach Grand Canary, as the main island is called, by charter airline. A priest in Denmark has operated a huge charter operation for years, which annually ferries thousands of Danes on holiday to and from Grand Canary.

Other, smaller charter lines fly into Los Rodeos Airport on Santa Cruz de Tenerife, one of the six smaller islands. Spantax Airlines, a Spanish charter carrier, is one example, and on December 3, 1972, it chartered one of its Convair 990-A Coronados to a group of Bavarian bus operators on holiday. The group, sunburned and happy, boarded the plane on a cloudless afternoon, and the takeoff was smooth and uneventful.

But 1,000 feet into the air, an engine of the Convair suddenly burst into flames. The plane wavered, stalled and then plunged to earth, exploding mightily upon impact.

Military personnel and civilians from a military base near the crash site reached the conflagration of twisted steel in moments. There was one survivor, a woman, who succumbed to her burns four hours later. One hundred fifty-five persons perished in the crash.

SPAIN
CANARY ISLANDS
SANTA CRUZ DE TENERIFE
March 27, 1977

In the worst accident in the history of civil aviation, two charter Boeing 747s, one belonging to KLM, the other to Pan Am, collided on the ground in a fog at Los Rodeos Airport on Santa Cruz de Tenerife on March 27, 1977. A combination of delays caused by a terrorist incident and the misunderstanding of the Spanish-speaking airport controllers by the Dutch crew of the KLM jet culminated in the death of all 249 aboard the KLM plane and 321 aboard the Pan Am 747.

Ironically, the deadliest accident in civil aviation history took place on the ground. And it involved two Boeing 747s that never would have been there in the first place had it not been for a small, fanatical group of militant separatists who had set off a small bomb near the florist's shop in the Las Palmas Airport earlier that day.

A Pan American Airways charter jet from Los Angeles via New York was taking passengers to Las Palmas to hook up with a Mediterranean cruise, and a KLM charter jet loaded with Dutch tourists had been diverted to Santa Cruz de Tenerife because the Las Palmas Airport had been shut down. Santa Cruz de Tenerife is ill equipped and unsuited for jumbo jets, or for most aircraft, for that matter. It is noted for the sudden fogs that close in on it at various times of the day and the year and for the dearth of equipment it possesses.

Still, it was probably the difficulty the Dutch crew of the KLM 747 had in understanding the Spanish-accented English of the control tower that was ultimately responsible for this terrible tragedy, which ultimately claimed the lives of 570 people.

The planes had been delayed on the ground nearly two hours when both were finally cleared for takeoff. There was one main runway at the airport and a small holding area at the end of the runway.

Finally, at 4:40 P.M., the KLM jet was given clearance to taxi to the end of the runway first, with the Pan Am jet behind it. When the Dutch craft reached the end of the runway, it would turn around and take off, while the Pan Am jet would taxi off and wait in the small side space. There would be plenty of time for both maneuvers.

It was extremely foggy, and the Pan Am jet had difficulty keeping the KLM jet in sight. Eventually, its taillights disappeared entirely into the mist. The Pan Am pilot reached the turnoff when suddenly he saw the lights of the KLM bearing down on him, at full throttle, in a takeoff. "What's he doing? He'll kill us all!" shouted the Pan Am pilot, as he frantically tried to pull his plane off the runway and into a field of high grass.

It was too late. The KLM jet smashed full force into the Pan Am 747 and then went on to tear through it. Both planes exploded in a giant roar of igniting jet fuel. Black clouds rose in the air, and the cries of the dying pierced the blackness like knives. Blood was everywhere.

Every single person on the KLM died. Because the Pan Am craft had partially left the runway, those in the front of the plane survived. "It exploded from the back," said Lynda Daniel, a 20-year-old Los Angeles college student. "We were sitting next to the emergency exit and it blew off. Most of the people sitting in the first six seats made it. People started climbing over me, and I saw flames, so I decided to get out."

Fire consumed most of both of the jets, and some of the bodies were burned beyond recognition. All 249 aboard the KLM jet were dead. Three hundred twenty-one died on the Pan Am 747 in a tragedy that never should have occurred.

SPAIN
GRANADA
October 2, 1964

No explanation has been found for the mysterious plunge into the Mediterranean near Granada of a Union Transports Africain Douglas DC-6 on October 2, 1964. All 80 aboard perished.

The Douglas DC-6 was a reliable plane, a piston-powered workhorse that carried thousands of passengers during the last days before jets took over the industry. Its efficiency and its fuel capacity allowed it to make intercontinental flights. Still, the DC-6 had its problems. Its engines were enormously complicated, using up most of their own power in the supercharger needed to force air into their hungry induction systems and depending for their successful operation on intricate cooling systems and electrical equipment. The technology of the modern jet resolved these problems.

Still, in 1964, DC-6s flew well and often. One such plane was maintained by the French-owned Union Transports Africain, and on October 2, 1964, it was filled with 73 passengers and seven crew members on their way from Palma, on the island of Majorca, to Mauritania, in West Africa. The aircraft took off from Palma, headed west and south near Granada and then abruptly lost altitude and dove into the Mediterranean.

Ships from several countries searched throughout the area where other ships had seen the DC-6 descend, but neither people nor wreckage was ever found, nor was the reason for the crash ever determined.

SPAIN
IBIZA
January 7, 1972

Pilot error was blamed for the smashing of an Iberia Airlines Caravelle into a mountainside on the island of Ibiza on January 7, 1972. One hundred seven died.

Ibiza is one of the gems of the Balearic Islands, located in the western Mediterranean off the coast of Spain. Although Majorca is the best known of this string of three islands (Minorca is the third), easily accessible in less than an hour by air from Valencia and Barcelona, Ibiza is the preference of artists and tourists who prefer splendid isolation and Roman and Phoenician ruins. A mere 221 square miles of terraced towns, it attracts discriminating tourists particularly during the winter holiday season.

On January 7, 1972, 98 Spanish tourists, fresh from spending the Christmas and New Year's holidays on Ibiza, boarded Iberia Airlines flight 602, a Caravelle, for their short return to the mainland. The flight took off from Ibiza Airport at approximately 2:00 P.M. Ten minutes later, it smashed into a mountain in the Atalayasa range, a semicircle of mountains rimming the city.

Although the precise cause of the accident was never discovered, the plane's pilot, José Luis Ballester, apparently misjudged his climb after takeoff and failed to clear the top of the mountain ridge. Wreckage and bodies were strewn over a one-mile radius. There were no survivors. All 107 aboard died in the crash and its consequent fire.

SWITZERLAND
BASEL
April 10, 1973

Pilot error caused the crash of a BEA Vanguard charter jet into a mountainside south of Basel, Switzerland, on April 10, 1973. One hundred seven died; 39 survived.

Mrs. Brenda Hopkins was very active in the Axbridge, England, Ladies' Guild, and she thought it would be a wonderful and original idea to organize a one-day air excursion to Switzerland. Some of the women could go to Lucerne to shop; others could visit the Swiss Industries Fair in Basel.

Seventy guild members liked the idea and joined the 60 or so other passengers on a chartered four-engine Vanguard turboprop that left Bristol for Basel bright and early on the morning of April 10, 1973.

It was snowing in Basel but not very hard. It was possible to make an instrument landing at the Basel airport, and officials there gave the pilot clearance to do so. He circled the airport once, then passed directly over it and made a wide circle southward. The control tower tried to reach him, but there was neither radar nor radio contact.

The reason was a calamitous one: The plane had slammed into a wooded hillside 1,800 feet high near the village of Hochwald, eight miles south of Basel. First, the wing had snagged on the branches of trees clinging to the side of the hill, and that had flipped it over on its back. Then, skidding through the three-foot-deep snow, it had disassembled, coming apart in several sections.

The tail portion remained relatively intact, and it was here that the 39 survivors of the crash were located. They huddled together in the shelter of their piece of an airplane, singing hymns and staying close, until rescuers arrived.

A farmer first saw the wreck and phoned police, who asked him and his fellow farmers to clear a path to the plane with their tractors. The farmers complied, allowing rescuers finally to reach the site of the wreck. Swiss army helicopters were unable to locate the wreck through the snow.

A local doctor and some farmers did what they could for the survivors and the injured. For the dead, some of whom were still hanging from their seat belts in the upside-down pieces of the plane, nothing could be done. One hundred seven people, most of them British mothers and housewives out on a once-in-a-lifetime one-day outing, were dead.

TAIWAN
TAIPEI
February 16, 1998

A China Airlines Airbus A300, on a scheduled flight from Bali to Taipei, crashed in fog and light rain 200 yards from Chiang Kai-shek Airport's northern runway on a road the pilot apparently mistook for the runway. Two hundred and seven were killed in the crash—196 on the airliner and nine on the ground.

The China Airlines Airbus, approaching Chiang Kai-shek Airport in Taiwan on the evening of February 16, 1998, was loaded with vacationers from Bali and several high-level Taiwanese bank executives returning from a conference in Bali.

There was fog and light rain surrounding the approach to the airport, and the pilot aborted his first approach, telling the control tower that he had trouble distinguishing the runway in the fog. He was given clearance to try again, and on the second try made a fatal error. A large road, running in the same direction as the northern runway, appeared to the pilot to be the runway itself.

At 8:09 P.M. residents in homes lining the road in a semirural area several hundred yards from the airport were startled by the thunderous roar of a descending jet, not a few hundred yards away but directly over them and falling fast. In an instant the plane plowed into the homes lining the road, shearing off their upper stories before setting them on fire and catching fire itself as it exploded upon impact. It plunged into the road, ripping itself asunder as fire consumed the plane. All 196 passengers and crew were killed. Nine others on the ground, including a two-month-old baby and a 10-year-old boy who was taken from the charred remains of his home and who died later in a hospital, were killed by the explosion and fire.

The sparsely populated area, home to fish farms, small factories, rice paddies and warehouses, was decimated and leaped into flame instantly. Firemen appeared swiftly and extinguished the flames within an hour.

Rescuers combed through the ruins but found only charred bodies and body parts along the road and throughout the area. Only the tail of the jet was recognizable. Pieces of it were everywhere. Searchlights illuminated a life raft wrapped around a tree stump. Scattered parts of the fuselage, yellow insulation, seats and pieces of rubber were strewn among clumps of dirt. The smell of burning rubber and jet fuel was everywhere.

"He tried to pull up but he couldn't," one observer on the ground told reporters after the crash. "I heard a blast, and was scared to death. Part of the house started to fall down," said Chen Ah-mei, who was forced to crawl out of the ruins of her home on her hands and knees. "It happened so fast—noise and fire," added an elderly farmer who ran to the scene as soon as he saw the flames.

"Visibility was extremely bad. The pilot said he was having trouble seeing the runway as he made his approach and asked to come around for another try," airline spokesman Liu Kyo-chien said. "Immediately after he asked for another try, the pilot lost contact with the tower." Struggling with the controls, the pilot finally managed to pitch the nose up 40 degrees and gain 1,000 feet in altitude, but the plane went into a total stall and dove into the road.

Tsai Tuei, director of the Taiwan Civil Aeronautics Administration, resigned to, as he phrased it, "take moral responsibility" for the crash, which was the worst in the airport's history.

TAIWAN
TAIPEI
October 31, 2000

Singapore Airlines Flight 006, a Boeing B-747 flying from Singapore to Los Angeles with a stopover in Taipei, turned onto a runway under construction at Chiang Kai-shek Airport on the night of October 31, 2000, in typhoonlike weather conditions and crashed into construction equipment. Eighty-three of the 179 aboard died.

In the world of commercial aviation, no person or airplane seems to be absolutely free of the possibility of error or breakdown. The safest airline with an impeccable record of safety is vulnerable. The most experienced and careful pilot can make a mistake.

Singapore Airlines enjoyed an international reputation for safety, passenger care and experience, and, although typhoon Xian Sang was nearing Taiwan and Chiang Kai-shek International Airport at Taipei on the night of October 31, 2000, there seemed to be nothing to worry about. The 159 passengers and 20 crew members boarding or who had just landed from Singapore on the Boeing B-747, headed for Los Angeles, were confident that all would be well.

The winds were high and the rain was blinding. There were some afterward who thought that the airline had no business flying in such weather, but all agreed that the signal for takeoff was within the parameters of international standards for flying.

The flight left gate B-5 at a few minutes before 11:00 P.M. and taxied to taxiway NP, which ran parallel to runways 05L and 05R. Runway 05R had been closed since September 13 for construction, and the notice had been announced to all the carriers using the airport. It was littered with concrete blocks and construction equipment, including heavy earth movers. For some reason it was lit on the night of October 10.

Runway 05L was clear and empty, and it was to that runway that the tower expected to plane to go, although heavy rain prevented the tower operators from seeing the exact whereabouts of the Singapore Airlines Boeing 747. The captain reached the end of the taxiway and made a 180-degree turn—but onto the wrong runway. Confused by the lights and the extremely poor visibility, he turned onto 05R, the runway ripped apart by construction.

After an approximate six-second hold, Flight 006 started its takeoff roll. Only three and a half seconds later, at exactly 11:18 P.M., the aircraft hit one of the earth movers. Almost instantly the plane collapsed onto the runway, broke asunder and burst into flames. It continued its forward slide, hitting concrete blocks and other equipment for approximately 6,480 feet beyond the runway threshold.

Rescue workers and firemen raced out into the worsening storm. The jet's full fuel tanks had exploded, sending pieces of the plane in several directions, but, miraculously, the rear of the aircraft was spared the worst of the fire. It was from there that escape chutes were deployed, and more than 40 of the passengers and crew slid down them to safety. The entire cockpit crew escaped from the nose portion of the jet, and more severely burned passengers were taken from the wreckage by rescue teams and transferred to hospitals in Taipei.

The rescue and salvage efforts continued all night and into the following morning, when the last of the bodies were removed from the charred aircraft. Relatives and friends of the dead passengers were directed to a makeshift

All that is left intact is the tail section of the Singapore Airlines Boeing 747-700 which collided with construction equipment as it blundered onto the wrong runway at Chiang Kai-shek International Airport on the storm-swept night of October 31, 2000. (AP/Wide World Photo)

morgue set up in an old hangar. Body bags were lined up next to empty wooden coffins; Buddhist monks chanted and Roman Catholic clergy and other religious leaders held vigil. It was an eerie, peaceful sight while the typhoon raged outside, forcing the airport to close down completely.

Investigation of the crash began immediately. The pilot, C. K. Foong, a Malaysian with a long and spotless record, told investigators that he had seen an object on the runway shortly before the crash and had tried to avoid it but could not. This was partially refuted by Kay Young, the managing director of Taiwan's Aviation Safety Council, who reported that the aircraft was traveling straight ahead when it made first impact.

Eventually, as the facts were assembled, it became clear that a combination of impossible visibility and crew error led to the death of 83 passengers and crew aboard Singapore Airlines Flight 006. The pilot particularly came in for heavy criticism. Although the driving wind and rain could have been partly responsible for his missing two signs that clearly indicated the number of the runway down which he apparently took the plane, he was forced to take heavy blame for missing a routine preflight briefing report that warned of the hazards on the runway under construction.

As this is being written, the case of the pilot and two co-pilots had been moved to the criminal court system of Taiwan. If convicted, the three would receive up to five years in jail on manslaughter charges. Although it acknowledged pilot error in the tragedy, Singapore Airlines continued to reiterate that it was a mystery why an experienced captain made such a "dreadful mistake."

THAILAND
SUPHAN BURI
May 16, 1991

On the night of May 16, 1991, a Lauda Air Boeing 767–300 carrying 233 passengers and crew smashed into a jungle mountain shortly after taking off from Bangkok airport on a flight destined for Vienna.

Thai police sergeant major Charan Palung was putting in another routine night shift in his police station in Suphan Buri province, close to the Myanmar border and roughly 100 miles northwest of Bangkok, on May 16, 1991. Only the sound of the jungle accompanied his shifts. But on this particular night, shortly before 11:00 P.M. the jungle sounds were suddenly augmented by the roar of an airplane flying low.

Charan Palung dashed outside, looked up and became one of the few witnesses to see a Lauda Airlines Boeing

767-300 with 223 passengers and crew aboard suddenly explode in an immense fireball as it plunged into the jungle. From where he stood, it was obvious that the plane had plowed into a hillside near Huai Khamin. Within moments rescuers that included more than 100 police officers, local residents and members of a Thai rescue team chopped and battered their way to the crash site.

It was a horrific scene. The mangled remains of the airliner were still burning when the first rescuers arrived. Plastic chunks of the plane, seat cushions and twisted pieces of metal were strewn about. Most pathetic of all, tattered and bloodstained clothing hung from numerous trees.

The flight, headed for Vienna, had originated in Hong Kong with 125 passengers. In its stopover in Bangkok, 88 more came aboard. It left Bangkok airport on time at 10:45 P.M. The American pilot, Thomas Welch, had logged hundreds of flying hours, the last of them with Lauda, Austria's largest charter airline, owned and run by Niki Lauda, the former world motor racing champion, but this flight would be a tragic punctuation, preceded by a fight for survival.

A day after the crash, when it was apparent that there were no survivors and before a detailed inspection of the wreckage could begin, theories about the reason for the tragedy ran rampant. In Hong Kong Francz Karner, the Vienna-based sales manager for the airlne, released a public statement that claimed that a bomb probably caused the explosion, because the crew reported no mechanical problems after takeoff. Asked for proof, he could offer none. Other "experts," seizing upon this theory, reminded reporters that during the Persian Gulf War, Western intelligence pinpointed Bangkok as a possible staging site for terrorism.

Somnuk Keetket, the governor of Suphan Buri province, was more circumspect. The possible cause could be one of three, he told the press: an exploding engine, a bomb placed aboard the plane or lightning.

Eleven days after the crash, the flight data recorders were opened for analysis. On June 2 the results were announced: A computer malfunction had switched one engine into reverse shortly after the plane took off from Bangkok. Investigators examining the left engine of the plane found approximately 80 percent of the left wing heavily burned.

Austrian Transport Minister Rudolf Streicher painted a picture of the pilot and co-pilot struggling with the controls, trying to disengage the computer and engage manual control over the engine, but to no avail. "It can be concluded," the minister continued, "that one of the two engines that were computer controlled during the ascent was suddenly switched to reverse. The pilots tried to solve this with the aid of the flight manual, but were unable to do so. The plane became unpilotable, stalled, and broke apart."

After the stall, the aircraft went into a steep high speed dive, broke apart at 4,000 feet and crashed into the jungle. Following the crash Boeing made modifications to the thrust reverse systems in all of its planes.

UNITED STATES
ALASKA
ANCHORAGE
June 3, 1963

No official reason was given for the mysterious crash of a Northwest Airlines DC-7 charter carrying U.S. military personnel over the Pacific Ocean near Anchorage, Alaska, on June 3, 1963. All 101 aboard died.

A military charter flight of Northwest Airlines took off from McChord Air Force Base in Washington State at 7:30 A.M. on June 3, 1963. Its destination was Anchorage, Alaska, and it carried 101 passengers and crew members.

At 10:06 A.M., Captain Albert F. Olsen, the pilot of the flight, requested permission from the air station at Sandspit, British Columbia, to change altitude from 14,000 to 18,000 feet. This sort of request is usually weather related, and it seems logical to assume that Captain Olsen and his flight were encountering some turbulence.

The tower at Sandspit radioed back, advising him to maintain his present altitude until they could get clearance from Anchorage. Their communication with Anchorage revealed that there was a Pacific Northern Airlines plane already in the vicinity at that altitude, but the Northwest flight was given permission to ascend to 16,000 feet. Sandspit attempted to reach the Northwest flight to advise them of this but met with nothing but silence.

The radio operators repeated their attempts to rouse the flight, but to no avail. They contacted the Pacific Northern Airlines flight to try to reach the military charter. They had no better luck.

On the ocean below, a fishing boat trolling off the Alaska coast picked up a faint distress signal. It radioed back to Anchorage, and rescuers immediately set out to comb the area for wreckage. Six Coast Guard cutters, three Royal Canadian Air Force planes and 12 U.S. planes converged on the area near the fishermen.

Finally, one of the RCAF pilots reported seeing debris floating in the sea off Graham Island, where the Sandspit air base was located. By the time the Coast Guard cutters reached the site a 55-knot wind had whipped the sea to a wild froth, and fog and darkness had begun to close in. No survivors were found. All 101 perished in this mysterious crash at sea.

UNITED STATES
ALASKA
JUNEAU
September 4, 1971

Alaska Airlines flight 1866, a Boeing 727, crashed into Mount Fairweather, near Juneau Airport, during a heavy storm at midnight, September 4, 1971. All 109 passengers and crew aboard died.

A raging storm roared around the Juneau Airport at midnight on September 4, 1971. Alaska Airlines flight 1866, a Boeing 727, requested permission for an instrument landing. Unfortunately, the Juneau Airport's instrument landing capabilities were elementary and inadequate for the weather conditions that night. There was no localizer, a device that lines up an incoming airplane with the center of the runway, and there was no glide slope device to warn the pilot that his craft had veered off the assigned approach and was heading in another, dangerous direction.

That is exactly what happened with Flight 1866. Coming in for an approach, it was apparently blown off its course and slammed into Mount Fairweather, near the airport. The plane exploded into flames on impact. None of the 109 passengers and crew members survived.

UNITED STATES
ARIZONA
GRAND CANYON
June 30, 1956

Human error on the part of an air controller caused the first midair collision in U.S. commercial airliner history and the worst air disaster to date when a United Airlines DC-7 collided with a TWA Super Constellation over the Grand Canyon on June 30, 1956. All 128 aboard both planes died.

The first midair collision of commercial airliners in the United States took place in a thunderstorm over Arizona's Grand Canyon on the afternoon of June 30, 1956. The enormity of the tragedy and the senseless set of circumstances that led up to it did have a positive result: the creation of the Airways Modernization Board and, a year later, the passage of the Federal Aviation Act, which set up the Federal Aviation Administration (FAA) as a separate entity.

If the two planes, a TWA Super Constellation and a United Airlines DC-7, had been covered by radar, the collision might not have occurred. But at the time both were operating beyond the boundaries of the existing radar control units and thus were relying purely on radio contact with nearby airports to maintain safe altitudes.

TWA flight 2 and United Airlines flight 718 took off from Los Angeles International Airport three minutes apart. The TWA flight, piloted by Captain Jack Gandy, was a half hour late for its scheduled trip to Kansas City. It was assigned a flight altitude of 19,000 feet. The United flight was on time, headed for Chicago and given a flight level assignment of 21,000 feet. Essentially, then, the two would be flying over the same route for most of their trip but separated by 2,000 feet—a safe distance.

Near Salt Lake City, Gandy radioed Los Angeles and requested permission to climb to 21,000 feet. The 19,000-foot level kept him in clouds and turbulence, and he wanted to rise above it. Ground control in Los Angeles radioed Salt Lake City, which warned that the United flight was at 21,000 feet and very near the Constellation.

Los Angeles radioed the TWA flight, ordering it to maintain its level of 19,000 feet.

And then an incredible sequence of events took place. Captain Gandy radioed Los Angeles again, requesting a "1,000-foot top." This meant that the Constellation would be able to fly 1,000 feet above the cloud cover, which would determine the altitude. Gandy must have realized the danger in this request; the ground controller in Los Angeles also must have realized it. But, incredibly, permission was given to TWA flight 2 to climb to a 1,000-foot top.

And that was the fatal decision that would put both planes on a collision course. Had the sky remained clear above the cloud cover, they might have seen and missed each other. But at 10:30 A.M. a thunderhead poked its head above the carpet of clouds below. Both planes entered the gray cloud, and somewhere over the Painted Desert the United Airlines DC-7 plowed into the side of the TWA Constellation, blowing it apart.

Damaged beyond control, the DC-7 pitched over into a steep dive, its crew shouting into the radio, "Salt Lake! United 718! We're going in!" The plane plowed into two enormous buttes—Chuar and Temple—in the Painted Desert, exploding on impact.

The planes and all of their combined 128 passengers and crews were fragmented and scattered for miles. Rescue helicopters, arriving hours later, found no survivors and no intact corpses. Those in the TWA Constellation had been incinerated; those in the United DC-7 had died on impact.

The new regulations drawn up by the newly formed FAA after the crash were better than nothing but hardly effective, considering what had happened to bring them about. A continental control was instituted on December 1, 1957, effective above 24,000 feet. (Remember, this collision occurred at 21,000 feet.) This meant that all aircraft would be under positive control whether they were flying on or off the airways. The regulation was mandatory under conditions of restricted visibility but could be ignored during good weather.

UNITED STATES
FLORIDA EVERGLADES
May 11, 1996

Valujet flight 592, a McDonnell Douglas DC-9, on a flight from Miami, Florida, to Atlanta, Georgia, on May 11, 1996, encountered fire and was returning to Miami when it crashed into the Florida Everglades. All 110 aboard were killed.

For months, five cardboard boxes of old oxygen generators, which were normally housed in the panels above the passenger seats in the planes of Valujet Airlines but which had been removed because they had reached their expiration dates, sat on a shelf in Valujet's maintenance contractor's hanger in Miami, Florida. Incorrectly packed, incorrectly labeled, they would, early on Saturday, May 11, 1996, be loaded onto a truck and driven to the gate area where Valujet 592 was waiting to depart for Atlanta.

A Valujet worker signed for the boxes, which were then loaded into the jet's forward cargo hold, some of them on top of three inflated airplane tires also belonging to Valujet and headed for recycling in Atlanta. They had been delivered with a form that described the oxygen generators as empty and as Valujet property being returned at the airline's request.

And thus would begin the chain of events that would eventually bring down Valujet 592, a McDonnell Douglas DC-9, in a fiery crash into the Florida Everglades, killing all 110 aboard. It would be a rare case of a maintenance crew on the ground being responsible for an accident that occurred in the air, and it would result in the grounding of Valujet and its departure from airline travel.

The takeoff from Miami, although slightly delayed by earlier weather disturbances, was normal. It was a daily two-hour flight that the low cost airline was able to fly full most days, and this was one of those days. Captain Candalyn (Candi) Kubeck, one of commercial aviation's first woman pilots—and the first to die in a crash—steered the DC-9 north northwest, toward Atlanta. It was a clear, cloudless afternoon.

Then, suddenly, 10 minutes out of Miami, catastrophe struck. The transcript of the unfolding of the drama in the jet retrieved from the cockpit voice recorder tells the tale. It begins with a sound that was later identified as an explosion in the forward cargo hold:

2:10:07: Pilot: What was that?
2:10:08: Copilot: I don't know.
2:10:15: Pilot: We got some electrical problems.
2:10:17: Copilot: Yeah. That battery charger's kickin' in. Ooh, we gotta —
2:10:20: Pilot: We're losing everything.
2:10:21: Tower: Critter five-nine-two, contact Miami center on one-thirty-two-forty-five. So long.
2:10:22: Pilot: We need, we need to go back to Miami.

Now, shouts from the passenger cabin are heard on the tape:

2:10:25: Female voices in cabin: Fire, fire, fire, fire.
2:10:27: Male voice: We're on fire. We're on fire.

There is more shouting and confusion, then:

2:10:41: Tower: What kind of problem are you having?
2:10:44: Pilot: Fire.
2:10:46: Copilot to tower: Uh, smoke in the cockp— smoke in the cabin.

More confusion and sounds. Then the voice of a flight attendant on the intercom to the cockpit is heard.

2:10:58: Flight attendant: O.K. We need oxygen. We can't get oxygen back here.

The plane has descended from 7,000 to 5,000 feet, and the copilot has asked for the nearest available airport. The tower answers:

2:12:58: Contact Miami approach on, correction, you, you keep on my frequency.

All through this, there are sounds of shouts and screams that indicate that the fire and smoke are raging through the passenger cabin.

At a little after 2:13, the recording stopped, indicating that the pilot and copilot had lost all electrical controls. And at 2:14 precisely, Valujet 592 fell in a twisting, turning almost vertical dive into the Florida Everglades, sending a mushroom cloud 100 feet into the air.

A flight instructor and his student, flying in a small plane roughly two miles away, witnessed the crash. "When it hit the ground, the water and dirt flew up," the instruc-

Airboats were the only early rescue vehicles possible when Valujet Flight 592 spiraled into the Everglades in Florida on May 11, 1996. The muck of the swamp swallowed the plane and all aboard. (AP/Wide World Photo)

tor later told authorities. He immediately radioed Miami and circled the point, roughly 20 miles west northwest of Miami, until rescue helicopters reached the scene.

"The wreckage was like if you take your garbage and just throw it on the ground. It looked like that," the instructor added, and, asked if there were any survivors, he answered, "It was terrible. Nothing could have survived that. It was in a 75-degree bank angle going down."

The Everglades, a freshwater marsh reaching south from Lake Okeechobee, covers 5,000 square miles and is a wilderness populated by bass, turtles, hawks, herons, panthers, alligators and poisonous snakes. The muck of the area seemed to swallow up all but some scattered pieces of wreckage and bodies that floated on the surface of the water.

There was no dry land at the site and hence no place to bring in grappling equipment to salvage the wreck or search for bodies. Coast Guard helicopters hovered over the site, and airboats were sent in to assess the possibilities of rescue. By late afternoon, a staging area was set up on a track of dirt road a half mile from the crash, but only the airboats and helicopters could get close to it. By nightfall, under searchlights, the Coast Guard admitted, "We don't see any large pieces of the jet. There are no signs of survivors."

Gingerly, the next day rescuers began the long, laborious task of finding and piecing together the wreck of the jet. A depression in the muck defined the point of impact of the plane, but the Everglades had closed in over it. Three to five feet of silty water overlaid the floor of muck that in places was 25 feet deep, and all of this was the habitat of alligators and poisonous snakes.

The engines were the first of the wreckage to be located, followed by the tail assembly. Navy divers, trained in salvage, were brought in, and two days later a Dade County policeman, Elio Gonzalez, searching the scene for body parts, stepped on the black box flight recorder.

A method of searching the area of mud that was sometimes under water a few inches deep and in others a few feet deep was set up: Teams of divers were ferried to the site by helicopter and boat and lined up five abreast, shoulder to shoulder, to wade through small, 10-foot wide areas on a search grid. Behind them an escort in an airboat with an automatic weapon looked for alligators. The searchers wore anticontamination coveralls under wet suits and worked in shifts of about 20 minutes in the 90-degree heat. Their search turned up mostly body parts that were retrieved and catalogued by the Dade County Medical Examiner's Office.

Now, the NTSB arrived and began checking passenger and cargo manifests. The oxygen generators were loaded with two chemicals that gave off oxygen when mixed and that, most importantly, produced heat of up to 500 degrees in the process.

The flight data recorder revealed that the plane was ascending at a normal climb power to 10,628 feet when what the NTSB inspector termed "an indication of an anomaly" occurred at three and a half minutes before the crash. "The first indication was a drop in indicated altitude and air speed," he continued. "The altitude dropped 815 feet, and the airspeed dropped 34 knots." The flight data recorder stopped recording 50 seconds before the crash, indicating a loss of complete power within the plane.

The analysis of wreckage and documents reached the FAA, which announced on May 15 that Valujet had not been authorized to carry as cargo the chemical devices that were packed in the cardboard boxes in the forward cargo compartment. Dubbed by the FAA hazardous material, they required special packing and special FAA authorization, neither of which Valujet had. Because of the enormous heat they generated and the fact that they were loaded on inflated tires, the possibility of either an explosion or an encompassing fire became nearly a certainty.

Searchers at the site now turned up several pieces of the plane that were covered with soot and a few that showed heat damage. Eventually, an aluminum seat from the passenger section was found, a twisted mass of melted metal, indicating a terrible fire in the passenger compartment.

Accusations began to fly between Valujet and Sabretech, a Phoenix company that was contracted to do maintenance for Valujet. The boxes containing the canisters were incorrectly labeled empty, thus implying that they had been discharged and were not hazardous materials, according to Valujet, and so the clerk that allowed them aboard was simply acting on the information he was given by Sabretech. Sabretech countered that they were acting upon Valujet's orders.

Now, heavy-duty helicopters were lifting large chunks of the plane. Metal caps from oxygen generators were found embedded in a partly burned tire that had been carried as cargo. Thus, investigators were able to link this with evidence of a strong fire or even an explosion in the cargo hold just before the crash, and this pointed strongly to the oxygen generators and tires as the leading suspected causes.

Finally, on May 26 the cockpit flight recorder was recovered along with portions of the passenger cabin walls that showed smoke damage and an overhead baggage compartment that was covered with soot.

The unidentified sound on the cockpit recorder gave credence to the theory that one or more of the tires in the cargo hold exploded from the heat and fire caused by unemptied oxygen generators activating. The crew acted immediately to land the plane, but the best of their efforts were to no avail. In its final report, the NTSB said ". . . the loss of control resulted either from flight control failure or incapacitation of the crew due to extreme heat and smoke." Besides Candi Kubeck, who was the first female commercial airline captain to be killed in a crash, San Diego Chargers running back Rodney Culver, his wife and songwriter Walter Hyatt also perished in the fiery catastrophe.

On June 17, 1996, the FAA shut down Valujet, citing "serious deficiencies" in its operation. A specific charge made by the FAA was that "the airline has not demonstrated that it has an effective maintenance control system" to oversee its subcontracted maintenance work. The use of old aircraft that sometimes had large maintenance problems was also cited as a reason for grounding the airline.

Furthermore, the agency found "a significant decrease in the experience level of new pilots being hired by Valujet as well as other positions such as mechanics and dispatchers," as well as accidents and incidents involving inexperienced pilots, inadequate maintenance and insufficiently trained flight attendants, brought about by the company's rapid expansion from two to 48 planes.

Valujet vowed to resume service, but it did not. Its failings, the FAA and other agencies affirmed, were unique and not endemic to the entire low-cost airline industry. They cited the excellent and outstanding safety record of Southwest Airlines to bolster their evaluation.

UNITED STATES
ILLINOIS
CHICAGO
May 25, 1979

Faulty maintenance by American Airlines mechanics caused the worst disaster in U.S. aviation history. Mishandling of engine repair caused the left engine of an American Airlines DC-10 to fall off in flight near Chicago on May 25, 1979, and the plane lunged to earth, killing all 277 aboard and two on the ground.

The worst disaster in U.S. aviation history occurred on the afternoon of the beginning of the Memorial Day holiday, May 25, 1979. American Airlines Flight 191, a DC-10 bound for Los Angeles and loaded with 277 passengers and a full complement of jet fuel, took off from runway 14 of Chicago's O'Hare Airport at 3 P.M., nearly on time. The weather was clear and the winds mild.

And then, those who had come to see their friends and family off to the West Coast looked on in horror as the plane reached an altitude of 400 feet, wavered, rolled to the left, stalled and then plunged into the small abandoned Ravenswood Airport, near O'Hare. There was an enormous explosion of jet fuel, and a pyre of flame and black smoke rose high enough to be seen in downtown Chicago's Loop, 10 miles away. The heat from the blaze was hellish. It would be hours before rescue workers could penetrate the intense heat given off by the flames, much less touch the white-hot metal of what was left of the airplane.

The passengers and crew did not have a chance of survival. Every one of them, plus two men working nearby on the ground, were either incinerated or blown to pieces by the impact of the plane hitting the ground. A mobile home park located near the crash site suffered fire damage when flaming pieces of the plane were flung through the air and landed on the roofs of its homes. A block away, a Standard Oil Company gasoline storage facility stood unharmed. If the plane had hit it, there would have been no stopping the fire.

It would be later that night before the remains of those killed in the crash would be removed and daybreak the

The worst disaster in United States aviation history in the making. American Airlines Flight 191 rolls sharply to the left moments after losing its left engine over Chicago's O'Hare Airport. (National Transportation Safety Board)

next day before the last of them would be carried off by ambulances serving duty as hearses.

At first, the reason for the crash eluded the experts. Why should an airliner just nose over and dive to earth in the midst of a perfectly normal takeoff? An eyewitness, Winann Johnson, described what happened: "I saw this silver cylinder thing fall from the plane onto the runway," she told reporters. "It burst into flames and then smothered real quickly."

A federal judge immediately impounded all DC-10s in the United States, including those of foreign airlines. The industry erupted in indignation, but the court acted in the interest of public safety, until it could be determined if this was a problem endemic to the DC-10.

Further investigation concluded that the left pylon, which supports the left turbofan engine under the wing, ripped away as the aircraft was lifting into the air.

Investigators maintained that the pylon also had ripped out hydraulic lines. As soon as this happened, the left wing's leading edge flaps, or slats, retracted prematurely. Since slats provide lift at low takeoff speeds, the loss of the left slats caused the left wing to stall and drop while the right wing still had normal lift.

The reason for the malfunction of the pylon was traced to American Airlines' maintenance procedures. Mc Donnell Douglas recommended that the engines be removed in a two-step procedure: first the engine, then the pylon. But American Airlines mechanics, untrained in this procedure, repeatedly removed the whole assembly with a forklift.

So every single American Airlines DC-10 then in operation was in danger of repeating the tragic experience of flight 191. Immediate inspections and corrections were made worldwide, and eventually the DC-10 was pronounced safe to fly. The lesson learned was at a monstrous cost.

UNITED STATES
INDIANA
SHELBYVILLE
August 9, 1969

Student pilot error and the lack of a transponder were responsible for the midair collision of an Allegheny Airlines DC-9 and a Piper Cherokee over Shelbyville, Indiana, on August 9, 1969. Eighty-three were killed.

All of the regulations in the world probably could not have prevented the midair collision between an Allegheny Airlines DC-9 and a Piper Cherokee over Shelbyville, Indiana, on August 9, 1969. The fear that haunts both commercial airline pilots and passengers was realized when a plumber, on his first solo flight, came propeller to fuselage with a commercial aircraft coming in for a landing.

The DC-9 was covered by radar at Weir Cook Airport in Indianapolis as it headed in on a calm, sunlit summer afternoon. It could not have been a more routine landing. The single-engine Cherokee carrying the student pilot did not, however, carry a transponder, which would have immediately alerted the tower of its whereabouts.

As the DC-9 circled for its final approach, the Cherokee approached it from the side. The airliner almost cleared the path of the private plane, but not quite. The Cherokee smashed into the DC-9's tail, knocking out its controls and sending it into a dive. Both planes were flying too low to regain altitude and safety. Each smashed into the ground, killing everyone aboard both craft. Eight-three people lost their lives in one of the most unexpected and unpredictable air crashes in U.S. aviation history.

UNITED STATES
MARYLAND
ELKTON
December 8, 1963

A combination of lightning and jet fuel with a high flash point caused the explosion of a Pan Am Boeing 707 in the skies over Elkton, Maryland, on December 8, 1963. Seventy-three people died.

A series of circumstances that converged on the night of December 8, 1964, resulted in a fatal tragedy for everyone aboard Pan Am Flight 214, en route from San Juan International Airport to Philadelphia via Baltimore. Most of those aboard were heading home after vacations in the West Indies. Seventy-one of the flight's original 144 passengers deplaned in Baltimore. They were the lucky ones.

The Boeing 707 assigned to flight 214 was the oldest jet in service at that time, but it showed no signs of metal fatigue. It was, however, burdened with a far more lethal problem. Like other commercial airplanes during the 1960s, it used cost-effective JP4 fuel, a mixture of gasoline and kerosene with a high flash point.

There are three kinds of fuel commonly used in civil aviation. Gasoline is used only for piston-engine planes and is highly volatile. Turbine aircraft can use kerosene, similar to domestic paraffin. It has low volatility and is the most widely used jet fuel. The third kind is JP4, the kerosene–gasoline combination called "wide-cut gasoline." It has a flash point of minus four degrees Fahrenheit, a phenomenally low ignition level, which means that it will ignite even though the outside temperature is well below freezing.

The Pan Am Boeing 707 was equipped with fuel vents to equalize the pressure on the inside of the tanks with the pressure on the outside. In 1959 a TWA Constellation blew apart over Italy while it was flying through an electrical storm. A later investigation revealed that the explosion occurred near a wing-tank fuel vent.

At the end of May 1963, seven months before Pan Am Flight 214 took off from San Juan, Lockheed engineers sent a report to the FAA noting that "lightning can ignite an inflammable mixture spewing out of the vent. It would not be necessary for the lightning to actually strike the vent. It could strike anywhere on the aircraft and bleed off into the path of the flammable vapors."

There was plenty of lightning in the air over Philadelphia and Baltimore on the night of December 3, 1963. It had been an unusually mild December, and a cold front moving through the area was setting off huge downpours and fierce lightning. Commercial aircraft were stacked in holding patterns all up and down the East Coast of the United States.

Pan Am flight 214 took off from Baltimore at 8:25 for the short flight to Philadelphia and almost immediately entered a holding pattern. It circled over New Castle, Maryland, in a configuration that would take it directly over Elkton. It was told to hold at 5,000 feet, an altitude that was buffeted by high turbulence and icing.

Shortly thereafter, the Pan Am jet radioed for permission to break the pattern and land at Philadelphia. It was advised to wait until the weather, which was improving, cleared somewhat. The Pan Am crew responded with patience.

Nothing happened for a few moments, and then, at 8:58 P.M., a distress call from the Pan Am Clipper erupted over the radios of every liner in the vicinity and every control tower below: "Mayday . . . Mayday . . . Mayday . . . Clipper two one four out of control . . . here we go."

A thousand feet above, a National jet radioed, "Clipper two fourteen is going down in flames."

It had been struck by lightning, and the lightning had ignited the fuel and blown the left wing off the plane. All control had been lost.

Gerald Cornell, a resident of Elkton, later reported to the CAB: "The sky lit up like a tremendous bolt of lightning . . . then there was a loud explosion, like thunder outside of my house . . . then we saw the plane burst into flames and fall apart."

Other eyewitness reports, collected during the post-crash investigation, were similar. Joseph Dopirak saw a falling jet "coming straight down with only one wing." It blew into a thousand pieces when it struck the ground, he added. Jerry Greenwald, a skater, was close enough to see the one-winged craft plummeting through the sky with bodies tumbling out of it.

Pieced together by the investigation (the flight recorder was useless—it had struck the ground with such force that it had blown apart), the sequence of events unfolded thus: A massive lightning strike occurred first in the left reserve tank and spread to the center tank and the right reserve tank, causing multiple explosions, which in turn tore off the entire left outer wing, resulting in an immediate loss of control. The explosions of fuel continued, scattering bodies and pieces of the aircraft all over the countryside.

As a result of this tragedy, a lightning protection committee was formed in the FAA to study lightning protection systems and their feasibility. Quietly, the use of JP4 fuel was banned on the fleet of planes reserved for flights of the president of the United States and his staff and replaced by the safer but costlier JP5.

UNITED STATES
MASSACHUSETTS
BOSTON
October 4, 1960

A flock of birds sucked into the turbines of the engines of an Eastern Airlines Electra caused it to lose control and crash into Boston Harbor on October 4, 1960. Sixty-one died; 10 were rescued.

Flocks of birds are a recurrent and unexpected danger to jets. Groups of birds soaring around airports can be, and frequently are, ingested into the engines of jetliners. In jet travel's early years, this occurrence would bring an engine to a complete stop, thus threatening takeoffs and landings. Modern jet engines, manufactured after the mid-1970s, are designed to handle "bird strike," as the phenomenon is now called, without failing.

But jet travel was in its infancy on October 4, 1960, when Eastern Airlines Flight 375, a propjet Electra flying to Philadelphia, took off from Boston's Logan Airport. It was a clear late afternoon, and the plane lifted off at 5:48 P.M., all four of its engines in prime working order. Its takeoff pattern took it over Boston Harbor, where it began a slow turn.

At that moment, flight 375 encountered a flock of starlings. The birds were sucked up by the turbines of the engines, stopping the port inboard engine and causing it to catch fire. The aircraft, out of control, dropped like a punctured parachute into the waters of Boston Harbor at the Pleasant Park Channel.

The waters were choppy that evening, with a strong set, but there were five yachts anchored nearby. Divers from these boats went into the water and managed to rescue 10 passengers, still strapped in their floating seats. None of the other 61 passengers and crew members survived.

At first, FAA investigation focused on the Lockheed Electra, a trouble-plagued plane, but close scrutiny of the wreckage revealed one of the first examples of "bird strike" as a hazard to safe jet travel.

UNITED STATES
MASSACHUSETTS
BOSTON
July 31, 1973

Malfunctioning instruments and delayed rescue attempts may have accounted for the deaths in the otherwise unexplained crash of a Delta Airlines DC-9 on landing at Boston's Logan Airport on July 31, 1973. Eighty-eight died.

Not wanting to miss an important business meeting because of weather delays, Charles R. Mealy summoned a flight attendant on Delta flight 723 to his seat as the aircraft, a DC-9, was taxiing out on the fog-shrouded runway at Manchester, New Hampshire. When the flight attendant

Pieces of a Delta Airlines DC-9 dot the landscape following the unexplained crash of the liner on its approach to Logan Airport in Boston, on July 31, 1973. (National Transportation Safety Board)

arrived, Mealy announced that he had to leave the flight. "I want to get off this plane" were his exact words, as reported later to local reporters.

The flight attendant informed the captain, who turned the plane around and delivered Mealy to the terminal. An hour later, Mealy would be one of two survivors of Delta Airlines 723, bound from Manchester to Boston on the morning of July 31, 1973.

The weather was still but cloudy, the fog banks reaching almost to the ground. The aircraft took off without incident from Manchester and made the short flight to the airspace over Boston. But at 11:08 A.M., while attempting to make an IFR, or instrument landing, at Boston's Logan Airport, the jet crashed in a huge, thunderous ball of fire, killing 88 passengers and crew and leaving only one person—Air Force Sergeant Leopold Chouinard—alive. Chouinard would survive, but without either of his legs.

The cause of the crash would never be determined, but one factor revealed in the subsequent FAA investigation might have contributed toward it, and another might have worsened it. First, the particular Delta DC-9 involved in the crash had a history of malfunctioning instruments. It was entirely possible that the pilot was receiving wrong information or no information at all from some of his instruments as he relied entirely on them to come into Logan Airport at zero visibility.

Second, the Logan tower was apparently unaware that the flight had crashed until six minutes after it had happened. Those precious six minutes of rescue time might have saved more than one life, although Harris A. Cusick and Geoffrey F. Keating, two airport construction workers who happened to be in the vicinity of the crash site, rushed immediately to it but found only a horribly burned Chouinard alive. The FAA did conclude, however, that had firefighters been on the scene earlier and extinguished the roaring, lethal blaze, more might have been saved.

UNITED STATES
MASSACHUSETTS
OFF NANTUCKET ISLAND
October 31, 1999

Thirty-three minutes after taking off from New York's John F. Kennedy Airport on October 31, 1999, EgyptAir flight 990, traveling from Los Angeles to New York to Cairo, plunged into the Atlantic Ocean off the coast of Nantucket Island. All 217 aboard were killed.

EgyptAir flight 990 from Los Angeles and New York to Cairo was two hours and 20 minutes late leaving New York's John F. Kennedy Airport in the early morning of October 31, 1999. The delay was due to the late arrival of the plane from Los Angeles, which meant that there was a certain fatigue and restlessness among the 199 passengers and 15 crew members. However, they were in experienced hands. The pilot was Hakim Rushdi, who had logged more than 10,000 hours of flight experience preceding this flight. The first officer of the relief crew was Gamil al-Batouti, another experienced pilot who was just months from retirement. A major, flyer and flight instructor during the Arab-Israel war in 1967, he had risen through the ranks in EgyptAir to the plum assignment of the Los Angeles–New York–Cairo run, but only as a first officer. For all his experience, he had never reached the status of captain, although he was treated with great respect and deference by his fellow officers, who even called him "captain."

The takeoff in clear weather from Kennedy was uneventful. The jet reached its cruising altitude of 33,000 feet by 1:40 A.M., and the passengers proceeded to settle down and catch up on the sleep they had been denied by the delay from Los Angeles. At that moment, Batouti entered the cockpit and insisted that the co-pilot take a break and get some sleep, and he settled into the co-pilot's seat. The captain headed out over the Atlantic, turned on the autopilot, and, at 1:48 A.M., rose and said to Batouti, "Excuse me, Jimmy, while I take a quick trip to the toilet."

"Go ahead, please," Batouti responded.

Less than 30 seconds after the captain left the cockpit, a chain of events was set in motion that would shorten EgyptAir 990's flight to just 33 minutes, produce two minutes of horrendous, wrenching terror for all but possibly one aboard and launch years of international contention and mystery.

At 1:48 A.M. and 40 seconds, Batouti said, faintly, "I rely on God." This was followed by a long series of unexplained clicks and thumps.

The impact-crushed wreckage of EgyptAir flight 990, grappled from the floor of the Atlantic off Nantucket Island, is spread out for analysis in a hanger at Quonset Point in North Kingston, Rhode Island. (AP/Wide World Photo)

At 1:49 and 45 seconds, the autopilot was shut off, presumably by Batouti. Three seconds later Batouti again said, "I rely on God."

Five seconds later, the engines were reduced to idling speed, and the plane's elevators began pushing the plane into a dive. Three seconds after this, Batouti began saying "I rely on God" in rhythmic repetition.

Twelve seconds later, the cockpit door burst open and the captain came in, asking, with great agitation. "What's happening? What's happening?"

Batouti repeated "I rely on God" 11 times until the captain again said, "What's happening?"

The plane was in a steep dive of 66 degrees, and its speed had increased rapidly to nearly supersonic levels—exceeding .86 Mach.

Now, the jet's elevators, which control its upward and downward motion and are designed to operate in tandem, split in different directions, the pilot's for a nose up position and the co-pilot's for a nose down position, which possibly indicated a fight for the controls between the pilot and Batouti.

At this point someone cut off fuel to the plane's two engines, shutting them down. "What is this? What is this?" the captain asked. "Did you shut the engine?"

Within seconds the captain ordered, "Shut the engines."

The dive was continuing straight for the sea. Batouti answered, "It's shut."

The speedbreak, a device used to slow the plane, was applied by the captain. For the next few seconds, he implored, "Pull. Pull with me. Pull with me. Pull with me." They were the final words on the cockpit recorder.

The plane had descended to 16,700 feet and now, gradually, it began to climb at 650 mph to 24,000 feet. And then it heeled over and began a second dive, straight down. At 10,000 feet it broke apart, and its various pieces and occupants crashed into the Atlantic Ocean 60 miles southeast of Nantucket Island.

The United States Coast Guard immediately set off from Nantucket to the crash site, pinpointed by the disappearance from radar of the flight. They found clothing, passports, life rafts, airplane seats, lifejackets and parts of the aircraft, but no survivors. All 217 aboard had died, cruelly and undoubtedly in extreme terror.

Now the long salvage job began. Most of the airplane rested in 270 feet of ocean, and large navy salvage ships were brought in to raise the heavier parts while navy divers and robotic submarines retrieved body parts and the black boxes.

As the NTSB investigation began, diplomatic problems immediately rose. EgyptAir insisted that the cause of the crash must have been mechanical failure of the aircraft.

Their insistence arose from speculation, once the cockpit recorder was retrieved in November, that Gamil Batouti had committed suicide, taking 216 others with him to a watery grave. The initial findings came from the repetition of *Tawkalt ala Allah*—"I rely on God"—by him throughout the crisis in the cockpit.

"This is our gate to Islam to say these words," Dr. Maher Hathout, the imam who presided over a prayer service for Batouti at the Islamic Center in Los Angeles, said, "and we'd like them to be the last words we say in our lives. This means that the person is religiously conscious and if he is so, he will never commit suicide because suicide is a major sin in Islam. I think it is cruel, I think it is an insult to the family, and I think it is very insulting to his soul, especially if he's innocent." Batouti's nephew Walid, in response to media reports that his uncle might have been depressed at not having made full pilot before his impending retirement, answered that his uncle was very proud of his record as a pilot, including being an instructor for the Egyptian Air Force. *Al Ahram*, Egypt's largest circulating newspaper, stated in an editorial that the suicide theory was a threat to national pride.

Soon after the crash the National Transportation Safety Board (NTSB) was poised to hand the investigation over to the FBI. Apparently, they had determined that the crash was caused deliberately. The hand-off never occurred, but the FBI did conduct an investigation of Mr. Batouti's supposed shenanigans at the Pennsylvania Hotel in New York. Their findings included reports that he had exposed himself to two teenage girls, stalked some female hotel guests and made romantic approaches toward hotel maids.

However, the hotel never reported him to police and never barred him from the premises. Shaker Kelada, president for safety at EgyptAir and a member of the crash investigation team, asserted that the FBI files contained "largely unsubstantiated" reports and had nothing to do with the crash.

In fact, glowing testimonials about Mr. Batouti poured forth from Egypt, particularly stories about how he would spend long layovers providing for his youngest daughter, who was suffering from lupus and was on her way to being cured in the United States.

Political pressure from both the Egyptian and U.S. administrations (Egypt was one of the United States' closest allies) convinced the NTSB to continue testing for possible mechanical problems with the plane. Privately, however, some sources close to the investigation told ABC News that they had wondered aloud whether the extra checking was a waste of time and money, because recovered and assembled data showed nothing mechanically wrong with the Boeing 767. Egypt, however, insisted that evidence showed that rivets in the control system of the plane's elevators were sheared in opposite directions. This could, Egyptian investigators insisted, indicate a possible jam of the controls that could have pushed the plane into a dive. Furthermore, after the preliminary release of findings by the NTSB in August of 2000, Egyptian inspectors released a statement that "facts do not support a so-called deliberate act [by Batouti]."

The Egyptian Civil Aviation Authority employed a psychiatrist to analyze the emotional state of Batouti from the sound of his voice on the cockpit recorder and concluded that he was acting coolly, probably in concert with the pilot to save the plane from crashing. EgyptAir issued a release stating that when captain Anwar asked Batouti to shut off the plane's engines, the first officer responded "It's shut." "Shutting off fuel to the engines is the first step in any attempt to re-start them," continued the release.

In opposition to this, former NTSB managing director Peter Goelz, who was on the investigation team, said that Batouti's recorded comments and actions suggested he was *not* working to save the plane. "The guy did not seem to be trouble shooting when the plane was in a dive," said Goelz. "The action did not really start inside the cockpit until the captain came back in. If I would be in a zero-gravity dive and I didn't know what was going on," continued Goelz, "you'd find me pulling back on the stick, breathing heavy, pumping against the column, going through procedures and probably shouting for help. It didn't appear to me like he was doing any of that."

And so the contentions continued. Obviously ill at ease, NTSB chairman Jim Hall agreed to further investigation of possible mechanical problems. Finally, the NTSB released its report, which indicated that EgyptAir Flight 990 crashed because of an intentional act.

On January 26, 2001, EgyptAir's lawyers accepted liability for the crash, which opened the way for the families of about 100 of the passengers to collect damages. However, the airline stated that it would reserve the right to sue other parties to share the costs and that it was not accepting the blame for the crash. This, in effect, cut short an effort by lawyers for the families to show that Batouti had committed public sex acts at the Pennsylvania Hotel and that he was mentally unbalanced. This did not, however, bury the contention between Egyptian and American investigators, and that contention, as of this writing, continues to shroud the precise cause of the horrific descent of Flight 990 in doubt.

UNITED STATES
MICHIGAN
SOUTH HAVEN
June 24, 1950

The cause of the crash of a Northwest Airlines DC-4 into Lake Michigan on June 24, 1950, in heavy weather has never been determined. All 58 aboard died.

Like so many airline disasters, the cause of the worst commercial air crash to that date has never been explained. A stormy night. A calm radio transmission. And then silence. The scenario has been played over and over, and each replay brings renewed efforts to improve communication and safety. The "black box"—the voice recorder that freezes the concluding moments of an airliner's life—has aided today's crash investigators, but there was no such voice recorder aboard a Northwest Airlines DC-4 traveling through heavy weather from New York to Minneapolis and thence to Seattle on June 24, 1950.

The storm was particularly fierce over Michigan, and the captain of the aircraft, Robert Lind, requested permission to drop from a turbulent 3,500 feet to what he hoped would be a calmer 2,500 feet. His request was denied. Traffic was particularly heavy at 2,500 feet, dangerously so.

Captain Lind accepted the turnaway and radioed his position. He was over Lake Michigan, near the shore village of South Haven. Then there was silence. Some time shortly after that message, the DC-4 and its load of 58 passengers and crew plunged into the waters of Lake Michigan.

A score of planes and rescue boats, including the destroyer escort *Daniel A. Joy* were sent to the location. It would be the *Daniel A. Joy* that would spot the telltale oil slick that indicated the aircraft's fate. A week later, divers would bring up all 58 bodies and the twisted fuselage of the Northwest airliner.

UNITED STATES
NEW JERSEY
COAST
April 14, 1933

Human error in piloting and design caused the airship Akron *to crash in a thunderstorm off the coast of New Jersey on April 14, 1933. Seventy-three were killed; three were rescued.*

The history of lighter-than-air travel is not a pretty one. A noble experiment that has now been refined, its early years were marked with incompetence, trial and tragedy. France (see p. 11) and Germany (see p. 27 and below) suffered grisly calamities in their race to put faster and more efficient lighter-than-air craft into the skies. And two of the United States' early efforts at utilizing and improving on the designs of Baron von Zeppelin crashed disastrously.

First there was the *Shenandoah* (see p. 68). Then there was the *Akron,* a ship with a load of troubles. From its maiden flight in August 1931, problems had arisen with the *Akron's* structure and instrumentation. On a transcontinental trip to California from its home base at Lakehurst, New Jersey, in 1932, two girders had unexpectedly and unexplainably collapsed, thus reducing the rigidity of the ship's hull. The $5 million 758-foot-long leviathan (the largest ship in the air at that time) then limped to San Diego, where it killed two sailors at its landing post. A sudden gust of wind lifted it, carrying away the men who were trying to tie it down. The two lost their grips on the ropes and plunged to their deaths.

Even under ordinary conditions, the *Akron* did not behave properly. Its instruments malfunctioned; struts and braces gave way regularly. It was a tragedy waiting to happen, and, with an assist from human failure, it fulfilled its promise on April 14, 1933.

Assigned by the U.S. Navy to take radio compass dimensions along the New England coastline, the *Akron* was ordered aloft on a day when wind and fog had grounded all commercial passenger planes in the Northeast. There were violent thunderstorms off the coast, and the *Akron* drifted straight toward one of the worst of them. Even though the *Akron* was filled with helium and not the highly flammable hydrogen that had claimed earlier airships, it was still vulnerable to updrafts and lightning strikes on its skin.

According to the later inquiry, two misjudgments were made at this point—one by the *Akron*'s commanding officer, Captain McCord, and one by his adjutant, Lieutenant Commander Wiley. First, Captain McCord, upon seeing a giant thunderhead looming up before them, ordered a course change by 15 degrees. The order was evidently misunderstood to be 50 degrees, enough of an error to send the *Akron* and its crew directly into the storm rather than around it.

The ship plunged forward, sucked by the winds of the thunderhead into its very center. Within moments, its altitude had dropped from 1,600 to 700 feet, and this precipitated the second, fatal mistake. Commander Wiley ordered the ship's water ballast to be dumped into the sea, reasoning that this would give them the necessary altitude.

It did, but far too suddenly. The ship bounced upward, severing its rubber cables and smashing its controls. It was helpless in the elements, and it dove, nose forward, into the sea, smashing apart upon impact and splaying pieces of itself and its crew into the water.

Rescuers were dispatched immediately, among them the crew of a J-3 nonrigid blimp. But the same weather system that destroyed the *Akron* also forced the J-3 to crash into the sea, drowning two of its crew of seven.

Boats rescued the remainder of the J-3's personnel and three men from the *Akron,* including Commander Wiley. Seventy-three others, including Rear Admiral Henry V. Butler, perished in the crash.

UNITED STATES
NEW JERSEY
LAKEHURST
May 6, 1937

Exhaust sparks igniting escaping gas, static electricity caused by a sudden rainstorm or sabotage caused the fiery crash of the giant dirigible Hindenburg *as it floated near its mooring mast in Lakehurst, New Jersey, after a triumphant transatlantic voyage. Thirty-one died; 61 survived.*

For millions, the lost era of lighter-than-air passenger flight will forever by symbolized by the catastrophic conclusion to the stunning career of the world's largest and fastest airship, the *Hindenburg*. The subject of countless books, studies and a feature film, the fiery crash of this 804-foot giant of the skies forms a part of the fabric of every television retrospective of the 1930s—indeed, of every electronic or print recapturing of that age. The tortured voice of Chicago radio station WLS's announcer Herbert Morrison recording those last cataclysmic moments has been rebroadcast countless times: "Oh! oh! oh! It's burst into flames! . . . Get

The German dirigible* Hindenburg *burns at its mooring post in Lakehurst, New Jersey on May 6, 1937. This spectacular finale to the age of lighter-than-air transportation was blamed on exhaust sparks igniting escaping gas, static electricity caused by a sudden rainstorm or sabotage. (Smithsonian Institution)

this, Scotty! . . . Get out of the way, please . . . this is terrible, oh my . . ." It was the true end of an era, and a dream.

In the beginning, when the first ships were designed and built, it was believed that airship transatlantic travel would become faster than that of transatlantic ship travel, but no less luxurious. The *Hindenburg* was the apotheosis of the dream, the true super airship, with its nonstop range of 8,000 miles, a 97-passenger capacity in spacious staterooms, lounges and dining rooms, 61 crew members, four 1,100-horsepower Mercedes-Benz engines and 16 gas compartments holding seven million cubic feet of—hydrogen.

And this was the puzzling and eventually fatal issue that gnawed at the dream: Why, when helium, which was so much safer, saner, available and widely used, did the *Hindenburg's* manufacturers and its directors choose to employ the far more dangerous and explosive hydrogen?

The surface answer has to do with maneuverability. Hydrogen-filled airships are easier to maneuver than helium-filled ones. But there were deeper reasons, too, that had to do with the sort of pride that presupposes immortality, and the *Hindenburg* was a proud ship. Its 36 passengers on the fateful trip from Europe to America of May 6, 1937, paid premium prices of $400 each for the privilege of crossing on this state-of-the-art, luxurious leviathan. And this trip must have been particularly luxurious. The ratio of crew members to passengers was almost two to one.

Although its speed of 85 miles per hour allowed it to cross the Atlantic in one-fifth the time of surface liners, the *Hindenburg*, like any lighter-than-air craft, was vulnerable and subject to weather conditions and air currents. A series of squalls and crosswinds had hampered the crossing of May 6, 1937, and Captain Max Pruss radioed ahead to Lakehurst, New Jersey. Instead of arriving at their estimated hour of 5:00 A.M., the ship would dock at sunset.

The army of reporters, cameramen, spectators and passengers arriving to board the *Hindenburg* for its return trip to Europe probably breathed sighs of relief and caught a few extra hours of sleep.

And so the ship sailed serenely westward, a craft that had proven itself to be speedy, reliable and safe. It had made 10 transatlantic crossings in 1936 alone, and it was the pride of Germany's emerging Third Reich.

At 3:12 P.M. the *Hindenburg* hove into view near Lakehurst. But as preparations were being made for its docking at the huge mooring mast, surrounded by a landing crew composed of sailors and marines, a sudden rain squall came up. Captain Pruss and his second officer, Captain Ernst Lehmann, thought it wiser to wait out the squall.

By 6:00 P.M. the skies had cleared, and the *Hindenburg* began its majestic descent to the mooring mast. The trapdoors in its bows flipped open, flinging its landing cables groundward. Some of the passengers crowded at the windows of the passenger gondolas; others relaxed in the lounges.

But Captain Pruss was bringing the *Hindenburg* in too fast. He ordered the two rear engines to full reverse and at the same time released 1,000 gallons of water ballast. At exactly 7:23 P.M. Chief Boatswain's Mate R. H. Ward, who was part of the landing crew on the ground, looked up and saw a slight fluttering of the fabric near the number-two gas cell, toward the stern of the ship. Gas was obviously escaping, and there were exhaust sparks erupting from the suddenly reversed engines nearby.

Whether it was this that caused the hydrogen in that gas cell to ignite and explode, or static electricity caused by the sudden rainstorm or, as some darkly hinted, sabotage, the catastrophe spread like a Sierra Nevada forest fire in fast motion.

The tail section of the *Hindenburg* detonated with a ground-shaking impact. Superheated flames shot fiery fingers up into the sky and toward the ground. Like a runaway river of fire, the flames ran forward, engulfing the giant ship in seconds as, one by one, the liner's 16 gas cells exploded. The metal frame, heated to the melting point, drooped by the tail and began to tear apart. The nose poked skyward for a moment, then reversed itself and headed toward the ground.

The approximately 1,000 spectators and press people waiting for the majestic ship to dock could not believe their eyes. WLS's Herbert Morrison provided the most dramatic and vivid coverage of the catastrophe and was later dismissed by his station for failing to maintain his equilibrium: "Here it comes," he exulted, as the ship neared the mooring mast. "And what a sight it is, a thrilling one, a marvelous sight. The sun is striking the windows of the observation deck on the westward side and sparkling like glittering jewels on the background of black velvet . . ."

And then the far less poetic reality: "Oh, oh, oh! It's burst into flames . . . Get this Scotty . . . get out of the way, please, oh, this is terrible, oh my, get out of the way please! It's burning, bursting into flames and is falling just short of the mast. Oh! This is one of the worst . . . Oh! It's a terrific sight . . . Oh, all the humanity . . ." and Morrison burst into tears.

The searing, white-hot heat from the burning hydrogen reached out and tightened the faces of the horrified spectators. On the ship, everything from pandemonium to stoic acceptance was occurring. Some crew members and passengers leaped from the now-descending ship to the landing cables. Very few made it; more fell to their deaths.

The longer passengers and crew stayed aboard the collapsing giant, the more severely burned they became. Those who leaped at the last minute were seen to be human torches, their clothes burned away, their hair on fire. One German businessman turned his back on the smashed windows of the gondola, through which his flaming companions were leaping, and walked into the interior wall of flame that was now turning the liner into a twisted mountain of molten steel.

As the *Hindenburg* settled to earth, collapsing like a spent balloon, sailors rushed into the still-molten hull, finding and dragging out possible survivors. Some passengers and crew leaped for the ground at the last minute and suffered only broken bones. Others did not survive even the short fall. Some perished in the flames, burned to an unrecognized mass. Others survived to die of their burns later in ambulances or hospitals.

It would be 10:45 P.M. that night before the wreckage would cool enough for some sort of accounting. Of the

92 people aboard, 61 survived. Thirty-one died, as did the dream of luxury transcontinental travel in lighter-than-air craft.

UNITED STATES
NEW YORK
BROOKLYN
December 16, 1960

Human error on the part of both flight crews and controllers was deemed responsible for the midair collision of a United Airlines DC-8 and a TWA Super Constellation over New York City on December 16, 1960. One hundred twenty-eight aboard the two liners and eight on the ground in Brooklyn died.

Almost five years after the terrible midair collision of a TWA Super Constellation and a United Airlines DC-7 over the Grand Canyon (see p. 47), two planes representing the same airlines were involved in another appalling midair wreck, this time over New York City.

A light snow was falling on the morning of December 16, 1960. A 17-inch snowfall had blanketed New York City a few days before. The temperature was 33 degrees.

United Airlines flight 826, a DC-8, was coming in from Columbus and Dayton, Ohio, headed for Idlewild (now called Kennedy) Airport. TWA flight 266, a Super Constellation, was arriving from Chicago, headed for LaGuardia Airport. Because of the intermittent snow, both flights were on IFR (instrument flight rules) and were being monitored by New York center radar—a system installed as a result of the Grand Canyon collision.

At about 10 A.M., the TWA flight was cleared to descend to 9,000 feet. Once TWA captain David Wollman reached that altitude, he was told to contact LaGuardia approach control. This meant that each flight was now being monitored by different facilities.

At roughly the same time, the United flight, captained by R. H. Sawyer, was being given a revised route by New York center to Preston, a radio check point, with instructions to descend from 11,000 feet to 5,000 feet. "Looks like you'll be able to make Preston at 5,000," said the New York center.

"Will try," United answered.

"If holding is necessary at Preston S.W. one minute pattern right turns . . ." radioed the center.

"Roger, we're out of 6 for 5."

"826, Roger, and you received the holding instructions at Preston. Radar service is terminated. Contact Idlewild approach control."

"Good day," was the cutoff from United.

It was the LaGuardia approach control that first noticed something threatening on its radar screen. At 10:30, it radioed the TWA flight, "Unidentified target approaching . . . six miles."

FAA tests in 1958 determined that pilots could not distinguish traffic until it was three to five miles distant. The

Wreckage is strewn over Sterling Place, Brooklyn, after the midair collision of a United Airlines DC-8 and a TWA Super Constellation on December 16, 1960. One hundred twenty-eight aboard the two liners and eight on the ground died. (National Transportation Safety Board)

same tests, quoted in "Captain X" 's study of airline safety, *Safety Last,* concluded that a pilot flying at 600 miles per hour will travel 920 feet before he can definitely determine if an object he spots out of the corner of his eye is a cloud, a speck of dirt on the windshield or another aircraft. According to the study, "2,680 feet will be used up while he decides whether to climb, dive, or turn left or right; . . . 4,792 feet will slip by before he can make the plane respond to his actions. A jet pilot would need a bare minimum of 9,584 feet before he could spot a target and maneuver to miss it, under excellent conditions: clear day, eyes focused for distant vision, no distractions, and a well-rested crew."

"Unidentified object three miles . . . two o'clock," continued the approach control at LaGuardia.

"Roger," acknowledged the TWA flight.

At 10:33, the United flight radioed approach control at Idlewild: "Approaching Preston at 5,000."

Approach control advised him to maintain altitude and that little or no delay was expected. This was followed by a weather report and landing instructions.

There was no mandatory acknowledgment from the United flight, possibly because at this moment, Captain Sawyer realized that they were not over Preston at all but over Staten Island, New York, and headed on a direct collision course for the TWA Constellation.

A moment later, the United DC-8 slammed into the Constellation, its wing ripping a jagged hole in the other plane's fuselage, which in turn ripped off one of the DC-8's jet engines. The Constellation spiraled to earth, crashing into Miller Field, a small military airport on Staten Island. Fortunately, there was no one on the ground below, although everyone aboard the Constellation was killed instantly.

Meanwhile, the DC-8 limped toward LaGuardia, hoping to be able to make an emergency landing there. But it never made it. Over Brooklyn, it lost altitude and plunged into a thickly populated part of that borough. The plane first hit the snow-covered roof of a four-story apartment house and then bounced down to Sterling Place, where parts of the fuselage split apart and skidded down city streets. The major part of the plane smashed into the ironically named Pillar of Fire Church, setting it and 10 nearby buildings ablaze.

Pieces of the flaming wreckage, with charred corpses still strapped in their seats, were deposited in streets adjoining the main conflagration. One 11-year-old boy was rushed to Methodist Hospital, where 10 surgeons fought to save his life but failed. There were no survivors. One hundred twenty-eight people aboard the two airliners and eight on the ground in Brooklyn died.

Charges and countercharges punctuated the follow-up investigation, which fixed the blame on the deceased crew of the United DC-8. The CAB determined that the pilots had failed to report that one of their two key radio navigation instruments was not functioning. Therefore, they were unable to report the plane's exact position. Furthermore, they had raced through the five-mile buffer zone designed to separate LaGuardia and Idlewild traffic.

But there were other factors at work: A trainee had been working the LaGuardia radar at the time. The United flight had not been carefully monitored. And there was no direct phone communication between the two approach-control facilities at Idlewild and LaGuardia.

Lawsuits totaling $77 million were filed against the two airlines and the FAA by the relatives of the victims. The suits were settled for about $29 million, with United taking 60% of the responsibility, the FAA 25% and TWA 15%.

As in the aftermath of the earlier TWA-United midair collision, new safety measures were instituted. Speed for aircraft entering the terminal areas was decreased. Extra controllers were added at busy airports.

A new phone system was installed between the two approach centers, and a system of positive radar handoffs became mandatory. This meant that a controller could not relinquish his or her guidance of a flight until the receiving controller had the blip on his or her radar scope and that blip was positively identified.

Today, transponders and three-dimensional radar have decreased the incidence of midair collisions. But it took this horrendous accident to prod airports to put them into use.

UNITED STATES
NEW YORK
COVE NECK, LONG ISLAND
January 25, 1990

A Federal Transportation Board inquiry determined that faulty judgments by the crew were responsible when an Avianca Boeing 707 ran out of gas and crashed into the residential area of Cove Neck, Long Island, on January 25, 1990. But a series of factors were involved in the tragedy, which resulted in the death of 73 aboard the jet. Eighty-five survived.

The ancient adage about a nail in a horseshoe eventually becoming responsible for the loss of a war was illustrated in the fatal crash of Avianca flight 52 on the stormy night of January 25, 1990. The word "emergency" was not uttered by the co-pilot when the Boeing 707, critically short of fuel, was being held in a circling pattern before landing at New York's John F. Kennedy Airport. Words the copilot did use—"priority" and "low on fuel"—should have indicated an emergency. But the controller, testifying at postcrash hearings, stated that the official word "emergency," which would have triggered an immediate response and brought the flight in safely, was never heard. So, instead of being brought in, the flight was given directions that would exhaust its fuel and cause its crash.

Still, airplane crashes almost always involve many factors. Those in effect that night included foul weather, a shortage of experienced controllers, an unrealistic estimate of the number of aircraft that could be handled at Kennedy given the weather conditions, a failure by a local controller to hear all of the co-pilot's transmissions, the tentativeness of the co-pilot's messages to controllers, the difficulty of communication between flight crews and controllers speaking different languages and the decision of the pilot to abort a landing pass. All of these factored into the incident, and any combination of two or more of them might have caused that tragic loss of life.

At 7:33 A.M. on January 25, 1990, the Central Flow Control Facility in Washington, D.C., warned airlines that it was going to be a bad day for flying on the East Coast. Gusty winds and marginal flying weather would force ground delay procedures at both New York's Kennedy and Boston's Logan airports, the advisory warned.

At 9:30 A.M. Central Flow Advisory 21 was issued, stating that at Kennedy a ground delay program would be implemented at 2 P.M., the time at which peak traffic would begin. Because of low ceiling and visibility, Kennedy would be "reducing the acceptance rate [of aircraft] from 38 to 30 an hour." On a good day, Kennedy is capable of accepting 60 flights an hour, but since it is an international airport, and international flights are supposedly exempt from acceptance-rate reductions, this seemed to be a generous figure for the conditions.

And it proved to be so. By the time the Avianca flight approached Kennedy, planes were being held up to 80 minutes on the ground and in the air. Despite the fact that the Avianca flight was an international one, it, too, was put into a holding pattern.

The wreckage of the Avianca Boeing 707 that crashed in Cove Neck, Long Island, on January 25, 1990, is flung over a large part of this thickly populated suburban community. Fortunately no one on the ground was killed. (National Transportation Safety Board)

The flight, with 158 aboard, had originated in Bogotá, Colombia, and had refueled at Medellín. It had battled strong winds as it made its way up the east coast of the United States.

A few hours earlier, an air controller supervisor at Kennedy had recommended that because of the driving winds and rain, the number of planes allowed to land per hour be further reduced, from the optimum 60 to 22. He was overruled, and the number was set at 33.

Thus, at 8:07, when the Avianca flight entered Kennedy airspace, it was stacked in a holding pattern off the coast of New Jersey about 40 miles south of Kennedy. At that time it was being monitored by a regional control center responsible for traffic in and around the New York area.

By 8:45 the plane had made three wide circles, and its fuel was getting low. Regional controller Philip Brogan contacted the flight and told it that it would be held at least until 9:05. The pilot and co-pilot of the Avianca flight conferred, and the co-pilot radioed back, "I think we need priority." Brogan inquired about their alternate airport and how long they could hold. The co-pilot replied that their alternate was Boston but that it was full of traffic. And then, the co-pilot added the phrase that should have tipped off the controllers that an emergency was in the making: "It was Boston but we can't do it now, we . . . run out of fuel now."

Brogan immediately handed off the plane to local controllers to guide it in its approach to Kennedy. But the handoff controller, Jeffrey Potash, was on the phone coordinating the transfer, and he did not hear the comment about running out of fuel. Nothing was said about it to Scott Machose, the local controller.

Had he known this, Machose would undoubtedly have brought the flight in immediately. But instead, he placed the plane in yet another hold, called a "spin," over Long Island. He would keep the flight in this pattern for six and a half minutes, while more precious fuel was burned.

Meanwhile, a thick fog had settled in over Kennedy, making its runway lights invisible from 1,300 feet away.

At 9:07, 77 minutes after it had first entered its holding pattern, the Avianca flight was cleared for landing. It was 27 miles from the airport. The crew was confident that they would make it.

At 9:09, the co-pilot told the pilot, "They are giving us priority."

"Tell me things louder because I'm not hearing it," complained the pilot.

At 9:20, the co-pilot informed the captain: "Yes sir, we are cleared to land."

The plane was flying at 150 miles per hour, and it was just four minutes from a safe landing. "All set for landing," the engineer assured the pilot.

The pilot nosed the plane toward the invisible runway. At 9:23 the Avianca's warning system tripped, setting off a horn that warned that it was descending too steeply. "The runway. Where is it?" the pilot asked the co-pilot.

"I don't see it. I don't see it," he answered. The pilot pulled the nose of the plane up, aborting the landing. "Tell them we are in emergency," he said to the co-pilot.

"Once again, we're running out of fuel," the co-pilot radioed to ground control. He failed to use the critical word "emergency," which triggers immediate clearance.

"Declare emergency," said the pilot again. "Did you tell him?"

"Yes, I already advised him," responded the co-pilot.

"Did you already advise him we don't have fuel?"

"Yes sir. I already advise him . . . and he's going to get us back."

The pilot asked for a bearing. The co-pilot gave him a wrong bearing, taking the plane another precious minute out of its way.

Now, at 9:26, the controller told the crew to fly out another 15 miles northeast before coming back for a second approach. "Is that fine with your fuel?" he asked the co-pilot.

The co-pilot paused a moment and then answered, "I guess so. Thank you very much."

And that was the last, fatal decision. Five minutes later, as the plane began to line up its approach over Long Island's north shore, the engineer announced, "Flame out. Flame out on engine number four."

Four seconds later, the second of the plane's four engines stopped.

"Show me the runway," the pilot called.

"We just, ah, lost two engines and, ah, we need priority please," radioed the co-pilot to the ground.

"You're 20.6 nautical miles from runway 22 left, and cleared for instrument landing," answered ground control.

But it was too late. The remaining two engines had stopped, and the 707 glided silently for almost two minutes before it slammed into a hillside behind a huge estate in Cove Neck, Long Island.

With a roar, it split apart, spewing baggage, bodies and pieces of the plane over the landscape. Local fire departments rushed to the scene, and all through that long and rainy night rescuers toiled, bringing 85 survivors—some of them horribly injured—out of the mangled wreckage. Seventy-three, including the entire cockpit crew, died in the crash.

UNITED STATES
NEW YORK
LONG ISLAND
July 17, 1996

At approximately 8:20 pm on July 17, 1996, TWA flight 800, bound from New York's Kennedy Airport to Charles de Gaulle airport in Paris, exploded just off the coast of Long Island and plummeted 13,000 feet into the Atlantic Ocean. All 230 aboard were killed.

On February 6, 1997, 163 people who lost someone in the July 17, 1996, crash of TWA Flight 800 braved a snowstorm to travel to an isolated hanger on the now deserted former site of a Grumman Aviation testing facility in Calverton, Long Island. They had been invited by the FBI and the National Transportation Safety Board (NTSB) to view the reconstruction of the plane from pieces that had been fished from the waters off East Moriches, Long Island, in the months after the crash.

Some brought flowers; all brought memories and grief. "When you look at the empty seats that are charred and busted up and mangled, you immediately run through your mind, these seats were full at one time," said Stephanie Maranto, whose brother, Jamie Hurd, had perished in the tragedy. "They were full of life and people and children and husbands and wives, and it's upsetting to see them sitting there like that, completely empty and just broken to pieces. It's a very quiet, cold and eerie feeling."

Another family member, Joseph Lychner, added a chilling footnote. "I was hoping for a clear indication that my family died instantly," he told reporters on the scene. "Unfortunately," he concluded, "because of where they were sitting on the plane, that was not clear at all to me." And so dissatisfaction and lack of emotional closure lingered with the survivors of this, one of the most horrendous—and possibly avoidable—crashes in all of U.S. airline history.

Twenty-five is young for a man but old for an airplane, particularly one that flies constantly for an airline. The Boeing 747-100 that was TWA's flight 800 from Kennedy Airport to Paris on July 17, 1996, was the 153rd Boeing 747 made. It came off the assembly line in Everett, Washington, on July 15, 1971. Its technology was essentially the same as that of the first 747 flown in 1969.

In the 25 years of its life, this 747 had made 16,000 flights, and in the two weeks preceding the crash had flown 100,000 miles, in 24 transatlantic flights. The day before it had made the nine hour, 45-minute flight from Athens to New York easily.

However—and this was a big however—the ground crew at Kennedy noted that the Athens to New York leg had drained the center fuel tank down to 50 gallons. Because the tank would not be needed for the shorter trip to France, it was not refilled. All of the 30,000 gallons of jet A kerosene was pumped into the jet's six wing tanks. Filling the center tank would increase the plane's weight, making the flight more expensive to operate.

Pieces of TWA Flight 800 were raised from the bottom of the Atlantic Ocean off Center Moriches, Long Island, N.Y., in the summer of 1996, then transferred to flatbed trucks at the nearby Hampton Bays Coast Guard Station. (AP/Wide World Photo)

In business terms, TWA had a point. The 433 seats of the jumbo jet were carrying only 176 fare-paying passengers, plus 54 TWA employees and their families flying free. Some of the paying customers had been transferred, as had the cockpit crew, from flight 848, destined to fly direct to Rome. The passengers would continue to Rome from Paris on the same plane.

The flight loaded on time, but because security precautions involving unidentified baggage had been increased after the 1988 bombing of Pan Am flight 103 over Scotland (see p. 128), an unidentified bag (later identified) halted the passengers at the gate for an hour. During that time, the air conditioners, located under the nearly empty center fuel tank, were operating at full capacity, thus pumping heated air from their motors into the tank.

Finally, at 8:18 P.M., Captain Ralph Kevorkian began to move the jumbo jet toward its takeoff runway. The takeoff was smooth, and within minutes the plane was over Jamaica Bay, headed east toward Europe at 287 mph.

Pumps in the wing tank were now turned on, and fuel began to flow at a rate of approximately 11 gallons a minute through a large aluminum pipe that ran along the back wall of the plane's center tank. It would run through this pipe for only another five minutes and eight seconds—the duration of flight 800.

The air speed of the plane was increased to 368 mph as it climbed to an altitude of 13,000 feet, where it leveled off. Three minutes later, air traffic control gave the pilots clearance to climb to 15,000 feet.

Underneath part of the passenger cabin, the 12,890-gallon center fuel tank was heating up. The remains of the kerosene from the Athens flight were being turned into a fog, which was becoming steadily more volatile with every increase in temperature. And then, just as TWA flight 800 reached 13,700 feet, the mixture of heated air and heated kerosene in the tank exploded with a terrible thunder.

This writer, who lives only a few miles from the point at which the plane crashed into the Atlantic Ocean, was fixing dinner when suddenly two monstrous explosions in quick succession shook the house, rattling dishes and knocking hanging pictures askew. Looking through a window I saw what must have been the last glimmer of a bright flame, like an exploded star, in the southwest sky.

The thunder had come from a force estimated later at 100 tons of pressure, which cracked TWA flight 800 in two and disintegrated its entire forward support structure. The cockpit was decapitated from the rest of the plane. In the still, if only momentarily, intact passenger cabin, pressure suddenly dropped, and the air in the cabin rushed out through fissures that were racing up the cabin walls at more than a mile a second, shredding the fuselage as they went.

Below the passengers' feet, pressure from the explosion in the tank lifted the floor in the area between the wings, and seconds later a 10-foot hole appeared. The passengers seated in this part of the plane were sucked through the opening in the bottom of the plane and fell two and a half miles through the sky into the Atlantic Ocean off the coast of Long Island at East Moriches.

The plane then disintegrated, and the rest of it began to fall in bits and pieces. Cables snapped, explosive thuds continued and glass shattered.

When the nose of the plane was severed, the center of gravity shifted, tilting the tail downward. The engines, still running, drove the passenger cabin into a brief climb until the outer ends of both wings snapped off, and the long descent began.

Now, 24,000 gallons of fuel spun outward from the ruptured gas tanks and atomized into explosive kerosene clouds that turned into immense fireballs. The entirety of both wings cracked and snapped off the fuselage, so that those passengers who were still strapped into their seats and survived the trauma of the explosion could see the sky through the front of the severed cabin.

Tongues of fire and disintegrating debris rained from the sky. The tail section gained speed as it fell and hit the surface of the Atlantic at 400 mph.

Two aircraft pilots saw the explosion of flight 800. Captain David McClaine, at the controls of Eastwind SB 507, saw it from 10 miles away and immediately contacted Kennedy. "We just saw an explosion out here, Stinger Bee 507," he shouted.

"Stinger Bee 507, I'm sorry I missed it . . . did you say something else?" the controller asked.

"Ah —" the shaken McClaine continued, "we just saw an explosion up ahead of us here, somewhere about 16,000 feet or something like that. It just went down into the water."

Two hundred feet above Gabreski Airport in Westhampton, Air National Guard pilot Captain Christian Baur, who was piloting a Sikorsky Pave Helicopter for the 106th Rescue Wing of the New York Air National Guard, saw the flash. At first he thought it might be the C-130 rescue plane that was flying a practice run nearby, but when he looked closer, he saw that it was another plane and that pieces of it were falling near him. He warned the C-130 to stay its distance. "Stay back, there's still a lot of falling debris," he said over his radio. Later, he told investigators that in the middle of the debris he had seen a person "falling right through the debris like a sack of potatoes."

Now the sea caught fire as an oil slick the size of two football fields ignited, sending flames 30 feet into the air. Coast Guard cutters, helicopters and local boatsmen converged on the scene, looking, against hope, for survivors. There were none.

A nightmare of assumptions and missteps ensued. An immediate assumption was that a missile, fired from land on Long Island or possibly from a small boat at sea, had brought down the airliner. The FBI acted upon this assumption and the further assumption that friendly fire had come from an area south of the plane's flight path called Whiskey 105E. On July 17 extensive military maneuvers were under way in Whisky 105E that involved Navy ships, submarines and aircraft.

"Something happened out there," an FBI spokesman told reporters that night. "We know that. If it was an act that was perpetrated by terrorists, we will find them. We will find the cowards."

Meanwhile, as investigators from the NTSB were on their way from Washington and the Coast Guard and other rescuers turned to fishing bodies and pieces of bodies from the Atlantic, the relatives and friends of the doomed passengers aboard flight 800 were kept waiting in a hotel near Kennedy. It would be 10 hours before TWA sent an executive to make apologies.

As the days and days of salvage and the search for the black boxes and the fuselage continued, politicians and personalities muddied the waters. Journalist Pierre Salinger announced that he was in possession of secret documents proving that a U.S. Navy missile had been responsible for the downing of the flight. His secret documents, it turned out, had been on the Internet for six weeks and were a hoax.

New York governor George Pataki, acting on information given to him by the FBI, announced at a memorial service on the beach near the accident scene that there were bodies trapped in the wreckage and that they would be brought up soon. His information, it turned out, was based on speculation reached after Navy divers spotted the main wreckage field.

Congressman Michael Forbes announced that one of the black boxes had been found before it had been found. Told of his error, he refused to back down. "It's been located," he told reporters. "I don't know if it's actually been pulled up at this point. . . ."

It would be a week before Navy divers located both of the black boxes, but they were largely uninformative to all but the NTSB investigators, who now found themselves at odds with the FBI, who was still looking for evidence of a missile or a bomb.

The NTSB knew that the airliner had broken up in flight, and their theory was bolstered by the discovery of fire-damaged parts of the center fuel tank. An earlier accident, while not as severe as this one, seemed to bolster their theory even more: In 1990 the center wing tank on a Phillipine Air Lines 737 had exploded, killing eight people. Other explosions, one in 1970 aboard a DC-8 in Anchorage, Alaska, and another in New Haven, Connecticut, of an Allegheny Airlines plane, all seemed to strengthen the explosion theory.

Still, the FBI pressed forward in its hunt for evidence that a bomb exploded on the plane. Rumors of evidence of plastic explosives in the wreckage surfaced, then died. The source of the explosives the FBI was tracking was traced to the military ships that carried the wreckage from ocean to shore.

The fuselage was reassembled in Calverton, and the NTSB pressed forward in its efforts to find the source of the explosion. Meanwhile, Boeing began a media campaign to foster the company's claim that the plane had not self-destructed. "The chances are zero of that happening," their engineers told reporters.

The NTSB was analyzing the wiring in Boeing 747s, and it found that much of it was old and that much of it was packed tightly together, so that if one worn wire shorted and flamed, it could affect the other wires. A pack of these wires ran through the center fuel tank, and the center fuel tank was poised above the air conditioning units, which were using the tank to drain off excess heat in their motors. Now it began to come together. The NTSB was startled to learn that the military had long since corrected wiring deficiencies that still existed in commercial airliners. It would be 1998 before the FAA would order airlines to modify their wiring through fuel tanks. They were given three years to comply.

Two months earlier, investigators had found a small black mark on a fuel probe in the reconstructed flight 800. It was identified as a bit of copper-sulfur created over time by the interaction of the metal conductor with the sulfur in jet fuel.

Then, in the fall of 1998, Tower Air maintenance people found evidence of a short circuit in a center fuel tank in one of their 747s. When they removed the probe, they found the same black deposits. They forwarded this information to the NTSB.

Further probing discovered that there had been 16 airplane explosions in the 35 years preceding TWA flight 800's catastrophic eruption. All of them had been traced to ignition problems, but none had been linked to trouble in empty fuel tanks.

As early as the 1960s, "inerting" systems to battle fires had been tried. The theory was that if one robbed a potential fire of oxygen, it would not burn, nor would it explode. The insertion of nitrogen was tried, which worked, but it was deemed too cumbersome and expensive.

However, in the 1970s another less expensive and cumbersome system of inerting using the gas generating

Smith Point Park, the nearest beach to the location of the horrendous explosion and crash of TWA Flight 800, became a memorial site for those who lost friends and family in the calamity. (AP/Wide World Photo)

system already aboard airplanes was discovered and made practicable. In the so-called OBIGGS method, nitrogen for inerting in fuel tanks was created by separating the nitrogen from engine air. By 1981, the Air Force began to incorporate this system in all of its military aircraft. But, astonishingly enough, the FAA exhibited only passing interest in it, as the airplane manufacturers complained that it would be too costly.

It would be the summer of 2001, five years after the terrible tragedy of flight 800, before a proposal for inerting was again raised, and once again the FAA did not recommend that it be used. Again, the airline and airplane manufacturing industries reasoned that it would be less expensive to have an accident and pay off the lawsuits entered into by the victims' families than to install inerting equipment that could preclude another such tragedy. The morality of placing profits above human lives was questioned in several newspaper editorials, but as of this writing, the FAA still has not required inerting in commercial aircraft, ones that, like TWA flight 800, carry millions of paying—and trusting—passengers every year.

UNITED STATES
NEW YORK
NEW YORK
July 28, 1945

Pilot misjudgment was responsible for a B-25 Mitchell bomber crashing into New York's Empire State Building on July 28, 1945, in a heavy fog. The crew of three and 11 workers in the building were killed.

One of the more bizarre airplane crashes in the history of aviation occurred on a foggy Saturday morning in July 1945. It involved the collision of a U.S. Army Air Corps B-25 Mitchell bomber with New York City's Empire State Building. The experts at the time said the odds of this sort of accident happening at all were 10,000 to 1. But it happened.

The morning of July 28, 1945 was a murky one in New York City. A heavy fog had rolled in from the Atlantic Ocean overnight, and the sun had not yet burned it away.

Twenty-seven-year-old Lieutenant Colonel William F. Smith, Jr., a highly decorated West Point graduate, was flying a B-25 on a short trip from New Bedford, Massachusetts, to New York's LaGuardia Airport. The plane was carrying a crew of three and one passenger. At approximately 9:30 A.M., LaGuardia ground control redirected him to Newark, New Jersey. LaGuardia was socked in with fog, ceiling zero and forward visibility about three miles. "The Empire State Building is not visible," added the ground controller, in an unconsciously prophetic after-thought.

Smith and his crew headed for Newark in a path that would take them squarely over the island of Manhattan. Fifteen minutes after receiving his wave-off from LaGuardia, Smith found himself a mere 500 feet above Rockefeller Center's towers, more than 40 stories high. According to observers on the ground, he climbed abruptly out of that danger and disappeared into the mist.

But apparently Smith did not climb high enough. Two minutes later, the bomber, flying at 225 miles per hour, smashed squarely into the north side of the 78th and 79th floors of the Empire State Building—the tallest building in the world at the time. The fuel in the engines exploded on impact, and the plane's two propeller-driven engines scissored their way through the concrete walls of the building. One engine emerged on the south side of the skyscraper, became airborne for a brief moment over 33rd Street and descended in a long loop through the skylight of the penthouse sculpture studio of Henry Hering. The other flaming engine demolished the elevator door on the 79th floor, snapping the cable of the elevator car. One woman was riding in it, and she began the long plunge to the bottom. Fortunately, the elevator's automatic braking device engaged, and she was saved from a plummet to the basement. But just as she stopped, part of the engine hit the top of the car, caving it in.

The plane broke into several pieces. Part of a wing careened a block east onto Madison Avenue. Pieces of the fuselage dug into the walls of nearby skyscrapers as if they had been shot through a high-powered machine gun.

If this had happened on a weekday, the Empire State Building would have been thickly populated with close to 50,000 office workers and tourists. But it was Saturday, and only a handful of workers were at their desks in the 79th-floor offices of the War Relief Services of the National Catholic Welfare Conference. Eleven of them were killed. Some of them burned to death; others were crushed. One man was flung through a window. His charred body would be discovered on a 72nd-story ledge of the Empire State Building. Twenty-five were badly injured. All three servicemen on the plane died violently.

Firemen appeared on the scene quickly and doused the blaze. For weeks, until war-appropriated material could be secured for repairs, a gaping hole 18 by 20 feet existed in the side of the Empire State Building. A tarpaulin was used to cover it, but it was continually blown loose.

The woman who fell 79 stories in the elevator shaft survived thanks to the quick thinking of Coast Guard hospital apprentice Donald Maloney, who just happened to be strolling by the Empire State Building. He ran to a drugstore and commandeered syringes, hypodermics and other medical necessities. He then dashed into the building, climbed down through the crumpled roof of the elevator cab, administered a shot of morphine to the injured woman and saved her life.

UNITED STATES
NEW YORK
QUEENS
February 3, 1959

Failure to read new instrumentation correctly resulted in the crash during an instrument landing of an American

Airlines Lockheed Electra at New York's LaGuardia Airport on February 3, 1959. Sixty-six died; seven survived.

"Undershoot," the technical term for a landing attempt in which the aircraft touches down on the ground before reaching the runway, does not necessarily result in a catastrophic accident. However, runway 22 at New York's LaGuardia Airport begins on the bank of the East River. Even a minimal undershoot there can be at least wet and at most calamitous; it was the latter that occurred there on the fog-shrouded evening of February 3, 1959.

The American Airlines flight, which originated in Chicago, was required to make an instrument landing because of a heavy blanket of fog and rain that had all but closed down LaGuardia. Veteran pilot Albert DeWitt was at the controls of the aircraft, a brand new Lockheed Electra propjet. The 12-day-old plane contained the latest in instrumentation, including a new altimeter designed for precision but configured differently than the altimeters DeWitt and other veteran pilots were used to relying on in instrument landings.

It was the misreading of this new altimeter that was later considered to be a possible cause of the crash. Considering the fact that the last radio transmission from DeWitt contained no inkling of trouble, it was concluded in the follow-up investigation that the captain misread his altimeter and approached too low. Since LaGuardia lacked a backup radio beam to warn off undershooting pilots, this misreading would not have been challenged.

For whatever reason, five minutes after clearance the American Airlines Electra disappeared from the radar and dove into the waters of the East River at 135 mph. The slow speed allowed part of the fuselage and the tail section to survive the plunge, and those who were in the rear lounge or the rear seats found themselves with enough time to escape.

A businessman and two flight attendants were among a small group of seven that pulled themselves up onto the still-floating tail section. Some others, including flight engineer Warren Cook, managed to escape from the shattered plane underwater. Despite extreme injuries, Cook helped other passengers swim to safety.

Immediate rescue came from a tugboat, the *H. Thomas Teti Jr.*, which was towing several barges. Captain Samuel Nickerson, in command of the tug, summarily cut his barges loose and made for the wreck, taking the seven survivors aboard.

They were the only ones to live. Sixty-six others died.

UNITED STATES
NEW YORK
QUEENS
November 12, 2001

Slightly more than three minutes after taking off from New York's JFK Airport at 9:14 A.M. on November 12, 2001, American Airlines flight 587, an Airbus 300, plunged into the residential area of Belle Harbor in Queens, killing all 260 aboard and five on the ground.

Belle Harbor is a quiet community in the city borough of Queens, across the East River from Manhattan and on the Rockaway peninsula, a small tongue of land that juts out into the Atlantic Ocean. Its residents are middle-class workers, and a stunning percentage of the New York City firefighters and police officers who lived there had died in the September 11, 2001, tragedy at the World Trade Center (see p. 147). And so, at approximately 9:37 A.M. on another sunlit autumn day, November 12, 2001, a number of its residents near the juncture of 101st Street and Newport Avenue, felt a horrible sense of déjà vu, as the thunder of a rapidly descending jet smashed through the early morning calm, followed by flames and multiple explosions as it sideswiped several houses and crashed directly into a home. Milena Owens, who lived two blocks from the crash site, later told a *Newsday* reporter, "I heard the explosion and I looked out the window and saw the flames and the smoke and I just thought, 'Oh, no, not again.'"

Earlier, at 9:14 A.M., American Airlines flight 587, bound for the Dominican Republic with a full load of passengers, took off 74 minutes late from runway 31 at New York's JFK International Airport, banked left over Rockaway Bay and headed south. One hundred and seven seconds into the flight, as the Air Bus A300 was climbing toward the 13,000-feet cruising altitude for which it had been cleared, the plane shuddered, probably from ground turbulence. Fourteen seconds later, it shuddered again, more seriously. And then, a few seconds later, the A300 was knocked sideways, to the right, viciously. This was more than mere ground turbulence.

Two minutes ahead of American flight 587, on the same runway, a Japan Airlines 747, had taken off and was a little over 4 miles ahead of the American flight on the same path. All jets produce what pilots call "wake turbulence," a minitornado in the form of a wake, produced by the jet's engines. This sort of trailing vortex can be dangerous for small planes, and larger planes are separated by 4 miles to assure that there is no problem.

Ordinarily wake turbulence would have caused no difficulties for the American Airlines flight, but as the flight data recorder revealed later, the plane began to yaw violently at 2,900 feet, at the moment it entered the 747's turbulence. Captain Edward States and First Officer Sten Molin fought to stabilize the crazily fishtailing plane, which swung wildly to the right, then to the left, then to the right again.

Perhaps they fought too hard and put too much strain on the tail section. The 177.5-foot-long fuselage of the Airbus A300, because of its length, can act like a long lever, producing immense pressures on the tail assembly. It was apparently too much stress. "A brand new tail would have broken," said one investigator after the crash, and this one did. It and its rudder broke off from the fuselage, then the right engine tore away from the wing, and after this, the left engine ripped off and plummeted earthward.

The plane was helpless and rendered uncontrollable. "The aerodynamic forces must have been preposterously wild," former NFTSB investigator Chuck Leonard told reporters later. Within seconds, the plane heeled over, the nose dropped, and it fell like a bomb, straight down from 2,900 feet.

"I heard a plane that didn't sound right. It was rumbling," said Susan Locke, who worked near the crash site. "I looked out of the window, and I saw the silver front of the plane nose-dive to the ground."

Kevin McKeon and his family had just finished breakfast, and he was walking through the kitchen to head out to work when the plane hit his house. "It felt as if the walls were blowing off," he told reporters later. The impact of the crash tossed his daughter and him into the backyard and his wife into the living room.

Part of the jet's wing plummeted into the McKeon's basement, and simultaneously, a piece of one engine ripped through the back of the house and slammed into the garage, which was set afire.

"Oh God, it was terrible," related Elaine Dolan, who was coming home after walking her dog on the beach. "There were flames. There were flames coming out of the plane and then it fell."

The airbus made a direct hit on one house, flattening it and setting it afire, along with three others. At least 10 houses were flaming infernos by the time the fire department arrived. The entire area from Beach 131st to Beach 128th Streets, bounded on the south by Newport Avenue and on the north by Cronston Avenue was ablaze. Most of the firefighters who responded to the tragedy had just buried colleagues who had perished in the collapse of the World Trade Center towers.

Firefighters found both of the engines, a block apart, one on a boat parked in a driveway, the other in a gas station, where it mercifully missed the pumps. The tail assembly was fished out of Rockaway Bay later that afternoon by the Coast Guard.

Not only was this the second worst crash in U.S. airline history, after the 1979 Chicago crash of American Airlines flight 191 (see p. 50), but also it was a traumatic shock to the populace of a city still trembling from September 11. The first reaction was that this was a second terrorist attack, and all New York City airports—LaGuardia, JFK and Newark—were immediately shut down. The Lincoln and Holland tunnels, the Port Authority bus terminal, the George Washington Bridge, and the Outerbridge, Goethals and Bayonne bridges were closed. The Empire State Building was evacuated and shuttered, and all entrances and exits to the United Nations building were sealed off.

However, as the morning wore on, it became apparent that this was not a terrorist attack but a tragic, terrible accident that killed all 260 of those aboard the jet, as well as five people on the ground. By 1:30 P.M., all of the closings had been rescinded.

It would be days before the smoldering wreckage of the plane and the demolished houses cooled. Bodies, body fragments, personal belongings, pieces of the plane were removed and turned over to the NTSB for investigation.

Nothing mechanical was found to be wrong with the airplane, and as the investigation continued into the new year, it became more and more certain that pilot error had been responsible for the crash. In trying to regain control of their plane, the pilot and copilot had overcompensated. They had caused the side-to-side jolts. The rudder, according to the NTSB, was jammed at a deflection of a full 10 degrees. Many pilots, interviewed by the media, said that in a situation of that nature, a rudder should be deflected, or moved only a few degrees.

As of this writing, the investigation continues into one of the worst and most ill-timed of U.S. crashes of commercial airliners.

UNITED STATES
NORTH CAROLINA
HENDERSONVILLE
July 19, 1967

Failure to heed instructions by the pilot of a small plane was blamed for the midair collision of a Cessna 310 and a Piedmont Boeing 727 over Hendersonville, North Carolina, on July 19, 1967. Eighty-two died.

Fear of falling from an airplane is part of the fear of flying. Like the dream from which the sleeper awakes just before he or she hits the earth, it haunts those who think of flying as an unnatural way to pass between two points. But rarely is this fear realized.

A tragic exception took place on July 19, 1967, over Hendersonville, North Carolina. On that day, a Piedmont Boeing 727 took off from the Ashville-Henderson Airport with no difficulty and with no warning of danger from ground control. On time and on course, the liner continued to climb toward its cruising altitude.

At the same time, a twin-engine Cessna 310 carrying two businessmen bound for Ashville and piloted by veteran charter flyer David Addison was headed toward the airport. Addison was 12 miles off his flight path, and ground control at Ashville-Henderson radioed to him to turn north. This he apparently failed to do, for barely three minutes later the Cessna plowed directly into the Piedmont 727.

The Cessna blew up immediately, atomizing itself and its occupants. The airliner, a gaping hole in its forward fuselage, lurched forward for a few feet, and then the jet fuel in its wing tanks ignited, and it, too, exploded. Cargo and passengers were flung free of the plane and rained like grisly confetti onto the countryside. Corpses landed in trees and on roads. Two crashed through the roofs of houses; one fell onto the pavement near the gas pumps of a service station. The plane itself dove into a patch of woods, setting the trees on fire. Eighty-two people on both planes died. There were no survivors.

UNITED STATES
OHIO
AVA
September 3, 1925

Political considerations apparently outweighed weather and safety factors when the Shenandoah, *America's proudest lighter-than-air ship, was forced by local votegetters to fly in bad weather over Ohio. On September 3, 1925, the ship went down in a pasture and broke apart, killing 14 crew members. Twenty-eight survived.*

The early days of dirigible travel are dotted with terrible tragedies. Every major ship—the *Akron* (see p. 55), the *Dixmude* (see p. 11), the *Hindenburg* (see p. 56) and the *Shenandoah*—crashed, with a consequent loss of life.

All were tragic, but the crash of the *Shenandoah,* America's prize entry in the international long-range and high-speed dirigible derby, was the one most tainted by controversy. "The Secretary of the Navy wanted to play politics by sending the ship over the Middle-Western cities," complained the embittered widow of the *Shenandoah's* commander, F. R. McCrary. "My husband was very much opposed to making the flight at this time because of the weather conditions he knew so well. He asked officials at Washington to delay the flight until a better season." The famous Colonel Billy Mitchell charged that the *Shenandoah's* fatal journey was nothing but a propaganda trip, and zeppelin captain Anton Heinen dubbed it "murder."

There was some basis to the charges. September, despite its unpredictable weather in the Midwest, was the month of country fairs, and there were a number of politicians from these states who were facing reelection. They apparently felt that a majestic flyover by the *Shenandoah* would help them with their local constituencies.

Patterned after the German L-49, the 680-foot-long ship was built in 1919 and, like the *Akron,* was plagued with problems almost from its first launching. The problems were not exclusive to the *Shenandoah;* all lighter-than-air ships were, ultimately, expensive, impractical and dangerous. According to pioneer aviator Laurence La

Political considerations forced the Shenandoah *to fly in bad weather and thus crash on September 3, 1925, near Ava, Ohio. The wrecked carcass of the once-proud airship was picked clean by souvenir hunters the following day.* (Smithsonian Institution)

Tourette Driggs, quoted by Jay Robert Nash in his book *Darkest Hours:* "When the heavy winds blew, she could neither leave nor enter her hangar. When ordinary storms broke about her, she was in peril. When no mooring mast was handy, she required 500 men to catch her and hold her to earth. When rain drops clung to her envelope, she feared to attempt landing under their weight until they evaporated. When every condition was favorable, she sailed through the skies majestically—*but to what purpose?*"

Not a high recommendation for the pride of the U.S. Navy, which for safety reasons was filled with helium instead of the hydrogen that would explode and incinerate the *Hindenburg* and its crew and passengers in 1937. Ironically, the *Shenandoah* encountered its first problems at the very mooring mast at Lakehurst, New Jersey, that would be the scene of the *Hindenburg* disaster.

On January 16, 1924, a sudden storm, with winds up to 65 mph, raged through Lakehurst while the *Shenandoah* was moored. The constant pounding of the winds eventually loosened the ship's mooring cables, and it broke loose with a skeleton crew aboard. The crew acted swiftly, redistributing ballast and fuel just in time to avoid a plunge into some spiky pine trees beyond the airfield. Buying time, they then took the *Shenandoah* up to balmier altitudes and waited until dawn when they would be able to land the craft safely.

Later that same year, the *Shenandoah* made a safe and highly publicized long-distance flight of 9,317 miles back and forth across the United States. Benign weather, good luck and the skill of Commander Zachary Lansdowne turned the trip into a triumph, establishing a speed record of 19 days, nine hours for the round-trip from Lakehurst to Los Angeles and back.

The September 1925 voyage was another experience altogether. The commander and everyone concerned were aware of two negative factors: The weather was bad and perhaps dangerous, and the flight was only really necessary to midwest politicians. Nevertheless, the navy men obeyed orders and left Lakehurst on September 2 bound for the fairgrounds of the Midwest.

By 4:00 A.M. on the third, they had reached Ohio and were immediately beset first by huge headwinds and then by a twister line squall. The turbulence tore at the *Shenandoah* rocking it, plunging it earthward and knocking it skyward. Lansdowne fought the weather with all of the skill he could muster. He dipped the *Shenandoah's* nose earthward to provide a better, less-battering aerodynamic flow; he maneuvered the rudder and elevators and gave its engines full power, but the ship was at the complete mercy of the up-and-down drafts and lateral winds that shot it upward from 2,000 feet to nearly 7,000 feet.

Now the *Shenandoah* began to come apart. Pieces of its envelope shredded, and the engines stopped. It was carrying far too much ballast to stay for more than a moment at 7,000 feet, and it began to dive precipitously. In an attempt to lighten the ship and stop the fall, the commander ordered that everything that could be torn loose from the interior of the *Shenandoah* be tossed overboard.

Lieutenant Charles E. Rosendahl took six men with him, and they clambered to the keel of the ship to release helium with hand valves. But no sooner had they arrived at their destination than the ship began to break apart. It ultimately split in two. First, its control compartment and its commander plummeted earthward, where all would be crushed inside the gondola.

Then, liberated from the weight of the gondola, the nose shot up to 10,000 feet, with Rosendahl and his men clinging to struts and grillwork. They continued to work frantically, releasing helium until the nose drifted, like a parachute, toward Sharon, Ohio, 12 miles from the point at which the gondola had carried Lansdowne and 10 of his men to their deaths.

While they were on their way down, Rosendahl effected one of the most spectacular rescues in the annals of aviation. Lieutenant J. B. Anderson, ordered by Commander Lansdowne to release a ballast of 800 gallons of gas, was crossing a catwalk when the ship broke apart and plummeted earthward. Anderson lost his balance and dove for a strut. He was hanging from this when the nose portion sailed close. Rosendahl lassoed Anderson with a rope just as the lieutenant lost his hold of the strut and was about to fall. Several men held onto the rope; it played out and held fast as Anderson fell the length of it. He hung suspended for a heart-stopping moment and then was hauled safely to the nose section by Rosendahl and his men.

Once the nose portion landed in an orchard owned by farmer Ernest Nichols, the sailors and Nichols tied it to several trees and a fence. Borrowing the farmer's shotgun, Chief Machinist's Mate Shine S. Halliburton shot holes in the remaining helium bags to prevent the nose portion from becoming airborne again. Meanwhile, the tail section, with 25 men in it, floated down into a nearby field.

The *Shenandoah* lay in pieces, its former glory gone forever. Fourteen crew members were dead; 28 survived. By the next day, souvenir hunters would loot the wreck of the belongings of the dead and even haul off portions of the ship that might have aided in an investigation of the crash.

UNITED STATES
PENNSYLVANIA
ALIQUIPPA
September 8, 1994

USAir flight 427, a Boeing 737-300 bound from Chicago to Pittsburgh, suddenly rolled over and dove a mile into a wooded ravine near Aliquippa, Pennsylvania, on September 8, 1994. All 132 aboard the flight died.

USAir flight 427, a Boeing 737-300 traveling from Chicago to Pittsburgh, was on schedule in the early evening of September 8, 1994. The weather was clear, and the wooded hillsides of western Pennsylvania were still easy to see shortly before 7:00 P.M. as the plane neared its final destination.

USAir had had its share of fatal accidents—four in as many years—but the National Transportation Safety Board had given the airline a clean bill of health, noting that there was no established pattern linking the four fatal flights. There also was no intimation of anything wrong as USAir flight 427 prepared for its final approach to Pittsburgh.

Then, at roughly seven miles from the Greater Pittsburgh International Airport, at an altitude of 6,000 feet, something went terribly wrong. The voice recorder in the cockpit recorded an alien sound, a "whoomp, whoomp" followed by a voice saying, "Jeez. What was that?" A moment later, either the pilot, Peter Mermano, or the copilot, Charles B. Emmett, exclaimed, "Oh, God."

The tower responded with landing directions to maintain its altitude. The next and last words from the cockpit were "Traffic emergency!"

On the ground Richard Trenary and Gerald Taylor were sitting in Mr. Taylor's yard in Hopewell Township, near the small, 13,000-person town of Aliquippa. "We heard this noise," Mr. Trenary later said, "and it didn't really sound like a jet. I thought it was a small plane because the engines were making a popping sound, like a muffled backfire over and over."

At the same instant, the crew of the jet was struggling with the plane, trying to regain control as it abruptly rolled to the left, and, for an instant, flew upside down. In the next moment it went into a nosedive and fell at 500 mph from a mile in the sky for 23 seconds until it smashed into the earth in a wooded ravine, driving its cockpit 20 feet into the ground and exploding.

"It looked like somebody was holding it by the tail and dropping it straight down," Mr. Trenary observed. "There was no doubt in my mind it was coming down. It was a matter of a second from then until it hit the ground. It looked to me like he was trying to turn the plane, like maybe he was going to try to land on Route 60. It wasn't a diving crash, it was a nose dive straight down—a gigantic ball of fire. I would say it was 10 times the size of a house. We heard a real loud, heavy thud. Then once it hit the ground, it exploded immediately."

Mr. Trenary ran to call the police, and Mr. Taylor made his way through the heavy underbrush to the crash site. "It wasn't too nice," he told reporters later. "Pieces of everything—. . . The only thing that looked like a plane was something leaning against a tree. There were chunks of metal all over the place and legs and pieces of people. There was no cabin, nothing that I could recognize. It was just a bunch of mangled junk thrown all over the place and a large hole with nothing in it. A large hole where the plane hit and there was nothing in it. Everything was just thrown everywhere. I walked around there hollering, and there wasn't anybody alive, to my estimation. There was a whole lot of fire."

Rescuers arrived within minutes, but it was clear that of the 132 passengers and crew aboard the plane, there were no survivors.

"It was a gruesome sight, just gruesome," Hopewell Township police chief Freddy David said. "There were body parts, pieces of seats, luggage for 200 yards. The biggest piece of the airplane I saw, it was only a little bigger than a car door. I would never have known it was an airplane."

Dr. Michael Zernich, who heard of the crash on his car radio and rushed to the scene, confessed that "It was the most horrible scene I've ever seen in my life. You can't describe it. The pieces—the pieces were unidentifiable. It's just indescribable."

For weeks, NTSB inspectors would comb the smoldering crater and the nearby woods for reasons that the plane abruptly rolled to the left and then dove straight for the earth. There had been a series of warnings regarding a rudder problem that had plagued Boeing 737s. On 737s and other planes like them, a malfunctioning rudder can orient the wings in a way that causes one to lift and the other to sink. Ordinarily, a pilot can compensate, but in some cases, government and industry researchers had found, a pilot might bank sharply to one side and lose control of the plane.

Inspections by the NTSB discovered that just before the nose dive, the plane had been banking 14 degrees to the left. But as it began to come out of the turn, the bank decreased. Then, suddenly, it increased to as much as 100 degrees, which meant that the plane had rolled over completely and was upside down. However, this particular 737 had been repeatedly inspected in the months preceding the crash—the last time about 100 hours before—to make sure it did not have a rudder problem. And none in the preliminary investigation was found.

Probers found the rudder of the exploded craft, and it appeared to be in a position for a modest right turn. No conclusions were drawn, and attention turned to the right engine. It could have become detached in flight, some investigators theorized, or the problem could have been in the thrust reversers. One of the six activating devices for the thrust reverser on the right engine was in an extended position, but that might have happened in the crash.

Still, this seemed to be the most probable cause of the catastrophe. Thrust reversers deflect the flow of air through the plane's engines, helping it to slow down after landing. When they are deployed, a cowl on each side of the engine extends to let air out through the engine's side, thus dramatically reducing forward thrust. The hydraulically operated pistons, or activators, in the cowls control the thrust reverser mechanism. Their function, then, is a powerful one, and most theorists focused on finding the pieces of the cowling of the right engine for study. It was further discovered that the inboard thrust reverser on the right wing of the jet had been reported causing difficulty. On June 3, a portion of the mechanism had been replaced, and there had been no subsequent reports of difficulty.

Before September ended, government inspectors discovered the rest of the spoilers and inspected the flaps on both wings. They determined that neither was at fault. The mystery deepened. Weather was not a factor; there was no evidence of a collision with a small plane or a bird and no traces of a bomb or sabotage.

Then, on September 29, *The Seattle Times* published the findings of FAA reports from 1974 through August 1994. They revealed 46 problems with the rudder system on 737 aircraft, 21 of which the airlines attributed to hydraulic fluid leaks in the power control unit. Of the 21 incidents, 11 had resulted in unscheduled landings, two involved emergency descents and one resulted in an aborted landing approach. In three cases pilots returned to the gate before takeoff.

The inspectors focused on the rudder assembly and discovered the true cause of the tragedy of USAir flight 327. The control valve of the main rudder had jammed. This had caused a reversal of controls on the rudder, and the rudder had moved into a direction opposite to that commanded by the pilots. They had lost control of the aircraft—thus the "Oh God" and "Travel emergency" calls—and it had rolled over and settled into its 23-second nosedive to earth.

UNITED STATES
TEXAS
DAWSON
May 3, 1968

No explanation has been given for the explosion and crash of a Braniff Lockheed Electra near Dawson, Texas, on May 3, 1968. All 88 aboard died.

The explosion and crash of a Braniff Lockheed Electra in a rainstorm near Dawson, Texas, has never been explained. En route from Houston to Dallas on the night of May 3, 1968, the plane radioed no distress signal, nor was there any indication that the crew was aware of any mechanical difficulty. And yet, somewhere outside Dawson, the plane simply exploded and plunged to earth, killing all 88 persons aboard.

There was some speculation that the endemic problem of engine mount defects that had caused several earlier crashes of the four-engine, turboprop Electras (see pp. 52 and 65) might have been responsible. But no huge midair explosions occurred in those accidents; wings had been torn off by the loosening of the engine mounts. And no evidence of a bomb was discovered in the wreckage.

Thus, the Braniff crash of May 3, 1968, would remain forever a mystery.

UNITED STATES
UTAH
BRYCE CANYON
October 24, 1947

Faulty design was determined to be the culprit in the crash of a United Airlines DC-6 on October 24, 1947, as it was descending for a landing at Bryce Canyon Airport, Utah. Fifty-two died.

A design fault caused the calamitous crash of a United Airlines DC-6 on October 24, 1947, just short of the runway to Bryce Canyon Airport, Utah. One of the DC-6's fuel vents had been located near a heater air scoop. Fuel being automatically transferred from one tank to the other during flight often tended to spill, and fuel vapor could then make contact with the electric coils of the cabin heating system. The result: an inevitable fire. One occurred aboard the United DC-6 and another, less than a month later, aboard an American Airlines DC-6 over Gallup, New Mexico.

The latter fire would not result in a crash. The fire aboard the United Airlines plane would, and the drama of its unfolding was spelled out by Captain E. L. MacMillan as he called in at various dangerous moments to air control at Salt Lake City.

The flight took off on time from Los Angeles and was two hours into its journey to New York when the fire was discovered in the cargo compartment. Heavy clouds of thick smoke were pouring from the hold into the passenger cabin, and crew members with fire extinguishers were trying to control it when MacMillan radioed his first message of distress. "We have baggage fire aboard," MacMillan said. "We are coming to Bryce Canyon. We have smoked-filled plane. Unable to put fire out yet."

Five minutes later, the crew had controlled the fire, but there was no way of telling how much structural damage it had caused. "The tail fire is going out," radioed MacMillan. "We may get down and we may not."

With these chilling words as a directive, Bryce Canyon Airport was prepared for a crash landing. The sky was clear, the runway was made ready and rescue apparatus stood by.

"We may make it. Think we have a chance now," said the captain. "Approaching the strip."

It all seemed easy. But then there was a shout from MacMillan: "The tail is gone!" he yelled. And that was the last word the airport would receive.

Now out of control, the DC-6 was on a glide path that, had it remained intact, would have taken it safely to the Bryce Canyon Airport runway. But fifty feet before the beginning of the runway, there was a small hill. If MacMillan had been able to keep his craft aloft for only a few seconds more, he would have cleared it.

He did not. The DC-6 plowed into the top of the hill and exploded. Its four engines shot 50 feet ahead, skidding to a stop along the runway. Some passengers were flung from the wreckage, others burned within it. Rescuers rushing to the scene could save no one. All 52 aboard died.

As a result of the crash of the United DC-6 and the subsequent fire aboard the American airliner, all DC-6s were grounded by the order of the CAB in November 1947, and the design flaw was corrected.

UNITED STATES
VIRGINIA
RICHMOND
November 8, 1961

Ancient equipment and a failure to profit from the past were the almost certain causes of a catastrophic crash of an Imperial Airlines Lockheed Constellation near Richmond, Virginia, on November 8, 1961. Seventy-seven died. Two survived.

In the 1950s and 1960s, the U.S. Army made it a practice to employ the lowest-bidding commercial carriers to ferry its personnel around the United States. However, low bidding in the charter airline business can be dangerous, and Imperial Airlines, one of the army's winners, was a dangerous carrier. Its record was appalling, its equipment outdated.

A quick glance at the airline's record with the army alone should have convinced both the FAA and the U.S. government that its contract with Imperial should have been long since abrogated.

In 1953, one of its DC-3s crashed near Centralia, California, killing 21. Eighteen of the victims were soldiers.

In 1959, Imperial was fined $1,000 by the FAA for loading 30 marines into an unairworthy C-46.

Earlier in 1961 three of Imperial's four pilots had had their licenses lifted by the FAA for "flying their aircraft under conditions dangerous to servicemen aboard."

Imperial's entire fleet consisted of four moth-eaten, patched-together planes, one of which, a 1946 model Lockheed Constellation, was pressed into service to ferry 74 army recruits from various locations to Fort Jackson, South Carolina.

Early on November 8, 1961, the Constellation began its journey at Newark Airport, where 30 recruits climbed aboard. Its first stop was Wilkes-Barre, Pennsylvania, where 31 more boarded. A final 13 were picked up at Baltimore, Maryland, for the last leg of the flight to South Carolina.

Shortly after leaving Baltimore, the ship's captain, Ronald Conway, began to have trouble with the Constellation. By the time he radioed Byrd Airport at Richmond, Virginia, to request permission for an emergency landing, two of his four engines had stopped.

The airport gave him clearance, and he circled for an approach. Partway through the circle, a third engine quit, but he was a seasoned pilot and could probably have made a landing with only one engine running if another malfunction had not suddenly occurred. The nose landing gear refused to lock.

Still, Conway might have brought it in. But on his second pass at the runway, the plane made a sudden lurch and heeled over into an abrupt dive toward the earth. It hit a forest, traversed a swamp and exploded, killing all 74 soldiers and three of the crew members. Conway and flight engineer William Poythree managed to escape through the cockpit hatch and were thus the only survivors of a crash that never should have happened.

UNITED STATES
WASHINGTON
MOSES LAKE
December 20, 1952

Pilot error was blamed for the crash, on takeoff, of a U.S. Air Force C-124 over Moses Lake, Washington, on December 20, 1952. Eighty-seven were killed; 44 survived.

Pilot error was judged responsible for the crash of a U.S. Air Force C-124 double-decker transport on the morning of December 20, 1952, at Larson Air Force Base, near Moses Lake, Washington.

The plane, carrying 131 servicemen, took off with apparent ease and climbed into the cloudless sky. But at a crucial moment in the takeoff, the pilot apparently failed to unlock the rudder and elevators from the plane's automatic locking gear. A knob on the throttle pedestal is ordinarily moved by the pilot on takeoff from the locked position to the unlocked position—a matter of two notches. If the knob is not moved, the plane can fly but cannot maneuver. For some reason, the knob was not moved to its proper notch, and two minutes after it took off, the plane plummeted to earth, crashing into a snow-covered field and igniting immediately.

Emergency crews were dispatched forthwith to the fiercely burning wreck, and rescuers in asbestos suits carried survivors from the conflagration. No one in the nose section lived; 87 died, either in the crash or at a nearby hospital. Forty-four were rescued, but some of them were crippled for life.

U.S.S.R.
KRANAYA POLYANA
October 14, 1972

Pilot error was responsible for the crash of a Soviet Aeroflot Ilyushin-62 on approach in bad weather to the Kranaya Polyana airport on October 14, 1972. One hundred seventy-six were killed.

The weather was abominable over Moscow on October 14, 1972. A heavy mist broken by intermittent periods of rain reduced visibility to practically nothing. Not only that, the instrument-landing apparatus at Sheremetevo International Airport was temporarily inoperative. The following afternoon, a British European Airways flight would be diverted to Stockholm when the pilot decided that the combination of bad visibility and an out-of-order landing system was a situation he was unwilling to battle.

The same decision should have been made the night before by the pilot of an Ilyushin-62 reportedly chartered from the Soviet airline Aeroflot by the state travel agency, Intourist. At 9:00 P.M. that night, he approached the airport and requested clearance for landing. He was granted it but reminded that there were no instrument-landing capabilities.

The flight had left Paris at noon that day with a full passenger load that included 102 French citizens. Early that evening, all but one of these French passengers had disembarked at a stop in Leningrad. Thirty-eight Chileans, five Algerians, three Italians, two Lebanese, a Frenchman, a Briton and 111 Russians were still aboard when it departed for the last leg of its flight to Moscow.

The pilot circled the field and came in for a landing. It was impossible to see the runway. He circled twice more and aborted each time. Finally, on the fourth try, he apparently felt that he had the runway in sight, but he was tragically wrong. He was three miles away from the airport, on the outskirts of the small village of Kranaya Polyana. The plane struck the ground with enough force to explode and send fiery sparks 100 feet into the misty night sky.

Everyone aboard—176 passengers and crew members—died, making this the worse commercial airline crash in the world to that date. The Soviet news agency Tass stubbornly refused to release details of the crash at first. Soviet authorities even tried to misdirect Western newspeople to the village of Chernaya Gryaz, seven miles from the site of the crash. But because there were foreign nationals killed in the crash, the Soviets finally released the statistics and a terse story about the disaster.

VENEZUELA
LA CORUBA
March 16, 1969

Pilot error was blamed for the crash on takeoff of a VIASA DC-9 at La Coruba, Venezuela, on March 16, 1969. Eighty-four died in the airplane; 76 were killed on the ground.

Maracaibo, a city of 800,000 in the northeastern part of Venezuela, is that country's oil capital. As such, it attracts a large number of foreign businesspeople. At 11 P.M. on March 16, 1969, 46 Americans were among those on board a VIASA DC-9 en route from Caracas to Miami that took off from Maracaibo's Grano de Oro Airport. The flight lasted only two minutes, and it would never rise higher than 150 feet.

According to witnesses, the plane wavered for a moment and then glided toward the ground. On its descent it struck a high tension wire and exploded, sending flaming wreckage into the suburb of La Coruba, just outside Maracaibo. Five square blocks of the town were obliterated by fire and falling parts of the plane. The entire neighborhood resembled a war zone.

Fire and ambulance crews immediately dashed to the scene of the crash. All 74 passengers and 10 crew members aboard the plane perished. There was no hope whatsoever for their survival.

The carnage was appalling. Rescue workers sifted through the wreckage, looking for survivors, sending the injured off to hospitals and caring for those who were less seriously hurt. Roman Catholic priests roamed through the crash area trying to console relatives of the victims and performing last rites for the dead. Boy Scouts went from house to house asking for sheets in which to wrap the dead.

Seventy-one died on the ground; five more would die later in hospitals from their injuries. A total of 160 people perished in this crash, the worst in commercial aviation up to that time. The reason was given as "pilot error."

WEST INDIES
GUADELOUPE
June 22, 1962

An Air France Boeing 707 crashed on landing at Guadeloupe on June 22, 1962. The cause of the crash was never ascertained. All 113 aboard died.

The year 1962 was not a good one for Boeing 707s. No less than five of them went down in 1962, killing 456 persons. The 707, a modification by Boeing of its KC-135 tanker, was first flown on December 20, 1957, and was received with great enthusiasm, despite the fact that the early models gained a reputation among pilots of flying "like civil bombers." They burned fuel heavily, but their pure-jet engines gave them a 25% speed margin over Viscounts and Electras, their chief competition, which made them ideal for transcontinental and transoceanic travel.

In 1962 Air France had a bad record for safety. In the period from 1950 to 1968, 812 Air France passengers had been killed. Like the Boeing 707, Air France resolved its deficiencies and developed into a model of safety. But not, unfortunately, before both were involved in a tragic crash on the French Caribbean island of Guadeloupe.

The island's airport, like most in the West Indies, is anything but spacious, and landing on it involves both skill and luck. Ringed by mountains, it demands a steep descent and constant attention on the part of the crew.

On June 22, 1962, an Air France 707 approached the airport at Guadeloupe for a routine landing. It smashed squarely into Dos D'Ane (the Donkey's Back), a rugged mountain peak overlooking the airport's runways. All 113 aboard were killed in the resultant explosion and fire.

No specific reason was ever discovered for the crash. Attention focused on the 707 in the FAA investigation, but no design flaws that could have led to the tragedy were found.

YUGOSLAVIA
LJUBLJANA
September 1, 1966

Pilot error was the reason given for the crash on landing at Ljubljana, Yugoslavia, of a charter Britannia Airways 102 on September 1, 1966. Ninety-seven died; 20 survived.

One hundred twelve British tourists chartered a turboprop airliner from Britannia Airways in September 1966 for a vacation in Yugoslavia. Because of pilot error, only 20 of them would survive the trip.

The aircraft made the journey from London's Heathrow Airport without trouble. The weather and the approach to the airport in Ljubljana, a city of 200,000 north and west of Zagreb and Belgrade, were clear. But the flight was obviously off course when it approached the landing field, and controllers advised the pilot of his error. He was 110 yards off course and 600 feet too low.

The crew apparently received this information and failed to act on it. The plane descended on the wrong path and at the wrong altitude. Then, according to one of the survivors, Arthur Rowcliff, "The plane slowed down. Then it started to vibrate. A few seconds later we crashed, bounced back in the air and finally fell down. We were thrown clear with our seats."

The aircraft had plowed into a ridge full of fir trees, leaped into the air and then plummeted to earth, where it burst into flames. Those who, like Rowcliff and his family, were thrown clear survived. The others perished in the inferno that erupted when the plane struck the ground. The ultimate death toll would be 97.

ZAIRE
KINSHASA
January 8, 1996

An overloaded Russian Antonov AN-32B operated by Africa Air failed to achieve a takeoff from N'Dolo Airport in Kinshasa, Zaire, on January 8, 1996, and plowed into a crowded marketplace, where it killed 225 people.

There are numerous private plane companies in Africa, and one of them, Africa Air, flew Russian Antonov AN-32B jets with Russian crews in 1996. Sometimes monitored, sometimes not, these flights were frequently flown on the knife edge of legality.

The Antonov AN-32B that left N'Dolo Airport in Kinshasa at 12:40 P.M. on January 8, 1996, was definitely off the knife edge and into illegal territory when it began its takeoff from the airport. Its certification to fly safely had been revoked, and the crew of six did not have authorization to fly.

Nevertheless, the overloaded cargo plane rolled down the single runway of N'Dolo Airport on the sunny afternoon of January 8. The crew struggled with the overweight plane, straining to get it off the ground, but it never truly left, and its engines began to belch flames and black smoke. The pilot nevertheless kept the throttle at full speed, and the inevitable happened: The plane skidded off the end of the runway and, traveling at full ground speed, plowed into the crowded marketplace of Simba Zikidi, a congregation of merchants and their customers packed together in a space punctuated by corrugated iron and wooden shacks.

The flaming aircraft would continue on for 330 feet, smashing the market stalls into kindling, crushing or burning to death at least 223 men, women and children and injuring hundreds more.

A fire crew from the airport rushed to the scene and extinguished the fire on the airplane and in the market. Meanwhile, with tears streaming down their faces from the acrid smoke and the terrible sights they were witnessing, hundreds of rescuers worked to pull the badly mutilated bodies of the marketgoers from the wreckage of the plane and the market.

Four of the six crew members aboard the plane survived the crash and were taken by Red Cross personnel to a nearby clinic for treatment of their injuries. An angry crowd gradually gathered outside the clinic and soon tried to break in to drag the crew members out and lynch them.

The police finally intervened and led the four crew members off to jail to await prosecution for the deaths of 225 people on the ground in the market. A police search for the remaining two crew members failed to turn them up, either as injured or dead.

N'Dolo Airport was shut down until further notice, Africa Air went out of business and the crew members were brought to trial and eventually extradited to Russia, where they disappeared from public sight.

CIVIL UNREST AND TERRORISM

THE WORST RECORDED CIVIL UNREST AND TERRORISM

CIVIL UNREST

* Detailed in text

Armenia/Turkey
- * Armenian massacres by Turks (1895–1922)

Austria
- Vienna
  - General strike following Nazi acquittal for political murder (1927)

Burundi
- Marangara
  - * Tribal confrontation (1988)

China
- Beijing
  - * Massacre in Tiananmen Square (1989)
- Northern Provinces
  - * Boxer Uprising (1900)

Czechoslovakia
- Uprising crushed (1977)

Egypt
- Insurrections over exorbitant taxes (189 B.C.)

Europe
- * The Holocaust—attempted genocide of European Jews by Nazis (1939–45)

France
- Paris
  - * St. Bartholomew's Day massacre of 2,000 Huguenots (1572)
  - Massacre of Champs de Mars (1791)
  - * Reign of Terror and White Terror (1793–95)
- Rouen
  - Burning of Joan of Arc at stake (1431)
- Vassy
  - 1,200 French Huguenots slain (1562)

Germany
- Alsace
  - First *Bundschuh,* or peasants' revolt (1493)

Great Britain
- England
  - Nationalist uprisings in north and west crushed by William I (1068)
- London
  - Evil May Day riots—60 hanged on Cardinal Wolsey's orders (1517)
  - * Gunpowder Plot—Guy Fawkes arrested in cellars of Parliament (1605)
- Manchester
  - Peterloo massacre (1819)
- Sheffield
  - * Soccer riot (1989)

Hungary
- Budapest
  - * Uprising crushed (1956)

India
- Amritsar
  - * Massacre (1919)
  - Riot between Sikhs and Hindus (1984)
- Bombay
  - * Hindu-Muslim riots (1992–93)
- Mandai, Tripura; Assam
  - * Tribal massacre (1980)
  - * Ethnic violence (1983)

Ireland
- Irish railway strike (1920)

Japan
- Edo
  - Yetuna, new shogun, overcomes two rebellions (1651)
- Satsuma
  - Revolt crushed (1877)

Northern Ireland
- Violence begins (1971–72); 467 Irish killed in 1972

Poland
- Warsaw
  - Massacre when Russian troops fire on demonstrators (1861)

Rome
- Gaius Gracchus killed in riot; his reforms abolished (121 B.C.)
- * Revolt of slaves and gladiators under Spartacus; crushed by Pompey and Crassus (71 B.C.)

Russia
- St. Petersburg
  - * Decembrist Revolt crushed (1825)

Saudi Arabia
- Mecca
  - * Stampede in tunnel connecting Mecca and Mina (1990)

South Africa
- Matabele
  - Revolt against British South Africa Company; crushed by Starr Jameson (1893)

United States
- California
  - Los Angeles
    - * Race riots in Watts (1965)
    - * Riots after acquittal of LAPD officers accused of beating Rodney King (1992)
- Colorado
  - Sand Creek
    - * Massacre of Cheyenne and Arapaho Indians (1864)
- Illinois
  - Chicago
    - * Race riots (1919)

Chicago (continued)
Strike against Republic Steel (1937)
* Police riot at Democratic National convention (1968)
Massachusetts
Boston
Boston Massacre between civilians and troops (1770)
New York
New York
* Draft riots (1863)
Ohio
Kent
* Killing of Kent State students by Ohio National Guard (1970)
Virginia
Southampton
Revolt of slaves led by Nat Turner (1831)
Washington, D.C.
* Bonus Army march (1932)

TERRORISM

Belgium
Red Brigade tries to kidnap and kill Alexander Haig, NATO commander (1979)
Brazil
Rio de Janeiro
* First of diplomatic kidnappings: Charles Elbrick, U.S. ambassador to Brazil (1969)
Comoro Islands
Moroni
* Ethiopian Airlines Boeing B-767 hijacking (1996)
East Africa
Nairobi, Kenya/Dar es Salaam, Tanzania
* Bombing of American embassies (1998)
France
Paris
Orly Airport bombed by Armenians (ASALA) (1983)
Germany
Lufthansa hijacking (1977)
Munich
* Olympic Games massacre (1972)
Neo-Nazi plants bomb at Bierfest (1980)
West Berlin
La Belle discotheque bombing (1986)
Great Britain
England
Birmingham
Provisional IRA sets series of bombs (1974)
Brighton
IRA bomb almost wipes out Margaret Thatcher and entire British cabinet (1984)
London
* Harrods attacked by car bomb planted by Provisional IRA (1983)
Scotland
Lockerbie
* Pan Am jet blown up (1988)
Greece
Athens
TWA plane from Tel Aviv attacked by National Arab Youth for the Liberation of Palestine (Libyan sponsored) (1974)
India
Bombay
Airliner explodes; Sikh terrorists suspected (1985)
JAL hijacking (1977)
Iran
Tehran
* U.S. Embassy held hostage (1979–81)
Ireland
Irish Sea
* Air India Boeing 747 from Toronto to London blown up and crashes into Irish Sea (1985)
Israel
Entebbe Airport; first defeat for international terrorism (1976)
Jerusalem
* Two bombings in public gathering places (1997)
Netanya
Bomb attack by Islamic Jihad (1995)
Sinai Desert
* Israeli Phantoms shoot down Libyan Boeing 707 (1973)
Tel Aviv
* Lod (Lydda) Airport; first transnational terrorist attack, between Palestine Liberation Front and Japanese Red Army (1972)
Car bomb attack on bus (1994)
Nail bomb attack outside Dizengoff Center (1996)
Italy
Bologna
* Neo-fascist terrorists bomb central train station (1980)
Rome
First Palestinian hijacking (1968)
NALYP bombing and hijacking of Pan Am Airliner (1973)
Rome and Vienna, Austria
Coordinated attacks on El Al check-in desks (1985)
Vatican City
Pope John Paul II severely wounded by Turkish Grey Wolves; Bulgarian Secret Service charged with complicity (1981)
Japan
Tokyo
First of Japanese Red Army's international actions (1970)
Hijacking of JAL plane to North Korea (1970)
* Sarin gas poisioning in subway system (1995)
Jordan
Dawson's Field
* Five planes hijacked; leads to Black September formation (1970)

Lebanon
Beirut
U.S. Embassy destroyed by car bomb (1983)
* Marine barracks bombed (1983)
* TWA Flight 847 hijacked (1985)

Malta
Hijacking; 58 killed after Israeli commandos rush plane (1985)

Mediterranean Sea
* *Achille Lauro* hijacking (1985)

Netherlands
Rotterdam
Fatah blows up fuel tanks (1971)

Saudi Arabia
Mecca
Muslim extremists kill 150 in Grand Mosque (1979)

Spain
Majorca
Lufthansa flight hijacked by PFLP/Baader-Meinhof (1977)

Sweden
Stockholm
West Germany embassy blown up by Rote Armee Fraktion (1975)

Syria
Damascus
* Bombing (1981)

United States
California
Berkeley
Patricia Hearst kidnapped by SLA (1974)
Colorado
Littleton
* Massacre in Columbine High School (1999)
New York
New York
FLN bomb exploded in Fraunces Tavern (1975)
LaGuardia Airport bombing (1975)
* Bombing of One World Trade Center (1993)
* Plot to blow up bridges, tunnels, buildings (1993)
New York and Washington, D.C.
* Suicide attacks of World Trade Center and Pentagon (2001)
Oklahoma
Oklahoma City
* Bombing of Alfred P. Murrah Federal Building (1995)
Texas
Waco
* Standoff between federal law enforcement officers and Branch Davidians (1993)
Washington, D.C.
* Muslim hostage taking (1977)

CHRONOLOGY

CIVIL UNREST

* Detailed in text

189 B.C.
Egypt; insurrections

121 B.C.
Rome; Gaius Gracchus killed in riot

71 B.C.
Rome; revolt of slaves and gladiators

1068
England; nationalist uprisings

1431
Rouen, France; burning of Joan of Arc

1493
Alsace, Germany; first peasants' revolt

1571
May 1
London; Evil May Day riots

1562
Vassy, France; slaying of French Huguenots

1572
August 24
* Paris; St. Bartholomew's Day Massacre

1605
November 5
* London; Gunpowder Plot

1651
Edo, Japan; shogun overcomes two rebellions

1770
Boston; Boston Massacre

1791
Paris; Massacre of Champs de Mars

1793–95
* Paris; Reign of Terror and White Terror

1819
Manchester, England; Peterloo Massacre

1825
December 14
* St. Petersburg, Russia; Decembrist Revolt

1831
Southampton, Virginia; revolt of slaves led by Nat Turner

1861
Warsaw, Poland; massacre of demonstrators

1863
July 13–15
* New York City; draft riots

1864
November 29
* Colorado; massacre of Cheyenne and Arapaho Indians

1877
Satsuma, Japan; revolt

1893
Matabele, South Africa; revolt against British South Africa Company

1895–1922
* Armenia/Turkey; Armenian massacres

1900
June
Northern Provinces, China; Boxer Rebellion

1919
April 13
* Amritsar, India; Amritsar Massacre
July 27–August 3
* Chicago, Illinois; race riots

1920
Ireland; Irish railway strike

1927
Vienna, Austria; general strike

1932
May 20–July 28
* Washington, D.C.; Bonus Army march

1937
Chicago, Illinois; Republic Steel strike

1939–45
* Europe; the Holocaust

1956
November 4
* Hungary; Hungarian uprising

1965
August 11–29
* Los Angeles; race riots

1968
August 25–30
* Chicago, Illinois; police riot at Democratic National Convention

1970
May 4
* Kent, Ohio; killing of Kent State students by Ohio National Guard

1971
Northern Ireland; violence begins

1977
Czechoslovakia; general uprising

1980
June 7–8
* Mandai, Tripura, India; ethnic violence

1983
February 18
* Mandai, Assam, India; ethnic violence

1984
Amritsar, India; riot between Sikhs and Hindus

1988
August 14–21
* Burundi, Africa; tribal confrontations

1989
April 15
* Sheffield, England; soccer riot

April 18–June 4
* Beijing, China; massacre in Tiananmen Square

1990
July 2
* Mecca, Saudi Arabia; stampede in tunnel connecting Mecca and Mina

1992
March 2
* Los Angeles; riots following acquittal of LAPD officers accused of beating Rodney King

December–January 1993
* Bombay, India; Hindu-Muslim riots

1993
February 28–April 19
* Waco, Texas; standoff between federal law enforcement officers and Branch Davidians

TERRORISM

* Detailed in text

1968
July 22
Rome; first Palestinian hijacking

1969
* Rio de Janeiro, Brazil; first diplomatic kidnapping

1970
September 6–8
* Dawson's Field, Jordan; five planes hijacked

1971
March 4
Rotterdam, Netherlands; fuel tanks blown up

1972
May 3
* Tel Aviv, Israel; first transnational terrorist attack

September 5
* Munich, Germany; Olympic Games massacre

1973
February 21
* Sinai Desert, Israel; Israeli jets down Libyan 707 airliner

September 5
Rome; Pan Am airliner bombed

September 7
Athens, Greece; TWA airliner blown up

1974
February 5
Berkeley, California; Patricia Hearst kidnapped by SLA

November 21
* Birmingham, England; IRA bombings

1975
December 24
New York City; FLN bomb explodes in Fraunces Tavern

1976
June 27
Entebbe Airport, Uganda; first defeat for terrorists

1977
March 9–11
* Washington, D.C.; Muslims take hostages

September 28
Bombay, India; JAL hijacking

October 13
Majorca, Spain; Lufthansa hijacking

1979
June 29
Belgium; Red Brigade attempt on Alexander Haig

November 4
* Tehran, Iran; U.S. embassy held hostage until 1981

1980
August 1
* Bologna, Italy; neo-fascist bombing of train station

August 1
Munich, Germany; neo-Nazi bomb at *Bierfest*

1981
May 13
Vatican City; attempt on Pope John Paul II's life

November 29
* Damascus, Syria; bombing

1983
July 15
Paris; Orly Airport bombing

October 23
* Beirut, Lebanon; U.S. Marine barracks bombed

December 17
* London; IRA Christmas bombing of Harrods

1984

October 12
Brighton, England; IRA attempt on Margaret Thatcher

1985

June 14–18
* Beirut, Lebanon; TWA flight 847 hijacked

June 22
* Irish Sea; Air India 747 blown up

October 7–9
* Mediterranean Sea; *Achille Lauro* hijacking

November 23
Malta; Egyptair hijacking

1986

April 5
West Berlin, Germany; La Belle discotheque bombing

April 15
Libya; U.S. aircraft attack Libya

1988

December 21
Lockerbie, Scotland; Pan Am flight bombed

1993

February 26
* New York City; bombing in basement of World Trade Center

June 23
* New York City; plot to blow up bridges, tunnels, buildings

1994

October 19
Tel Aviv, Israel; car bomb attack on bus

1995

January 22
Netanya, Israel; bomb attack by Islamic Jihad

March 20
Tokyo, Japan; sarin gas poisoning in subway system

April 19
* Oklahoma City; bombing of Alfred P. Murrah Federal Building

1996

April 4
Tel Aviv, Israel; nail bomb attack outside Dizengoff Center

November 23
* Moroni, Comoro Islands; Ethiopian Airlines Boeing B-767 hijacking

1997

July 30/September 4
* Jerusalem, Israel; two bombings in public gathering places

1998

August 7
* Nairobi, Kenya/Dar es Salaam, Tanzania; bombing of American embassies

1999

April 20
* Littleton, Colorado; massacre in Columbine High School

2001

September 11
* New York City/Washington, D.C.; suicide attack of World Trade Center and Pentagon

CIVIL UNREST AND TERRORISM

There is a single, dark thread that runs through and binds together the two categories of this section, and the name of it is motivation. Each of these similar undertakings—civil unrest and terrorism—is motivated by a belief.

That belief may be as simple as a fancied slight or as complex as a philosophy; as closely held as a catechism or as widely held as a form of government, a system of laws or an ordering of ideas. The point of the matter is that in each of the incidents described or listed in this section, the action taken was done with the purpose of either *overthrowing* a particular ideology or political system or *promulgating* a particular ideology or political system.

As the *threat* of terrorism turns into the *fact* of terrorism, certain subtleties arise. Whereas to some observers or victims, a set of fighters or bombers or practitioners of destructive suicide may become freedom fighters struggling against an oppressive regime, to other observers or victims, the same set of fighters or bombers may become terrorists intent on carrying out fanatic intentions. Somewhere, within the thickets of perception, is the truth, but the thickets keep getting in the way—of diplomats, warriors, the men and women at the tops of governments that make war, and the refugees from those conflicts.

This difference in perception has existed for centuries, but it has seemed to increase in frequency and ferocity as the 20th century has given way to the 21st. It was and continues to be at the heart of the worst of contemporary Middle East conflicts and the catastrophes therefrom. To Palestinians, Israel is an occupier of their land that has, through systematic starvation, maltreatment and invasion, pushed the Palestinian population to the brink of despair, thereby fostering terrorism. To Israelis, Palestine is a terrorist state, bent on destroying Israel and all that it stands for. As years of conflict have spun out, the perceptions of each side have solidified into doctrine, and the possibilities of disaster rise with each attack and reprisal.

In the 1980s, the mujahideen were, to the U.S. government, freedom fighters struggling against the occupation of Afghanistan by Soviet Russia, much as were the contra rebels in Nicaragua fighting the Soviet-backed Sandinista government. But as the freedom fighters in Afghanistan morphed into the Taliban and became the oppressors of the Afghanistan populace and the guardians of Osama bin Laden and the al-Qaeda network, they became the enemy and as guilty of terrorism as the multinational terrorists of the al-Qaeda network, to which they now pledged their allegiance.

On either side of the India-Pakistan border, perception has brought two nuclear powers to the brink of war. Many in Pakistan believe that the Muslim combatants against Indian troops in a disputed section of Kashmir are freedom fighters. In India the nearly universal perception is that they are terrorist militants, supported by the government of Pakistan. For a long while, the United States backed India, but once the war on terrorism began and Pakistan became a vitally important member of the coalition fighting Osama bin Laden, the Taliban and al-Qaeda, its backing softened.

And so it goes. Yesterday's enemies can become today's friends; the friends of the past can become the enemies of today. And in this climate, terrorism, which knows no borders, flourishes, causing greater and greater cataclysms, and all in the name of noble causes—some economic, some social, and most religious. To some, fanatic loyalty to a cause is their only anchor in a world that seems to be more and more adrift. Thus, civil unrest. Thus, terrorism.

What qualifies these events as disasters is that no matter the purpose, nothing was achieved through them. They were either failures, or they brought about the reverse of their intention. The world, or *their* world, was made worse for their actions. The collective loss of life in the American Revolution is not a disaster; the collective loss of life in the Decembrist Revolt of 1825 is, because it failed to accomplish its objective. The loss of life in any airplane crash is catastrophic, but it acquires a new dimension of tragedy when it is the result of a terrorist's bomb.

Who would have thought, after the horror of World War II, the opening of a pit of inhumanity into which supposedly civilized societies descended, that much of the same world would seem to learn practically nothing from the experience? A mere 50 years—half a century—later, religious and ethnic wars were begetting acts of terrorism and genocide that resembled what the world thought it would never experience again.

Ethnic rivalries led to massacres in the Balkans; Catholics and Protestants killed each other in Northern Ireland; Muslims and Hindus died in clashes in India; and terrorism ran rampant between Israelis and Palestinians.

Peace pacts were brokered and broken, despite their glorious names: the Good Friday Accord; the Oslo Agreement. Within astonishingly short periods of time, they became merely names for noble wishes.

And then, the deadliest, most vicious and fanatical of all terrorist attacks occurred on September 11, 2001, when four commercial airliners and their passengers were commandeered by suicide bombers and turned into missiles that brought down New York City's World Trade Towers, smashed in a wall of the Pentagon, and killed more than 2,000 people in the space of an hour.

And at this point, terrorism crossed the line into war, which was, perhaps, the inevitable result. It was what some of the leaders of some of the many Middle East terrorist organizations had wished for and tended toward: a confrontation between the West and particularly its leader, America, and their extreme—and, to many moderate Muslims, perverted—form of Islam. Thus, although acts of terrorism against military targets such as the U.S. Marine barracks in Beirut and the U.S.S. *Cole* in Yemen are not covered in detail in this section because they were both military targets, they became, in retrospect, preludes to the September 11 holocaust and are thus mentioned in that entry.

Political and civil unrest often results in mass assassination. It either involves large groups of people who have been whipped into a fanatic frenzy by leaders who appeal to their dedication to a cause, or it is the massing of the forces of a particular *government* with a particular point of view against a mass of people with an opposing viewpoint. Extended to an international status, it becomes war. Confined to a specific location, it is defined as riot. Extended within the borders of one specific country, it becomes civil war.

The events in this section are, with two exceptions, confined to riots. The remaining two events, included because they are of such horrific dimensions that they cannot be denied space, are two attempts at genocide: the Armenian Massacres (see below) and the Nazi Holocaust (see p. 91).

CIVIL UNREST

ARMENIA/TURKEY
1895–1922

Religious intolerance was the core cause for the 27-year attempt by Turkey to commit genocide upon Armenia and its populace. Two million Armenians were massacred, and Armenia as a country was eliminated from the map of the world.

Genocide is one of humankind's lower forms of activity, and one of the most dramatic exemplifications of this was the horrendous massacre, from 1895 to 1922, of Armenians by the Turkish government. During those 27 years, two million Armenian men, women and children were murdered, often after prolonged and barbaric torture. Others were driven across deserts or to ports of debarkation. The purpose was to totally annihilate the Armenian minority in Turkey as a holy necessity. Armenians were Christians; Turks were Muslims, and it became a holy war—traditionally the most savage sort of conflict.

Founded by Haik, a descendant of Noah, Armenia originally occupied the land at the source of the Tigris and Euphrates rivers in Asia Minor. Eventually, it became known as an incorporation of northeast Turkey, the Armenian Soviet Socialist Republic, and parts of Iranian Azerbaijan.

Long a disputed territory that was fought over by Persia, Russia and Turkey, Armenia became the scene of turmoil and oppression for centuries. The Turkish Ottoman Empire invaded Armenia in the 15th century and held all of it by the 16th century.

Although Armenians became successful merchants in Turkey, they were always an oppressed minority because of their religion. Saint Gregory the Illuminator established Christianity in Armenia in the third century, and the autonomous Gregorian church became the centerpiece of Armenian culture and belief. Thus, this country without a portfolio was also an island of Christianity in a vast sea of Muslims, and it was this religious identity that the Ottoman Empire, under Sultan Abd al-Hamid II, used as its reason for launching the first volley in an attempt to exterminate all Armenian infidels from what once was the Armenian Empire.

The Sassoun massacres in January 1895 were merely the first step in a 27-year-long genocidal campaign. Over three years, 300,000 Armenians perished either by the efforts of government troops, starvation or disease. Troops would swoop down on Armenian settlements with the orders "Exterminate, root and branch. Whoever spares man, woman, or child is disloyal."

Thus, when the troops entered a town, they butchered all Armenians without discrimination. Women were raped and then killed. According to the *New York Times* on January 1, 1895, a priest in one village was taken to the roof of his church, hacked to pieces and then set afire. A large group of women and girls was herded into the church, raped and then locked in as soldiers set fire to the church.

In Moosh, Alyan and 14 other villages in the Sassoun district, 7,500 Armenians were butchered in the grisliest of ways. Some escaped into the hills, but starvation eventually drove them back to the villages, where they were set upon by waiting soldiers. Fires were built, and three- and four-year-old children were tossed, alive, into them.

The priests of the church were particular targets for the soldiers. In Ashpig, Der Bedrase, the priest of Geliguson, was stabbed by 40 soldiers wielding bayonets, and his eyes were dug out before he was tossed into a shallow grave he himself was forced to dig. Der Hohannes, the priest of Senmal, faced an even more gruesome ordeal. According to an eyewitness: "The soldiers took out Der

A mother grieves over her dead child in a field near Aleppo, in the midst of the Armenian massacres. Two million Armenians were killed over a period of 27 years by Turks, who eventually erased Armenia from the map of the world. (Library of Congress)

Hohannes's eyes, seized his hands, and compelled him to dance. Not only was he deprived of his beard, the insignia of his priestly office, but the cruel creatures took along with the razor some of the skin and flesh as well. Having pierced his throat, they forced him to drink water . . . It flowed from the ghastly cut, down on either side. His head was kicked this way and that, as if a football. Human flesh taken from some of his mangled people was put into his mouth. He, too, was pitched into the ditch with more than two score of men that had the promise of safety if they would cease resistance and surrender."

This was only the beginning. In April 1909, hundreds of thousands of Armenians were butchered in the Massacre of Adana.

By 1913 mass deportations were organized, and tens of thousands of men, women and children were made to march across deserts without food or water. Along the way, they were whipped, bludgeoned, bayoneted and torn limb from limb. Women were raped in front of their husbands and children and then murdered and tossed by the side of the road.

In Marash and in Zeytoon, there were uprisings of Armenian youths, but they were summarily crushed. If soldiers did not kill the Armenians, mobs did, with shovels, axes and blacksmith tools.

In 1915 the deportations increased. Tens of thousands of Armenians were driven ruthlessly from one city to another and back again. On August 7 the prisons in Zeytoon and Fundajak were thrown open, and Armenian prisoners, chained together, were led through the streets to their slaughter. Some were hanged from scaffolds in the center of various villages; the rest were marched to the foot of Mount Aghur and shot.

Not all Armenians were slaughtered or deported. Some were saved for slave labor. Twelve thousand of them worked on the beds of various railroad lines around northeastern Turkey. Overseen by German officers (World War I was now in progress, and Turkey was Germany's ally), these men, women and children were rationed a loaf of bread a day and some water and counted themselves lucky. They were, at least, alive.

Concentration camps of Armenians living in tents sprang up in the countryside, and on June 14, 1916, another mass deportation imprisoned or killed thousands more. Hungry, thirsty, naked, dirty and near death, these Armenians were relentlessly tortured and then killed. When the survivors were led on a deportation march, they were frequently separated from their families. Those who became exhausted fell by the side of the road, where shooting had come to be a kind fate.

Abraham Hartunian, a pastor who survived despite having an eye gouged out and being shot twice in the hip, wrote of this deportation:

> Corpses! Corpses! Murdered! Mutilated! . . . Stepping over them like ghosts of the dead, we walked and walked . . . Armenians were being massacred on the way between Baghtche and Marash . . . Here were the bodies of those driven out before us and shot, stabbed, savagely slaughtered [but] the previous convoys had experienced more.
>
> The men in our group who struck the eyes of the *zaptiye* [Turkish police] were separated, taken a little distance away, and shot. Everyone expected his turn to come next. The old man whose young son had died in Baghtche was walking along beside me with his daughter-in-law and two small grandchildren . . . But now, unable to walk, he was getting in the way of those behind. A *zaptiye* saw him. He came and kicked him and, dragging him out of the group, tripped him into a ditch nearby and emptied his gun into his breast.

At various places along the march, Muslim mobs from nearby villages waited with guns, axes and sacks. Told they would be blessed by Allah if they robbed and killed Armenians, they did.

"Night fell," wrote Hartunian in his memoir, *Neither to Laugh nor to Weep,*

> and the prettier women were taken aside and raped. Among them was an extremely beautiful girl, about twenty-five years of age . . . one after the other, the *zaptiye* . . . raped [her] and then, killing her, threw her mutilated corpse to one side because they could not agree who should have her.
>
> Many women were stripped naked and lined up, and their abdomens slashed one by one, were thrown into ditches and wells to die in infinite agony. The *kaymakum* of Der-el-Zor, holding a fifteen-year-old girl before him, directed his words to a murderous band and then, throwing her to the ground, clubbed her to death with the order, "So you must kill all Armenians, without remorse."
>
> Convoy after convoy was driven night and day unceasingly, robbed, raped, then brought to the edge of

streams and forbidden to drink at the point of the gun. Under the burning sun, thousands perished from hunger and thirst.

Many were gathered in one place and burned alive. One of these, left half dead and later rescued, told me that for days she had remained with the corpses and had lived eating their flesh.

The chronicle of horror was endless. Finally, in 1919, when British troops entered Armenia, an end to the massacres seemed to be in sight. But in one final genocidal sweep, the Turks massacred thousands of Armenians as the British troops were landing. British forces did little to stop these raids, and the remaining Armenians began to lose hope again.

In 1920 the French occupied Turkey. Even so, in that year alone 15,000 Armenians were annihilated in Marash; 160 Armenian girls were taken from an American girl's seminary in Hadjin, raped in the Turkish harems and then massacred; 3,000 Armenians on the road from Marash to Adana were buried in snow and died and further massacres were planned under the eyes of the French.

The 1920 massacres were as brutal as any that had gone before. According to Hartunian, "Children were ripped open before their parents, their hearts taken out and stuffed down their mothers' throats. Mothers were crucified naked to doors, and before their very eyes their small ones were fixed to the floor with swords and left writhing."

By the middle of 1920, the Turks were in full revolt against the French and massacred Armenians at will. Open warfare erupted. Turks burned Armenian homes and businesses; Armenians burned Turkish mosques. The Armenians were eventually overcome, and the Turkish government confiscated the houses, vineyards and fields of dead or fugitive Armenians.

At the end of the year, the Treaty of Sevres was signed, restoring Armenia as a sovereign state. Most Armenians who could, left. There was no guarantee that the massacres that had raged for nearly 27 years would not begin again.

BURUNDI
MARANGARA
August 14–21, 1988

Long-standing animosity between the central African tribes of the Tutsis and Hutus resulted in mass slaughters between August 14 and 21, 1988. Five thousand people, mostly women and children, were killed; thousands were wounded, many seriously. Forty-seven thousand refugees crossed into neighboring Rwanda; nearly 150,000 were made homeless.

For three centuries, highly charged, emotional confrontation has existed between the central African tribes of the Tutsis and the Hutus. The Tutsis arrived from Somalia and Ethiopia in the 16th and 17th centuries and established themselves as a kind of feudal aristocracy over the Hutus. Tall, cattle-raising people, they set up a ruling regime that denied the short-statured Hutu farmers equal rights. Belgium, which later ruled Burundi as a colony, exacerbated the problem by allowing the Tutsis to dominate education, government and the army. When Burundi achieved independence in 1972, war broke out between the two tribes.

Between August 14 and 21, 1988, the Hutus, armed with rocks and knives, attacked Tutsi villages. The reprisals by the Tutsi army were swift and devastating. Hutu villages were burned to the ground. Women and children were shot, mutilated and beaten. Five thousand victims on both sides of the conflict were killed.

As a result of the violence, 47,000 refugees poured across the border into neighboring Rwanda, and the government of Burundi estimated that nearly 150,000 were made homeless by the slaughter. Hospitals were filled with the wounded, most of whom were suffering from infections resulting from wounds that went untreated for weeks while they hid from soldiers in the underbrush. The result was a multitude of amputations, some of them on children only two years old.

The government of Burundi applied to the United Nations for $15 million in aid, but with the government's own army responsible for the massacres, the world body took a negative view of the request.

CHINA
BEIJING
April 18–June 4, 1989

Pro-democracy demonstrations begun by several thousand students in Tiananmen Square, Beijing, on April 18, 1989, swelled to one million in mid-May, during a visit by Mikhail Gorbachev. On June 3–4, army troops sent by the government massacred over 1,000 students and workers, injured more than 10,000 and crushed the uprising.

Life in the present-day People's Republic of China is as controlled as if the government were a puppet master and its populace puppets. Government officials assign people jobs, determine where they may live and decide how many children they may have.

Thus, the eventual magnitude of the protest demonstration that filled Beijing's Tiananmen Square throughout most of the spring of 1989 must have come as a startling surprise to both the Communist government of Deng Xiaoping and the demonstrators themselves. Not that gatherings of protesters have no precedence in China. Ironically, it was dissenting students, on May 4, 1919, that played a key role in the founding of China's communist movement, and it was student demonstrations that followed the death of Zhou Enlai (Chou En-lai) in 1976 that helped Deng Xiaoping ascend to power.

In 1976, tens of thousands of mourners opposed the ascension to leadership of the so-called Gang of Four after the deaths of Mao Zedong (Mao Tse-tung) and Zhou Enlai, and Deng, accused of being behind the protests, was purged. Three years later, in 1979, the same protesters pasted posters on what became known as Democracy Wall, and they were encouraged by Deng, who was making a comeback. But after he and his allies secured power, he reversed himself and clamped down on them.

Deng was a conservative force. Student protests and pro-democracy marches in 1986, spreading from the provinces to Beijing, became a factor in the downfall of Hu Yaobang, whom Deng accused of failing to stem the protests.

By 1989 Deng was in his 80s and surrounded by like-minded, conservative Communists. In the age of Gorbachev and glasnost, China was ruled by neo-Stalinist hardliners who practiced a policy of restraining criticism, stifling independence and maintaining a party monopoly on power.

There had been no more dramatic assertion of this than the 1987 ouster of Hu Yaobang, the leader of the Chinese Communist party and the heir apparent to the aging Deng Xiaoping. His removal came about because of his advocacy of intellectual freedom in China, an advocacy that launched a wave of public support from university students throughout China.

On April 15, 1989, Hu Yaobang died, and his death, as had his public stands in life, led to yet another outpouring of mourning, support and protest from China's students.

In the predawn hours of April 18, more than 10,000 of them marched through the capital of Beijing, chanting democratic slogans, singing revolutionary songs and eulogizing Hu Yaobang. At the height of the demonstration, several thousand students peeled off and marched to the Communist party headquarters, where some of them attempted to force their way in to see the nation's leaders. Still others staged a sit-in in front of the Great Hall of the People, at one end of the square, chanting a series of demands, among them freedom of the press, a reappraisal of Hu, a repudiation of past crackdowns on intellectuals and a disclosure of the income and assets of China's leaders and their children. Late in the day, several officials emerged from the hall and accepted a list of demands, assuring the students that they would study them. Thus, the Beijing spring began in a friendly, if public, fashion.

By Friday night, April 21, the government had grown perceptibly nervous and issued a ban on all public demonstrations. In defiance of the ban, more than 100,000 students gathered in Tiananmen Square. By the next morning, thousands of them had set up an encampment, thus foiling government attempts to close off the area and prevent a mass rally.

Later that morning, as the students chanted democratic slogans, party leaders arrived at the Great Hall for memorial services for Hu Yaobang. In the streets students were conducting their own memorial by singing the "Internationale," which begins, "Rise up, you who refuse to be slaves. . . ." During the day, attracted by cries of "Beijing citizens, follow us!" workers began to join the students in the square.

By Sunday the 23d, protests had ignited in other parts of China. In Xian, according to the official New China News Agency, protesters, after watching the official memorial service on television in the public square, attacked the provincial government headquarters, injuring 130 officers and burning 20 houses, setting fire to 10 vehicles and attacking a tourist bus loaded with terrified vacationers. In Beijing the demonstrators were far more peaceful, and most left the square by the morning of the 23d, vowing to boycott classes until May 4, the 70th anniversary of the famous 1919 demonstrations.

But by Thursday the 27th, despite the presence of police barricades and army troops in trucks, they were back again, 150,000 strong, marching for 14 hours. This time, at least 75,000 workers joined in, and thousands more lined the route of the march. Workers applauded the students from the street, waved encouragement from office windows and sent food and drinks to the demonstrators. It was the first time the prodemocracy movement had spilled over from students and intellectuals to the ordinary workers of China, and the demonstrators escalated their demands to include populist themes such as increased funding for education and an attempt to control inflation.

Toward the end of the demonstration, crowds burst through the last police barricade surrounding Tiananmen Square. Nearly 1,000 soldiers in 20 trucks were surrounded by mobs who clambered onto the hoods and sides of the vehicles and appeared, according to Western newspeople, about to lynch the terrified soldiers.

But now a distinctive characteristic of the student demonstrations began to assert itself. Students, waving their identification badges, came to the soldiers' rescue, clearing a way for their retreat and shouting "Brothers, go home and till your fields!" No such gentleness came from the rulers of the country, who in the next two days adopted a stiff line of resistance to the demands in the street.

But the threats seemed to have no effect. A day after the initial demonstrations, the government offered to hold a dialogue with the students, provided they returned to their universities "at once." In the eyes of the students, the pronouncement amounted to a rejection, and new demonstrations spread to Shanghai, Nanjing and other provincial cities.

The impasse continued through early May, and on Saturday, May 13, 2,000 students began a hunger strike in Tiananmen Square, vowing to remain there until their demands were met. On May 14 an unannounced Politburo meeting reportedly endorsed the moderate line of the Communist party leader, Zhao Ziyang, who hinted at the possibility of greater democracy in China.

It may have been a ploy to clear the square for the May 15 visit of Mikhail Gorbachev, who was coming to seal the reconciliation of the world's two largest communist nations. The demonstrators now became an embarrassment to Deng and his hardliners. A welcoming ceremony for Gorbachev to be held in Tiananmen Square had to be rescheduled, in an abbreviated form, at the airport after protesters refused to end their vigil and hunger strike in the capital's center.

During the night of May 14, the crowds in the square swelled to 80,000, and by the night of May 15, during a state dinner for Gorbachev, 150,000 rallied around the hunger strikers, who were now being tended by a team of medical students, distributing salt tablets and glucose. By morning more than 100 hunger strikers had lost consciousness and were taken to hospitals.

By Wednesday the 17th, an astonishing crowd of more than one million Chinese citizens had gathered in the streets of Beijing, bringing the capital and much of the country to a virtual standstill. As more ambulances carried more unconscious hunger strikers to hospitals, a high school teacher cried to a *New York Times* reporter, "Our hearts bleed when we hear the sound of ambulances. [The students] are no longer children. They are the hope of China."

This sort of escalation could no longer be denied by either the Chinese leadership or Gorbachev, who mildly endorsed the demonstrators' aims as he went on to Shanghai, where scores of thousands of people took to the streets to demand democracy and show their support for the Beijing hunger strikers.

Now, Prime Minister Li Peng came to the fore, appearing on television and warning that chaos in Beijing was spreading all over the country. It was time for it to stop, he added, but the million people in the streets of Beijing braved a driving rain to defy him.

To defuse the situation, Li and the Communist party leader, Zhao Ziyang, went to Tiananmen Square to visit some of the hunger strikers, whose ranks had swollen to 3,000. Furthermore, the government capitulated on a key demand by arranging a nationally televised meeting between Li and student leaders. The meeting was an emotional one in which a tearful Zhao Ziyang stated to the young protesters, "You've come too late. You have good intentions. You want our country to become better. The problems you have raised will eventually be resolved. But things are complicated, and there must be a process to resolve these problems."

To the young people the process was obviously in the streets, where cries for the ouster of Deng and Li Peng were rising and rings of students and protesters guarded the hunger strikers. A makeshift loudspeaker system had been erected, and a copying machine ground out stacks of pamphlets, which were eagerly scooped up by the surrounding crowds. An elderly police officer spoke to reporters. "The student movement is terrific!" he enthused. "If the government commands a crackdown, will I obey their order? No, I will go against it!"

On Saturday, May 20, that crackdown began. Martial law was proclaimed in Beijing, and 1,000 troops were dispatched to the city. At the same time, Deng asserted his authority and removed Zhao Ziyang from his leadership of the party for being too conciliatory with the students. In his place, Deng named Li Peng as both prime minister and party leader. Authorities shut off drinking water fountains in the square.

But the demonstrators held fast. Troops were met by thousands of citizens who rushed into the streets to block their progress. By midday one million people defied the martial law edict and choked the thoroughfares, surrounding army convoys and threatening them. Once more, students locked arms and protected the soldiers, while citizens pleaded with the army to leave. "You are my army," said a Chinese businesswoman. "You are our brothers and sisters. You are Chinese. Our interests are the same as yours. We believe you have a conscience. You must not crush the movement." Twenty-one army trucks were blocked by workers on the outskirts of the city.

Meanwhile, the government was gathering its forces. Deng stripped Zhao not only of his party leadership but of his right to order troop movements. Foreign television journalists were forced to stop their transmissions, although print journalists continued to roam through the crowds.

And the crowds grew, exceeding one million in Beijing. In Xian 300,000 demonstrators brought that city to a standstill. In Hong Kong 500,000 marched. In Shanghai protesters carried banners reading "Li Peng does not represent us" and "Li Peng, do not use the people's army against the people."

Rumors of an approaching brutal repression began to circulate among the students, and on May 26 many of them prepared their wills. On Saturday, May 27, some of the leaders called for an end to the occupation of Beijing's main square. Two weeks of living there had spawned serious health hazards. Garbage carpeted the premises; the refuse, reheated by the 90-degree spring heat, produced a stupendous stench. "It is very difficult to continue our sit-in," Wuer Kaixi, a student leader, said. "As leaders, we have responsibility for students' health and the difficulties are obvious. Hygiene is extremely bad and the food is insufficient." Student leaders proposed an evacuation of the square and called for large-scale demonstrations for greater democracy and the resignation of Prime Minister Li.

But two days later, on Monday, May 29, a crowd of nearly 100,000 workers and students cheered loudly as a 27-foot sculpture, constructed by art students and modeled after the Statue of Liberty, was dragged to the square in several pieces on tricycle carts, reassembled and put in place. The plans to leave had obviously been rescinded, if not totally reversed.

The military, now under the direction of Prime Minister Li Peng, regrouped. On Friday, June 2, 2,000 unarmed troops again marched on Beijing; again, tens of thousands of citizens turned them back. But this time the confrontation was more volatile. Eight hundred riot police officers fired tear gas to clear an area outside Communist party headquarters, and 30 people outside the Beijing Hotel were beaten.

But these were isolated incidents. In the rest of the city, soldiers, who seemed to be peasants from distant areas, looked unenthusiastic about their mission. Halted by the citizenry, they sat down along the side of the road and listened while students and workers talked earnestly to them. "You are the people's army," a young worker told several soldiers. "The students' movement is patriotic, and you mustn't use violence against it. Think about it." Most of the soldiers apparently did; some cried. It seemed as if once again, the students and workers had stopped the government.

But those would be the last peaceful moments in Beijing's Tiananmen Square. Shortly after midnight on Sunday, June 4, tens of thousands of seasoned, hardened army troops from the provinces converged on Beijing, supported by tanks and armored personnel carriers. As citizens rushed out to block their paths, they opened fire, killing the protesters and running over their bodies.

Horrified, the citizenry charged the soldiers with greater force, but they were crushed by onrushing tanks and shot and bayoneted by the imperturbable soldiers, who were apparently deaf to the pleas not to kill their fellow citizens.

In the main square, news had been received as early as Saturday afternoon that an armed force was on its way. In hopes of a nonviolent, peaceful end to the confrontation, students had purposely destroyed guns and bombs they had accumulated. "We dismantled the bombs by pouring out the gasoline," a student leader later told the *San Francisco Examiner.* "We wanted to avoid any chance that they would be used by criminals, or be treated as 'evidence' that the students had committed violent acts against the troops."

Student leaders broadcast warnings to those in the square that bloodshed might come and advised them to leave. Some did, but approximately 150,000 remained.

Shortly after midnight, the first two armored vehicles appeared, speeding down the side of the square. Thick formations of heavily armed, helmeted soldiers followed. Machine gun emplacements were set up on the roof of the History Museum. The students retreated to the Heroes Monument, near the art students' statue of liberty named the Goddess of Democracy and Freedom.

Negotiations were taking place between the student leaders and army commanders for a peaceful retreat of the students. But at 4 A.M., in the midst of these negotiations, the lights in the square were suddenly turned off. The talks continued for another 40 minutes in the dark, and then, abruptly, red flares flooded the area with light. Thousands of additional soldiers had gathered under cover of darkness, and now they rushed into the square, setting up more machine gun emplacements. Interspersed between them were riot police, carrying electric cattle prods and rubber truncheons. They waded into the mobs of students, beating them mercilessly, kicking them and then shooting them.

Armored vehicles roared into the square now, blocking it off entirely except for a small opening in the direction of the History Museum. The students were trapped. Soldiers continued to advance, smashing the students' broadcasting and printing equipment and dragging the students down from the steps of the Heroes Monument.

It was their only refuge; apparently, the army did not want to damage this national symbol with gunfire. Students retreated again and again to the monument. Soldiers and riot police followed, rushed them, clubbed them and drove them into the street, where they were machine-gunned or shot by automatic rifle fire. Wave after wave of gunfire and clubbing continued until some students, in desperation, tried to escape through the gap between the armored vehicles near the museum. But the gap was soon closed, and the vehicles ringing the square now charged the students, crushing them beneath their tracks and wheels.

Citizens and workers tried to charge the soldiers but were beaten or killed. By 5 A.M. Tiananmen Square had been swept clean of students and workers. Blood was everywhere, and bodies were strewn for blocks. Reporters saw people gunned down, beaten, crushed. As bullets careened over their heads, one hysterical student begged a *New York Times* reporter, "We appeal to your country. Our government is mad. We need help from abroad, especially America. There must be something that America can do." Another sobbed, "Maybe we'll fail today. Maybe we'll fail tomorrow. But someday we will succeed. It's a historical inevitability."

The wounded, the dying and the dead were brought to hospitals. In Beijing Tongren Hospital, one mile southeast of the square, a doctor told reporters for the *Times,* "As doctors we often see deaths. But we've never seen a tragedy like this. Every room in the hospital is covered with blood."

All day on Sunday, troops crossed the city, and the sound of gunfire constantly reverberated through its streets. Rumors sped through the city as rapidly as the military vehicles. The university was about to be assaulted. It had been set afire. There were 20,000 dead; 200,000 dead.

Some soldiers were killed. Thirty-one military vehicles and 23 police cars were burned. A soldier who had shot a young child was overpowered by a large crowd in the Chongwenmen district early Sunday morning, hanged from a bridge and then burned.

But the casualties were primarily among the citizenry. A 24-year-old government official, fleeing from a barrage of bullets near the square, happened upon some wounded and bleeding victims. He stopped to help them. An army officer slammed his pistol into the side of his head. "Don't stir or you will be dead," he said, and a dozen soldiers ran up and began to beat the young samaritan with bricks and rifle butts. "I never thought they could be so brutal," he told a *Times* reporter later at the Union Medical College Hospital. At the same hospital, a doctor showed a reporter bullet holes in the side of an ambulance. Late Sunday afternoon, shoppers on a major side street in Wangfujing, one of the major shopping districts, were shot and killed when troops unexplainably opened fire on them. Altogether, more than 1,000 students, workers and unwary citizens were killed and 10,000 were injured in one of the bloodiest peacetime massacres in modern history.

To this day, the communist government of China states that it did not happen. The official version ignores the bodies, the blackened, twisted remains of barricades, the residue of blood on the Heroes Monument, the eyewitness reports of international reporters. A 42-year-old factory worker who described the slaughter to ABC News was later shown on state television, his head bowed, confessing that he never saw anything and that he was a counterrevolutionary. "I apologize for bringing great harm to the party and the country," he added.

Further television pictures showed soldiers peacefully cleaning up the debris in the square. Three hundred people were killed, many of them soldiers, according to the official version. "Not a single student was killed in Tiananmen Square," said an army commander. Later, the government amended his remarks to admit 23 student deaths.

Worldwide repugnance produced indignant public statements from the government of China. It attacked the United States for interfering in China's affairs when President George Bush protested the killing of civilians by suspending arms sales and visits by military officials between the two countries.

By Wednesday, June 7, the army was still in control in Beijing and still flexing its muscles. It fired into two diplomatic compounds that housed thousands of foreign diplomats and journalists. Stony-faced troops then surrounded the compounds for two hours, after which the United States embassy sent its marine guards to help evacuate Americans who wanted to leave. The exodus of foreign nationals rose to flood stage, although diplomats remained in the city. Shortly after the incident at the compounds, the army began to evacuate Beijing.

Tension and unrest remained in the capital city. More than 1,000 students were arrested; state television news characterized them as "thugs" and assailed them for supporting the "counterrevolutionary rebellion." Trials were held; scores of students and workers were sentenced to public executions.

Purges occurred throughout the country. Other executions took place in Shanghai. The world was horrified; China remained defiant, chastising its critics for meddling in its internal affairs. A cloak was drawn around the country. Dissidence was muffled, then eliminated. And the world turned its attention elsewhere.

In December 1989 President George H. W. Bush, apparently trying to maintain the open lines of communication between China and the United States established by former president Richard Nixon, tried to reestablish economic ties with Beijing by sending national security adviser Brent Scowcroft on a secret mission to the Chinese capital. Congress learned of the visits and objected to the mission. Some members sensed the behind-the-scenes influence of Nixon and Henry Kissinger, his former secretary of state.

In May 1991 President Bush again tried to reestablish most favored nation trading status for China, with no assurances of an improvement in human rights. Congress again refused to support him, but in March 1992 they gave in to Mr. Bush and awarded China most favored nation status. Meanwhile, Deng and his hardliners held China tightly under their control, maintaining it as the last neo-Stalinist stronghold in the world.

CHINA
NORTHERN PROVINCES
June 1900

Long-festering hostility between Chinese conservatives and foreign partitioners climaxed in the disastrous Boxer Rebellion of June 1900. Thousands of Western missionaries and residents were killed, hundreds more were injured, and China was left vulnerable to Western powers.

Dismembered bodies are grisly evidence of executions in Canton during the Boxer Rebellion, a grim conflict between Chinese conservatives and Western partitioners. *(Illustrated London News)*

A contemporary Chinese print depicts the Boxer Rebellion of 1900. (Library of Congress)

The Orient was opened in the 16th century to Western interests, but only temporarily. Within a very short time, the door was slammed shut to Western commerce and reform, and it would be the beginning of the 19th century before that door was forced open again by the British.

Canton, China, was opened to trade in 1834, and this was Britain's opportunity to establish a large foothold in China. In 1839, when China took action to enforce its prohibition of opium importing and proceeded to destroy supplies of opium belonging to British exporters in Canton Harbor, Great Britain instigated the so-called Opium War. The British used modern firearms, and the Chinese were quickly defeated.

Over the next 20 years Western interests took hold in China. Hong Kong became a British colony. Seventeen ports, including Foochow and Shanghai, were opened to Western trade, and France, Russia, Germany and the United States participated in widening the treaty, opening more "treaty ports" and establishing residences for diplomats, businessmen and missionaries.

This was not done without fierce resistance from the Chinese, who fought back periodically. In 1859 Chinese troops attempted to block the entry of diplomats into Beijing British and French forces not only reversed this, but they also occupied all of Beijing and burned the imperial summer palace to the ground.

By the 1890s China was further weakened by its unsuccessful war with Japan and the partitioning of China into foreign spheres of influence. But anti-Western, antiforeign sentiment was as pervasive as it had been in the 16th century.

Some of this was focused in a powerful secret society called Ho Ch'uan, which in Chinese means "righteous, harmonious fists." In English, this translated, much more prosaically, into "Boxers."

It was no small neighborhood club. Its adherents were fierce and dedicated, and by 1898, there were 140,000 members who had little love for the foreigners whom they saw as usurpers of their country's economy, land, culture and pride.

Ruling at this time was the dowager empress and regent T'zu Hsi. She was a strict adherent to the old values of China and no friend of the West. In 1875 she named her infant nephew Kuang Hsu to the throne, even though he was not a direct heir. Then, in 1898, she resumed the regency herself after Kuang Hsu attempted to institute political reforms of which she did not approve.

The dowager empress tacitly approved of the growth and aims of the Boxers, and, given support by the war party at court, they began in late 1899 to conduct raids against foreign missions and Chinese Christians. These forays were waged in the northern provinces—Chihi, Shanxi and Shantung, in Manchuria and in inner Mongolia. Western interests had built railroads in these provinces, and some of the earlier actions were directed against the railroads and Western landlords.

As the Boxer Rebellion—as it was eventually known—began to grow, it became more deadly and widespread. Western missionaries were massacred and their dwellings burned to the ground. The land around Beijing became a battleground. Telegraph wires from Beijing to Pau-ting-Fu were cut; bridges were destroyed. Finally, French, British, Russian and German troops took up positions, and the Boxers massed for an attack. Meanwhile, in the Imperial Palace, the Ultra Conservative party, headed by the dowager empress, opposed Prince Ching and his Moderate party, who argued that the Boxers should be repressed.

During the first week of June 1900, the Boxers struck, 140,000 strong, and occupied Beijing. For eight weeks, they attacked missionary outposts, foreign missions and installations. They were most vicious when encountering Chinese Christians. A Professor Headland of Beijing University, speaking to the *New York Times* on June 4, said, "I don't believe the Boxers intend to kill any foreigners unless they get mixed up in fights. They want to carry off some, perhaps, in order to get a ransom for them; but they are intent on killing off the native Christians."

For most of the summer of 1900, the Boxers held the northern provinces of China, and foreign interests were muted, if not murdered. And then in August, as they had on several previous occasions, a force of British, French, Russian, American, German and Japanese troops retook the provinces, killing great numbers of Boxers and driving the rest into hiding.

There would be sporadic raids on missions for another two years. But the Boxer Rebellion would be, for all intents and purposes, crushed, and the Western powers would exact more economic and trade advantages from China, plus $333 million in reparations and an agreement to allow foreign troops to be stationed in Beijing. Japan wanted considerably more from the nation that had allowed and encouraged the Boxers to rebell, but disagreements among the Western powers, plus the intervention of the United States, which advocated an end to the further partitioning of China, prevented these wishes from being fulfilled.

The Boxer Rebellion, then, was worse than a failure. It left China in debt and a subject nation to foreign powers.

EUROPE
1939–45

Adolf Hitler's determination to exterminate "inferior" races and establish a master race resulted in the six-year-long "Final Solution," or Holocaust, as it was eventually called. Between 1939 and 1945, Nazis systematically murdered five million European Jews, three million Russians and two million Slavs.

In 1939 there were 10 million Jews in Europe. In 1945, at the end of World War II, there were fewer than five million. To Adolf Hitler, the chancellor of the Third Reich, they, along with the Slavs of eastern Europe, were *Untermenschen,* or subhumans, and therefore destined for either slave labor or extermination.

The Nuremberg trials revealed that, in defiance of the Geneva Convention, the Nazis routinely killed prisoners of war, but none so thoroughly, gruesomely or energetically as Russian and Slavic POWs. Two million Russian prisoners of war died in German captivity from starvation, exposure and disease. The remaining million have never been accounted for; at Nuremberg, a good case was made that most of them either died from the above causes or were exterminated by the S.D. (S.S. Security Service).

But this was a small effort compared with the grand plan of the Third Reich, the elimination of all European Jews, a genocide of a magnitude even larger than that of the Turks during the Armenian massacres (see p. 83). It would be an act of unparalleled barbarism, possibly the most inhuman act in the history of the world.

The so-called Final Solution had its roots in a speech made by Adolf Hitler before the Reichstag on January 30, 1939, in which he stated, "If the international Jewish financiers . . . should succeed in plunging the nations into a world war the result will be—the annihilation of the Jewish race throughout Europe."

As early as 1939 systematic incarceration had already begun, with the *Einsatz* groups, organized by Heinrich Himmler and Reinhard Heydrich. These specially trained squads followed the German armies into Poland and rounded up Jews, locking them in ghettos. Two years later, between June 1941 and June 1942, at the beginning of the Russian campaign, this was escalated to extermination. Jews and Soviet commissars were rounded up in each village that was conquered; they were ordered to dig mass graves and then remove their clothing. They were shot at the edge of the grave and shoveled into it. More than 500,000 Jews in White Russia were killed this way.

Later the method was refined. Otto Ohlendorf, testifying at Nuremberg, described it thus: "The *Einsatz* unit would enter a village or town and order the prominent Jewish citizens to call together all Jews for the purpose of 'resettlement.' They were required to hand over their valuables, and shortly before execution to surrender their outer clothing. They were transported to the place of executions, usually an antitank ditch, in trucks. . . ."

In the spring of 1942, Himmler ordered the method of killing to be changed from shooting to gassing. For this, gas vans were constructed by two Berlin firms. Resembling closed trucks, the vans were loaded with up to 25 persons, ostensibly to be taken away for "resettlement." The motor was turned on, and the gas—carbon monoxide—was directed into the van.

However, these vans proved too small to handle the massacres that Hitler and Himmler envisioned, and so the Final Solution evolved to its final, most efficient and horrible phase. According to the Nuremberg statistics, all of the 30-odd principal Nazi concentration camps were really death camps, where millions were killed by torture, starvation or planned execution.

It was at the extermination camps, called the *Vernichtunglager,* where the killing was most widespread and gruesome. The most renowned of these was Auschwitz, in Poland, where four enormous gas chambers with adjoining crematoria killed up to 6,000 Jews a day.

Following the earlier methods of the *Einsatz* units, the Jews rounded up by the Nazis in occupied countries were ordered to gather their valuables and report for resettlement. They were packed like animals into freight cars and transported to the death camps at Auschwitz, Treblinka, Belzec, Sobibor and Chelmno in Poland, where extermination was carried out by gassing, or to smaller installations such as Riga, Vilnius, Minsk, Kaunas and Lvov, where it was done by firing squads.

In either case, "selection" would take place at the railroad siding at which the prisoners debarked from the train. According to Auschwitz commandant Rudolf Hoess, "We had two S.S. doctors on duty . . . to examine the incoming transports of prisoners. These would be marched by one of the doctors, who would make spot decisions as they walked by. Those who were fit to work were sent into the camp. Others were sent immediately to the extermination plants. Children of tender years were invariably exterminated since by reason of their youth they were unable to work."

Even those who were to be killed immediately were deluded into thinking they were being "resettled." Some were given lovely picture postcards marked "Waldsee," inscribed with a reassuring but viciously ironic message:

> We are doing very well here. We have work and we are well treated. We await your arrival.

"At Auschwitz," continued Hoess at his trial at Nuremberg, "we endeavored to fool the victims into thinking that they were to go through a delousing process. Of course, frequently they realized our true intentions and we sometimes had riots and difficulties. Very frequently women would hide their children under their clothes but of course when we found them we would send the children in to be exterminated."

The grim charade continued until almost the last days of Auschwitz. The gas chambers were disguised as plain-looking buildings fronted by green lawns punctuated by lush and colorful flower beds. At the entrance, which bore a sign that read "BATHS," an orchestra of young women prisoners played Offenbach and Lehar.

The prisoners were ordered to remove their clothing and were sometimes even given towels. Once 2,000 of them were within the "showers," the doors were slammed shut and barred. It was at this moment that all those inside knew they had been deceived and would die, and so the stampedes would begin. Mountains of humanity would pile up at the door, and many of the doomed would be trampled or clawed to death in this first desperate surge toward freedom.

Meanwhile, on the roof of the gas chamber, the S.S. would be at work. Concealed in the lawns and flower beds were the mushroom-shaped lids of the vents that ran into the chambers. Orderlies opened the vents and stood ready with amethyst crystals of hydrogen cyanide or Zyklon B, which had originally been manufactured as a strong disinfectant and were now supplied to the extermination camps by Tesch and Stabenow of Hamburg and Degesch of Dessau, two German firms that had received the patent from I. G. Farben. The former supplied two tons of the crystals a month, and the latter three-quarters of a ton.

A Sergeant Moll would give the order, *"Na, gib ihnen schon zu fressen,"* ("All right, give them something to chew on"). The crystals would be poured into the openings, and the openings would then be sealed.

The executioners watched the death throes of those inside through heavy glass portholes. According to Gerald Reitlinger, "they piled up in one blue, clammy, blood-spattered pyramid, clawing and mauling each other even in death."

It took as long as 30 minutes for the killing process to be consummated, and then pumps drew out the poisonous fumes. The huge door leading into the "baths" was opened, and Jewish male inmates, the *Sonderkommando* who were given adequate food and promised their lives, were let in to do their work. Protected with gas masks and rubber boots, they wielded heavy water hoses. Their first task was to wash away the blood and defecation. Then they went through the bodies, extracting gold teeth and hair. Once this had been accomplished, they loaded the naked and mutilated corpses onto lifts or railroad wagons that took them to the furnaces, where they were burned, and their ashes were either scattered into the Sola River or sold as fertilizer.

The gold fillings were later melted down and shipped, along with other valuables taken from the Jews at the railroad siding, to the Reichsbank, where, under a secret agreement between Himmler and the bank's president, Dr. Walther Funk, they were deposited to the credit of the S.S. in an account given the cover name "Max Heiliger."

The gas chambers and crematoria were the result of a disgusting partnership between German industry and the German military. There was spirited bidding to win the contracts for the crematoria, and the firm of I. A. Topf and Sons of Erfurt, manufacturers of heating equipment, won.

Even the efficient construction of the crematoria was no match for the speed at which the Third Reich carried out its extermination policy. In 46 days during the summer of 1944 at Auschwitz alone, between 250,000 and 300,000 Hungarian Jews were put to death. Midway through this period, the gas chambers fell behind, and mass graves were again dug by the victims before they were shot and thrown into them. The bodies were set on fire, after which bulldozers were used to cover over the mass mausoleums. The Russians who liberated the camp and confiscated its records estimated that close to four million died at Auschwitz alone, but later estimates reduced the figure to approximately one million.

Combined with a systematic and almost unimaginably barbaric series of "medical experiments"—operations without anesthesia to test pain thresholds, high altitude experiments, freezing experiments—and the stripping of skin from the bodies of the dead or dying to be tanned and formed into lamp shades and other decorations for the S.S., plus the systematic killing of all of the inhabitants of various places in Poland and Russia—the Warsaw Ghetto and the village of Lidice, for example—this attempted genocide and massacre of the innocents has no parallel in all of history. That it occurred in the 20th century is a discouraging affirmation that man seems to be eternally capable of inflicting the worst and most senseless disasters upon his fellow man.

FRANCE
PARIS
August 24, 1572

The continuing confrontation between Catholics and Protestants in France in the 16th century culminated in a massacre of 2,000 French Huguenots gathered in Paris on August 24, 1572, to celebrate both St. Bartholomew's Day and the wedding of Henry of Navarre and Margaret of Valois.

One of the bloodiest and most barbaric incidents in the so-called Wars of Religion that raged through France from 1562 to 1598 took place on St. Bartholomew's Day, August 24, 1572, in Paris and later throughout France. Ostensibly it was an attack upon Huguenots for practicing their Protestantism, but palace intrigue and politics were also involved that day.

The Protestant reform movement began in France at the start of the 16th century, but it was given a tangible symbol in 1559, when the first French national synod was held and the Presbyterian church, modeled after John Calvin's reform in Geneva, was founded. The adherents of

Protestantism in France were then known as Huguenots—from the German word *Eidgenossen*, meaning sworn companions or confederates. The confederacy extended across class lines but failed to mute the persecution of the Huguenots in France.

In 1560 the Conspiracy of Amboise brought about a fierce confrontation and heavy toll upon the Huguenots. The object of the plot was to allow the House of Bourbon to usurp the power of the Guise family, represented on the throne by Francis II. The plan was to march on the royal castle, abduct the king, and arrest Francis, duke of Guise, and his brother Charles, who was also cardinal of Lorraine.

The cardinal, however, got wind of the plot before it could be put into motion, and the rebel forces were set upon before they could organize themselves. A brutal slaughter followed, and for weeks the bodies of conspirators were hung from the castle and from every tree in sight. The Huguenots were enraged, and the first of the Wars of Religion, in 1562, was a direct result of this slaughter and its grisly aftermath.

By 1572 two of these civil wars had been fought, each ending in reconciliation. But in August 1572 the peace ended violently. That month, the Huguenot nobility was gathered in Paris to attend the wedding of Henry of Navarre (he would later become King Henry IV) and Catherine de' Medici's daughter, Margaret of Valois.

Catherine de' Medici and the duke of Anjou (later King Henry III), with the reluctant help of King Charles IX, tried, on August 22, to capture the duke of Coligny, the commander of the Huguenots in the second War of Religion and their most respected representative. The attempt failed, and Catherine and her cohorts then determined to kill Coligny and as many Huguenots as they could.

On August 24, St. Bartholomew's Day, while French Huguenots gathered in Paris to celebrate the day and the wedding, the soldiers of the king swooped down on them, massacring every Huguenot in sight. Leaders and ordinary citizens were cut down ruthlessly, and before the day was over, 2,000 Huguenots lay dead in the streets of Paris.

During the next few days the massacre spread to the countryside and to other cities in France, and within days the Huguenots regrouped, and the Third War of Religion began. Two more wars would be fought after this, and though the wars themselves would end in 1598, true freedom from oppression for Protestants in France would not come until 1905, when church and state were finally declared separate.

The brutal slaughter of French Huguenots in the streets of Paris, the famous St. Bartholomew's Day Massacre of August 24, 1572, is depicted in a period drawing. (New York Public Library)

FRANCE
PARIS
1793–95

Between 23,000 and 40,000 were executed during the double reign of terror that followed the French Revolution in 1793–95.

Excess may seem a tame word to define the French Revolution's Reign of Terror and its aftermath, the White Terror. But consider the statistics of the Reign of Terror: In Paris alone between May 1793 and June 1794, 1,251 people were executed, either by the guillotine or in mass drownings, called *noyades*. Between March 1793 and August 1794, when the Reign of Terror ended and the White Terror began, 16,594 death sentences were handed down by the Revolutionary Tribunal.

It was, in each case, retribution that set the sentences, and they arose from many complex causes. The French Revolution marked the end of feudalism in France, but it also meant the potential beginning of anarchy. In order to prevent this, the Committee of Public Safety was created in April 1793, and none too soon. In September 1793, spurred by hunger, thousands rioted in the streets of Paris, and there was a fear that these riots had been fomented by royalists and counterrevolutionaries.

Under the dominant guidance of Maximilien Robespierre, the committee announced its goals: to eliminate all internal counterrevolutionary elements (ever since the execution of Louis XVI in January 1793 royalists had been organizing to overthrow the revolution), to raise new armies (Austria and Prussia, soon to be joined by Great Britain, had taken advantage of a perceived weakened France and had gone to war against it) and to regulate the national economy (hungry people were rioting people, and civil order was necessary).

But the Reign of Terror, a retributive set of mock trials and multiple public executions carried out by Revolutionary tribunals, was counterproductive in its sweeping excess. Defense and preliminary cross-questioning of the accused were abolished, and juries could convict on nothing more than moral proof. And there were only two choices: acquittal or death.

The bases for conviction were general and chilling: "Those who have aided and abetted the plans of the enemies of France by persecuting and slandering patriotism, those who have sought to spread a spirit of discouragement, to deprave the morality of the people, to undermine the purity and energy of revolutionary principles, all those who, by whatever means and under whatever pretext, have attacked the liberty, the unity and the security of the Republic or have worked to prevent these from being established on a firm, lasting basis. . . ."

Small wonder, then, that at one time the Paris prisons held 8,000 prisoners, which in turn led to fears of a prison revolt. Thus, tribunals were held and executions were swift. According to Fouquier-Tinville, the public prosecutor at the Revolutionary Tribunal, "heads were falling like tiles."

Robespierre, however, was overthrown on July 24, 1794, and now the opposition took over. The Thermodorian Reaction, or White Terror, began, and by the first of March 1795 the Monteuil section of the National Convention demanded death for the death dealers of the Reign of Terror: "What are you waiting for, in order to cleanse the land of these cannibals? Do not their ghastly hue and sunken eyes proclaim enough their foster parents? Have them arrested . . . The Trenchant blade of the law will deprive them of the air they have too long infested."

On February 2, 1795, in Lyons, the first massacre of imprisoned ex-terrorists was carried out. Thousands more would be murdered, individually and in large numbers, while the convention looked on, mindful that inflation, famine and the numbing cold of the winter of 1795 were eating away at the loyalty of the people of Paris and the rest of France. Mass killings tended to keep them indoors and discouraged rioting.

But the White Terror continued through the spring of 1795. All over France prisons were broken into and prisoners massacred. At Tarascon in June, Jacobins were hurled into the Rhone from the walls of the Chateau du roi Rene, and there were other such atrocities committed in Salon, Nimes and Pont-Saint-Esprit. "Wherever you look there is throat-cutting," lamented a deputy of the convention.

Finally these excesses produced a reaction, and a calmer return to the values of freedom and the Rights of Man established by the revolution prevailed. The so-called War of the Black Collars in the summer of 1795 rounded up the executioners of the White Terror; far right members of the Council of Deputies were arrested; and a general amnesty was extended for "deeds exclusively connected with the Revolution." It had ended, but the terror on both sides had greatly weakened the purposes of the French Revolution.

GREAT BRITAIN
ENGLAND
LONDON
November 5, 1605

The presence of harsh penal laws against English Catholics in 1605 led to the ill-fated Gunpowder Plot to blow up Parliament on November 5, 1605. It failed; all of the conspirators were executed.

Anti-Catholic sentiment ran deep in England at the beginning of the 17th century, and it had official and royal sanction. There were harsh penal laws designed to all but prohibit the practice of Catholicism, and in protest against them a plot was originated in 1605 to blow up both Parliament and King James I. It would take place on November 5, the opening day of Parliament, which was normally given over to ceremony. The king would be in attendance.

Three young men, Robert Catesby, John Wright and Thomas Winter, originated the plan. They were soon joined by Christopher Wright, Robert Winter, Robert Keyes, Thomas Percy, John Grant, Sir Evirard Digby, Francis Tresham, Ambrose Rookwood, Thomas Bates and Guy Fawkes—the last a convert to Catholicism who served as a soldier with the Spanish in Flanders.

The plan was straightforward: blow up the entire government and set in place a Catholic monarchy. Like all grand designs, it was too good to be kept secret. At any rate, by the middle of 1605, when Thomas Percy had rented a subcellar under the House of Lords and the conspirators had stocked it with 36 barrels of gunpowder, overlaid with steel bars and firewood, the grand design was known throughout much of the Catholic community of London, including Henry Garnett, the superior of the English Jesuits.

Members of Parliament, however, did not know of the Gunpowder Plot until October 26, when Francis Tresham sent a letter to his brother-in-law, Lord Monteague, warning him not to attend Parliament on November 5. The planners might as well have announced it in the middle of Piccadilly. Lord Monteague informed his colleagues, among them the first earl of Salisbury, who, in short order, discovered the cellar and its lethal contents.

On the night of November 4, Guy Fawkes, who because of his military background had been elected to detonate the dynamite, crept quietly into the cellar to check the fuses and the powder. Soldiers were waiting for him and arrested him on the spot. He was taken to the Tower of London and under torture revealed the names of his co-conspirators.

Soldiers fanned out throughout London and began to make arrests. Catesby tried to fight his way through the arresting party and was killed. Percy, the renter of the cellar, was shot and mortally wounded while trying to flee from his captors. The rest were captured and either imprisoned, killed outright or sentenced to be hanged. Among those hanged in November 1606 were Henry Garnett (the superior of English Jesuits), Thomas and Robert Winter and Guy Fawkes.

Rather than making the lot of Catholics in England better, the Gunpowder Plot worsened their lives. Instead of erasing repressive laws against the practice of their religion, the aborted plot caused the enactment of harsher, more repressive ones. And to this day in England, Guy Fawkes Day is celebrated on November 5 with fireworks and bonfires and the image of Guy Fawkes hanged in effigy.

GREAT BRITAIN
SHEFFIELD
April 15, 1989

In the worst tragedy in the history of British soccer and one of the worst disasters in the history of sport, 95 soccer fans were trampled to death on the overcrowded terraces of Hillsborough Stadium in Sheffield, England, during the first six minutes of the April 15, 1989, match between Liverpool and Nottingham. More than 180 were injured.

No one can accuse the builders of British soccer stadiums of making them overly comfortable. Most of them are sandwiched into working-class neighborhoods, most were built during Victorian or Edwardian times. Many of them have turned ramshackle with age. All of them provide relatively cheap entertainment for those who neither can afford nor wish to spend their Saturday afternoons at the cricket grounds.

British soccer fans are also far from genteel. For five years following the tragic riots during the European Cup final between Liverpool and Juventus at Heysel stadium in Brussels, in which 39 fans—most of them Italian—were killed and more than 400 injured, British fans were banned from European competition.

On the afternoon of April 15, 1989, there was no rioting between opposing fans during the opening moments of the Football Association cup tournament game between Liverpool and Nottingham in Sheffield's Hillsborough Stadium. There was much pushing and shoving, as there always was. And there was nearly criminally poor judgment on the part of the officials, who allotted 6,000 fewer seats to Liverpool fans than to Nottingham fans despite the undisputed fact that the Liverpool club had far more supporters than did Nottingham.

Nevertheless, the Liverpool fans who were swarming through the gates to the terraces—concrete steps behind the goal line, where fans stand throughout the game—were willing to endure a little crowding to cheer on their team. As Brian Wolfson, a Liverpool native who runs London's Wembley Stadium, put it: "You have a culture, the camaraderie of the terraces. It is directly behind the goal, where the action is, where you can throw yourself around and come away drained, as if you had played yourself."

As a safety precaution and to keep the fans from fighting one another and spilling onto the playing field, officials had erected fortified walls and high steel frame fences around the terraces. When play began the section was dangerously overcrowded. And then, as play heated up, the fans surged forward to cheer their team on. Those at the gates shoved ahead to enter the section. Those in the upper terraces pressed against those in front of them, and a terrifying domino effect began to build. The spectators in front were flung forward against the iron barricades and fences. The crush increased, and within moments, despite their screams, 95 fans died of trampling or suffocation, some impaled on the fence, some crushed underfoot, some asphyxiated. More than 180 suffered painful injuries.

It was the worst soccer tragedy in the history of British soccer, surpassing the death toll of 66 people in 1971 when crowd barriers collapsed during a match in Glasgow, and only surpassed in the history of sport by the catastrophic riots on May 24, 1964, when fans from Peru and Argentina fought one another during an Olympic qualifying match. In that horrible debacle 300 fans died and 500 were injured.

The Hillsborough tragedy brought about increased security and some physical improvements in the facilities of the United Kingdom's soccer stadiums. But most experts, PMs and officials agree that the next soccer tragedy is an incident waiting to happen.

HUNGARY
BUDAPEST
November 4, 1956

Hungarian freedom fighters, fired by the success of the Polish uprising and a Russian-appointed regime, declared independence on October 23, 1956, with a student uprising. On November 4, 1956, Russian tanks reclaimed Budapest and Hungary. Twenty-five thousand Hungarians were killed in the fighting; thousands more were killed in the ensuing executions.

Today it seems almost inconceivable that tens of thousands of Hungarians lost their lives and their hope in one week of fierce fighting against the Stalinist regime that had held Hungary captive since World War II. Nor does it seem possible that both the United Nations and the Western powers—themselves involved in the "Suez crisis"—could stand by and watch as a bloodbath of staggering proportions took place, while the victims pleaded over their radio station for help from the rest of the world.

Yet it happened, from Tuesday, October 23, 1956, when the first student uprising occurred, until Sunday, November 4, 1956, when Russian tanks reclaimed Budapest. The world had a chance to help, but did not. Perhaps it was bad timing. Events choose their moments, and this was not the most propitious moment for Hungary to find its freedom.

The impetus for the Hungarian uprising came with the death of Joseph Stalin in March 1953, which was surrounded by a maelstrom of plots and threats. Stalin's tyrannical hold on Russia and the lands it had acquired at the end of World War II was so complete and so ruthless that it carried with it the seeds of an inevitable revolt.

There were demonstrations in East Germany and Czechoslovakia, and in Yugoslavia there was Tito, who preferred his own brand of communism to that of the Stalinist loyalists. So in May 1955 Nikita Khrushchev formed the Warsaw Pact, a treaty of mutual friendship, cooperation and mutual aid uniting the satellites as an answer to NATO.

But within the Warsaw Pact there was discontent. In Poznań, Poland, on June 28, 1955, a strike occurred. It was put down ruthlessly. Russian tanks and troops arrived, surrounded the city for two days, killed 113 people and broke the strike.

Almost simultaneously Anastas Mikoyan arrived in Hungary, where a group of intellectuals were already expressing discontent about life under Stalinism. Matyas Rakosi, the head of Hungary's Central Committee, was notoriously repressive, employing the AVH, or Secret Police, to enforce his edicts and intimidate the Politburo. He was opposed by Imre Nagy, a moderate. If Mikoyan had, on that visit to Budapest, replaced Rakosi with Nagy, the Hungarian Revolution might never have taken place. He did not. He replaced Rakosi with Erno Gero, Moscow's handpicked man and one of Rakosi's henchmen.

Meanwhile, crisis erupted in Poland. Riots over the Poznań killings sprang up in Warsaw. The army and Polish students faced each other. To defuse the situation, Khrushchev met with Polish leaders and announced that the Soviet government would allow a form of communism to exist in Poland that was not precisely Russian. The Hungarians thus learned that it was possible to stand up to the Russians and win. They congratulated Poland and made plans for their own revolution.

From October 19 through 21, 1956, Hungarian students and intellectuals escalated their demands for the withdrawal of Soviet troops from Hungary. Meetings were held. A large student demonstration was scheduled for Tuesday, October 23 in Budapest. It was first forbidden; then the prohibition order was withdrawn.

The demonstration took place, and its makeup would have made Lenin smile. It was a spirited alliance of workers and intellectuals—just the mix that Lenin said was indispensable to a revolution.

By that night the revolution was well under way. The intellectuals, students and workers wanted Nagy; Gero was determined to remain in place. The demonstrators in the streets had created a 16-point manifesto, and they went to the radio station to request that these 16 points be broadcast. According to a UN report filed later, an army major volunteered to present the paper to the head of broadcasting, but as he approached the main entrance to the building he was gunned down by police.

And so the bloodshed began. Tear gas was lobbed into the crowd, and AVH men, the Hungarian equivalent of the KGB, opened fire, killing a number of people and wounding more. Tanks arrived, but the commander in charge informed everyone that he was a worker and would not participate in a massacre.

The crowd was now armed with machine guns and rifles, driving vans and trucks taken from factories. On Dozsa Gyorgy Street an immense bronze statue of Stalin was hauled down with the help of metalworkers using blowtorches. Red stars and other Communist emblems were shot off buildings, and Russian bookshops were looted.

By the next day Nagy had been installed as prime minister, but the Russians were still in control. Russian tanks reinforced the AVH, which had taken up positions around the city.

The Hungarian Army soon joined the street demonstrators. Most important, a heavily decorated war hero, Colonel Pal Maleter, ordered to lead a formation of five tanks against the insurgents, made a fateful decision. "Once I arrived there," he later said, "it quickly became clear to me that those who were fighting for their freedom were not bandits, but loyal sons of Hungary. As a result I informed the Minister of Defense that I was going over to the insurgents."

On the 25th a huge group of demonstrators advanced on Parliament Square demanding Gero's resignation. They were unarmed. Russian tanks and AVH men opened fire on them, killing an estimated 600 unarmed civilians. It was a ghastly massacre, and the Russians replaced Gero as first secretary of the party with Janos Kadar.

The fighting in the streets increased. After the Parliament Square massacre, AVH men were hunted down and strung up on trees. Sometimes they were found to be carrying their pay—10 times that of a worker—and the money was then pinned to their bodies.

A revolutionary cabinet was now formed, with Nagy at its head. By Sunday, October 28, a cease-fire was negotiated, and the Russians appeared to be allowing Nagy and his followers to assume control. It was, to the jubilant insurgents, Poland all over again.

On Monday, according to the United Press, "Soviet tanks crunched out of this war-battered capital [Budapest] . . . carrying their dead with them. They left a wrecked city where the stench of death already [rose] from the smoking ruins." An announcement was made that the AVH would be abolished. Nagy set about tying the various strands of the revolution together. Hungary was free, despite the fighting in the streets.

Meanwhile, in the outside world, Great Britain and France were making plans to take the Suez Canal by force. On the morning of October 31, the news reached Hungary. Nagy fell into a mild depression. "God damn them!" one of the ministers exploded. "Aren't we going to put out feelers to the Western powers *even now?*"

"Certainly not *now,*" Nagy replied.

The Russian withdrawal from Hungary seemed to be inexplicably stalled. Reports from the countryside told of tanks stopping and soldiers grouping. Communications circulated that trainloads of soldiers estimated at more than 75,000 men accompanied by 2,500 tanks were moving across the frontiers from Russia, Romania and Czechoslovakia. The new, free government of Hungary sent telegrams to the Kremlin questioning this apparent violation of the October 28 agreement of conditions for a cease-fire. By Saturday, November 3, a Russian military delegation arrived at Parliament to negotiate the withdrawal of Soviet troops.

Joseph Cardinal Mindszenty, who had been arrested and convicted of conspiracy in December 1945, was freed, and he gave a radio address. At 8 o'clock that evening, General Maleter drove to Tokol, outside Budapest, to renew negotiations at the Russian military headquarters.

He would never return. Shortly after midnight, General Serov, chief of the Soviet Security Police, would arrest the entire Hungarian delegation.

At 5 A.M. on November 4, 1956, Imre Nagy went on Hungarian radio. He was broadcasting from the Parliament building, where he had spent the night. His words contained a heartrending urgency:

"Attention! Attention! Attention! Attention!

". . . This is Imre Nagy speaking, the president of the Council of Ministers of the Hungarian People's Republic. Today at daybreak Soviet forces started an attack against our capital, obviously with the intention to overthrow the legal Hungarian democratic government.

"Our troops are fighting.

"The government is in its place.

"I notify the people of our country and the entire world of this fact."

Free Radio Kossuth would continue to broadcast bulletins to the world throughout the day. The reports would increase in intensity and despair. Gyula Hay, the playwright and friend of Nagy, broadcast the most impassioned one:

"This is the Hungarian Writers' Association speaking to all writers, scientists, writers' associations, academies and scientific organizations of the world. We appeal for help to all intellectuals in all countries. Our time is limited. You know the facts. There is no need to review them. Help Hungary! Help the writers, scientists, workers, peasants and all Hungarian intellectuals. Help! Help! Help!"

Kadar had gone over to the Russians and now announced a breakaway government. Nagy took refuge in the Yugoslavian embassy with his other cabinet ministers.

Heavy artillery opened up on the city. Soviet tanks rolled into Budapest and rumbled through its streets. When sniper fire came from a building, they blasted the entire building to oblivion. The Hungarian News Agency painted the picture: "People are jumping up at the tanks, throwing hand-grenades inside and then slamming the driver's windows. The Hungarian people are not afraid of death. It is only a pity that we can't stand for long."

The fighting roared on for three days and nights, sputtered and then died. The Soviet tanks had completely retaken Hungary, and all the cries for help had gone unanswered. At exactly the moment that Soviet tanks entered Budapest, British and French paratroopers were dropped at Port Said, Egypt, and America and the United Nations were busy bringing about a cease-fire there.

In a later interview on television, President Dwight Eisenhower said, "The thing started in such a way, you know, that everybody was a little bit fooled, I think, and when suddenly the Soviets came in strength with their tank divisions, and it was a *fait accompli,* it was a great tragedy and disaster."

It certainly was for the Hungarians. The government of Janos Kadar took over and negotiated amnesty for Nagy and his associates. They were loaded into a bus that was to take them to their homes. But before it could leave the Yugoslavian embassy, the bus was boarded by Soviet military personnel, who commandeered it and took it to the headquarters of the Soviet Military Command. The two Yugoslavian diplomats who were to accompany the former Hungarian officials to safety were ordered to return to the embassy, and Nagy and his associates were arrested and imprisoned outside the country at Sinaia in Romania.

In June 1958 those who had not died in captivity were executed: Nagy, General Pal Maleter, the journalist Miklos Gimes and Nagy's secretary, Jozsef Szilagyi. In total, 2,000 were executed and 20,000 were imprisoned after the uprising. An estimated 25,000 died in the street fighting.

The border with Austria remained open for a few weeks after the revolution, and 200,000 refugees streamed across it. Housed in camps and shelters in Austria, they moved on to whatever countries in the West would accept them. Among the refugees were some of the finest minds in Hungary.

Shortly after this, barbed-wire enclosures went up, and the Iron Curtain was firmly redrawn around Hungary. It would be 35 years before it would be torn down again, this time through diplomacy.

INDIA
AMRITSAR
April 13, 1919

Several thousand Indians, gathering in defiance of a British ban on public meetings, were fired upon in the Sikh Holy Shrine of Jalianwala Bagh on April 13, 1919, by soldiers under the direction of Brigadier General R. E. H. Dyer. Three hundred seventy nine Indians died in the barrage; 1,200 were wounded. None was armed.

Amritsar, a city in the northwestern Indian state of Punjab, was for centuries a center of the Sikh religion, a healing spiritual philosophy designed to reconcile Hindu and Muslim concepts. At the center of the city, in the middle of a placid lake, the Golden Temple still stands, the holiest place in the world for Sikhs.

Thus it is particularly poignant and ironic that one of the two most devisive and bloody events in modern Indian history (the other was the Indian Mutiny of 1857) should take place in Amritsar. Nothing would be quite the same after the Amritsar Massacre of 1919. Great Britain, for all that it had accomplished and all it had promised, would never be counted trustworthy by Indians again, and Mohandas Gandhi's star would rise precipitously as a result of the slaughter.

As early as 1861 the first steps toward Indian independence were taken when Indian councillors were appointed by the British colonial government, and provincial councils with Indian members were formed. In 1885 the Indian National Congress was established, and two factions split it immediately. One, led by Gopal Krishna Gokhale, was in favor of a dominion status for India; the other, led by Bal Gangadhar Tilak, agitated for complete independence.

With the coming of World War I, both factions united behind Britain, but as the war wore on, famine, an influenza epidemic and a feeling that independence was a distant dream began to pervade India, and the two potential ruling parties started to regard Britain as an enemy to be overthrown.

During this period Britain dangled the carrot of eventual self-government with the Montagu declaration in 1917 and the Montagu-Chelmsford report of 1918. But it did little to settle the unrest, and in 1919 the Rowlatt Acts were passed, which provided for the arrest and imprisonment of political agitators without trial.

It was like throwing gasoline on a fire. India drifted between self-determination and authoritarian rule, and in the middle there was rioting, particularly in the Punjab. On April 10, 1919, in Amritsar, two nationalist leaders were arrested and deported. The crowds began to grow in the streets of the city. One group tried to enter the European cantonment but was turned away. They began to riot.

General R. E. H. Dyer, under the orders of Sir Michael O'Dwyer, one of the architects of the Rowlatt bills, restored order, and all public meetings and assemblies were declared illegal.

In defiance, a public meeting of several thousand Indians was held on April 13 in the Jalianwala Bagh, an enclosed park. Hearing of this, General Dyer took 90 Gurkhas and Baluchi soldiers and two armored cars to the Jalianwala Bagh. He blocked the only exit from the park with the armored vehicles and, without a word of warning, began to fire into the densely packed crowd.

One thousand six hundred five rounds were pumped into the screaming, panicked mass of men, women and children. Officially, 379 were killed and 1,200 injured, but the Hunter Report (see below) eventually set the casualty figures at 1,200 killed and 3,600 injured. General Dyer finally withdrew, but he kept his armored cars in place, preventing medical personnel from entering the park to tend to the wounded or survivors from carrying the wounded out.

The brutality continued. More fierce demonstrations erupted. The following day a mob rioting and burning at another spot was bombed and machine-gunned from the air by British forces. On April 15 martial law was declared in Amritsar, and it would remain in place until June 9. During this time Indians were forced to crawl on all fours past a spot on the street where a woman missionary had been attacked. Public floggings were ordered for such offenses as, according to the Hunter Report, "the contravention of the curfew order, failure to salaam to a commissioned officer, for disrespect to a European, for taking a commandeered car without leave, refusal to sell milk, and for similar contraventions."

In October 1919 the British government convened the Hunter commission of inquiry to investigate the massacre. The panel was composed of four British and four Indian members. However, three of the British members were civil servants, and all four of the Indians were moderates.

General Dyer appeared before them, and he was absolutely unrepentant. "If more troops had been at hand," he testified, "the casualties would have been greater in proportion. It was no longer a question of merely dispersing the crowd but one of producing a sufficient moral effect from a military point of view not only on those who were present, but more especially throughout the Punjab."

The panel condemned him and his actions, but in mild terms such as "unfortunate" and "injudicious." Mohandas Gandhi set about putting into practice *Satygraha,* or nonviolent civil disobedience. It became the policy of the Indian National Congress and it in effect ended British control of India. Great Britain would no longer be able to rule by force, despite the fact that it would repeatedly jail

Gandhi and incite those other Indians who did not follow his philosophy of nonviolence to riot.

Ultimately, Gandhi would prevail in most of his goals—except in the uniting of Hindus and Muslims, which the British creation of Pakistan successfully buried—and the British would leave. By that time Dyer would be removed from his command but lauded by much of the British press and people.

Still, even the staunch supporters of Dyer and O'Dwyer, who had once cried, "There is another force greater than Gandhi's soul force!" would have to admit that the Amritsar Massacre was a turning point. From that moment on the struggle for Indian independence would admit no compromise, and the good faith of British concessions would nevermore be believed.

INDIA
BOMBAY
December 6, 1992–January 1993

A month of riots between Hindus and Muslims that began with the destruction of a mosque in Ayodhya on December 6, 1992, and that finally centered in Bombay killed 1,770 people and injured thousands more.

The partitioning of the subcontinent of India in 1947 into the primarily Hindu country of India and the almost totally Muslim country of Pakistan sparked extensive and bloody riots between Hindus and Muslims, and, eventually, the assassination of Mahatma Gandhi, the foremost force in India's fight for freedom from British Colonial rule, and, ironically, a foe of partition. After 1947 Hindus and Muslims began to intermingle more in India than in Pakistan, whose governments have remained steadfastly Muslim. By 1992 about 82 percent of India's 870 million people were Hindu, but more than 100 million Muslims, along with smaller numbers of Sikhs, Christians and Parsis, populated the country.

An uneasy relationship existed between Hindus and Muslims, trembling on the knife edge of hostility. One unfortunate incident had the potential for enormous and tragic consequences.

On December 6, 1992, that incident occurred. Mobs of Hindu extremists in the city of Ayodhya, on the northern plains of India, destroyed a 16th-century mosque and replaced it with a makeshift Hindu temple. As news of the destruction spread, so did religious rioting and clashes with security forces. Throughout the entire country skirmishes between Hindus and Muslims erupted during the entire month of December.

By January 1, 1993, 1,200 people had been killed, 225 of them in Bombay, the capital city of Maharashtra province, an essentially Hindu stronghold. Two days later the fury seemed to have been spent, however, and government forces began to disperse.

But it was only a momentary pause before the bloodiest riots yet. On January 6 warfare suddenly exploded again with a far more brutal edge and intensity. Hindu mobs armed with daggers and swords, rocks, iron rods, homemade bombs and guns began to attack Muslim homes and businesses, burning the homes and businesses and slaughtering their occupants. Muslims fought back, and the battle was joined. By January 10, 1993, scores of fires raged through Bombay as smoke from burning lumber yards, shops and a tire store blackened the sky. Sudharkarrao Naik, the chief minister of Maharashatra, appealed for peace on television and simultaneously petitioned the central government for troops to control the situation.

The need for troops was apparent: Hindu policemen in too many cases, despite shoot-to-kill orders, stood by and watched as Hindu mobs set fire to Muslim businesses and homes and fired only on Muslims, who in turn battled the police with guns, stones and bottles. Meanwhile, firemen were unable to extinguish the multiplicity of fires. Hindu mobs threatened them and in some cases stoned or shot firemen trying to do their job.

At first, the army was given orders only to stage shows of force without firing their weapons, but they were constantly overwhelmed and eventually began to shoot into crowds. Mayhem pervaded Bombay for the first two weeks of January. Terrified, thousands of people crowded into railroad stations, fleeing to home villages to escape the danger. The city was under siege, and by January 11 it became apparent that the violence was well organized by criminal gangs as well as by supporters of different political parties.

"They are all criminals," police commissioner Srikant Bapat told local reporters. "These are incidents of madness. I do not view them from the point of politics. This is not a law-and-order problem. It is caused by religious fundamentalism of every kind." In the first week of rioting, 3,000 arrests were made, and 1,500 of those arrested were jailed as criminals under preventive detention laws.

The new rules of force given to the army seemed to calm some of the rioting, but only temporarily. As police moved in with tear gas and troops followed, mobs scattered but reformed elsewhere, and fighting continued unabated. More firemen were killed by shots from the mobs.

By the end of nine days, the death toll had risen to 558, and the fighting, which in the beginning was largely confined to the slum areas of Bombay, where poor Muslims conducted business and lived, now began to spread to middle- and upper-class areas. Menacing Hindu groups began to visit apartment buildings searching for Muslims. Janitors began to remove the names of all residents from building lobbies. In other apartment complexes apartment owners paid thousands of dollars in protection money to keep mobs away.

"Fundamentalists on both sides are equally culpable, and Bombay is being held ransom by them," former judge Bhakdawar Lentin announced on January 14 as he and others demanded the resignation of the government of Maharashtra.

The stench of uncleared garbage filled many roads and lanes now, and the charred hulks of scores of taxis torched by mobs continued to line the roads. By January 15 more than 40,000 people had fled their homes and were living in refugee camps, and more than 50,000 people had crowded through Bombay's Victorian railroad station bent upon fleeing the city. As the fighting seemed to subside, more and more bodies were found in lanes, streams and even sewage pipes. The streets of Bombay were deserted, but even as more than 200 men and women from both faiths carrying white flags walked for three hours through some of the worst-hit areas singing songs of unity, political leaders threw gasoline on the flames.

Bal Thackeray, chief of the Shiv Sena, a Hindu militant group suspected of the destruction of the mosque and of leading the riots in Bombay, issued a statement through the Indian press. "Muslims have been constantly provoking Hindus," he said, "hence, the current riots. At no place were Hindus aggressors. They only acted in self-defense. Hindus cannot be held responsible for the current riots."

His words were met with demands by Muslims for his arrest, but the Indian Congress Party demurred. In a statement from a member who refused to be identified, their position was made clear: "[The arrest] cannot be contemplated now," he said. "The situation is still very delicately poised and could get out of hand if such a move takes place."

Still, the terror continued. "It is a pogrom of Muslims, it is a nightmare," said Alyque Padamsee, a Muslim who ran an advertising agency and was also a playwright and actor.

Finally, Shiv Sena spokesman Pramod Navaikar admitted that Shiv Sena members were heavily involved in the rioting. "Our boys were involved," he said, "but for every 5 Shiv Sainiks on the street, there were also 20 antisocial elements involved."

Sakhruddin Khorakiwala, a prominent Muslim industrialist who held the largely ceremonial post of sheriff of Bombay, said that in other parts of the world the attacks would have been described as "ethnic cleansing." And other Muslims noted that it would be difficult to control younger Muslims who wanted to retaliate. It became apparent as the rioting quieted that some slum landlords had organized violence in some areas to clean out squatters and unwanted tenants.

A sense of despair descended over Bombay. Some Muslim leaders warned that Bombay could be turned into a Beirut, with endless clashes between religious groups. And that thought seemed to sober both sides into stopping the violence.

"If the country forsakes the path of secularism, it will break," Prime Minister P. V. Narashimba Rao said after surveying the damage in the once bustling city. A day later the Indian cabinet resigned to give Prime Minister Rao a free hand in reshuffling his government.

When the carnage finally calmed 1,770 people had been killed and thousands more injured in one of the worst religious riots in modern history. Nani Palkiwala, one of India's best known constitutional experts, summed it up disturbingly: "Neither Hindus or Muslims, but Indians, many of whom have nowhere else to go, have suffered most," he said. "It's a question of who and which community will be next."

INDIA
MANDAI, TRIPURA; ASSAM
June 7–8, 1980;
February 18, 1983

Driven by Hindu-Muslim hatred, tribal youth organizations conducted raids on Indian villages, massacring Muslims. In the two worst incidents, on June 7–8, 1980, and February 18, 1983, 5,500 Bengali immigrants were massacred; 1,000 were wounded; 300,000 were made homeless.

The most savage and disastrous confrontations between people have been ethnic and religious ones. Consider the Holocaust of World War II (see p. 91), the Armenian massacres (see p. 83) or, closer to home, the Chicago race riots (see p. 109) and the Watts riots (see p. 104). When a cause is perceived to be a religious one, cruelty can apparently be justified and swallowed up in the cause. "Men never do evil so fully and so happily as when they do it for conscience's sake," said Pascal, and this has never seemed so true as in the confrontations between those driven by differing beliefs.

When Great Britain partitioned India in 1946, isolating its Muslim population in Pakistan and its Hindu population in India, it sowed the seeds of religious confrontation. Hindus in Pakistan were driven out by Muslims and settled in the northeastern provinces of West Bengal and Bangladesh, which was, until 1971, East Bengal. The Bengalis, concentrated mostly in the adjoining states of Tripura and Assam, were predominantly Hindus. The native tribal people who lived there were mostly Muslims.

Thus, the Bengalis were looked upon as foreigners, immigrants and Hindus, a combination the tribespeople regarded with bigoted distaste. In 1980 the situation erupted when tribal youth organizations, dedicated to the expulsion of "foreigners," began systematic massacres.

The first such massacre to reach the attention of the world occurred in Mandai, in Tripura state. Adjoining the border of Bangladesh, the village had long been a Bengali dwelling place; some of its inhabitants had lived there a generation, ever since the partition. Others were new immigrants.

On June 7 and 8 tribal gangs swooped down on the village and laid waste to it. Armed with guns, spears, swords, scythes and bows and arrows, they emptied houses, chased men, women and children out into the flatlands around the village and slaughtered them, crushing their heads and severing their limbs. Children were run through with spears. Three hundred fifty people were killed in this massacre, which left not one person alive in the village of Mandai. As news of the massacre began to emerge from India, officials in Tripura admitted that nearly 700 people had died in the past year after similar raids, which had rendered 200,000 homeless and had necessitated the setting up of 100 camps to house them.

The Communist-led government of Tripura, which had displaced Prime Minister Indira Gandhi's Congress Party in the 1977 elections, was charged with incompetence and blamed for not controlling the situation. The Tripura government countercharged the New Delhi government with indifference.

Two weeks later, on June 22, Tarun Basu, a reporter for the New Delhi weekly current affairs journal *Contour,* uncovered a staggering fact: Four thousand Bengalis had been slaughtered by tribesmen during the month of June, and the government of Tripura had burned many of the bodies to cover up the slayings. This resulted in an immediate tightening of security for the Bengalis that lasted for three years. Then, an important state election was scheduled to elect a 126-member state legislature and 12 members of the national parliament from the neighboring state of Assam. In this case, the settlements were Muslim and the tribespeople Hindus. This time, an added political dimension, a motive to prevent Muslims from electing Muslims to the legislature and parliament, was added.

The village of Bhagduba Habi in the center of the state was typical. There, Bengali-speaking Muslims who went to vote found themselves surrounded by hostile Assamese tribespeople who turned them back from the polling places. Those who did manage to vote were found murdered the next day.

And then the worst massacres occurred. On Friday, February 18, the violence was particularly vicious. In Bhagduba Habi, old people, women and children were chased into the outlying fields, tortured and killed. Reporters arriving the next day counted 157 bodies, mostly those of children, lined up in rice fields being readied for mass burial.

On that same day, 17 Muslim villages were attacked in a 20-square-mile area about 50 miles northeast of Gauhati, the state capital. Out of a population of 12,000, 1,200 were slaughtered. Retaliatory attacks against Hindu villages swelled the figure to 1,500. Police, trying to restore order, killed 127. Eight hundred wounded were treated in hastily set up camps.

Once again, the government of India stepped in. In Gauhati, the capital of Assam, the high court released four prominent leaders of the anti-immigrant movement, hoping to defuse the situation.

With thousands of voters boycotting the voting and thousands more prevented from voting, the Congress Party of Prime Minister Indira Gandhi swept to a landslide victory in the state legislature. Troops would maintain peace, but Indira Gandhi's term in office would be short. One year later she would be assassinated.

ROME
71 B.C.

The gathering of freed slaves and gladiators attempting to sever themselves from Roman rule and escape to the Alps was crushed by the Roman generals Crassus and Pompey in 71 B.C. Six thousand slaves and gladiators were crucified.

In Ancient Rome slaves were usually enemy soldiers who had been defeated and captured in battle. One of these was Spartacus, a Thracian captured when Rome defeated and annexed Thrace during the first century B.C.

Taken to Capua, Spartacus was installed in the gladiator school there. But in 73 B.C., leading 78 men armed with kitchen knives, he escaped from Capua and established an army of runaway slaves that would eventually number 100,000.

Their first attacks were ragtag ones made with makeshift swords. Compensating for a lack of organization and skill, they excelled in fierce resolve. As time went on they defeated more and more Roman legions, appropriating their weaponry and improving their organization. Slave prisons were invaded, and their inmates joined the swelling ranks.

The Roman Senate first sent small armies headed by praetors into the field. They were roundly defeated. The Senate then dispatched consular armies, and they too were beaten. It was time to unite several armies under one consul. Pompey was abroad fighting in Spain. The second choice was Marcus Crassus, better known for his real estate astuteness than his fighting skills. But Crassus was ambitious, and he brought a strong organizational hand to his resolve to crush the slave revolt. This was a distinct advantage over Spartacus's command. His army was effective but unruly.

It had been Spartacus's aim to fight his way north from Capua and then leave Italy and strike out for the Alps. He wanted nothing more than to return to Thrace. His army laid waste to southern Italy and Campania and developed a taste for plundering. They were in no hurry to leave.

It would be a fatal error. The Roman army under Crassus was given time to organize and arm itself and to force the slave army into a formal battle, a situation both Crassus and Spartacus knew the slave army could not win.

The battle took place near Rhegium, in the toe of Italy, where Crassus trapped the slave army. It was a rout. A heavy snowstorm provided cover for a third of the slaves, who managed to flee from this battle. The remaining two-thirds broke into two fleeing armies, and Spartacus met Crassus on the field near Lucania. He died in a hand-to-hand battle with the Roman consul.

Meanwhile, Pompey had returned from Spain and caught the rest of the escaping slave army. His troops annihilated them, and those they did not kill in battle they

The crushing of the uprising of the slaves and gladiators and the death of their leader, Spartacus, are depicted in a 19th-century drawing. New York Public Library (H. Vogel)

crucified. Six thousand slaves on crosses lined the highway from Capua to Rome.

Ironically, the armies of Pompey and Crassus found 3,000 Roman prisoners in the abandoned camp of the rebels. They were unharmed.

RUSSIA
ST. PETERSBURG
December 14, 1825

The Decembrists, a secret society of army officers who had served in Europe and were influenced by Western liberal ideals, revolted against Czar Nicholas I on December 14, 1825. The coup failed, and several hundred officers were killed.

There is a certain mystical fascination about the reign and life of Czar Alexander I. It was he who defeated Napoleon in 1812. It was he who advocated a benign, liberal treaty with France afterward. It was he who then, after 1812, began to subscribe to a sort of general Christian ideal but, in contradiction to its teachings, began to suppress any liberal movements in Russia, calling them "threats to Christian morality." He supported Metternich in crushing all national movements and, under the influence of Juliana Krudenar, created the Holy Alliance to uphold Christian morality in Europe.

In Russia he established military colonies and paraded them as Christian enclaves. They were actually little serfdoms in which the common soldiers were treated like chattel.

Alexander I died in 1825, or possibly he didn't. Rumors maintain that he actually went to Siberia and became a hermit. In 1926 his grave was opened, and it was found to be empty. The mystery remains unsolved to this day.

When Alexander disappeared in 1825, the throne passed unexpectedly to his brother Nicholas I. Nicholas inherited all the problems set in place by the repressive measures of Alexander. One of these was the formation of secret societies challenging Alexander and his fervent repressions.

One of the secret societies was called the Decembrists, and they were a unique group. Composed of army officers and aristocrats who had fought Napoleon and had thus spent time in the rest of Europe, they were consumed by new, liberal ideas of existence and government. They advocated the establishment of representative democracy but disagreed about the form it should take in Russia. Some supported a constitutional monarchy; others wanted a democratic republic. They were not the ideal group to stage a rebellion, but the disappearance of Alexander I and the ensuing confusion offered them an opportunity they chose not to refuse.

It seemed that the assumed death of Alexander, who had remained childless, would necessarily result in the assumption of the throne by the next in line, Constantine. But, unbeknownst to all but a very few, he had renounced the throne in 1822. The confusion led the Decembrists to think that they could challenge the unpopular younger brother, Nicholas I, overwhelm him and demand that Constantine grant a constitution.

Badly organized but determined, they advanced upon Senate Square in St. Petersburg on December 14, 1825, the first day of Nicholas's reign. Fully armed and riding horses, the Decembrists formed themselves into a fighting force.

Nicholas attempted to negotiate. They refused. Artillery opened up on them, and the czar's cavalry charged. Improperly shod, the Decembrists' horses slipped and fell on the icy pavement of Senate Square. The artillery cut down more of them.

Within a short time the revolt was crushed. Five of the leaders were later executed; hundreds more were killed in Senate Square.

It had been an unsuccessful revolt, but its effects would be felt for years. There was an immediate police repression ordered by Nicholas I and inherited by his heirs. This inspired considerable revolutionary fever and activity. For a small effort, the Decembrist revolt of December 14, 1825, accomplished much and precipitated more.

SAUDI ARABIA
MECCA
July 2, 1990

Approximately 1,426 Muslim pilgrims were trampled to death in a stampede on July 2, 1990, in a tunnel connecting Mecca and Mina, Saudi Arabia, during the Muslim celebration of hajj.

Every year between 1 and 2 million Muslims travel to the holy city of Mecca, in Saudi Arabia, to celebrate hajj, or the holiday Eid Al-Adha (the Feast of Sacrifice). The rituals of the holiday commemorate the prophet Abraham's offering of his son in sacrifice to God, and sheep are slaughtered to emulate Abraham's oblation. The Islamic faith requires that each worshipper try to attend the pilgrimage, and in the early summer of 1990, more than two million worshippers crowded into Mecca and the nearby tent city of Mina.

In previous years terrorism, attributed to Iran or to radical Shiite Muslims, had marred the deeply religious celebration. In 1986 Saudis confiscated large quantities of explosives from Iranian worshipers. In 1987 Iranians engaged in pitched battles with security forces, and 402 Iranian pilgrims died in the conflict. Following that tragedy, the Saudis set a quota of 1,000 pilgrims per million of population for each Islamic nation, and the Iranians angrily boycotted the pilgrimage.

By June and July of 1990, terrorism, except for an occasional bomb planted or thrown by Shiite Muslims, had dissipated, but the crowds had not. The concluding days of the 1990 hajj were blisteringly hot. The temperature climbed to 112° F (44° C) on the morning of July 2, the day that pilgrims would trek to Jamarat al Akaba in Mina, one of the three "Satan's stoning points." Each pilgrim would cast pebbles at a rock pillar in a ritual that symbolized the faithful's struggle against evil.

At 10:00 A.M. thousands of pilgrims in traditional attire began the short journey from Mecca to Mina. They had a choice: either cross a pedestrian bridge or enter an air-conditioned tunnel 1,500 feet long and 60 feet wide cut through a mountain. Most chose the tunnel, but some trudged across the bridge.

Shortly after 10 o'clock a railing gave way on the bridge just above the exit of the tunnel in Mina. Seven worshipers leaning on the railing plunged 26 feet to their deaths among the slow-moving crowd of pilgrims exiting from the tunnel.

Foot traffic came to an immediate standstill around the seven bodies. But at the other end of the tunnel and in it, unaware pilgrims still pressed forward. Within a few minutes 50,000 of them entered a space that was built for a maximum of 10,000.

Then a power failure abruptly plunged the tunnel into total blackness, and the air conditioning stopped. In the sudden, eerie stillness the air seemed to drain away. Panic set in, and the crowd pressed forward mindlessly. The wave of humanity increased its speed and force, and individuals in it began to faint from the heat or lose their footing. More and more were trampled as thousands slammed against one another in a mindless stampede. Hundreds were crushed against the walls or one another. "It was terrible," a survivor later told Saudi television. "When one stumbled, scores trampled him and hundreds fell on top of them." Before the screaming, writhing mass of terrified and dying pilgrims could be calmed and led out of the tunnel, an astonishing 1,426 Muslims were trampled to death or died of suffocation.

To Western ears, the statement by Saudi King Fahd that "It was God's will, which is above everything. It was fate," sounded peremptorily heartless and needlessly cruel. He amended it for world consumption shortly thereafter, adding that he would work for better safety measures so that "God willing, we will see no tragedies in the coming years."

According to Islamic teachings, to die while on the hajj ensures immediate ascension to heaven. That fetid and horrible morning, 1,426 Muslims made the ascent.

UNITED STATES
CALIFORNIA
LOS ANGELES
August 11–29, 1965

Festering racial resentments, poverty and sultry summer weather converged to cause the racial riots of August 11–29, 1965, in Watts. Thirty-four died; 874 were injured; 3,800 were arrested; and there was $20 million in property damage.

There can be no doubt that the 1960s, remembered in retrospect as the age of the flower children, was also a decade of extreme violence—Vietnam; the assassinations of President John F. Kennedy, Robert Kennedy and Dr. Martin Luther King Jr.; and civil unrest (see also the police riot at the Democratic National Convention, p. 110, and Kent State killings, p. 116).

The summer of 1965 was a hot and sultry one in Los Angeles, and in the Watts neighborhood, an outwardly neat and well-kept suburban section of Los Angeles, it was seemingly serene. Roughly 20 square miles in area, Watts held about a sixth of Los Angeles County's 523,000 blacks.

Shortly before 8 o'clock on the night of August 11, 1965, at the corner of Imperial and Avalon streets, a white California Highway Patrol officer stopped Marquette Frye, who, with his brother Ronald, was driving erratically. Some 25 people watched while their mother, Mrs. Rena Frye, entered the scene and began to berate her son, who in turn began to berate the police.

The crowd grew and became involved. Stones were thrown. The police radioed for help. By 10 P.M. crowds were stoning city buses, and 80 police officers sealed off the 16-block area in an effort to contain the violence.

It was fruitless. Looters had already moved beyond the sealed-off area, and the Watts riots of 1965 had begun.

By the next night black youths had acquired firearms and were firing on police from the tops of buildings. Anarchy ruled the streets. Fires were started and fire fighters fired upon. White television crews were mauled and their equipment destroyed.

As the fever mounted over the next two days, roving bands of black teenagers assaulted cars containing whites. By August 13 four people had died, 33 police officers had been injured and 249 rioters had been arrested. And the fierceness and tempo of the riot were increasing.

No whites were safe anywhere near the Watts section of Los Angeles. Whenever a car containing whites entered the area, gangs of teenagers, egged on by shouts of "Kill! Kill!" descended upon it. Black police officers who tried to contain the crowds were jeered, called traitors and then stoned.

Robert Richardson, a black advertising salesman for the *Los Angeles Times* who had entered the area, wrote, "Light skinned Negroes such as myself were targets of rocks and bottles until someone standing nearby would shout, 'He's blood. He's a brother—lay off.'

"As some areas were blockaded during the night, the mobs would move outside, looking for more cars with whites. When there were no whites, they started throwing rocks and bottles at Negro cars. Then, near midnight, they began looting stores owned by whites.

"Everybody got in the looting—children, grownups, old men and women, breaking windows and going into stores.

"Then everybody started drinking—even little kids 8 and 9 years old."

And it was then that teenagers started to fan out into white neighborhoods up to 20 miles from the riot scene. One group of 25 blacks tossed rocks in San Pedro, in the harbor area; another appeared in Pecoima, a black community in the San Fernando Valley.

Los Angeles Police Chief William Parker did nothing to try to bring peace. His public statements, comparing the rock throwers to "monkeys in a zoo," only fanned the flames.

On the morning of August 13, 2,000 heavily armed National Guard troops converged on Los Angeles. Moving in with machine guns, they fired at rioters, who fled and then regrouped elsewhere. At one point a group of rioters charged Oak Park Hospital, where those who were injured and wounded in the riots were being treated.

The next day 20,000 National Guardsman were called up and began to penetrate the riot area. A curfew from 8 P.M. to sunrise was imposed on a 35-square-mile area surrounding the riot scene. Snipers shot from rooftops. Fires began to break out with increasing frequency. There was hardly an unlooted store in Watts or the surrounding area. Twenty-one people had been killed. Nineteen were rioters, one was a sheriff's deputy and one was a fireman. Six hundred had been injured.

Chief Parker appeared on television to assail black leaders in the community, calling them "demogogic . . . pseudo leaders of the Negro community who can't lead at all." There was some evidence that Black Muslims were encouraging the riot and egging teenagers on, but by and large black civil rights leaders from all over the country issued pleas for the violence to end.

There were not enough fire fighters and equipment to stop the continuing string of fires that blazed through the area. But 2,500 of the 15,000 National Guard troops in Los Angeles began to secure the riot area, and by late on August 14 the rioting had begun to subside.

The death toll rose that day from 22 to 31. One victim was a 14-year-old girl killed in a traffic accident while fleeing the scene of a looting; one was a five-year-old child shot by a sniper. Bricks, rocks and bullets continued to rain down from rooftops, some striking guardsmen and police, some striking black residents. The guardsmen used rifle fire, tear gas, machine guns and bayonets, and Governor Edmund G. Brown widened the curfew area to 50 square miles.

Meanwhile, violence broke out in Long Beach, south of Los Angeles. One policeman was killed and another wounded when they were ambushed by snipers. Troops were ordered into that city.

Governor Brown came to Watts surrounded by guardsmen and met with black leaders, trying to calm the atmosphere and effect a reconciliation. Gradually, an embittered calm descended over Los Angeles, and the gasoline bombs, rifles and rocks began to disappear.

A score of relief agencies entered the battle-scarred, smoldering area to begin rehabilitation after five days of riots. Slowly the troops were withdrawn, but the curfew remained in place. Racial tension throughout the city remained for a long time.

Two hundred businesses were totally destroyed; 500 were damaged; $200 million in property damage was estimated. Nearly $2 million in federal funds was allocated to aid in the rebuilding of a 45-square-mile area of Los Angeles. One thousand six hundred people were hired under the antipoverty program of Los Angeles County to aid in the cleanup.

In the last hours of the riots, a black woman was killed by a Guardsman, bringing the death toll to 34.

It would be a long path back for Watts.

UNITED STATES
CALIFORNIA
LOS ANGELES
March 2, 1992

On March 2, 1992, a wave of riots swept over Los Angeles, California, following the acquittal of four LAPD officers who were caught on tape beating Rodney King after stopping him for a speeding violation on March 3, 1991. Fifty-three died, thousands were injured, more than 7,000 were arrested and more than $1 billion in property damage occurred during the riots.

At 12:30 A.M. on March 3, 1991, a Hyundai belonging to Rodney King was spotted speeding on the 210 Freeway in Los Angeles, California, by California Highway Patrol officers Tim and Melanie Singer. The two radioed for backup and chased King at speeds of more than 110 mph. Fifteen minutes later, King's vehicle was cut off. Guns drawn, the Singers advanced on King while four Los Angeles Police Department officers—Sgt. Stacey Koon, patrolmen Laurence Powell, Theodore Briseno and Timothy Wind—grabbed King and threw him to the ground.

King fought an increasing circle of officers, but, as an 81-second amateur videotape shot by civilian George Holliday from his nearby apartment later showed, some of them kicked him, and three of them struck him more than 50 times with metal batons before finally handcuffing him. Sgt. Koon, the highest ranking officer, apparently did nothing to stop the beating. An ambulance finally took King to a hospital.

The following day Holliday gave his videotape to Los Angeles television station KTLA, which took the tape to LAPD headquarters, where senior officers viewed it. That night, KTLA broadcast the tape; CNN acquired a copy and broadcast it nationally. A day later all the major networks showed the startling scene of obvious police brutality on their evening news broadcasts. National outrage followed, and on March 7 Los Angeles Police Chief Daryl Gates announced that the officers involved in the beating would be prosecuted. Fifteen officers present at the scene were suspended.

A week later a grand jury returned indictments against Stacey Koon, Laurence Powell, Timothy Wind and Theodore Briseno, while Los Angeles Mayor Tom Bradley announced that a commission headed by Warren Christopher would evaluate the performance of the LAPD and investigate charges of racism on the force. The next day the mayor asked for the resignation of Police Chief Gates, but the chief refused to resign.

Believing that emotions were running too high in Los Angeles for the officers to receive a fair trial, their defense attorneys moved for a change of venue from Los Angeles County. The motion was denied, then reversed on appeal, and the trial was moved to the predominantly white and conservative Simi Valley, California.

A jury of 10 whites, one Hispanic and one Filipino-American was chosen, and the trial began. The defense focused on the arrest record of Rodney King and stated that the policemen believed he was acting under the influence of the animal tranquilizer PCP, which often causes violent and unpredictable behavior. His alcohol level that night was also presented. It was .19, more than double that allowed in most states before drunken driving charges are brought.

At 3:15 P.M. on March 2, 1992, the jury, after seven days of deliberations, acquitted Officers Koon, Wind and Briseno of all charges and announced that they were unable to reach a verdict on one charge against Officer Powell. The judge declared a mistrial in his case.

The news spread rapidly. Civil rights activists immediately denounced the verdict: For example, ACLU executive director Romona Ripston called the verdicts "a travesty of justice," and California State Senator Ed Smith told the United Press, ". . . the world saw the videotape and if that conduct is sanctioned by law in California, then we have to re-write the law." Los Angeles Mayor Tom Bradley, speaking on television, said, "We must express our profound anger and outrage [at the acquittal] but we also must not endanger the reforms that we have made by striking out blindly. We must demand that the LAPD fire the officers who beat Rodney King and take them off the streets once and for all."

It was like throwing gasoline on glowing coals. By late afternoon southeast and South Central Los Angeles had erupted in riots. Stores were set afire and looted; cars were set aflame; roving gangs of armed black youths roamed through predominantly black neighborhoods kicking in store windows and looting the contents of the stores. Senseless violence begat senseless violence. Gunshots rang out; snipers fired indiscriminately from rooftops; firemen were shot as they attempted to put out fires.

The rioting spread to Inglewood, then downtown Los Angeles, Pasadena, Hollywood and Koreatown. The entire city seemed to be engulfed in mayhem. In three days 3,600 fires were set. At one point the Los Angeles Fire Department received calls for three fires every minute. Eventually, the fires exceeded the fire department's ability to deal with them, and some burned unchecked until other departments, summoned from outside the city, arrived.

Most of Los Angeles had been turned into a war zone. Banks and businesses closed and were boarded up. Those that could withstand sledgehammers and fire stood. Others were transformed into blackened skeletons.

Roaming television news helicopters, a trademark of Los Angeles, caught scenes of terrible brutality. White motorists were pulled from cars and trucks, flung to the ground, kicked and beaten. One of these incidents, that of Reginald Oliver Denny, who made the mistake of stopping to address looters in the street, was caught and broadcast live by one of the news helicopters. Dragged from his truck and mercilessly set upon by a mob, he was left for dead. Paramedics arrived and took him to Daniel Freeman Hospital, where he underwent brain surgery to remove a blood clot.

His beating was only one of many as the rioting grew in intensity and scope. No longer blacks against whites, it became an outlet for white and Latino street gangs and anarchists. The looting became a game indulged in by entire families, who dragged every manner of luxury item from devastated stores.

Police called for reinforcements which arrived quickly. Thousands of extra police converged on Los Angeles to do battle with guerilla bands that formed and reformed, some engaging in gun battles with the police. Wholesale arrests were made, but the riot raged on, night and day.

On April 1 4,000 National Guardsmen, 4,000 regular Army and Marine troops with orders to "return fire if fired upon" and 1,000 federal law enforcement officers moved into the city, augmenting the local and state police force, which now numbered 4,000. By the following weekend, much of the violence had abated as a dawn to dusk curfew was put into effect by Mayor Tom Bradley.

But by this time, the fires of the Los Angeles riot had spread to other parts of the country. In San Francisco 1,400 people were arrested in rioting that engulfed the city's downtown area before a state of emergency and a curfew were established. In Las Vegas a mob of 200 went on a rampage, setting fires and engaging in snipering and drive-by shootings. The Nevada National Guard was called out to quell the riot.

Seattle, Washington, was struck by smaller mobs, who broke windows and looted numerous cars in the downtown area. In New York City blacks reportedly pulled two white men from a truck, stabbed one and beat the other. Two hundred protesters attacked the doors at Madison Square Garden but only caused property damage. Another crowd of 400 black youths roamed the city, smashing windows and stealing merchandise. Eighty arrests were made and several police officers were injured as businesses released workers early.

In Atlanta, Georgia, more than 300 rioters throwing rocks and bricks were met with tear gas from Georgia National Guard troops. And sporadic acts of violence, arson and property damage erupted in Tampa, Florida; Pittsburgh, Pennsylvania; Birmingham, Alabama; and Omaha, Nebraska.

When, in the middle of April, the riots quieted, then stopped, the death toll stood at 53. Thousands had been injured. More than 7,000 people had been arrested, and more than $1 billion in property damage had been sustained—considerably more than in the Watts riots (see p. 104).

A strained peace settled over Los Angeles as President George H. W. Bush announced that he had ordered the Department of Justice to investigate the possibility of filing charges against the LAPD officers for violating the federal civil rights of Rodney King. A federal grand jury returned indictments against the four officers, and on April 16, 1993, a federal jury convicted Koon and Powell on one charge of violating King's civil rights. Wind and Briseno were found not guilty. Civil rights leaders around the country appealed for calm, and their appeals were heard. No demonstrations followed the verdicts.

Rodney King filed two civil suits, one against the city of Los Angeles, one against the four officers. He was awarded $3.8 million in damages in the first suit and nothing in the second.

It was a sobering lesson from which some police departments—notably, those in New York City, Cincinnati and the state of New Jersey—apparently learned little. On the other hand, the success of Rodney King's lawsuits established one way, at least, to achieve retribution for perceived police brutality.

UNITED STATES
COLORADO
SAND CREEK
November 29, 1864

Colonel John M. Chivington's disdain for both Indians and treaties manifested itself in the massacre on November 29, 1864, of peaceful Cheyenne and Arapaho Indians. One hundred forty-eight Indians, mostly women and children, were killed, among them nine Indian chiefs; hundreds of Indians were wounded or mutilated; nine soldiers were killed; 38 soldiers were wounded.

There were many Indian raids along the Platte Road in Colorado in the 1860s. It was the main route of stagecoaches, and there were Cheyennes among the Sioux who were chiefly involved in the raids.

But the encampment of Cheyenne and Arapaho in Sand Creek, Colorado, southeast of Denver and 40 miles northeast of Fort Lyon, was a peaceful one. In fact, a verbal agreement was reached between them and territory governor John Evans. They would be under the protection of Fort Lyon, and a formal treaty signing was to take place in a very short time. Chief Black Kettle flew a huge American flag before his adobe hut at Sand Creek in celebration and expectation.

The camp consisted of 100 lodges of Cheyenne, under Chief Black Kettle, and 10 lodges of Arapaho, under Chief Left Hand. They were certain that any day word would arrive from Kansas that a treaty had been concluded.

However, anything but peace was on the mind of Colonel John M. Chivington. Colonel Chivington commanded a company of Colorado volunteers made up of the Third Colorado Cavalry, each of them "100 days men." They were not trained soldiers, but recruits made up of mining toughs, gamblers, "bull-wackers" and general frontier flotsam. There was no discipline; the men chose their officers by vote and then ignored them. They did not wear uniforms, and the one common thread that united them was a desire to kill Indians.

As early as April 1864, Colonel Chivington and his men had attacked Cheyenne indiscriminately. They had been ordered to go to Kansas and fight the Confederates. Colonel Chivington apparently did not want to do this. An attack on Indians would, he felt, make his company indispensable Indian fighters, too good to send East.

It was no problem for him. He hated Indians with an abiding passion. His standing order was to take no prisoners. Under his command, various companies made repeated raids on Cheyenne villages, killing the inhabitants. In one incident in May, 600 Cheyenne warriors met 100 soldiers led by Lieutenant George S. Eayre. The soldiers would have been killed to a man if it had not been for the intervention of Chief Black Kettle, who ordered his warriors not to fire.

But this had no effect whatsoever on Colonel Chivington's plan to exterminate Cheyenne, and so at the beginning of November, while Chief Black Kettle and his village waited patiently for the peace treaty to be signed, Colonel Chivington made his own plans to massacre every Indian in Sand Creek.

The Indians there were lulled into a sense of security first by Major E. W. Wynkoop and then by Major Scott J. Anthony, who treated them deferentially from their command posts at Fort Lyon. In fact, Major Anthony assured them that they could continue to live safely under his protection at Sand Creek until word came about the treaty.

Meanwhile, Colonel Chivington decided to strike one blow at the "Red Rebels" and endear himself to Colorado voters, whom he would be facing soon, after he was mustered out of the service. He began gathering his troops on November 20. On November 24 they had reached Booneville, a little settlement on the Arkansas River above Fort Lyon. On that day he ceased all traffic down the Arkansas River, even the mail, so that no one, not even Major Anthony, would know of his approach.

On the morning of November 28, he appeared at Fort Lyon, to the surprise of everyone. He surrounded the fort and stationed sentries with orders to let no one out. He then went into the fort and informed Major Anthony of his plan to attack the camp at Sand Creek. Major Anthony, who by now had become convinced that the camp was a peaceful one, objected. He knew the kind of men Chivington was commanding, and he knew that attacking the camp would precipitate Indian raids from less peaceful tribes. It was foolish, and he told Chivington so. The colonel was adamant and ordered Anthony and his men to accompany him in the attack.

Shortly after midnight on the 28th, the army, 1,000 strong, set out on horseback for Sand Creek, led unwillingly—but at the threat of the loss of his life—by Robert Bent, the brother of George Bent, a white man who was then living with the Cheyenne at Sand Creek.

At dawn of November 29, 1864, the main body of the troops attacked in a princer movement from both sides of the camp. Others made for the pony herds to the south of the camp. The sleeping Indians rushed from their huts in dazed confusion, which immediately melted into screams from the women and children and fierce war cries from the men, who dashed back into their huts for their weapons.

Chief Black Kettle grabbed the pole of the large American flag that flew outside his hut and called to his people not to be afraid. The soldiers would not hurt them.

His words died in the dawn as the soldiers opened fire. The Indians who escaped being hit by the cavalry's bullets scattered. Some dashed for the sand hills; others scurried down the dry creek bed. Two miles upstream, some dug holes in the ground to hide. Wherever they went, soldiers followed. Knots of people were cornered and slaughtered. Others stood and fought; still others were chased and gunned down.

The camp was now empty. The soldiers turned their full attention to killing any Indians running from the site. Black Kettle stood in disbelief before his hut until the camp was almost empty, and then he took his wife and started up the creek after his people. After a few steps his wife was hit. She fell, and he continued on until he found a group of Cheyenne hiding in holes they had dug in the side of the riverbank.

After dark he returned to where his wife had fallen and found her still alive. She told him, and a later board of inquiry, that the soldiers had fired nine more bullets into her as she lay on the ground. The peace commissioners in 1865 counted her wounds and confirmed her story.

By 5:00 P.M. the soldiers withdrew. If they had been better trained, they would have closed in and finished off the rest of the Indians. But being a mob, they feared getting close. Instead, they retraced their way along the creek, killing the wounded and scalping them. Little Bear, a friend of George Bent's, described the scene afterward: "After the fight I came back down the creek and saw these dead bodies all cut up and even the wounded scalped and slashed. I saw one old woman wandering about; her whole scalp had been taken off and the blood was running down into her eyes so that she could not see where to go."

Frontier art depicting the slaughter of Cheyenne and Arapaho Indians during the Sand Creek, Colorado massacre on November 29, 1864. (Currier and Ives)

All of that cold night, the surviving Indians from Sand Creek stayed in the hills. George Bent wrote later in his letters to Owl Woman, a Cheyenne, "The men and women who were not wounded worked all through the night, trying to keep the children and the wounded from freezing to death. They gathered grass by the handful, feeding little fires around which the wounded and the children lay; they stripped off their own blankets and clothes to keep us warm . . . Many who had lost wives, husbands, children, or friends went back down the creek and crept over the battleground among the naked and mutilated bodies of the dead."

Before dawn they made their way east toward the headwaters of Smoky Hill, 50 miles away, to another Indian encampment. Frightened that the soldiers would pick up their trail and slaughter them before they reached there, they sent some men ahead to get help. By nightfall a huge contingent of Indians with ponies met them and brought them to Smoky Hill. "As we rode into that camp there was a terrible scene," wrote Bent. "Everyone was crying, even the warriors and the women and children screaming and wailing. Nearly everyone present had lost some relations or friends, and many of them in their grief were gashing themselves with their knives until the blood flowed in streams."

Back at the camp, the soldiers plundered the Cheyenne lodges and captured the pony herds. The disciplined troops returned to Fort Lyon. The 100 days men were too busy plundering and squabbling over the ponies to move. Two days later, they headed for the Arapaho village to massacre its residents. But the Arapaho had long since been warned and had escaped to a place near the Kiowa and Comanche, south of the river.

Colonel Chivington went in triumph to Denver, where he and his men were received as heroes and acclaimed for their victory over Black Kettle and his "hostiles." One evening in a Denver theater, a band of soldiers stepped onstage and exhibited over 100 Cheyenne scalps and tobacco bags made of pieces of skin cut from the bodies of dead Cheyenne while the orchestra played patriotic airs.

Eventually, the truth of the Sand Creek massacre was heard in Washington, and Chivington was ordered to be court martialed. But he had long since been mustered out of the service and was beyond the reach of a military court. Colonel George Shoup, his adjutant and the leader of the 100 days men, was elected to the U.S. Senate.

UNITED STATES
ILLINOIS
CHICAGO
July 27–August 3, 1919

The famous Chicago race riots of July 27–August 3, 1919, were caused by ill feeling between blacks and whites after World War I, a small altercation on a Chicago beach and hot summer weather. By their end 35 were dead and more than 500 were injured.

July 1919 was a hot month in the Midwest. On the 27th citizens of the South Side of Chicago, Illinois, gathered on a South Side beach to escape the heat. The beaches in Chicago in 1919 were divided. Blacks swam on one side of the precaution line; whites swam on the other. That afternoon several blacks wandered across the barrier after some whites had amused themselves by throwing small stones at black bathers.

The confrontation escalated. Larger rocks were thrown. A black swimmer on a raft was struck by a rock that was flung with enough force to knock him into the lake, where he drowned. A white swimmer was hit by another rock, and he, too, drowned.

A melee broke out on the beach and spilled onto 29th Street. Within hours the pent-up energy that had been accumulating over a period of months, during which isolated bomb explosions, small fires, shootings and sporadic fistfights had punctuated the days and nights, erupted into full-scale violence.

There had been small, contained racial confrontations in a number of cities that summer: Washington, Detroit and New York had had their battles. Blacks who had fought in the war came home with a new sense of equality but discovered that there was neither equality nor justice in some hiring practices and wage policies. White laborers, perceiving blacks as willing to work for lower wages, were resentful.

In the early evening of July 27, the streets on the South Side, particularly along State Street, were packed with brawling whites and blacks. Rocks, stones and sniper bullets flew through the air. Fires were set, and fire apparatus was blocked from extinguishing them. Police poured into the area from other precincts; police were fired on and fired back. By midnight, the situation had turned ugly.

The death toll had risen to 14 by midnight of the first night; nine victims were white, five were black. Seventy-six had been injured.

The riots continued throughout the next day, and looting began. By nightfall it had spread into the Stock Yards district to 35th and Halstead Streets and all through the "black belt" of Chicago. Every available policeman was rushed to the scene; former soldiers and sailors were deputized; 3,600 men in four regiments of the National Guard and reserve troops were activated.

Rocks flew continuously through the South Side. Elevated trains ceased to run, and telephone wires were cut, isolating the area. Roving bands roamed both black and white neighborhoods, cornering lone blacks or whites and stabbing them on the spot.

By July 29 fighting had spread to the Loop, Chicago's business district, where two blacks were killed and a score of others were captured, kicked and beaten. The toll had climbed to 28 dead and more than 500 injured. By 7 o'clock that night, the rioting had reached the North Side of Chicago, where a crowd of Sicilians attacked a group of black families living in one tenement on West Division Street. They were met with a hail of bullets.

That same night an unidentified black man riding his bicycle on Lytie Street on the North Side was waylaid by a mob of whites who stabbed him and then shot him 16 times. When he fell dead from his bicycle, his body was splashed with gasoline and set afire.

At the same time, car men on the Chicago Transit System, unhappy over stalled contract negotiations, struck the system, immobilizing it and adding to the general pandemonium. Blacks were hanged in effigy; black policemen in full uniform fired on white policemen. Anarchy ruled.

The following day blacks, who constituted one third of the workforce in the stockyards, failed to report to work. Those whites who did report lay in wait for blacks.

The Stanton Avenue Police Station, in the middle of the "black belt," became an arsenal, as confiscated weapons piled up in it. By evening of the third day of rioting, the "black belt" was threatened with starvation when wholesalers refused to send delivery wagons of meat or groceries into the area.

The Loop was a maelstrom. Abandoned trolley cars were overturned; a black porter was cornered by a mob of whites at Wabash Avenue and Adams Street, where he was beaten, trampled and finally shot to death. Five hundred whites stormed the famed Palmer House, looking for black employees.

Meanwhile, Illinois governor Frank O. Lowden and city officials issued public statements saying that the rioting was under control. It would take another day of rioting for Mayor William Hale Thompson to finally order 4,000 troops into action. On the evening of July 30, the troops began to patrol the streets of the South Side while blacks fired on them and tossed rocks from side alleys.

Governor Lowden and city officials met that afternoon in the Blackstone Hotel. A mob of whites chasing a lone black man roared past the hotel, interrupting the conference.

That night whites infiltrated black neighborhoods, set fires and then prevented firemen from putting them out. One hundred twenty-three fires were reported on the night of July 30.

Blacks retaliated, invading white districts. A white woman was gunned down as she innocently walked near 47th Street and Indiana Avenue.

Finally, on July 31, 4,000 troops backed by 3,000 more in reserve began to restore order. Still, that day 32 fires were set between 7:00 A.M. and noon. In several cases streetcars were stretched sideways across streets, blocking firemen. A huge mob of whites with incendiary devices and a determination to "burn the Black Belt to the ground" were intercepted and turned back by troops.

In other sections of the city, troops thwarted street-corner lynchings. Hearing that the army was in control, a group of black workers attempted to return to their jobs in the stockyards. A mob of white workers met them, pursued them, wrestled them to the ground and began to beat them. The militia appeared on the scene, fixed bayonets and separated the groups, forming a cordon around the terrified workers.

Machine guns were set up at other strategic positions, particularly and ironically at the corner of Garfield Boulevard and Normal Avenue.

Finally, on Friday, August 1, as the Chicago car men voted to accept management's offer of 65 cents an hour and an eight-hour day, the rioting quieted. Six thousand troops still patrolled the South Side, but they were meeting fewer demonstrators. Under their protection fresh food, milk and ice once again flowed into the "black belt." Fifteen thousand black stockyard workers announced that they would return to work on Monday, August 4.

That Sunday, August 3, more violence erupted. Six blocks of black homes near the stockyards were destroyed by fire, and plans for blacks to return to work the next day were abandoned. Snipers hiding in an alley of the eastern section of the riot zone tried to assassinate Captain A. R. Wenhelm of the fourth Illinois Reserve Infantry. As he left his headquarters, snipers on nearby rooftops opened fire on him, and a small group of three blacks leaped out of a side alley and lunged at him and his adjutants with knives. Captain Wenhelm received a knife thrust in his side but escaped alive.

Finally, the next day, order was restored. Thirty-five had been killed; between 500 and 600 had been injured. It would be weeks before blacks would be able to return to work throughout the city, and years before any semblance of peace between blacks and whites would arrive in Chicago.

UNITED STATES
ILLINOIS
CHICAGO
August 25–30, 1968

Antiwar demonstrators who had gathered outside the Democratic National Convention on August 25–30, 1968, to protest the Vietnam policies of the Johnson administration were beaten and gassed by a police force that had gone momentarily berserk and had been ordered to enforce excessive repression measures by Mayor Richard J. Daley. Hundreds of demonstrators, scores of newspeople and bystanders and some delegates were injured.

In 1968, embittered and discouraged by public reaction to his Vietnam policies, Lyndon Johnson announced that he would not run for reelection. It had been a terrible year for him and for the country. The nation was sharply and irrevocably divided over American involvement in Southeast Asia. In April Dr. Martin Luther King Jr., an advocate of civil rights, had been assassinated. In June Robert Kennedy, openly challenging President Johnson's Vietnam stance and an announced candidate for the Democratic nomination for president, had been assassinated.

Eugene J. McCarthy, a solid opponent of the Vietnam War, was the voice of anti-war youth in the country, but it was generally agreed that the powers in the Democratic Party did not feel that he could challenge Richard Nixon for the presidency in 1968. They favored Hubert Humphrey, vice president under Lyndon Johnson and therefore an advocate of the Vietnam strategy that had divided the country so bitterly.

Thus, a large contingent of antiwar protesters journeyed to Chicago in August 1968 to make their voices heard and, it was hoped, influence the floor votes at the Democratic Convention.

Fearful of violence, Mayor Richard Daley, who was also the political boss of the Cook County Democratic Committee, ordered a mobilization of police forces to contain the demonstrators and prevent possible riots. In addition, he requested and received from Governor Samuel Shapiro 5,649 Illinois National Guardsmen to be stationed on round-the-clock duty. On top of this, 6,000 regular army troops received riot control training at Fort Hood, Texas, in an exercise called Operation Jackson Park, after the park in Chicago that was expected to be the gathering point for the student demonstrators. On August 25, 5,000 of these troops were flown to Chicago, where they were quartered at the Glenview Naval Air Station and the Great Lakes Naval Training Center outside the city. It was an awesome array of power for a peaceful nation to set up against a portion of its own citizenry.

Meanwhile, approximately 1,000 student protesters gathered, in a carnival mood, in Lincoln Park, on the fringes of one of the posher areas of Chicago. On Sunday night, August 25, at 11 P.M., the curfew hour on all public parks in Chicago, they were ordered out of the park by 400 policemen carrying tear gas launchers and rifles and wearing riot gear. The police drove the crowd into the downtown area of Clark Street and LaSalle Street. Traffic was disrupted, and the police waded into the mob of demonstrators, clubbing them with their nightsticks.

Claude Lewis, a black reporter for the *Philadelphia Bulletin,* was scribbling notes when a policeman approached him and demanded that he hand over his notebook. "He snatched the notebook out of my hand and started swinging away," Mr. Lewis later wrote. The first of many journalists, to be worked over by Chicago police, Mr. Lewis was treated for head lacerations at the Henrotin Hospital that evening.

By the next night 27 newspaper and television reporters had been roughed up by police despite the fact that they had displayed their press credentials. Delos Hall, a cameraman for the Columbia Broadcasting System, reported to the *New York Times* that he was filming police action when he was clubbed from behind, knocked down and then attacked by several more policemen. He was treated for a blow on the mouth and a cut forehead. James Strickland, a cameraman for the National Broadcasting Company, was struck in the face when he photographed Mr. Hall lying in the street.

At 12:20 on the night of August 26 in Lincoln Park, 300 policemen wearing Plexiglas shields fired tear gas into a crowd of nearly 3,000 youths who had erected a barricade of overturned picnic tables, upon which they had affixed Viet Cong, black anarchist and peace flags. In the crowd was poet Allen Ginsberg, who led 300 protesters in a gentle chanting of "Om," the mystic Sanskrit sound of peace and love.

This apparently deflected most of the police force. They waited to make arrests. Finally, Tom Hayden, one of the protest coordinators, was arrested for the second night in a row. All in all, some 150 protesters had been booked so far, and nearly 60 had been injured.

On August 27 at 12:30 in Lincoln Park, the police again moved into the mob of demonstrators that now numbered 2,000 and began to fire tear gas into it. A group of clergymen and demonstrators, gathered around a 12-foot cross that they had set up in the hope of conducting an all-night prayer vigil, were routed and clubbed.

The 2,000 made their way toward Michigan Avenue to Grant Park, where they merged with some 3,000 Yippies, New Leftists and adherents of the National Mobilization Committee to End the War in Vietnam. Grant Park was directly across the street from the Conrad Hilton Hotel, a center of activity for the Democratic National Convention and the headquarters of all of the major nominees.

August 28 was the climactic—and most brutal—day and night. Police were joined by National Guardsmen in the streets. The Democratic National Convention was nearing its most important business, and Hubert Humphrey was expected to win the nomination. A huge march was planned on the Amphitheatre that housed the convention. Already, news from the streets and some of the violence had invaded the floor of the convention itself.

Alex J. Rosenberg, a delegate from New York, was wrestled from the floor by an orange-arm-banded security guard when he refused to show his credentials. Once in the entryway, he was struck by a policeman and hauled away. Paul O'Dwyer, the Democratic candidate for the U.S. Senate, was roughed up by police when he attempted to intercede, as was Mike Wallace, the CBS television reporter, who was also struck on the chin and hauled from the hall.

Later that night Robert Maytag, a delegate from Colorado, interrupted the seconding of Hubert Humphrey's nomination for president by shouting into a microphone, "Is there any rule under which Mayor Daley can be compelled to end the police state of terror being perpetrated?" Cheers greeted the interruption while Mayor Daley sat impassively, and no move was made by the Democratic National Convention to deter the attempts to control demonstrators on the streets of Chicago.

Later, in his nomination speech putting George McGovern's name in contention for Democratic nominee for president, Senator Abraham M. Ribicoff of Connecticut said, "If Senator George McGovern were president, we would not have these Gestapo tactics in the streets of Chicago."

Impassive no longer, Mayor Daley and his supporters rose angrily to their feet and tried to shout down Senator Ribicoff, who turned to them and added, "How hard it is to accept the truth."

Outside the convention hall a steady crescendo of activity had been accumulating all day. That afternoon a gathering of approximately 15,000 young people had filled Grant Park and gathered around its band shell in a rally designed both to protest the violence in the streets and to prepare the demonstrators to march on the Amphitheatre.

There were skirmishes between police and protesters at the exterior of the gathering. At the band shell poet Allen Ginsberg again led the group chanting "Om," though his voice by now was cracked from chanting and swallowing tear gas; French author Jean Genet spoke to the crowd through a translator; authors William Burroughs and Norman Mailer exhorted them; comedian Dick Gregory, mounting the platform, said, "You just have to look around you at all the police and soldiers to know you must be doing something right." Entertainers Judy Collins and Peter, Paul and Mary led the crowd in folk songs of the resistance movement, and leaders of the protest led the mob out of the park.

Some groups, such as the Poor People's March, had permits, but more purposely did not. There was manipulation by the more militant leaders of more naive, young and marijuana-smoking youngsters.

At the Congress Street Bridge leading from the park to Michigan Avenue, police and Guardsmen opened up with tear gas and mace, attempting to hold the demonstrators within the park. But the numbers were overwhelming, and between 2,000 and 5,000 youths, led by David Dellinger, the national chairman of the Mobilization Committee to End the War in Vietnam, and poet Allen Ginsberg, headed south on Michigan Avenue toward the Amphitheatre.

The police, moving in in phalanxes and using their clubs as prods, broke up the march, sending protesters fleeing up side streets. Those who escaped the police charges faced a tank and National Guardsmen with machine guns. Newspeople were again clubbed to the ground as demonstrators chanted, "The whole world is watching."

Reverend John Boyles, the Presbyterian chaplain at Yale and a staff worker for candidate Eugene McCarthy, was hauled off to a patrol wagon and charged with breach of the peace. Speaking to a *New York Times* reporter afterward, Mr. Boyles said, "It's an unfounded charge. I was protesting the clubbing of a girl I knew from the McCarthy staff. They were beating her on her head with clubs and I yelled at them 'Don't hit a woman.' At that point I was slugged in the stomach and grabbed by a cop who arrested me." There were 178 arrests that night alone, and 100 persons, including 25 policemen, were injured.

Shortly after midnight an uneasy calm settled over the city, as 1,000 National Guardsmen arranged themselves in front of the Conrad Hilton Hotel and 5,000 demonstrators drifted back into Grant Park. A field piece was poised in the lobby; officials had prevented the National Guard from bringing bazookas onto the premises.

Blue police barricades were lined up on the streets, and several dozen people, many of them elderly, watched quietly as protesters and police, illuminated by television lights, chased one another in and out of the park. Suddenly, "for no apparent reason," according to reporters on the scene, the police turned on the spectators and charged the barriers, crushing the spectators against the windows of the Haymarket Inn, a restaurant in the hotel. The plate glass windows gave way, sending screaming women and children backward through the broken shards of glass. The police then ran into the restaurant and beat some of the victims who had fallen through the windows. As they were clubbing them, they arrested some of these bewildered and bleeding citizens.

"Outside," wrote Nora Sayre, who was caught in the crush, "people sobbed with pain as their ribs snapped from being crushed against each other. . . . Soon, a line of stick-whipping cops swung in on us. Voiceless from gas, I feebly waved my credentials, and the warrior who was about to hit me said, 'oops, press.' He let me limp into the hotel, where people were being pummelled into the red carpet, while free Pepsi was timidly offered on the sidelines."

In St. Chrysostrom's Church and the sixth-floor offices of the Church Federation of Greater Chicago, volunteer doctors treated the gassed, maced and injured demonstrators. A specially equipped van manned by students of Yale Medical School and the Columbia College of Physicians and Surgeons roamed the riot area to dispense first aid, but police harassed first aid teams and forced the van away by threatening to confiscate it. Dr. Albert S. Braverman, an internist from Manhattan who was helping the wounded, was himself a victim. "I was hit and pushed by a cop while I was coming back from dinner and while wearing my white coat and red cross," he told reporters for the *Times*. "When a friend said I was a doctor, the cop replied, 'I don't give a damn.'"

On the next night, August 29, another march was planned to the Amphitheatre. This time, several delegates joined the 3,000 marchers led by Dick Gregory (Dellinger was in jail, and Rennie Davis, his second, had had his arm broken the night before).

"Such blood . . ." wrote Nora Sayre.

> Broad bloodstreaks on the pavements showed where bodies had been dragged. . . . Each day, scores staggered bleeding through the streets and parks, reeling or dropping, their faces glistening with vaseline—for Mace. . . . With two doctors, I walk[ed] five blocks ahead of . . . Dick Gregory . . . to the Amphitheatre; we [saw] the tank with the machine guns that await[ed] them. We turn[ed] back to tell them, discovering that the empty alleys—where we'd planned to disappear if necessary—[were] now crammed with police and Guardsmen.

Armored personnel carriers and jeeps with barbed-wire barriers mounted on their hoods further blocked the way. The marchers were ordered to stop at 18th Street and Michigan Avenue on the advice of the Secret Service. Dick Gregory argued that he had invited the demonstrators to his home on 55th Street and wished to pass. He was denied his request, as were others behind him. He pushed past the barricades and was arrested, as were 422 others, including nine delegates. A steady shuttle of police vans ran between the street corner and a specially convened night court.

Ms. Sayre returned to McCarthy headquarters on the second floor of the Conrad Hilton Hotel. "As I watched the beatings and gassings from a second-floor McCarthy room," she wrote, "twelve policemen surged in, slamming down the windows, drew the curtains, and told us to turn away and watch the TV set, where Humphrey was starting to speak—'And that's an order.'"

Earlier, in the Convention Hall, Mayor Daley had mounted the podium to defend his police and his tactics. He mentioned that 51 police had been injured but failed to note that 300 demonstrators had also been injured. "The people of Chicago," he said, "will never permit a lawless, violent group of terrorists to menace the lives of millions of people, destroy the purpose of a national political convention and take over the streets."

At dawn police raided the headquarters of Senator Eugene McCarthy on the 15th floor of the Conrad Hilton Hotel, herded 30 McCarthy aides from several rooms into elevators and took them to the lobby. In the scuffle three McCarthy workers were injured seriously enough to require hospital treatment. One required 10 stitches in his head, another, six. The police reported four injuries to their ranks. No arrests were made and no charges filed, but Senator McCarthy postponed his departure to call a news conference to protest the police action, which was, authorities said, in response to complaints that objects were being thrown from the windows of McCarthy headquarters—a charge the aides denied.

That same morning Frank J. Sullivan, the Chicago Police Department's director of public information, called a news conference. He described the demonstrators as "revolutionaries" and called some of their leaders, including Tom Hayden and Rennie Davis, "Communists who are the allies of the men who are killing American soldiers.

"The intellectuals of America hate Richard J. Daley," he continued, "because he was elected by the people—unlike Walter Cronkite [the CBS anchor man at the convention]."

By August 1 the demonstrators had dispersed, as had the National Guard and police. The army personnel had never been called up, and they quietly boarded planes and returned to their camps. More than 71 percent of the Chicago citizens polled by local papers approved of the police handling of the demonstrators; many replied that they were certain the protesters were Communists, and J. Edgar Hoover announced an FBI investigation. Nevertheless, the rest of the nation and the world looked upon the riots differently. "The Chicago cops taught us that we were rubble with no protection or defense," wrote Nora Sayre. "In future, we can understand the ghettos' rage."

In September 1969 eight radicals and antiwar activists—David Dellinger, Rennie Davis, Tom Hayden, Abbie Hoffman, Jerry Rubin, Bobby Seale, John Froines and Lee Weiner—were tried in Chicago for conspiracy to riot. Bobby Seale, the head of the Black Panthers, was chained and gagged for outbursts in the courtroom. His case was separated from that of the others. Five of the remaining seven defendants were convicted of intent to riot and sentenced to five years in prison and fines of $5,000 each, the maximum penalties permitted. The riot convictions were eventually overturned on appeal because of improper rulings and conduct by the trial judge, Julius Hoffman. Contempt charges stemming from the trial were also dropped.

UNITED STATES
NEW YORK
NEW YORK
July 13–15, 1863

The July 11, 1863, Conscription Act, designed to replenish a depleted Union Army during the Civil War, allowed wealthy draftees to buy their way out of the draft for $300. It produced three days of rioting in New York City, from July 13–15, in which 2,000 rioters died, 10,000 rioters were wounded; 60 soldiers died; 300 soldiers were wounded; 76 blacks, turned on by the rioters, were reported "missing"; 18 blacks were hanged; and 5 blacks were drowned.

The Civil War was going poorly for the North in 1863, and in April of that year, President Abraham Lincoln announced that 300,000 men would be drafted according to the Conscription Act of Congress. The process would begin in New York City on July 11.

The act was a discriminatory one, pretending to draft all equally but actually skewed so that it drafted only the poor. If a man had $300, he could buy himself out of the army. And it was that provision that ignited riots in the North, the most notable and brutal of them the draft riots of New York City.

During the Civil War New York City had a population of approximately 815,000, 5 percent of whom were immigrants with no particular allegiance. More than 203,000 Irish had escaped the potato famine to live in extreme poverty in the slums of New York. Illiterate and hardly able to make a living, they were discriminated against in practically every sense. They would form the bulk of those to be drafted on July 11.

When the drums containing random numbers began to spin on the morning of July 11, discontent was not confined to Irish immigrants. There had been inflammatory stories in the papers raging against the unfairness of the draft; there was resentment in the slums toward the affluent and also toward blacks, since they were willing to work for even lower wages than those offered immigrants. There was also objection to the primary reason for the war, which was abolition.

And lower New York was a brutal place. Along Five Points, near its lower tip, the Plug Uglies, the Dead Rabbits and other gangs composed of men and boys living in subhuman conditions roamed and pillaged and killed. Any excuse would easily turn them out into the streets, and the draft was a better than ordinary reason for them to move uptown.

Just after dawn on Saturday, July 11, there were rumors that a political organization, the Knights of the Golden Circle, was going to take over the U.S. Arsenal at 7th Avenue and 35th Street. It never materialized.

Meanwhile, a large, hostile crowd gathered at Third Avenue and 46th Street, where the drum was spun, and 1,236 names were drawn. The crowd dispersed, muttering.

Sunday, July 12, was a day of festering for the poor and payments of $300 to the authorities for relief from service by the wealthy whose names had been pulled.

By Monday morning, July 13, the unrest had turned to action. A few fires had been set the night before, but they had been swiftly extinguished with no interference. But early Monday morning a mob began to move up the West Side toward the homes of the wealthy bordering Central Park. It gathered in a lot just east of the park at about 8 A.M. and then began to break into segments, snaking their way toward the draft lottery drum at Third Avenue and 46th Street.

The mob gathered around the draft office for six blocks in every direction and toppled carriages, forced pedestrians off sidewalks and knocked the top hats off unwary victims. Crudely written posters appeared reading NO DRAFT!

The Back Joke Gang, composed of members of Volunteer Engine Company No. 33, heard that their fire chief's number had been picked in the draft, and they were determined to invade the office and smash the drum. They were armed and forced their way into the building, smashed the drum and set fire to the structure (police escaped with the official records). When other fire companies arrived to extinguish the blaze, they were held off by the mob.

Superintendent of Police John A. Kennedy arrived in an open carriage to survey the scene. He was recognized, rushed by the mob, beaten mercilessly and tossed over an embankment. He was only saved from murder by sympathetic bystanders who convinced his attackers that he was already dead.

Now insurrections began all over the city. Police were attacked; precinct houses were raided. The so-called Invalid Corps of wounded soldiers was brought into action. Marching up Third Avenue, they were met with a barrage of bricks, paving stones and a dead cat or two. When one of their men was killed, they were ordered by their officer to shoot into the crowd. They did, killing a woman and six men. The crowd, infuriated, rushed them, shooting some of the soldiers with their own muskets. The battle had been joined. What had begun as a poor man's protest was now an armed fight against any sort of authority.

The city rallied, brought in National Guardsmen and deputized more than 1,000 citizens. Before it was over, 10,000 soldiers, cavalry, infantry and dozens of batteries of artillery would roar into New York to beat back the mob.

By noon the crowd had gathered leaders, and they decided that they needed firearms. They marched on the State Armory at Second Avenue and 21st Street, reserving the Union Steam works, an active munitions plant farther north, for later.

By 4 P.M. they were battering at the armory's main gates, while a contingent of police fired on them from inside the armory. Within minutes the mob had broken down the gates, and the police had retreated to the 18th Precinct Police Station at 22d Street and Third Avenue. The mob rushed the station and burned it, forcing the police to retreat to the Mulberry Street headquarters.

Inside the armory the mob armed itself. Fearing the return of the police, it locked and barred the door to the third floor drill room. It would prove to be a fatal error, for when the police returned and began to invade the building, some of the mob set fire to it. All of those on the third floor were incinerated when the ancient building went up like an incendiary torch.

In other parts of the city, the mob sought out the rich and blacks. Blacks were captured and lynched. Bodies—an average of three a day—swung from lampposts, trees and gateways.

Stores were looted and then burned. The provost marshal's office at Broadway and Ninth Street was set afire and its contents hacked up and thrown into the street. A mob moved out to burn Mayor George Opdyke's mansion on First Avenue and police headquarters on Mulberry Street.

A force of 125 policemen managed to beat back the assault on police headquarters. The mob was also turned away from Mayor Opdyke's home. Frustrated, it turned its

The confrontation between police and rioters in front of Horace Greeley's Tribune *during the draft riots in New York City, July 13–15, 1863. (Harper's Weekly)*

attention to the Colored Orphan Asylum, which housed 200 orphaned black children, all under 12 years of age. The superintendent gathered his charges, and they escaped out the back way of the building on Fifth Avenue between 43d and 44th Streets. When the rioters arrived, they found one girl cowering under a bed. They murdered her on the spot, chopped up the furnishings and set fire not only to the asylum but to several neighboring buildings as well.

By that night the mob held the town. In Printing House Square they marched on the *New York Tribune* and its editor and publisher, Horace Greeley. Greeley was chased down Park Row but escaped. When he returned to the *Tribune* building, he found that the police and soldiers had turned it into a fortress. Gatling guns protruded from upstairs windows, and a field howitzer was positioned in the main lobby.

Fires sprang up all over the city. Fire companies trying to extinguish them were shot at or pelted with debris. If a rainstorm had not suddenly come up at 11 P.M., the city might well have met the fate of Chicago earlier in the century (see p. 230).

During that night Governor Horatio Seymour conferred with the mayor, and both sent telegrams to the War Department in Washington requesting that New York regiments that were currently recovering from the Battle of Gettysburg be sent to the city immediately.

That night 2,000 soldiers, sailors and marines set up positions on the streets and in the buildings of New York City, manning an arsenal of firepower and ready for the worst.

Tuesday, July 14, was Bastille Day in France, and the streets of New York resembled Paris during the revolution. Street barricades were built from wagons, dismantled lampposts, street poles, boxes, barrels, kegs and looted furniture. The barricades blocked off First and Ninth Avenues and snaked through the streets from 11th to 14th Street.

The Union Steam Works now became the focal point for the mob, and police and army units gathered around it. Police broke heads, literally, with their truncheons; army units fired into the crowds, killing many.

Colonel H. J. O'Brien, who commanded a regiment of artillery, was particularly aggressive in his use of force. He committed murder en masse, firing his cannon into crowds and tearing them to pieces. Whether it was arrogance or ignorance or a taste for self-destruction that caused him to later go back to the scene of that slaughter, no one will ever know. He *did* go back, alone, on horseback, and was, of course, spotted and pelted with bricks.

Sensing danger, O'Brien dismounted and took refuge in a saloon at the corner of Second Avenue and 19th Street. But after taking on a little bravado fuel, he left the saloon and, brandishing a sword and a pistol, waded out into the midst of a hostile mob. Within minutes he was clubbed to the ground and kicked mercilessly. A rope was tied around his ankles, and the mob took turns dragging him over the cobblestones. Finding him still alive, the mob then worked him over with knives and left him in the middle of the street. Later that night he was discovered by another mob, dragged around again, tossed into his own back garden and assaulted with more knives. His body was discovered by his family that night.

The mob then went on to the Union Steam Works, which it captured, but it was eventually routed by 200 policemen. The dead piled up like cordwood around the factory and on the front walks.

Meanwhile, infantry companies ringed the Sub Treasury on Wall Street, and at the Brooklyn Navy Yard the gunboats *Gertrude, Unadilla, Granite City* and *Tulip* loaded their big guns and stood ready to defend the yard. At the Battery the ironclad *Passaic* rode at the ready.

Late Tuesday, Secretary of War Edwin M. Stanton informed city officials that five regiments of the Union Army in full war gear were being transported by steam cars and ferries to the city.

On Wednesday, July 15, in an effort to quell the riot, officials informed the press that the insurrection had passed its peak.

This announcement, plus the entry into the city that day of thousands of war-trained veterans, the 74th National Guard, the 26th Michigan, the 152d New York Volunteers and others, finally defused the draft rioters. The combined militia set up their positions and their howitzers, and they were a formidable force that soon demolished the remaining resistance in the city.

For three days the siege played itself out in a steadily decreasing spiral. Somewhere in the midst of it, someone at city hall issued a proclamation: "The Conscription Act is now suspended in New York and will not be enforced. The Board of Aldermen at a specially called meeting has voted $2,500,000 for all poor men to buy their way out of army service."

The mobs cheered and thought they had won. But within an hour the proclamation was declared a hoax, and after a few desperate forays the mob finally dispersed.

Martial law remained in place for a few days. The police combed poor areas to retrieve stolen items. The toll was never precisely counted, but it was concluded by the newspapers that 2,000 of the rioters had died and from 8,000 to 10,000 had been wounded. These figures were probably low, because many of the dead and wounded were carried off and hidden.

Only a few police were listed as dead. Some 50 to 60 soldiers were killed, although there was never an official listing by the War Department. Three hundred soldiers were wounded. Seventy-six blacks were listed as "missing," which probably meant dead. Eighteen were lynched. Five were drowned in the Hudson and East Rivers.

More than $5 million in property damage was reported, including the loss of the Colored Orphan Asylum, a Protestant mission, three provost marshall offices, three police stations, one armory and a score of factories and stores.

Later, armchair analysts dubbed the draft riots the "Roman Catholic Insurrection" because of the large number of immigrant Irishmen who were involved. But of course, it was not. It was an uprising against a conscription act that was perceived as being clearly discriminatory against the lower middle class and the poor.

UNITED STATES
OHIO
KENT
May 4, 1970

Three days of demonstrations by students of Kent State University protesting the U.S. invasion of Cambodia climaxed on May 4, 1970, in the killing of four students by Ohio National Guardsmen. Nine students were wounded, and the protest movement of college students in the United States lost its momentum.

The 1960s were a turbulent decade marked by the assassinations of three major American leaders, the most unpopular war in U.S. history and an enormous upsurge in activism among the youth of the country. While the drug culture, spearheaded by Timothy Leary, was urging the flower children to "tune in; turn on; drop out," organizations such as the SDS (Students for a Democratic Society) were urging the youths of America to stand up for their beliefs and not trust anyone over 30.

It was certainly a divided America. There were those who favored the war in Vietnam—a steadily decreasing number by 1970—and those who were vocally, vociferously against it. In 1968 President Lyndon Johnson had declined to run for president again in part because he did not want to defend further U.S. involvement in Vietnam.

But President Richard Nixon and his vice president, Spiro Agnew, were defenders of the war and foes of the antiwar demonstrators, whom they characterized as "malcontents," "bums," "hippies" and so forth, in spite of the fact that much of the American public had already wearied of a war whose moral justification seemed murky and whose end seemed to be nowhere in sight. Finally, in 1969, bowing to public pressure, the United States began to pull its troops out of Vietnam while continuing to bomb Communist strongholds in Cambodia.

It was a peculiar way to end a war, and antiwar protests continued on campuses across the United States. On April 8 there were student disturbances in Cleveland, Ohio, and 952 National Guardsmen were called out to quiet them. On April 16 and 17 there were student demonstrations in Oxford, Ohio, and 561 National Guardsmen were called out there. The next day, in Sandusky, more student demonstrations erupted and 96 National Guardsmen were called up. On April 29 there were rumors that the United States was about to invade Cambodia, and student demonstrations flared up all over the country. At Ohio State University in Cleveland, it turned into a riot, and 2,861 National Guardsman were called in to quell it. Obviously, Governor James A. Rhodes was not averse to calling up the National Guard to quiet civil demonstrations.

On April 30, 1970, President Nixon appeared on television and announced the U.S. invasion of Cambodia. College campuses all over the country erupted with angry student demonstrations.

Kent State University, in Kent, Ohio, was no exception. Although a small and relatively quiet campus, it had its chapter of SDS, and some of its buildings had been taken over by rebellious students. Yippie leader Jerry Rubin had addressed the students in early April. A year earlier the students had presented a list of demands to the administration of the university:

1. Abolish ROTC
2. Abolish the Liquid Crystals Institute
3. Abolish the Northeast Ohio Crime Laboratory
4. Abolish the Kent Law Enforcement Training Program.

On May 1, the day after Nixon's announcement, some students at Kent State held a campus ceremony in which they buried the Constitution of the United States. Not all of the campus population was happy about this, but some were fired up not only by it but by reports of huge demonstrations on other campuses against the Cambodian invasion.

By 8 o'clock that night, 1,000 of Kent State's 20,000 students had spilled into the streets of downtown Kent and were generally raising hell, climbing up on streetlights, stopping traffic and marching down streets and in and out of stores. Fires were set in the middle of streets. Forty-seven establishments had their windows broken. Mayor LeRoy Satrom, celebrating Law Day at a meeting in the Treadwell Inn in nearby Aurora, received the news badly. He drove swiftly back to Kent and conferred first with the town's police chief, Roy D. Thompson, and then clapped an 8:00 P.M. to 6 A.M. curfew on the town. Damage was first estimated at $100,000; then it was reduced to $50,000, then $16,000 and finally to $10,000. A few arrests were made, but the mayor was prompted to call Columbus and alert the National Guard. By 3:00 A.M. a National Guard officer was in town to assess the situation.

The next morning, May 2, the local *Kent-Courier* carried the following headline: "Nixon Hits Bums Who Blow up College Campuses." All day the city of Kent was bombarded with false fire alarms, bomb threats and violent rumors. At 5:30 that afternoon Mayor Satrom, afraid that the students would again invade downtown, called in the Ohio National Guard.

While all of this was going on, Robert White, the president of Kent State University, was in Mason City, Iowa, where he was scheduled to make a major address. It would be Sunday morning, May 3, before he would return.

That Saturday evening the students set out to burn the rickety old ROTC building on campus. A relic of World War II, it was scheduled for razing anyway. But to the student activists it symbolized U.S. involvement in Asia. By 8 o'clock a sizable crowd of around 600 students began to throw rocks at the building. Witnesses later testified that 10 or 15 policemen probably could have stopped matters then and there, but there were none visible.

By 8:30, after several attempts to kindle the building failed, two young men (some said of high school age) entered it and really set it ablaze. A half hour later the Kent volunteer fire department arrived, but students cut their hoses and roughed up some of the firemen, who then left the campus.

The crowd grew and other incidents erupted. A shed containing archery equipment was set afire. There was talk of setting the president's home ablaze, but it never happened.

Meanwhile two National Guard generals, Sylvester De Corso and Robert Canterbury, arrived in Kent, and 1,196 Guardsmen, taken off duty in Akron where they were monitoring a teamster's strike, were on their way. By 10 o'clock the Guard had arrived, and student rioters were again running amock in downtown Kent. Their actions had become more bizarre and indefensible. Arsonists tried to set fire to various buildings. Rocks were thrown at Guardsmen. The atmosphere turned ugly.

Under the protection of the National Guard, the firemen returned to campus, but by now the ROTC building was a smoldering wreck. Tear gas forced the students from downtown back to campus and their dormitories, and by midnight everything seemed to be under control. Thirty-one people had been arrested for violation of the curfew.

On Sunday morning, May 3, Governor Rhodes arrived in Kent to personally supervise the situation. He inflamed it, releasing public statements that accused student demonstrators of being "worse than the Brown Shirts and communist element and also the nightriders and the vigilantes. They're the worst type that we harbor in America."

The effect on the campus was depressing at best. There were rumors that the government was going to close the college down, and, indeed, it had been suggested.

But it did not happen, and that Sunday has been described as more like a carnival than anything else. Sightseers wandered onto the campus. Students and Guardsmen fraternized. One coed placed a flower in the muzzle of a Guardsman's gun.

But some members of the faculty were disturbed by the presence of the military on the campus and tried to have them removed. They did not succeed.

That night the turning point in the confrontation was reached. At 7:00 P.M. students began to gather in groups on the campus. Some Molotov cocktails were found by campus police, and the Guard took up positions. Confrontations occurred, and the Guard threw tear gas again, driving some students back but also driving some 200 of them downtown. There the students were met with an armed tank.

They retreated and might have gone back to the campus, but they met other students who had decided to sit down in the middle of town in defiance of the curfew.

Everything escalated. The victory bell on campus began to ring. Helicopters with searchlights flew back and forth over the campus, spraying it with intense light. The Guardsmen fixed bayonets. Student leaders demanded a meeting with President White and Mayor Satrom. Although there had been no guarantees made to keep police and National Guardsmen off campus, a student with a bullhorn assured the crowd that guarantees had been offered, and when the National Guard moved in with fixed bayonets and attacked seven students, the student leaders yelled, "We've been betrayed!"

By 11:40 the Guard had secured the campus. But it was an armed truce.

And then, May 4, 1970, dawned. By 11:00 A.M., after some classes had been dismissed early because of the tensions on the campus, students began to gather. All outdoor demonstrations had been banned, although no martial law had officially been declared by Governor Rhodes (he would declare it, retroactively, after the shootings).

A rally had been called by student leaders for noon. An officer in a jeep tried unsuccessfully to break up the rally peacefully. At 12:00 General Canterbury gave the order to prepare to disperse the students. There were 113 Guardsman on campus at the time. There were approximately 1,100 students.

The Guardsmen began to move out, firing tear gas canisters as they went. The students again broke and ran, hurling the gas canisters, rocks and epithets at the Guardsmen. Waves of soldiers and students swept back and forth across the campus. Tear gas filled the air.

Suddenly, at 12:24 P.M., one group of Guardsmen stopped, wheeled and aimed their rifles at the students, who were some 200 yards away and therefore could not have possibly harmed the soldiers. One shot rang out; there was a period of silence for two seconds, and then a fusillade erupted. Sixty-seven shots were fired in the next 13 seconds, and when it all stopped, 13 bodies were scattered on the grass and the parking area beyond. One was dead; 12 were wounded, some fatally; some would be crippled for life. The closest victim was 71 feet from the National Guard; the farthest, 745 feet.

No one knows who started firing. Several officers frantically stopped it, and the Guard was moved instantly back to the ROTC area.

Horror had replaced confrontation. Jeff Miller, a student, was lying face down, the back of his head blown off, his blood spreading over the parking lot. Allison Krause, Sandra Scheuer and William Schroeder had all been wounded seriously. Ms. Krause, denied oxygen by an ambulance attendant, died on the way to the hospital, as did Sandra Scheuer. William Schroeder would die in the hospital from a chest wound.

On the campus a terrible standoff developed. Three faculty members, Seymour H. Baron, Mike Lunine and Glenn Frank, became the peacemakers. They prevented an adamant general and a troop of frightened soldiers from causing more bloodshed; they talked reason back into the minds of the students, and a slaughter was averted.

Afterward, the reaction to the May 4 killings was sharply divided. There were those who were sickened by it. The Soviet poet Yevgeny Yevtushenko wrote a poem:

> Allison Krause, you were killed because you
> loved flowers
> . . . Ah, how fragrant are the lilacs,
> But you feel nothing.
> As the president said of you, you are a bum.
> Each victim is a bum. But it is not his fault.

On the other hand, a Kentucky mother whose three sons attended Kent State and were there during the shooting spoke candidly and chillingly to a researcher for James Michener, who was compiling a book about the incident:

MOTHER: Anyone who appears on the streets of a city like Kent with long hair, dirty clothes or barefooted deserves to be shot.

RESEARCHER: Have I your permission to quote that?

MOTHER: You sure do. It would have been better if the Guard had shot the whole lot of them that morning.

RESEARCHER: But you had three sons there.

MOTHER: If they didn't do what the Guards told them, they should have been mowed down.

PROFESSOR OF PSYCHOLOGY: (listening in): Is long hair a justification for shooting someone?

MOTHER: Yes. We have got to clean up this nation. And we'll start with the long-hairs.

PROFESSOR: Would you permit one of your sons to be shot simply because he went barefooted?

MOTHER: Yes.

PROFESSOR: Where did you get such ideas?

MOTHER: I teach at the local high school.

PROFESSOR: You mean you are teaching your students such things?

MOTHER: Yes. I teach them the truth. That the lazy, the dirty, the ones you see walking the streets and doing nothing ought all to be shot.

The heart went out of student protests that day at Kent State. The war would wind down. Vice President Agnew would resign in disgrace, and President Nixon, too, would be forced to resign as a result of the Watergate scandal. But a numbed silence, as of the grave, would settle over the campuses of the nation.

UNITED STATES
WASHINGTON, D.C.
May 20–July 28, 1932

World War I veterans, out of work during the Great Depression, marched on Washington, D.C., in May 1932 to demand that a bonus due them in 1945 be paid immediately. On July 28 an army unit led by General Douglas MacArthur cleared the veterans out and set fire to their encampments. One veteran was killed; scores were injured.

It was, for 20th-century America, the worst of times. The stock market crash had spiraled the country downward into depression, and by the summer of 1932, there were more than eight and a half million people out of work.

Many of them were veterans of World War I, men who had fought well and selflessly for their country. They had been promised a bonus totaling $2 billion to be paid to them in 1945, when, it was presumed, they would be old enough to need it (this was before Social Security). However, with the country and many of its people in dire straits in 1932, Representative Wright Patman introduced a bill in Congress that would pay these veterans their bonus then rather than in 1945.

But as the congressional session lurched toward its summer recess, the proposal seemed to rest dead in the water. At the end of May, various informal contingents of veterans from all over America began funneling into Washington hoping to prod Congress into passing the Patman bill.

They had their precedent. In the 1700s unpaid members of the Continental Army had laid siege to Philadelphia until Congress met their demands. In 1932 the leaders of what would come to be known as the Bonus Expeditionary Force felt that a peaceful, orderly settling into Washington would produce the same effect.

The first contingent of approximately 1,300 men arrived on May 28 in 16 trucks, which were festooned with American flags provided by the state of Maryland. The city of Washington had hot stew, bread, milk and coffee waiting for the arriving servicemen at a vacant government building on Pennsylvania Avenue. Superintendent of Police Pelham D. Glassford personally supervised the welcome, noting parenthetically that they would all have to vacate the premises within 48 hours.

But every day for weeks, more and more veterans poured into the capital as 48-hour deadlines were issued and ignored. It was all very low key and friendly, and General Glassford (he was a retired brigadier general and kept the title) managed to have army rolling kitchens set up to accommodate the new arrivals, who now occupied another vacant building and had begun to erect shanties on the banks of the Potomac. Eventually, their main camp would be located on the Anacostia River and would consist of a small village of slapped together lean-tos, shacks and tents.

Congress reacted with anxiety. Some senators and representatives were sympathetic but concerned about the source of the bonus; others were hostile enough to call the veterans Communists. (There *was* one contingent of Communists, but the other veterans disowned them, forbidding red flags to be flown and chanting "Eyes front, not left" as they sometimes marched from encampments to the Capitol.)

Senator Carl Lewis of Illinois delivered an address at the National Old Soldiers Home in which he fulminated, "If the veterans persist in terroristic tactics, they will endanger their chances of receiving any favorable treatment whatsoever."

In truth, the burgeoning bonus army could not have been more peaceful. Although some had engaged in violence in commandeering freight trains to get to Washington, once there they settled docilely in.

By June 4 there were several thousand veterans in Washington, and food supplies were running out. The District commissioners offered trucks to transport the marchers 50 miles from the city. The veterans politely refused.

By June 7 the army had swelled to 7,000, and a parade was planned to demand that the $2.4 billion bonus be paid. "If we have to stay until 1945, we will," said one veteran.

Members of the "bonus army" of World War I veterans gather outside one of their makeshift dwellings in Washington, D.C., in the summer of 1932. Shortly afterward, a U.S. Army unit led by General Douglas MacArthur routed the protesters and set fire to their encampment. (Library of Congress)

During the next week families began to arrive, and food was sent from the home states of many of the veterans' companies. Morale was high in the Expeditionary Force, but the patience of the federal government was dwindling. On June 9 Dr. William Fowler, the District health officer, and Dr. James G. Cumming toured the camp on the edge of the Potomac and concluded that the camps constituted the "gravest health menace in the history of the District of Columbia." Dr. Fowler called the situation "frightful," pointing to the presence of insects and the fact that many of the veterans were suffering from exposure and malnutrition.

Still, the army grew at the rate of 100 men an hour. By June 11 there were 15,000 ex-servicemen camped throughout Washington, and the District authorities confessed that they could no longer handle the problem. It was up to the federal government to take a hand.

June 17 was the day the Senate chose to debate and vote on the Patman bonus bill. The galleries were packed with veterans. The plaza before the Capitol building bulged with 10,000 men. For weeks they had petitioned and met with their representatives. There was no doubt in Congress about the determination of this army.

However, at 8:20 P.M. on June 17, the Senate voted down the Patman Bill 62 to 18. There was a collective groan from the gallery, and outside in the plaza, after the results of the vote were announced, there was a momentary ripple of anger and a movement toward the outer fringes of the gathering.

W. W. Walters of Oregon, the elected leader of the army, and his associates calmed the crowd, blowing bugles for formation to march back to the camps. "Go back to your camps," shouted Walters. "We are not telling you to go home. We are going to stay in Washington until we get the bonus, no matter how long it takes. And we are 100 times as good Americans as those men who voted against it. We are just asking you to obey the law and not antagonize the authorities."

On July 2 the army returned to the Capitol steps, protesting the adjournment of Congress without reconsidering the bill and vowing to remain until it was passed, no matter how long it would take. On July 4 it staged a parade down Pennsylvania Avenue.

As July wore on and the humid summer heat seeped down into the camps and buildings, some veterans returned home. But thousands stayed on. They had no place to go and no jobs. They decided to stick it out.

By the end of July, President Herbert Hoover, who had ordered the gates to the White House chained, now decreed that the bonus army be cleared from Washington.

On the morning of July 28, 1932, General Glassford and his police proceeded to march into the camps and gently remove the veterans, who resisted. It was obvious that the police could not do it alone, and shortly after 2 P.M. Hoover ordered a unit of the U.S. Army, under the command of General Douglas MacArthur, to clear the camps. MacArthur and his forces marched down Pennsylvania Avenue with four troops of cavalry, four companies of infantry, a machine gun squadron and several tanks.

The veterans cheered them, as did a large crowd of spectators gathered at the curb. Suddenly, the army unit charged. Cavalrymen rode pell-mell into the crowd, swinging sabers over their heads. Infantrymen tossed tear gas bombs. Men, women and children, veterans and spectators alike were trampled.

The troops moved on, scattering civilians and veterans, throwing tear gas indiscriminately. The peaceful protest had been transformed into a bloody confrontation. On the steps of one of the vacant veteran-occupied federal buildings, a policeman was hit by a brick tossed from an upper story. He drew his revolver and fired two shots into the crowd, killing one man, William Hashka, an unemployed veteran from Illinois. Other police officers drew their guns; General Glassford shouted to them to put away their firearms, which they did, but not before pointing them directly at him.

Meanwhile, the army unit, under the orders of General MacArthur, who was acting in direct opposition to his orders from the president, began to burn the shacks of the Anacostia camp. He encountered no resistance. Veterans helped his soldiers torch their own Hooverville. The Washington sky was a bright orange all night as the camps of the bonus army were destroyed, and their inhabitants wandered into the streets, joining the larger army of the unemployed of the Great Depression. By midnight the forced evacuation was complete.

President Hoover issued a statement in which he asserted: "An examination of a large number of names discloses the fact that a considerable part of those remaining are not veterans; many are Communists and persons with criminal records."

He tried to soften this by adding, "The veterans amongst these numbers are no doubt unaware of the character of their companions and are being led into violence which no government can tolerate."

But the violence had not occurred until the army ordered by Hoover himself arrived on the scene. It was a case of misjudgment, one among many that would, to a large degree, account for Hoover's defeat at the polls by Franklin D. Roosevelt in November 1932.

KEY TO MAJOR TERRORIST ORGANIZATIONS

AFRICA

South Africa

ANC (African National Congress)
Formed to fight for freedom from white domination in South Africa.

EUROPE

France

Action Directe
Anti-NATO group.

ALNC Armée de Liberation Nationale Corse (Corsican National Liberation Army)
Employed bomb attacks against France in cause of Corsican independence.

Germany

Baader-Meinhof Gang
Urban guerrillas named after Andreas Baader and Ulrike Meinhof, their leaders.

RAF (Red Army Faction)
Present form of Baader-Meinhof Gang, merged with second June Movement, named after date on which a student was shot in West Berlin demonstration.

Greece

November 17
Named in memory of date of student uprising in 1973 against colonels' regime.

Ireland

IRA (Irish Republican Army)
Traditional guerrilla and terrorist group dedicated to unification of Ireland.

PIRA (Provisional Irish Republican Army)
Extremist terrorist group within IRA that has taken over completely and is now solely responsible for terrorism in Northern Ireland and British Isles.

Italy

Red Brigades
Leftist terrorist group.

Spain

ETA Basque Homeland and Liberty
Separatist organization founded in 1959.

FAR EAST

India

All India Sikh Students Federation
Followers of Sant Jarnail Singh Bhindranwale, killed in Amritsar massacre.

Japan

JRA (Japanese Red Army)
Active at home and internationally with Palestinian groups.

LATIN AMERICA

Brazil

ALN (National Liberating Action)
Left-wing terrorist group based in Brazil.

MR-8 (October 8 Revolutionary Movement).
Left-wing terrorist group based in Brazil.

Colombia

M19 (April 19 Movement)
Colombian terrorist organization.

El Salvador

FMLN (Farabundo Martí National Liberation Front)
Antigovernmental groups in El Salvador.

MIDDLE EAST

Abu Nidal Faction
Breakaway from Fatah; based in Libya; responsible for attacks in Europe.

Al-Borkan (The Volcano)
Anti-Qaddafi group; the terrorist wing of the National Front for the Salvation of Libya.

Al-Dawa (The Call)
Shiites, mostly Iraqis hostile to Assad regime and supporters of Khomeini.

Al-Gama'a al-Islamiyya
Egypt's largest militant group, seeking to overthrow Egypt's government.

Al-Jihad
Close partner with Osama bin Laden. Seeks to overthrow Egyptian government. Responsible for assassination of Anwar Sadat in 1981.

Al-Qaeda
Osama bin Laden's worldwide organization dedicated to overthrowing all non-Islamic governments and expelling Westerners and non-Muslims from Muslim countries.

Fatah
Fatah is Yasser Arafat's power base in the Palestine Liberation Organization (see below) and devotes itself to what he refers to as "the armed struggle."

Force 17
Originally bodyguard of PLO leader Yasser Arafat; conducts raids on Israeli targets.

Grey Wolves
Turkish terrorist group; nationalist and fascist.

Hamas
Friendly with Iran and an outgrowth of Palestinian branch of Muslim Brotherhood. Pursues establishing a Palestinian state in place of Israel.

Hezbollah (The Party of God)
Lebanese Shiites; part of Islamic Jihad; supported by Iran in order to establish Islamic republic in Lebanon.

Iraqi Islamic Revolution
Tehran-based; opposes President Saddam Hussein of Iraq.

Islamic Amal (Islamic Hope)
Lebanese Shiites.

Islamic Jihad (Islamic Holy War)
Umbrella organization uniting Lebanese, Iranian and Iraqi terrorist groups dedicated to making war on West and installing Shiite Islamic Revolution throughout Middle East.

PLO (Palestine Liberation Organization)
Umbrella for various Palestinian groups, terrorist and otherwise; avowed spokesperson for Palestinians.

PFLP-GC (Popular Front for the Liberation of Palestine—General Command)
Operating under Syrian government orders, headed by Ahmed Jibril; totally terrorist organization.

Palestine Liberation Front
Splinter group of General Command, split itself into three groups: pro-Syrian (commander: Abdul Ghanem); pro-PFLP (commander: Talaat Yaquib); pro-Arafat (commander: Abu Abbas).

UNITED STATES

Alpha 66
Miami-based Cuban exile group, engaged in sabotage, assassination, invasions of Cuba.

CFF (Croation Freedom Fighters)
Separatist group.

FALN (Armed Forces of Puerto Rican National Liberation)
Nationalist guerrilla group.

JDL (Jewish Defense League)
Counterterrorist group attacking Arabs in United States.

Omega 7
Anti-Castro Cuban group based in United States.

SLA (Symbionese Liberation Army)
See Weathermen.

Weathermen (The Weather Underground Organization)
Terrorist organization that preaches solidarity with ethnic minorities.

TERRORISM

BRAZIL
RIO DE JANEIRO
September 9, 1969

Members of Brazil's ALN and MR-8, demanding the release of 15 political prisoners, kidnapped U.S. ambassador Charles Elbrick on September 9, 1969. The prisoners were freed, and so was Ambassador Elbrick.

The first in a long line of diplomatic kidnappings occurred on September 9, 1969, in Rio de Janeiro, Brazil, at 2 P.M., as U.S. ambassador Charles Elbrick's limousine slowly pulled up to his home on Marquis Street.

The kidnappers, members of Brazil's two most extreme left-wing terrorist groups, the National Liberating Action (ALN) and the October Eight Revolutionary Movement (MR-8), had been waiting all morning in two

Volkswagens. Now, the two Volkswagens lurched from the curb and blocked the street while three unobtrusive loiterers advanced on the limousine with drawn pistols.

They entered the car, held the ambassador at gunpoint and ordered the chauffeur to drive to a nearby secluded dead-end street. There the ambassador was hustled into a microbus that was parked, its engine running. The microbus sped off at breakneck speed for an unknown destination, leaving, on the seat of the Cadillac, a ransom note. It contained two demands: first, the publication of a 1,000-word document denouncing Brazil's military regime, and second, the release of 15 political prisoners. If the demands were not met within 48 hours, the note trumpeted, "We will be compelled to mete out revolutionary justice."

While the Brazilian government and American representatives mulled this over, they were directed to a second note that was left in a box at a church in downtown Rio. In Elbrick's handwriting it read: "Hurry to meet the conditions for my release."

The Brazilian government agreed to the demands. A few hours later, a note from Elbrick in a Rio supermarket announced that he would be freed as soon as the liberated prisoners arrived safely in Mexico City.

"If they get away with it in Rio," said one American Foreign Service Officer to *U.S. News and World Report,* "no diplomat will be safe anywhere in the world."

They did, and the prophecy would come true. Elbrick was returned safely, and the political prisoners were freed. The pattern had been set, and from 1969 onward, political kidnappings would continue to occur, often with more tragic results.

COMORO ISLANDS
MORONI
November 23, 1996

An Ethiopian Airlines Boeing B-767 en route from Bombay, India, to Africa's Ivory Coast was hijacked on the afternoon of November 23, 1996, shortly after the flight left Addis Ababa. The plane ran out of fuel and ditched in the Indian Ocean off Great Comoro Island. One hundred twenty-seven of the 175 aboard the plane were killed.

Ethiopian Airlines Flight 961, a Boeing B-767, left Addis Ababa on schedule on November 23, 1996, with a full load of 163 passengers and 12 crew members. Its final destination was Africa's Ivory Coast after stops in Nairobi, Kenya, and Brazzaville, Congo. It was the last leg of a flight that had originated in Bombay, India.

But 20 minutes after the plane left Addis Ababa, two passengers burst out of their seats and ran up the gangway to the cockpit. At first flight attendant Hiwot Tadesse thought they were having a fight, but when a third man followed the first two holding something in his gloved hands, she realized that the plane was being hijacked.

"I pushed back my trolley and told the other girl to stop serving drinks," she revealed later. "The terrorists said for everyone to be seated. They said they had explosives and they were going to blow up the airplane."

For four hours the escaped prisoners drank and threatened the rest of the passengers and crew aboard the plane. They wanted to go to Australia, they told the pilot, who patiently tried to explain to them that they did not have enough fuel to make Australia. Either ignoring or disbelieving him (their dialect encompassed English, French and Amharic, Ethiopia's main language), they insisted that he fly to Australia anyway.

And so the nightmare aboard the jet intensified. The hijackers beat the co-pilot and threw him into the passenger cabin, where the third hijacker, the one with the bomb, remained, glaring at the passengers to keep them in their seats. "He let us feed the children," Mrs. Tadesse said. When he looked down one aisle, she said, she would let passengers along the other aisle sneak to the bathroom or get a sandwich.

Meanwhile, in the cockpit one of the hijackers squeezed into the co-pilot's seat and played with the joystick, causing the plane to dip frighteningly. It produced gales of drunken laughter from the other hijacker.

Over and over the pilot reiterated that he did not have enough fuel to get them to Australia, but the hijackers continued their insistence, waving the ax and oxygen bottle they were using for weapons and threatening to beat him or kill him if he did not follow their directions.

Finally, at 3:10 P.M. the pilot made his first allowed announcement to the passengers. "We have lost one engine," he said. "We are running out of fuel. We are going to have a crash landing. Get ready."

The third hijacker ran up the plane's aisle shouting "Sit! Sit!" Ten minutes later, completely out of fuel, the jet glided over the Galawa Beach Hotel, located on the Indian Ocean near Morani on Grand Comore Island. When the plane hit the water just off the hotel's beach, Caroline Fotherby, the hotel's manager for water sports, was in her beach clubhouse doing clerical work. "I heard a plane approaching, but not loudly," she later said. "Then there was a crash. First, I thought it was thunder, but it was sunny. Then I thought it was the volcano here, but then my staff screamed 'Emirates!' [the name of an airline that flew to the island once a week] I turned round and saw the wing of a plane sticking out of the water, and a tail just a stone's throw from one of my dives in progress."

The plane skimmed the surface of the sea for a short distance, then the left wing dug into the water and the plane followed, breaking in two when it hit the ocean 500 yards off the beach at Le Galaw. Ms. Fotherby immediately called the hotel nurse and all her instructors by radio, and they rushed to the site in high-speed rubber boats, grabbing survivors where they could. A glass-bottomed cruise boat pushed off the beach and brought a dozen survivors to shore.

The nurse set up a clinic in an open-air restaurant, and eight vacationing French doctors and two from South Africa worked for hours in their bathing suits saving lives.

As the living were treated in the restaurant, the dead were set on top of old sheets and linens on the beach.

The survivors described their horror when they found they were strapped in their seats with waves pitching the plane back and forth. Those who struggled out of their seatbelts crawled through the twisted metal of the cabin to reach holes where they could see sunlight. Uninjured in the crash, many of them emerged with cuts as they crawled through the openings in the jet's fuselage. Once through, they clung to the wreckage, screaming for help as pieces of bodies and luggage floated near them.

Franklin P. Huddle, the United States consul general in Bombay, said later, "I thought I was dead when we hit the water. When the plane hit the water, it hit gently. There were a couple of good-sized lurches, but not too violent, and then a hard swerve." He and his wife managed to swim to a windsurfer, where they clung until they were rescued.

"It is a miracle for me," said Nagin B. Surti, another passenger from Bombay. He had climbed the overhead rack and squeezed through a hole in the roof the plane. "I had made my last prayer saying, 'Please, God, we are your human beings, please.'"

Hotel scuba divers who dove into the wreck described bodies trapped in their seats, sometimes held in by the life vests they had put on when the captain said they were going down. Ultimately, 127 passengers and crew died in the crash, many of them trapped and drowned.

The injured who were treated on the beach were transferred to the run-down hospital in Moroni, where they were jammed, eight to a room, on ancient iron beds. Some, like Mr. Huddle, were flown to the French Island of Réunion by helicopter. Both the pilot and co-pilot survived, as well as two of the hijackers, who were identified and immediately thrown back into jail.

EAST AFRICA
NAIROBI, KENYA/DAR ES SALAAM, TANZANIA
August 7, 1998

In two coordinated bombings blamed on Osama bin Laden, the American embassies in Nairobi, Kenya, and Dar es Salaam, Tanzania, were bombed on the morning of August 7, 1998. Two hundred and fifty-seven were killed and 5,538 were injured.

After the bombing of the American embassy and Marine barracks in Beirut, Lebanon, in the 1980s, United States embassies in certain foreign countries were discovered to be vulnerable to terrorist attacks. As a result of the Beirut bombing, the U.S. State Department ordered modifications and toughening of security standards worldwide.

But the funding did not meet the demands of the project. Over the years less than a third of the money that would have been necessary to meet the new standards was appropriated. Thus, low risk embassies like the ones in Nairobi, Kenya, and Dar es Salaam, Tanzania, received almost no funds at all, and work was only partially done.

In Dar es Salaam a nine-foot wall was built in 1988, but it was as close as 25 feet to nearby streets and buildings. The embassy in Nairobi was set back only 30 feet from the street on a busy downtown intersection.

When two bombs exploded before the two embassies within minutes of each other shortly after 10:30 A.M. on August 7, 1998, the proximity of the embassies to the streets of each city was thought to have made it easier for the bombers to do their work, which was devastating. A total of 257 people died, and 5,538 were injured in the twin blasts, which destroyed not only most of the embassies but other buildings as well.

In Dar es Salaam, the bomb, set off outside the embassy's wall, nevertheless created major damage to the face of the embassy and to other diplomatic facilities and residences in the immediate vicinity. The back of the embassy was peeled away, and staircases hung suspended in the air, leading nowhere. Concrete slabs blown from the building were strewn in crazy profusion on the grounds and the streets. Dozens of cars parked on the streets near the embassy were set afire or blasted apart. The American ambassador's residence, a thousand yards away and vacant at the time, suffered a collapsed roof and fragmented interior ceilings. A total of 12 people were killed and 85 were injured.

In Nairobi the devastation was considerably more because of the proximity of the bomb and the location of the embassy. There was massive, explosive damage to the interior of the building. Floor after floor was reduced to rubble. Windows, window frames and walls on the rear side of the building were reduced to twisted fragments. The force of the explosion sent glass flying, internal concrete walls buckling and furniture and fixtures soaring through the air, turned into lethal weapons as they were flung directly at the workers within the offices. Most casualties, in fact, resulted from these flying objects.

Next door to the embassy, the Ufundi Building collapsed entirely, trapping hundreds and killing many within its buckled structure. Flying glass from the Co-op Bank Building threatened and in some cases killed unwary pedestrians and motorists in the crowded streets near the embassy. Practically every building within a two- to three-block radius was damaged. In Kenya alone more than 213 people were killed and more than 5,000 were injured, some critically.

The suicide bombers of the Dar es Salaam embassy drove a pickup truck loaded with explosives up Laibon Road to one of the two vehicular gates. An embassy water tanker was stopped at the entrance toward which they were pointed and guards inspected it minutely for possible evidence of a bomb attached to its undercarriage. As the truly bomb laden truck drew up, one guard was bending down and focusing a mirror attached to a pole on the undercarriage of the water truck. The fuse in the bomb truck was set for 10:39 A.M., and as it was pulled up behind the water truck it exploded.

For weeks investigators linked the blast to the water truck, but examination of the direction of the explosion finally determined that it came from the truck directly behind it.

In Nairobi the entrance of the bombers was considerably more complicated and, eventually, violent. Shortly after 10:20 A.M. the guards at the rear of the embassy opened the fence gate and its drop bar to allow a mail truck to exit the embassy's garage. As they closed the gate and lowered the bar behind the mail vehicle, they noticed a truck pulling into the uncontrolled exit lane of the rear parking lot.

The heavily loaded truck maneuvered to the embassy's rear access control area, but just before it reached it a car came out of the Co-op Bank's underground garage, stopping it momentarily. The car backed up, and the truck proceeded to the embassy drop bar.

One of the terrorists demanded that the guards open the gates. They refused, and the terrorist on the passenger side of the truck drew out a gun and began shooting at the unarmed guards, who ducked for cover. A flash grenade was thrown by one of the terrorists, but it missed the guards and the chancery. Crouching behind a pillar, one of the guards took out his hand held radio and tried to contact the armed Marine Guard at the command post at the front of the embassy. The single radio frequency was occupied with other traffic. The other guard ran to a telephone, picked it up and tried to call the Marine post. The line was busy.

Embassy employees, hearing the gunfire, came to the windows to see what was happening, and just as they did, at 10:30 A.M., the truck exploded. The guards and every one of the employees at the windows was either killed or seriously injured.

The sheer magnitude of the explosion and its attendant casualties overwhelmed both the first rescue crews and the hospitals to which the dead and wounded were taken. Kenyatta National Hospital, the city's largest facility and the hospital to which most of the casualties were taken, soon began to run out of everything. X-Ray film, respirators, sterile gloves, pain-killing drugs and other basic supplies rapidly disappeared. "This place was filled with blood," Henry Njoroge, an X-ray technician, told reporters. "We treated so many who were unconscious and in shock, and we were shocked ourselves."

Hundreds of people appeared at the hospital and donated blood, food and blankets. But blood shortages continued in outlying hospitals. A bureaucratic snafu delayed medical supplies and food that were put aboard an

The tangled and burning wreckage of a truck and a car, exploded by terrorists outside of the U.S. Embassy in Nairobi, Kenya, smolder in the aftermath of the attack. (AP/Wide World Photo)

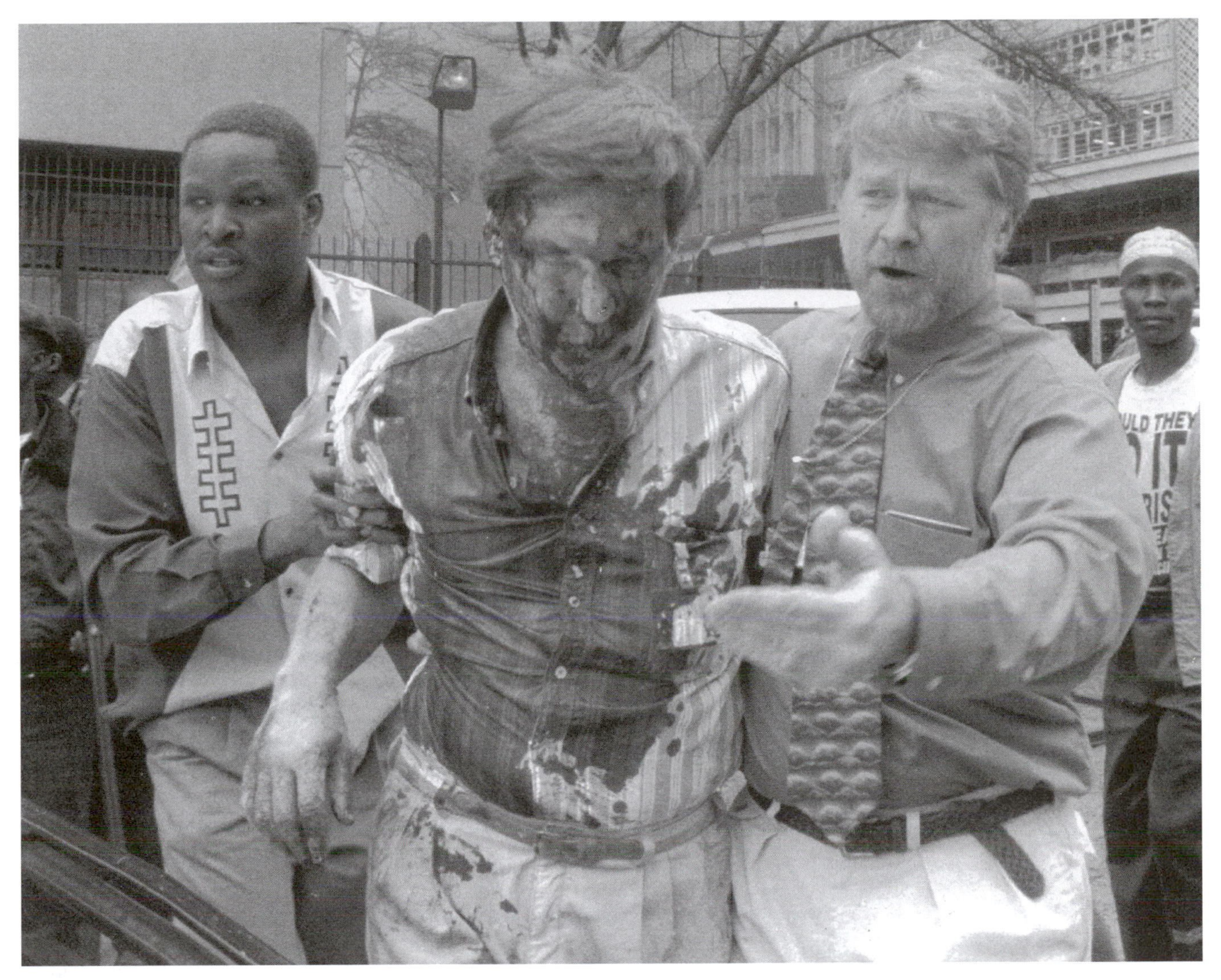

A rescue worker and a fellow employee help a severely wounded U.S. Embassy staff member to an ambulance following the blast that ripped apart one building and severely damaged the U.S. Embassy in Nairobi. (AP/Wide World Photo)

airplane from Washington, while another broke down in Sicily long enough for a cumulative outcry of accusations to develop that the Americans were treating their own and failing to rescue or minister to the thousands of Kenyans who were victims of the bombing.

Rescue work proceeded chaotically, too, with no real leaders until the arrival of a crack Israeli rescue team accompanied by equipment and dogs. They brought three cranes into play in the Ufundi Cooperative House to lift giant concrete slabs, then sent dogs trained to find bodies into the wreckage. That night at 10:45 P.M., they pulled a hysterical businessman from the ruins of the building. "We were kind of like psychologists," the captain of the rescue squad told reporters. "We had to calm him down and get him to do some things."

For days individual survivors were found, miraculously, in the wreckage of various buildings. But more bodies continued to add to the casualty figures.

Meanwhile, at Dar el Salaam an undamaged security video camera pointing at the spot at which the blast occurred promised to yield evidence. But it did not. There was no videotape in it or, in fact, in any of the other security cameras surrounding the embassy.

Inspectors arrived, and the search for the terrorists began. Evidence surfaced that the ambassador in Nairobi had pleaded with the State Department for increased security for the embassy and was refused. Finally, on August 13 the FBI found pieces of the truck that carried the explosives in Nairobi, and the bomb crater was scooped for samples of chemicals.

In the middle of August, American investigators in Pakistan arrested Mohammed Sadiq Howaida, who had flown from Kenya to Pakistan immediately following the explosions and who had attempted to enter the country with a fake passport. Mr. Howaida, later reidentified as Mohammed Saddiq Odeh, was a devotee of Osama bin

Laden, the Saudi-born extremist who had vowed war upon the United States and was currently in Afghanistan.

Three days later, after Mr. Odeh was transported to Kenya, Pakistani authorities arrested two more men who were trying to cross into Afghanistan at Torkham in the Khyber Pass and had been on the same flight that had taken Mr. Odeh from Kenya to Pakistan. Meanwhile, the U.S. attempted to press the Taliban militia in Afghanistan to expel Osama bin Laden so that he could be tried in the United States as a suspect in the twin bombings. The Taliban refused.

The story became more complicated. Evidently despairing of being able to try Osama bin Laden, America tried to kill him with missiles. On August 20, 1998, approximately 75 cruise missiles were launched from U.S. Navy ships in the Arabian and Red Seas. Most of them struck six separate targets within a camp near Khost, Afghanistan, dubbed by President Clinton "one of the most active terrorist bases in the world."

Another barrage of missiles struck a factory in the Sudanese capital of Khartoum. The factory, according to U.S. government officials, was being used to produce important components for making chemical weapons, particularly VX nerve gas.

The twin attack did not achieve its purpose and damaged relations with the Sudanese. Bin Laden had left his base in the camp a day before the missile launch. As for the factory in Khartoum, its identity as a producer of components of nerve gas was based upon faulty information; it was merely a pharmaceutical factory owned by Shifa Pharmaceutical Industries and dedicated to producing medicine and veterinary supplies.

Shortly after this, Mohammed Rashed Daoud al Owhali, the passenger in the van that approached the gate in Nairobi and the terrorist who threw the flash grenade, was brought from Pakistan to the United States to join Mr. Odeh for trial.

Mohammed Saddiq Odeh had confessed that he was trained at terror camps in Afghanistan by Osama bin Laden. By September authorities arrested Wadih el Hage, a Texas resident who had been Mr. bin Laden's personal secretary. Accused of making false statements to investigators, he was jailed but later freed.

The two conspirators received life sentences. A grand jury indicted Osama bin Laden on 238 counts of murder, and the United State offered a reward of $5,000,000 for information leading to the arrest and conviction of Mr. bin Laden. To this date it has not been collected, nor has the multimillionaire sheik been arrested.

GERMANY
MUNICH
September 5, 1972

Israel's Olympic team was captured in Munich's Olympic Village on September 5, 1972, by Black September terrorists seeking to free political prisoners. A bungled rescue attempt by German authorities ended in the deaths of all 11 athletes and three terrorists.

There will hopefully never be another session of the Olympic Games remotely like that which occurred in September 1972 in Munich, Bavaria.

One of the objects of the games was to contrast the Nazi Germany of 1936—the last time the games were held there—with the prosperous, democratized West Germany of the 1970s. Instead, they would be interrupted by the horrible massacre of 11 of Israel's top athletes.

Until the fearful events of September 5, the XX Olympaid had been an enormous success. More records had toppled than in any other Olympiad to date.

But while this was occurring, Black September, a fanatical splinter group in Fatah, the PLO fighting unit that drew its recruits from other Palestinian groups working under the PLO umbrella, was planning a dramatic kidnapping plot that would publicize its cause to the world. The week before the Olympics began, several Black September members, bearing a veritable arsenal of Russian-built Kalashnikov submachine guns, pistols and hand grenades, set out for Munich. Once in Munich, they spread out, and a number of them got jobs among the 30,000 temporary employees of the Olympic Village.

At 4:20 A.M. on the morning of September 5, 1972, two terrorists wearing sports warm-up suits and carrying athletic equipment scaled the six-and-one-half-foot fence surrounding the village. Two telephone linemen saw them but thought little of it. They were, as far as they knew, a couple of Olympic athletes who had broken the curfew and were sneaking back to their quarters.

In total, there were eight Black September members within the compound. Pausing momentarily outside the athletic quarters, they either blackened their faces with charcoal or pulled on ski masks and made their way to the Olympic Village apartments that housed 22 Israeli athletes, coaches and officials. Two of them knocked on one door, inquiring in German, "Is this the Israeli team?" Wrestling coach Moshe Weinberg opened the door a crack, saw the masked gunmen, flung himself against the door and shouted to his roommates to flee. Immediately, Weinberg was riddled by a burst of submachine gunfire through the door. He died on the spot.

Simultaneously, in the other apartments similar scenarios were being played out. In one, Yosef Gottfreund, an impressively tall wrestling referee, tried to hold off invading terrorists and was knifed to death.

Altogether, 18 Israeli athletes scrambled through windows to safety. Nine who did not make it were bound hand and foot in groups of three and pushed together onto a bed in one of the apartments. As hostages, they would be the bartering chips for the terrorists.

By 6 A.M. Munich police had been alerted, and 600 of them surrounded the area. An ambulance removed Weinberg's body, which had been dragged to a terrace and left by the gunmen.

Police Chief Manfred Schreiber attempted to brazen his way into the apartments. He was met by the group's leader wearing a white tennis hat and sunglasses. For a moment it seemed possible for Schreiber to take him hostage, and then, according to Schreiber, the man asked, "Do you want to take me?" He opened his hand and showed a hand grenade to the police chief. The terrorist's thumb was on the grenade's pin.

At 9 A.M. a message in English was tossed from a window. On it was a list of 200 prisoners currently held in Israeli jails. They included Ulrike Meinhof and Andreas Baader, the leaders of a gang of German terrorists who had robbed eight banks, bombed U.S. Army posts and killed three policemen before they were captured the previous June, and Kozo Okamoto, a Japanese Red Army terrorist who took part in the massacre at Tel Aviv's Lod Airport in which 26 people died (see p. 133). All were to be freed, according to the note, before the Israeli athletes would be released.

Furthermore, the Palestinians demanded that they and their prisoners be flown out of West Germany to any Arab nation except Lebanon or Jordan aboard three airplanes that would leave at agreed upon intervals. Officials had three hours to comply. If they did not, the hostages would be executed at the rate of two every thirty minutes.

International phone lines hummed. West German Interior Minister Hans Dietrich Genscher took personal charge of the negotiations, first offering an unlimited sum of money for the release of the hostages and then offering himself and other West German officials as hostages in place of the athletes. Although he was turned down, he was able to stall for time by stating that he was in touch with Israeli authorities. There were two extensions of the deadline, the first to 3 P.M., the next to 5 P.M.

Meanwhile, 15 volunteer police sharpshooters were brought into the area, and worldwide television coverage showed them crouching in readiness until German authorities realized that the terrorists could also tune them in, at which point they ceased the TV coverage.

The games were suspended at 3:45 P.M. that afternoon, after a request from Israel. By that time Willie Brandt, West Germany's chancellor, had made the decision to permit the terrorists to fly out of West Germany with the hostages. Speaking of the athletes, Brandt said to news-people later, "We are responsible for the fate of these people."

By 6 P.M. Genscher had run out of stalling tactics. He was told by Brandt that the Palestinians and their hostages would be taken to Munich's airport and flown out on a Lufthansa 727 jet to any place they named. The terrorists selected Cairo, and a 7 P.M. deadline was set. In actuality, the Germans were moving their sharp-shooters to Furstenfeldbruck Field, and the Arabs were planning a destination other than Cairo.

At 10 P.M., 18 hours after the initial assault, the terrorists herded their prisoners, tied together in single file and blindfolded, out of the building and into a German army bus, which drove them through a tunnel under the village to a lawn 275 yards away. The green expanse had been converted into a helicopter pad. There were three helicopters there; two took the terrorists and their hostages on the 25-minute ride to Furstenfeldbruck airport; the third went ahead, carrying German officials and Israeli intelligence officers.

The airport was ringed by 500 soldiers. But there were only five sharpshooters to pick off eight terrorists. The rest had unexplainably been left behind at the Olympic Village.

The helicopters landed. The terrorists leaped out and took the German crews hostage. They arranged them in front of their helicopters and proceeded to inspect the 727.

As they walked toward it, the police sharpshooters opened fire. The two Arabs guarding the helicopter crews were killed, and one of the pilots was wounded. One more guerrilla on the tarmac died. The leader dove under a helicopter, fired back and knocked out the floodlights on the field and the radio in the control tower. A Munich police sergeant was gunned down.

The battle would rage for an hour more. Five guerrillas, including their leader, would be killed, and three would surrender. And every one of the hostages would be killed. One group of four was burned to death when a terrorist tossed a grenade into the helicopter in which they were being held. The remaining five would be machine-gunned by their captors.

It would be four hours before the horrendous results of the failed ambush would reach the outside world. Reaction from the Arab world would be divided. Lebanon would offer condolences to Israel. Egypt would charge that German bullets killed them all.

In coming days, Israeli retaliation was swift and fierce. On the eve of Rosh Hashanah, Israeli jets struck Lebanon and Syria with the heaviest strikes since the 1967 war. Arab sources later said that 66 were killed in the raids by 75 jets. Israeli ground troops crossed the Lebanese border to battle commandos who had been mining roads in Israel. Syria put its army on alert. Meanwhile, in Libya Colonel Qaddafi conducted a martyr's funeral for the dead terrorists.

The three surviving Arabs would be tried and imprisoned, but in November they would be released in exchange for a Lufthansa airliner hijacked in Beirut by other Black September terrorists. In Tripoli Colonel Qadaffi would give the three a hero's welcome and parade.

GREAT BRITAIN
ENGLAND
LONDON
December 17, 1983

Continuing IRA terrorism attacks on British citizens resulted in the explosion of a car bomb planted by the Provisional IRA during the Christmas shopping season on December 17, 1983. Six were killed; 94 were wounded.

It was Dalthi O'Connell, one of the founders of the Irish Republican Army, who is credited with inventing the car bomb, the stock-in-trade of terrorists worldwide. By 1983 the Provisional IRA had much experience with setting car bombs, and it was one such device that accounted for the

Christmas carnage of Harrods department store in London on December 17, 1983.

There had been warnings issued via radio, television and newspapers of possible IRA bombings during the holiday season. Scotland Yard intelligence reports from Northern Ireland had warned of "a Christmas blitz."

At approximately 12:40 P.M. on December 17 the phone rang at the Samaritans, a voluntary charity organization in London. A distinctly Irish-accented voice announced, "Car bomb outside Harrods. Two bombs in Harrods."

The Samaritans called Scotland Yard, and at 1:15 a team of police, including animal handlers and trained dogs, arrived on the scene. They went to work in the store first, trying to trace down the interior bombs, while other police conducted a search of cars on the streets surrounding the giant store. The last Saturday before Christmas, thousands of shoppers jammed the store and the sidewalks surrounding it.

At exactly 1:20, a car not checked by the police suddenly exploded with a thunderous, earsplitting roar. Black smoke and shrapnel erupted as if they had been launched from a volcano. The concussion shattered windows for blocks and instantly killed five shoppers and a policeman. Other unsuspecting pedestrians and shoppers were injured, some horribly, by rainstorms of glass and metal.

Ninety-four would be injured; six would die. Those responsible for the bomb would never be captured.

GREAT BRITAIN
SCOTLAND
LOCKERBIE
December 21, 1988

No clear-cut responsibility was established for the midair terrorist bomb explosion aboard Pan Am Flight 103 from London to New York on December 21, 1988. Two hundred fifty-six people on board the plane and 11 on the ground were killed in the fiery crash.

Pan Am Flight 103, a Boeing 747, took off from London's Heathrow Airport 25 minutes behind schedule, at 6:25 P.M., on December 21, 1988. Aboard were 246 passengers and 10 crew members, among them 35 of 38 Syracuse University students who had been studying abroad and were returning home for Christmas and Brent Carisson of Sweden, the chief administrative officer of the United Nations' Council for Namibia. Carisson was flying to New York for the signing of an accord on Namibian independence.

Fifty-two minutes later, while Flight 103 was flying at an altitude of 31,000 feet over the small village of Lockerbie in the extreme southeren end of Scotland, a bomb, planted in a tape recorder and radio in the plane's luggage compartment, exploded. The main part of the airplane dropped like a flaming missile, landing near a gas station on the outskirts of Lockerbie and setting fire to the station, a dozen row houses and several cars that were on the A74 highway to Glasgow.

Other pieces of the liner and some bodies were strewn over the countryside in an 80-mile-long arc. It was the worst airline crash in British history and the worst single plane crash in Pan Am's history. The BBC broadcast horrendous pictures of raging fires, devastated houses and cars and shreds of aircraft wreckage. "The plane came down 400 yards from my house," said Bob Glaster, a retired policeman, to reporters. "There was a ball of fire 300 feet into the air, and debris was falling from the sky. When the smoke cleared a little, I could see bodies lying on the road. At least one dozen houses were destroyed."

The terrible part of the tragedy was that Pan Am and government agencies had been warned of the possibility of the bombing in ample time to prevent it and had been unable to accomplish this. One week before the bombing, the American embassy in Finland had received a notice saying that an unidentified caller had warned that "there would be a bombing attempt against a Pan American aircraft flying from Frankfurt to the United States." Flight 103 had originated in Frankfurt on a 727 with the same flight number. In London, at Heathrow Airport, passengers and baggage had been transferred to the larger 747 for the longer leg of the journey to New York.

Later investigation revealed that there was, indeed, increased surveillance of passengers boarding the craft and that when embassy personnel scheduled to board the flight were warned of the threat, many canceled their reservations. The general public was *not* warned, and this would make headlines in the United States, particularly in light of later disclosures (see below).

For many months, because of the plastic nature of the explosive, blame was directed toward two anti-Arafat Palestinian terrorist groups, the PFLP General Command, led by Ahmed Jabril, and the Fatah Revolutionary Council, led by Abu Nidal. Later investigation by Scotland Yard, however, led to the conclusion that the initial investigation linked the bombing to the Iranian callers, or, some speculated, at least a terrorist group sympathetic to Iran.

In November 1990 it was revealed that the U.S. Drug Enforcement Agency regularly used Pan Am Flight 103 to fly informants and suitcases of heroin from the Middle East to Detroit. Nazir Khalid Jafaar of Detroit was aboard this flight and involved in this operation.

Pan Am's baggage operation in Frankfurt, it was further revealed, was used to put suitcases of heroin on planes, apparently without the usual security checks, under an arrangement between the drug agency and German authorities. Thus, it was eminently possible that Jafaar, who was either an agent or an informer for the DEA, was the unwitting carrier of the bomb that destroyed Flight 103 and killed 256 people in the air and 11 on the ground on the night of December 21, 1988.

Later information refuted this and intimated that the bomb was planted by Libyan terrorists in retaliation for the 1986 attack on Colonel Muammer el-Qaddafi by American jets. This line of investigation ultimately led to the demand by Great Britain and the United States for the extradition for trial of two Libyans, Lamen Khalifa Fhimah and Abdel Basset Ali al-Megrahi. The two, the investigation concluded, had planted the bomb responsible for the crash of Pan Am 103.

The two powers took the charges to the United Nations Security Council in early 1992, asking for sanctions against Libya if Colonel Qaddafi did not turn over the two agents. In March 1992 the Security Council passed Resolution 731 ordering Libya to surrender the two men for prosecution in Britain and the United States and also surrender evidence that could be used against them.

Colonel Qaddafi refused, then agreed to turn them over to representatives of the Arab League, then changed his mind again. In late March the United Nations Security Council gave the colonel two weeks to conform to Resolution 731 or face sanctions that would cease air travel into and out of that country and severely reduce Libya's diplomatic presence in the rest of the world. Colonel Qaddafi's reply was to threaten to cut back on oil exports to various countries sympathetic to the UN resolution.

Even if the two terrorists were handed over, experts and family members of those who perished in the crash announced that this would be settling only a small part of the crime. International politics—specifically, the role of Syria in the Persian Gulf War—were preventing investigators from acting on what they knew about the entire operation, these critics and family members charged. President George H. W. Bush's statement that Syria had received a "bum rap" simply did not square with the facts, they noted.

In a statement to *New York Times* columnist A. M. Rosenthal on March 30, 1992, Steven Emerson, the Washington journalist who, with Brian Duffy, wrote *The Fall of Pan Am 103 in 1990,* said:

> The undisputed intelligence shows that Syria-based and -supported terrorists, led by Ahmed Jabril, head of the Popular Front for the Liberation of Palestine—General Command, planned and organized multiple airplane bombings against U.S., European and Israeli airlines in October 1988.
>
> The money and orders for the operation came from Iran, seeking revenge for the shooting down of the Iranian airbus that summer by the U.S. According to intelligence officials, Iranian officials traveled to Germany to oversee the operation and to personally witness the transfer of explosives and bombs.
>
> But the plan went awry when Syrian-based terrorists were arrested by German police in late October 1988. Jabril, who had received funding from Libya for at least the previous two years, handed off the operation to Libya, which had its own terrorist infrastructure in place.

Thus the sequence of events as reconstructed by international investigators: Iran bankrolled it, Syrian-based terrorists planned it; Libyans executed it.

The reason that Pan Am 103 was chosen? According to Vincent Cannistraro, who headed the CIA investigation of the crash until he left the agency in 1990, the Jabril group settled on Pan Am because its surveillance indicated that in Frankfurt the airline was not "reconciling" baggage fully. That is, it was not making sure that every piece of luggage "was identified directly with a passenger before being taken on board."

In January 2001, a Libyan man, Abdel Basset Ali al-Megrahi, was convicted of having planted the bomb that brought down flight 103.

IRAN
TEHRAN
November 4, 1979–January 20, 1981

The need for the Ayatollah Khomeini to galvanize anti-Western, pro-Islamic loyalty was the root cause of the taking of the U.S. embassy in Tehran and the imprisonment of 52 hostages for 444 days.

At a few minutes before 11 A.M. on November 4, 1979, 400 young Iranian "students," later learned to be members of the Revolutionary Guard, cut through chains that joined together the gates of the American embassy in Tehran, Iran. The Iranian guards stationed at the gates offered no resistance, and the invading crowd soon swarmed over the embassy compound.

It seemed at first to be a mirror image of other temporary embassy takeovers that had occurred in the world in the previous months, and President Jimmy Carter, spending Sunday at Camp David, Maryland, did not even return to Washington when informed of the break-in. But by the next day, it became apparent to the president and the world that this was a move without precedent. Bound and blindfolded hostages were paraded before angry crowds chanting death to the Great Satan, America.

A year earlier Shah Mohammad Reza Pahlavi had been deposed by revolutionary forces spurred on by Iran's spiritual leader, the Ayatollah Khomeini. The shah had fled to Mexico, and two weeks before the break-in at the American embassy in Tehran he had flown to New York, where he was undergoing treatment for cancer. The "students" demanded that the deposed shah be returned to Iran for trial.

The United States refused, and thus began one of the longest standoffs in history. For the next 444 days, a tug of war would pit Iran against the United States, with the hostages as the pawns in the game.

Two weeks after the takeover of the embassy, the militants released 13 hostages—eight black men and five women—who returned to the United States in time for Thanksgiving of that year. The shah's health improved, and Mexico, not wishing to become involved in the U.S.-Iran standoff, refused to let him return. He fled to Panama, where he remained in exile, despite Iran's demands.

That Christmas the White House tree remained dark out of respect for the hostages, and Americans, using the words of a popular song as their cue, began to tie yellow ribbons around trees, where they would shred and fade before the hostages were finally set free.

At year's end, 1979, three American clergymen were permitted to hold Christmas services for most of the hostages. The International Court of Justice at The Hague unanimously called for their immediate release, and the secretary general of the United Nations, Kurt Waldheim, went to Tehran to try to mediate the standoff. He was denied meetings with either the hostages or the ayatollah, and his car was mobbed and beaten on by demonstrators.

In mid-January American television crews and correspondents were expelled from Iran, but by late that month some hope was held out by the newly elected president of Iran, Abolhassan Bani-Sadr, who criticized the militants and promised to try to calm the situation. Later that month, six American diplomats who had been hidden in the Canadian embassy escaped using forged Canadian passports. It was the first good news from Iran in three months. In February Bani-Sadr announced that the hostages might be released without the return of the shah, but this hope was dashed quickly by the ayatollah, who refused to allow a United Nations commission to see the hostages.

As winter gave way to spring, President Carter's approval rating began to drop. He was up for reelection the following November, and the hostage crisis was eroding the president's chances for a second term in office. Something had to be done to break the deadlock.

On the day after Easter, President Carter formally broke off diplomatic relations with Iran, ordered all Iranian diplomats out of the United States within 24 hours, asked Congress to allow Americans to settle claims against Iran on the $8 billion in Iranian assets that the government had frozen following the embassy takeover and announced a trade embargo on Iran. Later that month he also banned travel to Iran by all Americans except journalists, who had been recently readmitted.

On April 25 the United States government launched a dramatic attempt to rescue the hostages. But in a tangle of confusion and mismanagement, the raid dissolved in ignominious failure. Three of the eight helicopters assigned to the mission dropped out with mechanical failure (they were the wrong kind of machine for the Iranian desert). Without them, the mission was canceled, but not before one of the remaining helicopters collided on the ground with a C-130 transport plane, sending both up in flames. Eight servicemen died in the fire, and the rest fled, leaving the charred bodies of their comrades in the sand of the Iranian desert 250 miles short of Tehran, their destination.

It was the worst fiasco since the CIA-bungled Bay of Pigs invasion of Cuba in 1961. President Carter would be permanently damaged by the incident. Secretary of State Cyrus Vance, who had opposed the raid from the very first, resigned and was replaced by Edmund Muskie.

In early July 1980 Richard I. Queen, one of the hostages who was suffering from multiple sclerosis, was released. This left 54 still in captivity. On July 27 the shah died in Egypt of the final effects of his cancer. Now the ayatollah went on Iranian radio and read off a new list of conditions: return of the shah's wealth, cancellation of U.S. claims against Iran, unblocking of Iranian assets frozen in America and a pledge by Washington not to interfere in Iranian affairs. It was a list designed to humble a major power before the resolve and strength of the ayatollah.

But the events of history blunted even the power of the spiritual leader of Iran. A full-scale war broke out between Iran and Iraq as the U.S. elections drew near. The hostages faded from the front pages of the world's newspapers. Behind the scenes the United States knew that Iran was strapped for spare parts and ammunition. It was also aware that Iran felt that it might be able to gain concessions from a president fighting for reelection.

The United States pledged neutrality in the Iran-Iraq war and hinted that if the hostages were freed, some Iranian assets would be unfrozen and more than $500 million worth of spare parts already purchased by Iran would be delivered.

There was a flurry of rumors, climaxing in the week before the election, that an agreement was imminent. But the ayatollah, as was his pattern, again dashed hopes on the eve of election day, 1980, which was the first anniversary of the hostage seizure. Jimmy Carter lost the election by a landslide to Ronald Reagan, and the day after, President Carter asserted that the 11th-hour developments in the hostage crisis had been the primary cause of his defeat.

During his final weeks in office, President Carter and his administration, using Algeria as an intermediary, haggled with Iran over the hostages. For the second year, he ordered the Christmas tree at the White House to remain dark, but, responding to a request from the hostages' families, he lit it on Christmas Eve for 417 seconds, one for each day of their captivity.

On Christmas Day three Iranian Christian clergymen and the papal nuncio held religious services for the hostages, and negotiations resumed. Iran conceded that its claim of $14 billion in frozen assets was high and accepted the U.S. figure of $9.5 billion, of which approximately $2.5 billion was subject to legal claims.

Iran's minister for executive affairs, in a speech to the Iranian parliament, probably put his finger on the real reason for the agreement. "The hostages are like a fruit from which all the juice has been squeezed out," he said. "Let us let them all go."

Thus, at 12:25 P.M. on January 20, 1981, just as the newly elected president, Ronald Reagan, was finishing his inaugural address, an Algerian Airlines 727 lifted off from Mehrabad Airport in Tehran with 54 hostages aboard. "God is Great! Death to America!" chanted the men who had brought the hostages to the airport. As the plane left Iranian airspace, nearly $3 billion of Iranian assets were unfrozen by the United States, and more was made available for Iranian repayment loans. The next day, $8 billion of Iranian assets would be funneled into the Bank of England in a special Algerian account accessible to Iran.

The purpose of the ayatollah had been served. He had humbled a great Western power and had arguably sent an American president down to defeat at the polls.

"With thanks to Almighty God," said Ronald Reagan at an inaugural luncheon, "I have been given a tag line, the get-off line, that everyone wants for the end of a toast or a speech, or anything else."

"I doubt that at any time in our history," said Jimmy Carter from his home in Georgia, "more prayers have reached heaven for any Americans than have those given to God in the last 14 months."

IRELAND
IRISH SEA
June 22, 1985

On June 22, 1985, a bomb planted by Sikh extremists exploded in an Air India 747 over the Irish Sea. All 329 aboard died in this, the first downing of a jumbo jet by a terrorist bomb.

On June 6, 1984, a violent confrontation took place between Sikhs and Hindus at the Golden Temple in Amritsar, in the Punjab region of India. One thousand two hundred were killed that day when the Indian Army raided the temple, among them Bhai Amrik Singh, a former president of the Sikh Student Federation, a militant terrorist organization that had been outlawed by the Indian government. Leaders vowed revenge, and although a lack of physical evidence precluded positive proof, it is generally accepted that it was the Sikh Student Federation that was responsible for the planting of a bomb that blew an Air India 747 to pieces over the Irish Sea on June 22, 1985.

The flight, bound for London, took off uneventfully from Toronto on the evening of June 21. At 8 A.M. the following morning, air controllers at Shannon Airport made contact with the crew of the flight as it entered Irish airspace. Clearance was given to proceed to London.

At 8:15 the airplane disappeared from Shannon's radar screens. No distress signal was radioed by the jet's captain, Commander Narendra. The flight merely disappeared in an instant.

Rescue boats and helicopters were dispatched immediately. It was a bone-chilling morning, with clouds at 500 feet and heavy rain, and at first it was thought that perhaps a freak of weather had caused the crash. But this was rejected summarily by rescuers, who noted that pieces of the airplane were scattered over a five-square-mile area, indicating that the plane had exploded long before it hit the sea. None of the bodies recovered from the water were wearing life jackets, indicating that the explosion occurred without warning, and no piece of wreckage was larger than 30 square feet, indicating that the detonation must have been enormous.

That very day the *New York Times* received a telephone call from a member of the 10th Regiment of the Sikh Student Federation, who claimed responsibility for the bombing. Their purpose was, in his words, to "protest Hindu imperialism." Similar calls were placed to other newspapers in Europe and India.

There were no survivors; all 329 aboard the jetliner were killed. It was the first jumbo jet downed by a terrorist bomb. No arrests were made, and no incendiary device was discoverd.

ISRAEL
JERUSALEM
July 30, 1997; September 4, 1997

Two terrorist bombings separated by five weeks killed 21 and injured more than 400 innocent shoppers and tourists in two public gathering places in Jerusalem in July and September of 1997.

Terrorism has become a way of life in the Middle East. Hardly a week or even a day goes by without the terrible occurrence of another incident. Two that were really part of one action will have to suffice to typify them.

The Oslo Peace Agreement between Israel and Palestine was supposedly in force on July 30, 1997, when two suicide bombers dressed in black suits carrying attache cases entered the old Mahane Yehuda market in western Jerusalem. At the height of market day, when the market was as crowded as it can get, the two detonated the bombs in their attache cases, blowing themselves apart, killing 17 and injuring 150 shoppers and shop keepers.

Moshe Rahamim, the owner of a shoe stand, spoke to reporters who descended on the carnage. He recalled a ball of fire boiling toward him, he said. "The explosion threw me to the side. People were all over the ground. One person was on fire, and I tore his clothes off. Another had a huge injury to his head and eye. I saw a woman—the whole stall fell on her back. We lifted it all. She had almost no blood. She was dead. The roof fell on people. They were screaming. It seemed like an eternity before the police came."

"I saw it! I saw people without hands, without legs, with lots of blood," a young Jewish boy, anxious to be heard, shouted at the reporters.

A wave of anger rushing toward fury washed over the survivors and witnesses to the mayhem. "Peace is a nightmare!" shouted one man. "Kill the Arabs!" chanted a group of young men, who might be called upon to do just that.

Ambulances arrived along with police, who formed cordons and shoved the clusters of spectators to the side every 10 minutes. Inspectors and the cordon of bearded religious men whose job it was to gather every shard of human remains for immediate burial arrived.

In the midst of the high emotion, some shopowners like David Bershel, who owned a cheese and olive shop, sighed in resigned acceptance. "Five years ago," he said, pointing at his shop, "this was a restaurant. An Arab woman blew herself up in the bathroom, evidently by mistake. A waitress was killed. This is how we live."

Shortly after the bombing, a leaflet appeared in Jerusalem claiming that Hamas, the militant Islamic organization, had done it to demand the release of Palestinian prisoners held by Israel. Prime Minister Benjamin Netanyahu went from place to place and microphone to microphone to denounce Palestine and its leader, Yasser Arafat, and to denounce the Oslo process and Americans who sent money to the Palestinian authority, while Palestinians blamed Mr. Netanyahu for creating the conditions for terror.

Five weeks later, in the middle of a sunny afternoon, three suicide bombers entered Ben Yehuda Street, a shady pedestrian thoroughfare in West Jerusalem. The area was lined with chic boutiques, punctuated by outdoor cafes and populated by Israelis and foreign tourists.

Once again, three men blew themselves up within moments of each other. Once again, carnage, rage and horror swept through a public gathering place. This time, four were killed and more than 180 were injured.

The bombs contained screws, nuts and bolts designed to maim, kill and injure as many within their range as possible. The bodies of the bombers were ripped asunder. Blood was spattered on the walls of buildings. Limbs and torsos were flung into the streets. The side alleys were choked with debris and the wounded.

This time, the identity of the bombers and the organization that had sent them was precise and undeniable. Shortly after the attack, a phone rang in the office of a news agency. The anonymous caller announced that there was a communiqué for reporters in Ramallah, a town that was—and still is—in Palestinian territory close to Jerusalem and from which terrorists operated.

This time the written boast of responsibility was more strongly worded than the one that had accompanied the bombing in July. A group calling itself the "Martyr's Brigade for Freeing Prisoners," a cell within the Qassam military wing of Hamas, reiterated its demands for the freeing of all Arab prisoners held by Israel. The demands also included the freeing of Sheikh Ahmed Yassin, the spiritual leader of Hamas, and carried a deadline of 9 P.M. on September 14, 1997. It went on to state that the attack was a "painful response" to an Israeli rocket attack on the Lebanese city of Sidon in August, in which several Palestinians were killed, and to the punitive actions taken by Israel in the West Bank. Finally, it denied that the group had an interest "in interfering in the collapsed political game which is led by America and its supporters in the region."

The reference was to the continuing negotiation for peace between Israel and Palestine, one that had resulted in the assassination of Prime Minister Itzahk Rabin, possibly

Ultra orthodox forensic volunteers clean up the carnage after two suicide bombers detonated themselves in the crowded Mahane Yehuda market in West Jerusalem early on a Sunday morning. It was typical of the ongoing terror attacks that reignited the continuing hostilities between Israel and Palestine. (AP/Wide World Photo)

Israel's most eloquent spokesman for peace in the region. His widely publicized handshake with PLO leader Yassar Arafat had enraged militants on both sides of the conflict, led to his death and had created a downward spiral of trust and an escalation of violence. Two more prime ministers would be elected to lead Israel in an increasingly hard-line stance, with provocation and retaliation the rule of behavior on both sides.

United States secretary of state Madeline Albright made a scheduled trip to the Middle East soon after the September bombing, and, by the end of the Clinton administration, genuine progress seemed to be made. Then the cycle of violence once more increased as hopes for peace drifted, until the gloom of hopelessness settled not only over the Middle East but the world in general.

"It's almost a sense that this is a fact of life we have to live with," Jonathan Shiff, an Israeli lawyer, commented to the press in September of 1997. "Yes, it is anger, but it's not directed anger."

Amos Oz, the famous Israeli writer, encapsulated the situation with succinct eloquence: "It is becoming clear," he wrote, "that beyond the struggle between Israel and Palestine an invisible conflict is taking place between, on the one hand, Israelis and Palestinians who want a peace based on a historic compromise made between the two peoples, and on the other, Israeli and Palestinian extremists, each of them wanting to eradicate the national existence of the other."

ISRAEL
SINAI DESERT
February 21, 1973

Fear that a Libyan jetliner had been hijacked and was flying a bomb aimed at Tel Aviv was the reason given for the downing of the jetliner by Israeli Phantoms on February 21, 1973. All 106 aboard died.

In late January 1973, Israeli intelligence received information that Palestinian guerrillas were planning a suicide mission in which a jetliner would be hijacked, armed with bombs and crashed into the heart of Tel Aviv.

In the early afternoon of February 21 of that year, Libyan Arab Airlines Flight 114, a Boeing 707 piloted by a crew loaned from Air France, took off from Tripoli headed for Cairo. Forty-five minutes before it was to land, it radioed Cairo that it was having radio trouble and had lost its way because of bad weather.

At 1:55 the plane, which had overshot Cairo considerably and had wandered into Israeli airspace over the Sinai Desert, was intercepted on Israeli radar. Proceeding on its course, it penetrated 50 miles into Israeli territory, flying over Israeli military concentrations and a military airfield along the Suez Canal.

Israeli authorities tried to contact the jetliner, but it did not respond, obviously because its radio was malfunctioning. Phantom jets took off from the military airfield and intercepted the jet, advising it to turn back. It again did not answer. Warning shots were fired near it. It still did not respond.

At 2:30 the radio aboard the jetliner finally began to function again. The captain radioed Cairo, stating that he was lost and surrounded by Israeli fighters. And then the radio contact abruptly stopped. The fighters homed in on the jetliner and shot it down. It fell to the earth in the Sinai Desert, killing all but 13 of its passengers. The 13, gravely injured, later died of their wounds, and the death toll rose to 106—every passenger and crew member aboard the aircraft.

Reaction was immediate and outraged. Israeli premier Golda Meir expressed distress and propitiation: "The Government of Israel expresses its deep sorrow at the loss of life resulting from the crash of the Libyan plane in Sinai and regrets that the Libyan pilot did not respond to the repeated warnings that were given in accordance with international procedure."

But even within the Israeli government, Israel Gahli dubbed the incident a "disaster," and ordinary Israeli citizens were horrified that their armed forces had shot down a civilian airliner. Conciliatory statements were released by many major governments and delivered to Libya. President Richard Nixon sent a message of condolence, as did UN secretary-general Kurt Waldheim. Libya denounced it as a "criminal act" and vowed revenge, and in August 1973 five passengers were killed and 55 wounded aboard a TWA plane arriving in Athens from Tel Aviv. Responsibility was claimed by members of the Libyan-sponsored National Arab Youth for the Liberation of Palestine.

ISRAEL
TEL AVIV
May 31, 1972

An agreement between the PFLP and the Red Army to continue terrorist attacks on Israel resulted in the first transnational terrorist incident at Lod (Lydda) Airport on May 31, 1972. Twenty-six unsuspecting travelers were killed; 76 were wounded.

"How does it happen," asked one dazed and bloodied survivor of the Lydda Airport Massacre of May 31, 1972, "that Japanese kill Puerto Ricans because Arabs hate Israelis?"

It was a microcosmic question that had no reason to be asked until that horrible day. Until that time each terrorist organization seemed to be autonomous. But some time in late 1971, members of Japan's United Red Army made contact with George Haddash, the leader of the Popular Front for the Liberation of Palestine, PFLP, and met with him in Pyongyang, North Korea. From there they traveled to Jordan, where, along with members of West Germany's Baader-Meinhof Gang, they underwent guerrilla training.

The Japanese Red Army, an ultra-leftist group, had lost support in Japan earlier that year when, after police had arrested hundreds of its adherents, including five of its leaders, the bodies of 14 young people were discovered. The 14 had been tortured to death for deviating from the Red Army's revolutionary line. Thus, the Japanese terrorist group sought to gain credibility and approval in the terrorist world by aligning itself with the PFLP.

Three of its members, trained in Lebanon, boarded an Air France jet in Paris on May 31, 1973. Passengers aboard Air France jets in 1973 felt safe. France practiced a friendly relationship with Arab countries, and at the beginning of May, although Asher Ben Nathan, Israel's ambassador to France, had called on Herve Alphand, secretary-general of the French Foreign Ministry, to plead for increased security measures on flights to Israel, Alphand had refused him, noting that France and the Arab countries were not enemies.

Air France Flight 132 arrived on time and without incident at Tel Aviv's Lydda, or Lod, Airport on May 31. Passengers debarked and proceeded to luggage conveyor belt number 3.

Three young Japanese tourists claimed their bags from the belt and then began to behave strangely. They removed their jackets and crouched to open their suitcases. When they straightened up, they were all holding Czech-made VZT-58 automatic rifles, which they immediately began firing, rapidly and indiscriminately. They fanned their weapons in wide arcs, mowing down the passengers near them, and then raised the barrels of the rifles to shoot those farther away.

Two of the gunmen then dashed for the tarmac, firing at two parked planes. One killed the other, nearly decapitating him with a brutal burst of automatic gunfire. The third terrorist leaped on the now blood-slick baggage conveyor belt, holding a grenade. He pulled the pin, slipped, fell on the grenade and was blown to bits by its explosion.

The one surviving gunman continued to shoot into the crowd until he ran out of ammunition. As he stopped to reload, an El Al traffic controller, Hanan Zaiton, leaped on him and beat him to the ground. Guards rushed to his aid and then hauled the terrorist, a 24-year-old college dropout named Kozo Okamoto, into a nearby office.

The airport was strewn with the dead, the dying and the wounded. Twenty-six people, including Dr. Aharon Katchalsky, one of Israel's leading scientists, and 14 Puerto Ricans making a pilgrimage to the Holy Land, were killed. Seventy-six others were injured. An hour after the killings, a spokesperson for the PFLP announced to local papers that it had recruited the Japanese fanatics "to kill as many people as possible."

It was barbaric and senseless, but it proved that there was no safe haven for the innocent who hoped to escape terrorism in the 1970s.

ITALY
BOLOGNA
August 1, 1980

To "honor" an accused neo-fascist bomber, neo-fascist terrorists detonated a bomb in the Bologna train station on August 1, 1980, at the beginning of the holiday season. Eighty-four died; 200 were injured.

August is the traditional holiday month in Italy and France, and on August 1, 1980, the Bologna train station was packed to capacity with vacationers and tourists. That morning a judge in Bologna announced that eight neo-fascists, among them Mario Tuti, had been indicted and would be tried.

Amid unsuspecting travelers in the Bologna train station, a group of terrorists planted a bomb equivalent in power to 90 pounds of TNT in a corner of the second-class waiting room. Shortly after 1 P.M. the bomb exploded with a thunderous roar, totally demolishing one wing of the massive train station. A restaurant, two waiting rooms and a train platform were flattened as the roof collapsed on them.

Mayhem followed. The screams of the injured and the shouts of rescuers and survivors ricocheted off the columns and walls of the ruins. It was the worst terrorist disaster in Italy's history, eclipsing the 1968 bombing of a Milan bank in which 16 people were killed and 16 injured. The toll in Bologna would be 84 dead and 200 injured.

On August 2, while smoke still filtered upward from a wing of the station that now had only two iron girders standing, 10,000 Bolognans turned a left-wing rally into a demonstration against terrorism, and labor unions held a rally in Rome's Colosseum in which they announced a four-hour strike.

That same day two calls from terrorist groups were received by police. One claimed responsibility for the blast by the Red Brigades, the far left wing terrorist organization that had kidnapped and killed former Prime Minister Aldo Moro. But the caller incorrectly described the time and location of the bomb.

The other caller claimed to represent the Armed Revolutionary Cells, a neo-fascist organization, eight of whose members were to be tried for the railroad bombing of the Bologna-Florence train six years before. The bombing, it was stated, was to honor Mario Tuti, one of the accused.

Police believed this caller and began an exhaustive investigation that would end a little over a year later, on September 12, 1981, in London, with the arrest of the terrorists who planted the bomb.

JAPAN
TOKYO
March 20, 1995

Fifteen people died and more than 5,500 were injured in a gas poisoning attack on the Tokyo subway system during the March 20, 1995, morning rush hour. The assault, which involved sarin, a nerve gas, was part of a guerilla war against the Japanese establishment by Aum Shinrikyo, a doomsday religious sect.

In June of 1994 seven people died of a mysterious gas poisoning in Matsumoto, in central Japan. People in their homes were poisoned by sarin, a nerve gas developed by the Germans in the 1930s and used by Japan against China in the 1940s. But the origin of the gas remained a mystery, and no arrests were made.

On March 5, 1995, 11 passengers in a subway car in Yokohama, the major port city adjacent to Tokyo, complained of dizziness and eye pain—two symptoms of sarin poisoning—and were hospitalized. Again, no trace of the source of the gas was found and no arrests were made.

Looking back after a catastrophe, the pieces of the jigsaw puzzle that go into making the ultimate disaster can be discovered, and these two seemingly unrelated incidents were later seen to be pieces in the puzzle, rehearsals for the horrific gas poisoning in the Tokyo subway during the morning rush hour on Monday, March 20, 1995. Fifteen were killed and more than 5,500 were injured in this deliberate act of domestic terrorism by Aum Shinrikyo, a doomsday religious cult.

Sarin is designed to disrupt the synapses between nerves, which in turn leads to suffocation from paralysis of the diaphragm and other muscles used in breathing. Odorless and viciously powerful, it can, in minute quantities, cause huge casualties. Its hideous power inhibited its use in World War II in a sort of nuclear standoff because both sides had developed it and knew its capabilities.

The method of contamination in the Tokyo subway on March 25, 1995, was primitive, simultaneous and effective. Individual cult members carried containers, sometimes disguised as lunch boxes, sometimes as beer cans or soft drink containers, onto subway cars on the Chidyoda, Hibiya and Marunouchi lines. At 7:03 A.M. gas was placed on the 7:09 A.M. train out of Abiko station of the Chiyoda line. At 7:43 A.M. a gas container was placed on the 7:43 A.M. out of Kita-Senju station of the Hibiya line and simultaneously another was set on the 7:59 A.M. out of Nakameguro station going in the opposite direction. And at 7:47 and 8:32 A.M. canisters were put on trains leaving the Ikebukuro station on the Marunouchi line.

Passengers, suspecting nothing, crowded into the world's most populated transit system minutes later. By 8:45 A.M. the entire system was contaminated. As trains pulled into stations, passengers staggered out onto the platforms and collapsed. They lay on the concrete, some bleeding from the nose and mouth, others with bubbles coming from their mouths. Still others attempted to negotiate the stairs to the street and collapsed on the stairs.

In the Tsukiji station, near the Tokyo fish market, passengers and workers on the subway stumbled out of cars at 8:15. Nausea, dizziness and blurred vision was universal. "I was losing my sight," a subway employee told Japanese television later, "I couldn't see." Sarin does that. It shrinks the pupils of the eye, causing temporary and sometimes permanent blindness.

At the Kokkai-gijidomae station, which serves the National Parliament of Japan, additional gas canisters were placed on the stairs leading from the station to the Capital Tokyo Hotel, next to the Parliament building.

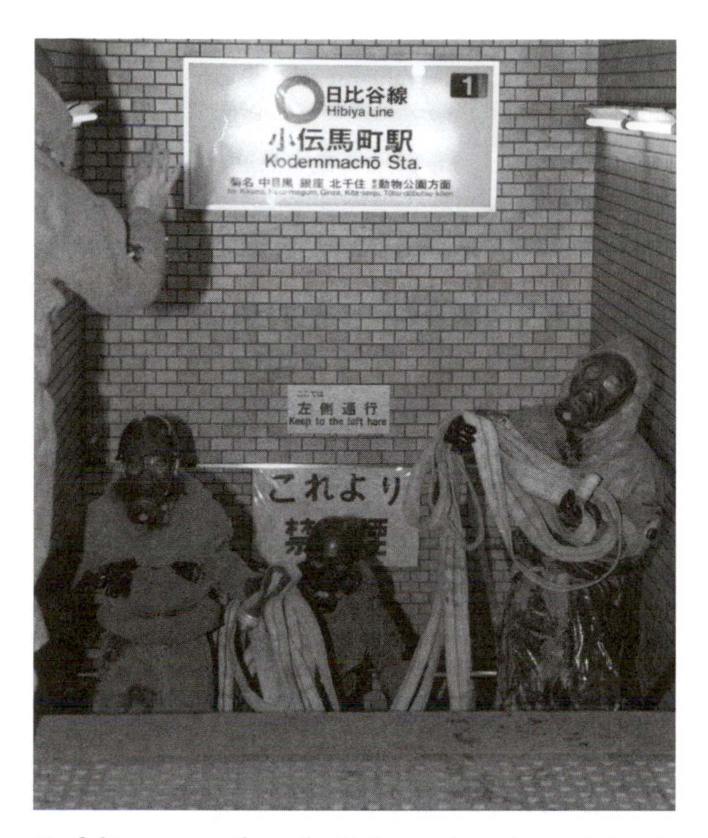

Firefighters emerge from the Kodemmacho subway station in Tokyo after cleaning up effects of the Sarin gas attack by the Aum Shinrikyo cult. A dozen people were killed and more than 5,500 others were sickened in the attack. (AP/Wide World Photo)

Ambulances converged on the stations, and attendants gathered up and took the dead and injured to Tokyo hospitals. Police followed, sealing off the stations and sending squads in gas masks and protective clothing to search for any passengers left behind.

Tents were set up on the sidewalks to shelter the wounded until ambulances could find them and take them to hospitals, and the military's chemical warfare team arrived with a vehicle designed to purify the air in the cars, the tunnels and the stations and simultaneously to sample the gas to discover its identity. It was, without doubt, sarin.

Now the Japanese Military Self-Defense Force mobilized a 140-member chemical warfare unit. Prime Minister Tomlicht-Muraya called an emergency cabinet meeting and ordered an all-out rescue effort and increased security for all railways, airports and seaports. "This is a case of organized, indiscriminate murder," Masashiro Terao, a senior police official announced at a hastily summoned news conference.

Like a stone dropped in water, waves of fear radiated out from the poison gas attack in the Tokyo subway to the rest of the country. Japan was, after all, a peaceful place. Self-confidence had helped to keep it that way, and this was a body blow to the nation's self-confidence.

Even as chemical experts were spraying huge quantities of a chemical with an odor like chlorine into the subways to neutralize what was left of the sarin, ordinary people were gripped by fear. This, they thought, was the sort of tragedy that could happen in New York, but not in Tokyo. And yet it had.

By March 22, two days after the attack, 2,500 police began to raid the offices of Aum Shinrikyo, whose guru, Shoko Asahara, had declared war against the Japanese government and forecast that the world would end in 1997. The cult, with offices also in Russia and New York, had 10,000 followers in Japan. Preaching a combination of Hindu and Buddhist theology, it was heavily financed and had been rumored to enslave and even murder some of its members who tried to leave the cult. Two days before the subway disaster police had raided its Osaka offices and charged those there with kidnapping a young man whose parents had retrieved him from the cult's headquarters and training compound in the tiny dairy town of Kamikuishiki, at the foot of Mount Fuji.

As long ago as 1993, neighbors had complained of noxious white fumes being emitted from a Tokyo building owned by Aum Shinrikyo, but police, without a search warrant, never investigated. And in the summer of 1994 residents of Kamikuishiki complained of eye and nose irritation. Several months later, by-products of sarin were found in the village. Now, in a concentrated raid on 25 sect offices, 2,500 police were making astounding and guilt-provoking discoveries. The group's leaders had fled the day after the subway attack in a convoy of luxury limousines traveling at speeds of nearly 100 mph.

The interior of the compound yielded an enormous cache: two tons of 10 kinds of chemicals, including 15 or more bottles labeled acetonitrile, a solvent for sarin, traces of which were found in subways after the attack. Other containers were found holding isopropyl alcohol and sodium fluoride, both of which can be used in making sarin. Still others were labeled sodium cyanide, a deadly poison. There was a magazine being prepared by the sect that warned that poison gas attacks or other calamities would kill 90 percent of the people living in major cities in the coming years preceding the end of the world in 1997. And there was the equivalent of more than $7,000,000 in cash and 22 pounds of gold ingots. In other parts of the compound, emaciated people who were being held without food or water against their will were discovered.

Shortly after the first police sweep, Shoko Asahara appeared on Japanese television on a tape submitted to him by television station NHK. He denied any links to the subway tragedy.

The police raids continued. In a warehouse owned by the sect in Kofu, a prefect capital not far from Aum's training center, 500 metal drums containing phosphorous trichloride, a key nerve gas ingredient, and other drums containing material to make nitroglycerine were found. On the training grounds they discovered a laboratory that contained large quantities of a substance used to treat nerve gas poisoning.

Once again, Mr. Asahara appeared on television. "We aim to set up a self-sufficient life-style, and for that we need lots of chemicals," he said, then ticked off their uses: to treat pottery, computer chips and plastic products and as agricultural chemicals.

At the end of March Takaji Kunimatsu, Japan's top police official, was shot and badly wounded on a Tokyo street. Two members of the Aum Shinrikyo sect were arrested and found with the guns that were fired at the official and some explosive chemical. By the middle of April, 60,000 police officers were stationed in train stations and other public places. Not only were authorities concerned with other efforts by the sect, but with copycat crazies. On May 6 just this happened. In the men's room of the Shinjuku subway station, two bags were discovered by a patron. One contained a half gallon of sodium cyanide in granulated form; the other contained a smaller amount of liquid diluted sulfuric acid. One of the bags had been set on fire, apparently so that the bags would both catch fire and the chemicals combine. If they had, they would have formed enough cyanide gas—the sort that is used to execute prisoners in the United States—to kill between 10,000 and 20,000 people in the subway system. Fortunately, four subway employees doused the flames with water before both bundles could catch fire.

Finally, on May 16, 1995, the police found Shoko Asahara meditating in a secret room within a building in the sect's headquarters. He was arrested and charged with murder and attempted murder.

As other leaders were arrested, confessions came forth. Masami Tsuchiya, the head of the sect's "chemical squad," acknowledged that he oversaw the manufacture of sarin nerve gas, most recently in January. Another captured member confessed to carrying the gas to a subway station. The notebooks of Yoshiro Inoue, the sect's intelligence chief, revealed detailed information regarding railway schedules on the lines on which the attack took place.

By the end of June, Shoko Asahara confessed to ordering the murder of a cult member trying to help others escape from the group. The time of his forecast for the end of the world, 1997, came and went. The cult, for all intents and purposes, was no more. Its Armageddon had come early.

JORDAN
DAWSON'S FIELD
September 6–8, 1970

A PFLP guerrilla group's demands for the release of Palestinian prisoners culminated in the 1970s' most spectacular terrorist incident at Dawson's Field in Jordan. Five civilian jetliners were hijacked; four were flown to Jordan and blown up. One hijacker was killed in the explosion, and 300 passengers were taken hostage.

In what began as the largest, most spectacular hijacking operation of the 1970s, four airliners from four countries were hijacked in one morning and early afternoon. The day was September 6, 1970, and all four hijackings took place shortly after each plane took off from its home airport. All were headed for New York.

The first of these, El Al's Flight 219 from Tel Aviv, a Boeing 707 with 148 passengers and 10 crew members aboard, took off from Amsterdam in the morning. Shortly after takeoff, two hijackers, Patrick Arguello, a Nicaraguan working for the PFLP, and Leila Khaled, a Palestinian, sprang to their feet shouting. Arguello dashed toward the cockpit door. Schlomo Vider, an El Al steward, pounced on him but was shot in the stomach.

The plane was thrown into a steep dive by its pilot, which flung the hijackers off balance and allowed one of the two Israeli security guards aboard to shoot and kill one hijacker, Arguello. Meanwhile, a passenger overpowered Khaled, and she was tied hand and foot with string and a necktie.

The jet diverted to London's Heathrow Airport, where it made an emergency landing. Vider, with three bullets in his stomach, was taken by ambulance to a hospital, and Khaled was led off to jail. The plane was cleaned up, the passengers reloaded and the flight proceeded to New York.

The other three hijackings were considerably more successful for the hijackers. Pan Am Flight 93, a Boeing 747 on the last leg of a flight from Brussels to New York with 152 passengers and a crew of 17 aboard was hijacked shortly after it left Amsterdam. Apparently confused, the hijackers allowed the plane to land in Beirut, Lebanon and then, after they conferred with PFLP "brothers" at the airport, allowed the plane to take off again and eventually land in Cairo. There the hijackers unloaded passengers and crew and blew up the $23 million craft. The passengers were evacuated the following day to New York and Rome.

Meanwhile, TWA Flight 741, a Boeing 707 on an around-the-world voyage with 141 passengers and a crew of 10 was commandeered shortly after it took off from Frankfurt and diverted to Dawson's Field, a former World War II RAF base in the Jordanian desert. Dawson's Field had been taken over as a "revolutionary airfield" by the PFLP.

Finally, Swissair Flight 100, a DC-8 with 143 passengers and 12 crew members, was hijacked 10 minutes after it took off from Zurich and was ordered to change its course to the Middle East. It set down at Dawson's Field shortly after the other jet.

A spokesman for the PFLP, speaking from Beirut, announced the reasons for the multiple hijacking: The American planes had been seized, he said, "to give the Americans a lesson after they have supported Israel all these years" and in retaliation for the U.S. involvement in peace negotiations in the Middle East between Israel and the Arabs. The Swiss plane was captured and held in ransom for the release of three Arab commandos convicted by a Swiss court for an attack on an Israeli airliner at the Zurich airport the previous December. The spokesman also gave the reason for the hijacking of the El Al jet: "We are fighting Israel; they are our enemy and we will fight them everywhere."

On the morning of September 7, the scene was an ominous one. The two planes were poised on the old runways, shimmering with heat. In between was a tent, the hijackers' field headquarters. Nearby was a water truck with a sign in Arabic reading "The Popular Front at your service." Around the periphery of the field were PFLP guerrillas armed with Russian Kalashnikov submachine guns, Katyusha rockets and jeeps with heavy-caliber machine guns.

And 250 yards away was the Jordanian Army, ringing the field in an impenetrable circle of tanks, armored personnel carriers, anti aircraft guns, communications jeeps, ambulances and fire trucks. Each was pointed directly at the planes. King Hussein of Jordan wanted nothing to do with the hijackers, nor did he want the world to feel that he was in complicity with them.

By noon the hijackers threatened to blow up the planes and their hostages unless the armor was withdrawn. The Jordanian Army backed off two miles and reformed its circle.

As the days passed and deadlines were made and then extended, the hijackers released women and children, conducted press conferences with appearances by hostages and eventually allowed the International Red Cross to fly in a planeload of relief supplies and a Jordanian Airlines toilet-cleaning vehicle to service the parked planes.

At midweek, on September 8, the PFLP hijacked another jetliner, a British Overseas Airways Corporation VC-10 carrying 117 passengers and crew. The plane was diverted to Beirut, where it refueled, and then flown to Dawson's Field, where it was forced to land and take up its position with the other two jets. It and its occupants would be held, the PFLP announced, until Leila Khaled was released.

By this time, five governments and the United Nations were involved in negotiations. As the days drifted on, token numbers of hostages were released. Sixty-eight were taken to Amman, Jordan, and flown to Nicosia, Cyprus, and London; 23 more were transferred to a hotel in Amman; 20 were allowed to fly to Beirut. The hijackers now demanded the release of an unspecified number of Arab guerrillas in Israel.

The parent of the PFLP, the PLO, was growing increasingly restive with the extremist tactics of the guerrillas and began to withdraw its support. Eventually, it would disassociate itself completely from the group, which continued to fire off demands including the release of six Arab guerrillas imprisoned in Germany, the return of two Algerians taken from a BOAC flight by Israelis and the release of 2,000 guerrillas held in Israel.

As the relationship with the PLO continued to deteriorate, the hijackers decided on a dramatic action: They freed 260 passengers, kept 40 and blew up all three jetliners. The hostages, some from each of the negotiating countries—the United States, Britain, West Germany, Switzerland and Israel—were to be held until prisoner exchanges could be arranged with each country.

Eventually, all hostages would be released, but not all Palestinian hostages. Leila Khaled would be returned; the Swiss and West German hostages would be traded; but that would be all.

Foreseeing the consequences of violating Jordanian sovereignty by the actions of George Haddash and his PFLP, PLO leader Yasser Arafat expelled the organization from the PLO for lack of discipline. King Hussein would eventually drive the PLO from its base in Jordan as a result of the incident, and Black September would be formed.

LEBANON
BEIRUT
October 23, 1983

A suicide bomber representing the Islamic Jihad drove an explosive-filled truck into the U.S. Marine compound in Beirut, Lebanon, on October 23, 1983. Two hundred forty-one U.S. Marines were killed; 58 French soldiers were also killed in a related attack.

Lebanon was a bloody battlefield in 1983. Beirut, formerly one of the most beautiful cities in the world, was on the way to becoming the demolished, smoking shell it is today. Rival religious factions roared back and forth across it and the rest of Lebanon, and both Syria and Iran financed and supplied some of these groups.

The Islamic Jihad, one of the most fanatical of terrorist groups, was founded at this time, and its chief support came from Syria and Iran, though it was headquartered in Lebanon. Its enemies were the United States, Western Europe, Iraq, Jordan and Egypt. A supercharged religious energy, as much as money and arms, fueled this fledgling terrorist organization. Some Middle East observers, among them correspondents Christopher Dobson and Ronald Payne in their study of terrorism, *The Never Ending War,* assert that it was the Israelis, who invaded Lebanon in June 1982 that gave the Islamic terrorists in Lebanon their "launching pad."

After a terrorist attack on its ambassador to Britain in London, the Israeli government ordered its armed forces into Lebanon ostensibly to root out the Palestinians. The attack worked, but it also strengthened the resolve of terrorist groups in that country.

Shortly after the Israeli invasion and while Israeli troops were surrounding Beirut, President Ronald Reagan sent U.S. Marines into that city to try to restore peace through a forceful presence. Aided by small contingents of French, British and Italian troops, they managed to maintain an uneasy truce.

But Shiite organizations, and particularly the newly formed Jihad, were planning the use of a new and gruesome weapon in the continuing escalation of terrorist attacks against the West and Arab enemies. On Monday, April 18, 1983, a truck loaded with explosives was driven onto the U.S. embassy grounds and detonated. Forty-five people were killed, including 16 Americans. It was the first such suicide bombing in the Middle East, but it would not be the last; nor would it be the most devastating.

That would come at 6:20 A.M., Sunday, October 23, when a pickup truck approached the south gate of U.S. Marine headquarters in Beirut. It was coming from the direction of Beirut International Airport and was noticed by the sergeant on guard duty. Inside the barracks some 200 marines were sleeping.

Suddenly, the driver of the truck gunned his motor, sending the truck barreling through a sandbag barrier and into the interior courtyard of the compound. Seconds later the driver tripped a switch and blew himself and the truck and most of the barracks into small slivers of metal, wood and flesh. The crater left by the bomb was 30 feet deep and 40 feet across, and the explosion had the force of a ton of TNT.

"I haven't seen carnage like this since Vietnam," said Marine Major Robert Jordan, who, with other officers who were quartered elsewhere, rushed to the scene to begin the grisly business of rescue and recovery of bodies.

A few minutes after the initial blast, another car bomb driven by a suicidal terrorist rammed into a building housing a company of French troops and exploded.

Machinery was moved into place to lift girders and concrete from wounded survivors. Lebanese and Italian soldiers aided in the rescue effort. Ships of the U.S. Sixth Fleet went on alert. Helicopters airlifted the wounded to the amphibious assault ship *Iwo Jima* and the battleship *New Jersey,* poised offshore near Beirut harbor. Muslim snipers on rooftops shot at the helicopters.

Two hundred forty-one marines were killed, nearly 100 were wounded, and 58 French soldiers were killed in the related bombing of their quarters.

Retaliation consisted of shells lobbed from the *New Jersey* into known terrorist strongholds. Israeli jets bombed the same targets. Intelligence forces from Israel, the United States, France and Lebanon sifted through the rubble. The 12,000 pounds of TNT and PETN, a plastic explosive that was used in the marine barracks bombing, and hexogen, which was involved in the explosion at the French barracks, plus a $50,000 money order to local mercenaries that was traced to Iranian diplomats clearly implicated both Syria and Iran.

Fourteen terrorists were finally named. They included Palestinians, renegade PLO members, professional terrorists and a fundamentalist mullah. The organization was being financed and supported by Shiite, Syrian and Iranian forces, but its network of supply and its operations were Byzantine in their complexity.

On December 12 the suicide bombers struck again, this time in eight locations in Kuwait. One was the American embassy, which was destroyed through the same method used in Beirut. The other seven locations included the headquarters of the Raytheon Company, an American outfit that was installing Hawk missiles in Kuwait, and the French embassy.

The peacekeeping force of American, French and Italian armed forces withdrew from Lebanon, leaving it to be destroyed by warring factions. Special barricades would be erected in front of the White House in Washington to guard against possible suicidal drivers of trucks loaded with explosives. Other government buildings would be provided with concrete pylons.

Apparently because of a dwindling supply of terrorists willing to commit suicide for their cause, this type of terrorist bombing temporarily stopped after this incident. But it had made its psychological point and had weakened the West in the Middle East.

LEBANON
BEIRUT
June 14–18, 1985

The Islamic Jihad was responsible for the hijacking of TWA flight 847 from June 14 to 18, 1985. Hundreds of passengers were terrorized, 39 men were held hostage and then released, and one U.S. Navy man was killed.

In 1970 the PFLP had staged a spectacular hijacking of five jetliners (see p. 136), and the world had trembled and negotiated. In 1985 the Islamic Jihad and its Shiite special squads decided that another hijacking was necessary to reaffirm the power and purpose of the Islamic revolution and its anti-Western orientation.

But between 1970 and 1985, the world changed. International agreements had, at least on the surface, largely forbidden hostage deals, although Islamic Jihad was well aware that clandestine dealing still existed. The arrangements that would eventually erupt the following November into the Iran-Contra scandal were already energetically under way in Tehran.

Still, the incidence of air piracy was distinctly down from 15 years before. In 1970 there were 91 airline hijackings; in 1984, 17. The reduction was partly due to increased airport security. However, security remained lax at two airports, Athens and Beirut, and it was here that the Islamic Jihad concentrated its efforts in 1985.

An Islamic Jihad plan to hijack two American airliners simultaneously was uncovered, but half of the plan was aborted when East German authorities picked up two passengers carrying explosives in their baggage.

TWA Flight 847 was, however, not so lucky. The flight originated in Cairo, stopped in Athens and was then to proceed to Rome, and, as a new flight, to Tel Aviv. Slightly after 10 A.M. on the morning of June 14, after a late arrival from Cairo, the flight became packed to capacity in Athens. There were 145 passengers aboard—120 Americans, a Greek pop singer, some Australians and 21 Catholic pilgrims.

A few minutes after takeoff, two Arab terrorists charged up the aisle, one waving a 9-millimeter pistol, the other brandishing two hand grenades, one in each hand. "Hijack! Hijack!" they screamed. "We have come to die!" A mace-like substance had been thrown prior to their charge, filling the cabin with choking smoke. One of them aimed a karate kick at Uli Derickson, the purser on the flight, slamming her against the cockpit door.

The hijackers, Ali Younis and Ahmed Ghorbieh, kicked in the cockpit door. One held the gun to the head of pilot John Testrake while the other pulled the pins on the twin grenades. Captain Testrake acceded to their demands to divert the liner to Beirut. The hijackers then proceeded to terrorize the passengers, beating some of them mercilessly and forcing the rest to assume agonizing postures in their seats.

At first Beirut refused permission to land, but Testrake convinced controllers and officials there that they were low on fuel and that the hijackers were hysterically out of control, plucking the pins from hand grenades and reinserting them at the last minute, brutalizing the passengers and committing other violent acts to convince all aboard that they were dedicated in their mission.

The plane landed in Beirut long enough to refuel and release 17 women and two children. In Algiers, the next stop, the terrorists released 22 more women and children. Once more, Flight 847 took off, heading back to Beirut. It was night by now, and the Beirut airport shut off its landing lights, hoping to deter the hijackers from landing.

They had already singled out a young U.S. Navy man, Robert Stetham, and had beaten him badly. Unless they got their way in two minutes, they shouted over the radio to the control tower, they would "let one American loose off the plane." The runway lights were turned on; the plane landed; the hijackers demanded that Shiite leaders be brought to the plane.

Again they were refused; again they threatened to kill one American, and this time they did. They shot Stetham in the head and dumped his body onto the tarmac. They would, they told the tower, begin to kill the other Americans one by one unless a rambling list of demands was met. Among the demands were the cutting off of oil sales to the West, the removal of all Arab money from Western banks and the release of all Shiites imprisoned in Kuwait and Israel.

It was insane, but so, apparently, were the hijackers, who, having found a translator in purser Derickson (one of the hijackers spoke German, and she was German), began to wage psychological warfare on the passengers. Ordered to point out the Jewish passengers, Derickson refused and was beaten. Later, the German-speaking hijacker proposed marriage to her. Shouting and screaming and kicking the passengers one minute, the hijackers would order the flight attendants to make omelettes for everyone the next.

Now, reinforcements arrived. Twenty-five armed Shiites rushed aboard the plane. Twelve remained on the tarmac.

Meanwhile, the U.S. Sixth Fleet was moved in close to the shore of Lebanon, and a 1,800-marine amphibious unit was drawn up in the same area. Squadrons of F-16 fighters were redeployed at bases in Turkey.

Late Friday night the hijackers ordered the plane to again take off and head for Algiers. There, the Greek government handed them Ali Atweh, a terrorist who had been separated from the original two hijackers at Athens and arrested. The deal was that he was to be traded for Greek pop singer Demis Roussos. But the hijackers reneged on the arrangement, freed a group of 10 Greek passengers and the flight attendants but did not include Roussos.

On Sunday morning, June 16, the plane again took off for Beirut. But by this time, Nabih Berri, the head of the Amal, which was the most respectable part of the Shiite organizations in Lebanon, offered to enter the picture as a mediator. His offer was accepted, and on Monday morning, June 17, Nabih Berri's militia freed the remaining passengers, with the exception of 39 American men, including the pilot and co-pilot. They were taken to three safe houses near Bourj el-Barajneh.

But four days of negotiations with Nabih Berri did not resolve the crisis. Finally, President Ronald Reagan was prevailed upon to contact President Assad of Syria. Appealed to as a head of state, Assad delivered an ultimatum to Hezbollah: Return the captives, or face a cutoff of supplies and communication from Syria.

The Israelis agreed to release some 500 Shiites, Islamic Jihad surrendered and on Friday, June 30, all 39 men were driven to Damascus, where they boarded a plane that flew them to Frankfurt, West Germany.

MEDITERRANEAN SEA

October 7–9, 1985

A perceived need by the PLO to call world attention to itself led to the bungled, improvised hijacking of the Italian cruise ship Achille Lauro *in the Mediterranean from October 7 to 9, 1985. One American was killed.*

The hijacking of the 23,929-ton Italian luxury liner *Achille Lauro* on October 7, 1985, was planned far in advance by its mastermind, PLF leader Abu Abbas. He and Yasser Arafat had come to the conclusion that, considering their recent failures in landing commandos in Israel, it was time for a major terrorist move that would garner world attention.

Thus, the *Achille Lauro* hijacking was planned in the summer of 1985. Four suicide terrorists were to book passage on this cruise ship that plied the Mediterranean, taking tourists from Genoa to Egypt and the Holy Land. The terrorists would board in Genoa and stay under cover until the ship reached the Israeli port of Ashdod. They would wait behind while the passengers debarked; then, under cover of night, they would leave the ship, infiltrate portside oil-storage tanks and an ammunition depot and blow them and themselves up. At least that is what Abu Abbas told the Middle East News Agency in Belgrade after the incident.

The liner was scheduled to sail from Genoa on October 3, with stops in Naples, Syracuse, Alexandria, Port Said, Limassol, Rhodes and Ashdod. Issa Mohammed Abbas, a relative of Abu Abbas, was placed in charge of the Genoa operation. Masir Kadia, one of Abbas's trustees, had already obtained bookings for the four suicide terrorists, Magied Youssef al-Molqi Hallah al Hassan, Ali Abdullah and Abdel Ibrahim. They were set to occupy cabin 82.

But from that point onward, the operation proceeded to fall apart. On September 25 three PLO terrorists shot three Israelis aboard a yacht moored in the harbor of Larnaca, Cyprus. On October 1 a squadron of Israeli planes flew 1,500 miles to Tunisia and, in retaliation for the September 25 incident, bombed the headquarters of the PLO, south of Tunis. Sixty-seven Arabs were killed, and all communication between PLO headquarters and Genoa was severed.

On that same day Italian authorities arrested Issa Mohammed Abbas and confiscated four Kalashnikov automatic weapons, eight grenades and some detonators, which had been hidden in the false bottom of his car's gas tank.

The four young terrorists were left on their own, without their leader and the rest of their supplies. They were armed with their own Kalashnikovs and 9-millimeter pistols, but they lacked the dynamite that would be needed to carry out their mission in Israel.

The *Achille Lauro* left Genoa on October 1. The four remained aboard, silent and unnoticed until October 7. The evening before, the ship had pulled into Alexandria, Egypt, and nearly 600 passengers had debarked to tour the pyramids. The ship left at 10 A.M. for Port Said, where it would retrieve the passengers who had left on the Egyptian tour and proceed with the rest of its itinerary.

When the *Achille Lauro* reached international waters, the four terrorists suddenly smashed into the dining room, firing machine guns and pistols. They held the crew of 80 and the remaining 427 passengers at bay, invaded the bridge and commandeered the vessel, ordering Captain Gerardo de Rosa to sail northward toward the coast of Syria.

But Syria would not allow the ship to dock, and on Tuesday, October 8, the terrorists broadcast a demand for the immediate release of 50 Palestinians held by the Israelis, or they would begin to kill the passengers one by one, beginning with Americans and Britons.

The only passenger to die was 69-year-old New Yorker Leon Klinghoffer, a crusty gentleman confined to a wheelchair who taunted his captors repeatedly. They shot him and threw his body and the wheelchair overboard.

Meanwhile, an international confrontation was rapidly forming. Italy dispatched paratroops to the British bases on the island of Cyprus, toward which the *Achille Lauro* was now steaming. Specially trained SEAL frogmen commando units of the U.S. Navy and Delta army teams were sent to Sicily.

Realizing that the incident had gone too far, Arafat ordered the operation to be aborted. Abu Abbas, who had planned it all in the first place, emerged as the negotiator, and within hours the hijackers announced that they were willing to free everyone upon landing in Port Said, Egypt.

Once the ship docked in Port Said, the hijackers and Abbas were whisked off to Cairo, much to the indignation of U.S. ambassador Nicholas Veliotes, who telephoned his colleagues in Cairo and gave curt orders: "You tell the foreign ministry that we demand they prosecute the sons of bitches."

This was not done. Instead Egyptian President Hosni Mubarak told U.S. sources that the hijackers had already left and were on their way back to Tunis. Intelligence sources, however, uncovered the real story. Preparations were still being made to put the five Arabs aboard an Egyptair Boeing 737 from Al Maza Airport near Cairo to Tunis. The United States utilized the time to scramble F-14 fighters aboard the USS *Saratoga* in the Mediterranean.

On the night of October 10, over international waters, four F-14s intercepted the Boeing carrying the hijackers and forced it to land at a joint Italian-NATO base at Sigonella, in Sicily. The high degree of secrecy necessary to carry out the counterhijack displeased Italy mightily, since it was not informed in advance of the U.S. action. A confrontation developed between U.S. and Italian military units at Sigonella, and Italian authorities won out, taking the four terrorists and Abbas to an Italian court.

Abbas was released immediately, again much to the outrage of U.S. authorities, and headed back to Tunis by way of Yugoslavia.

On June 18, 1986, the trial of the *Achille Lauro* hijackers began. Three of the planners of the incident, including Abu Abbas, were tried in absentia and sentenced to life imprisonment—sentences that would never be realized.

The other four were sentenced to prison terms ranging from 15 to 30 years. Magied Youssef al-Molqi was later tried for the murder of Leon Klinghoffer, convicted and sentenced to life imprisonment.

SYRIA
DAMASCUS
November 29, 1981

Anti-Assad Muslims' continuing terrorist attacks on Syrian installations were responsible for the detonation of a car bomb on the crowded streets of Damascus on November 29, 1981. Seventy bystanders and police were killed; scores were wounded.

The terrorist attacks of one faction against another have often been as vicious and devastating as those of terrorist organizations against a common enemy such as Israel or the United States. Syria's Muslim Brotherhood, which operated from Jordan, was relentlessly opposed to the ruling party headed by Syrian president Hafiz Assad. Drawn from the more fanatical elements of Syria's population of Sunni Muslims, it repeatedly denounced Assad's ruling clique of Alawites as nonbelievers and, over a five-year period from 1976 to 1981, assassinated hundreds of Alawite members of Assad's ruling Baath party.

In the early afternoon of November 29, 1981, a white Honda minivan pulled up in front of the Azbakiyah Recruiting Center in the middle of a middle-class residential district in downtown Damascus. Sensing trouble, police moved in immediately. The driver of the van, Yasin Ben Muhammad Sarji, a 19-year-old former vegetable vendor and a member of the Muslim Brotherhood, drew a pistol but was killed on the spot.

Within seconds the van exploded. Its 220 pounds of TNT blew a crater in the street, showered the area with lethal shards of metal and masonry and killed 70 innocent bystanders and police. Three buses burst into flames, and their passengers were counted among the dead.

President Assad blamed the West, describing the terrorists as "agents of Zionism and imperialism," but the Muslim Brotherhood soon drowned him out with its own stated reasons, "to further the Islamic revolution in Syria." It added unmistakably that the bombing was in retaliation for "massacres, arbitrary executions and assassinations against Syrian citizens in their country and abroad."

UNITED STATES
COLORADO
LITTLETON
April 20, 1999

On April 20, 1999, at Columbine High School in Littleton, Colorado, Eric Harris and Dylan Klebold, two Columbine High School juniors, killed 15 of their fellow students and teachers and wounded 25 in a bloody massacre.

There has always been bullying in schoolyards and fights among students. But from 1997 forward in an America seemingly obsessed with guns, the bullying and the fights

A student is helped from Columbine High School in Littleton, Colorado, after the shooting rampage of Eric Harris and Dylan Klebold on the morning of April 20, 1999. (AP/Wide World Photo)

A makeshift memorial, erected by students for their slain classmates, presides over a gloomy landscape following the killing of twelve students and a teacher at Columbine High School. (AP/Wide World Photo)

turned lethal. Arguments were being settled with guns rather than fists. Two students were killed in Pearl, Mississippi; two in Springfield, Oregon; three at West Paducah, Kentucky; five at Jonesboro, Arkansas.

But no school shooting was quite so terrible or sobering as the massacre at Columbine High School in Littleton, Colorado, on Tuesday morning, April 20, 1999. At 11:30 A.M. that day, two Columbine juniors, 18-year-old Eric Harris and 17-year-old Dylan Klebold, wearing long black trenchcoats, parked their cars in the school parking lot, and armed with two sawed-off shotguns, a 9 mm semiautomatic rifle, a semiautomatic handgun and a bag full of homemade bombs, began a rampage that would leave 25 seriously wounded and 15 dead, including the gunmen themselves.

As evidenced from computers, carefully kept diaries and the admissions of their friends, this assault had been long planned. The sad fact that the parents of neither boy had seen any obvious signs that it could occur would remain an unsolved mystery and the reason for several civil lawsuits.

Littleton, Colorado, is a middle-class suburb of Denver, with a population of 35,000. Columbine High School, which opened in 1973, had an enrollment of about 1,800 in 1999. Neither Harris nor Klebold were particularly interested students, and they were reportedly teased and tormented by some of the other students at Columbine. Much of their time was spent before their computers, and both had jobs at a local pizzeria.

They were also preoccupied with guns and seemed to possess more than the usual teenage taste for violence. Although there was a so-called Trenchcoat Mafia in the school, an outcast group that wore the costume and expressed hatred for blacks and Hispanics, the two were loners and not part of this group, although they did wear black raincoats to school on the day of the shooting.

Evidence later indicated that the planning of the massacre began at least two years before it occurred. Notes that Klebold passed to a girl more than a year before the slaughter were worded in gunslingerese: "He will always be under the gun and in the sights since he hit me in the face, even though it didn't hurt at all" read one of them.

In 1998 both Harris and Klebold were arrested for breaking into a van. Rather than chastening them, this seemed to steel their resolve, and they began to appear at gun shows and gun shops, inquiring about and admiring various weapons designed to kill as many people as possible. In January of 1999, 22-year-old Mark Manes, a former Columbine student with whom Klebold and Harris worked at the pizza shop, made a straw purchase at a gun show of a TEC-DC9 semiautomatic pistol, capable of holding ammunition clips of more than 30 rounds, for the two teenage terrorists.

In March five teenagers, including Eric Harris, were caught on surveillance videotape trying to buy an M-60 machine gun and a silencer-equipped assault pistol in the Dragon Arms gun shop in Colorado Springs. At another gun show that summer, 19-year-old Robyn Anderson, Klebold's prom date three days before the attack, made a straw purchase of two sawed-off shotguns and a 9 mm carbine rifle. Philip Duran, another former Columbine student and a fellow worker at the pizza parlor, had arranged the purchase of the semiautomatic by Manes and accompanied Harris, Klebold and Robyn Anderson to a shooting range in Douglas County, where the shotguns and TEC-DC9 were tried out. A videotape was made of this trip by Harris and Klebold. The TEC-DC9 would be fired 55 times on April 20, killing four and wounding two. It would also be the gun that would end Klebold's life.

A week before the massacre the two teenagers set up a video camera and recorded a chilling recitation of their plans. They thanked Duran and Manes for helping them get the semiautomatic pistol and urged the police not to press charges against the two.

"If they wouldn't have f------ helped us out, then we would have found someone else," Harris bragged. "We would have gone on and on. We would have found some way around it 'cause that's what we do."

Later, Kate Battan, the detective who first found the videotapes, would remark, "It just flabbergasted me that so much evil came out of these two teenage boys."

Apparently, their parents did not see this, nor any sign of their lethal plans, and did not question the presence of bomb-making materials in their sons' bedrooms.

At 11:20 A.M. on the morning of April 20, two of these bombs, made from propane barbecue tanks, gasoline cans and other fuel cylinders and wired to pipe bombs, which in turn were wired to alarm clock timing devices, were scheduled to explode in the school's kitchen. Both failed to ignite, thus saving the lives of the 660 students in the cafeteria at that moment.

Ten minutes later slightly more than half of the 1,800 Columbine students were either working, studying or eating in the cafeteria. The others were playing on the soccer field or eating their lunches on the lawn.

At that moment Harris and Klebold, cursing the failure of their first bombs and wearing black trenchcoats to conceal the weapons they were carrying, parked their cars in the school parking lot, unloaded other bags of bombs, walked through the parking lot, shot and killed two students who were standing by their cars and wounded eight others.

Five minutes later they entered the school through the back of the cafeteria and walked into the crowded room, guns blazing. The students screamed and scattered. Teacher Dave Sanders yelled at the students to "Get down! Get down!" The terrorists turned their guns on him, hitting him in the face. Bleeding abundantly, he was dragged by some students into an adjacent science room, where he bled to death. The terrorists cornered other students under tables, shooting some, ignoring others.

As a score of cell phones called 911, the two killers ascended to the second-floor library. The tape of the 911 dispatcher captured the scene through teacher Patti Neilson's phone. Neilson had been shot near the west doors earlier and had crawled, wounded, to the library, where she took charge of the terrified students huddling there:

> *Neilson:* The school is in a panic. And I'm in the library. I've got students, under the table. Kids! Kids under the table! Kids are screaming. . . . We need police here. . . . He's turned the gun straight at us and shot and my God the window went out. And the kid standing there with us, I think, I, he got hit. (Sound of gunshots) Oh God! Oh God!
>
> *Dispatcher:* I just want you to stay on the line with me, we need to know what's going on. O.K?. . . . Is there any way you can lock the doors?
>
> *Neilson:* Smoke is coming in from out there and I'm a little afraid. (Sound of gunshots) The gun is right outside the library door. O.K.? I don't think I'm going out there.
>
> *Dispatcher:* We have paramedics, we have fire and we have police en route.
>
> *Neilson:* I can't believe he's not out of bullets. He just keeps shooting and shooting and shooting.

The mayhem escalated. Harris and Klebold laughed and shouted, asking questions of students before they shot them, tossing pipe bombs and attempting to set off the crude larger bombs. They still did not ignite. But this did not discourage the shooters. They entered the library, where the worst carnage occurred. Leaving the library, they walked down the halls, banging on classroom doors, shouting "We know you're in there!"

Panicked students, some of them bleeding from wounds, some dying, tried to climb out of windows to escape. As a SWAT team arrived, a boy, bleeding profusely, climbed through a second story window and hung precariously from it. SWAT team members gathered underneath him, scaled part of the wall, and saved him.

Two final shots were heard in the library. Harris and Klebold had returned there, turned their weapons on each other and fired. They were the last two victims of the massacre.

Police evacuated the remaining students from the school, and at 12:30 the SWAT team entered it and swept the classrooms. It would be the rest of the afternoon before hysterical students, crouching in hiding places, often near dead and dying classmates, were entirely freed from the building.

At 10:30 P.M. that night a bomb, timed to go off in one of the cars in the parking lot, exploded. It was one of the only bombs to detonate. Two days later police would find the largest bombs in the school kitchen.

The tally began: 15 died and 25 were wounded in this, the worst and most senseless school shooting to date in the United States. The following investigation unearthed a terrible plot of which this massacre was supposed to be only the beginning. Robyn Anderson, Klebold's girlfriend, told police that the two had planned to kill 500 students at the school, commandeer an airplane and crash it into New York City, killing thousands more.

The outpouring of shared grief in the nation was massive. On the Sunday after the shooting, 70,000 people, including Vice President Al Gore, assembled in a parking lot outside a nearby shopping mall for a prayer service.

The National Rifle Association, which had scheduled its annual convention in Denver on April 21, was asked by Denver Mayor Wellington Webb to cancel it. The NRA came but stayed for only one day, during which its president, Charleton Heston, blamed loose national morals and a lack of school security for the tragedy. The U.S. Congress passed more restrictive gun legislation following the tragedy but failed to close the gun show loophole that had allowed the killers to obtain their weapons.

UNITED STATES
NEW YORK
NEW YORK
February 26, 1993

A horrendous bomb planted by terrorists from the Liberation Army tore through the basement of One World Trade Center on February 26, 1993. Six people were killed and more than 1,000 were injured.

Who knows specifically when and where a terrorist act begins? From some point either carefully guarded or difficult to discern, it grows slowly in the minds and plans of terrorist organizations and their leaders. The followers merely follow orders and bring the plans to fruition, as they did at 12:17 P.M. on February 26, 1993, when they ignited a bomb in the basement of One World Trade Center, in lower Manhattan. The bomb killed six and injured more than 1,000.

In the early 1990s in America, Sheik Omar Abdel-Rahman, the spiritual leader of al-Gama'a al-Islamiyya, a radical Muslim group seeking to overthrow Egypt's secular government and install an Islamic state, was considered to be the top man in the terrorist network in America. He resided in Brooklyn, and the followers—or at least admirers of the sheik—who would plant and detonate the bomb were all quartered in Manhattan, Brooklyn and Jersey City.

That snowy and awful day came as the climax of a long series of sometimes careful but more often clumsy preparations. There was much interweaving between these plotters and those who planned to turn much of the island of Manhattan into an inferno and a tomb. But for all intents and purposes, it is generally agreed that terrorism truly came to America on September 1, 1992. On that day Ramzi Yousef arrived in the United States, and, with skill and false papers, slipped through customs and the INS.

Yousef was an Afghan-trained terrorist with 12 aliases who knew how to build and plant explosives. He was installed in a Jersey City apartment by Mohammad A. Salameh, a mostly unemployed Jordanian of Palestinian descent. Joining them shortly was Nidal Ayvad, born in Kuwait, who had become a naturalized American citizen. He was valuable chiefly because he had a degree in chemical engineering from Rutgers University and a steady job.

In October he and Salameh opened a series of bank accounts to handle the expenses being paid by sources in, among other countries, Iran, Saudi Arabia and Kuwait. The precise source was never discovered.

Others who participated in the plot were Ibrahim el-Gabrowny, Ahmad Ajaj and Abdul "Aboud" Yasin. The leader of the mission was Mahmud Abouhalima, an Egyptian-born New York City cabdriver who had been a former aide and driver for Sheik Omar Abdel-Rahman and who had driven El Sayyid Nosair, Rabbi Meir Kahane's assassin, from the scene of the murder. Nosair was serving time in Attica prison for the crime, but it was later shown that he was in constant communication not only with these bombers but with the planners of the wider series of bombings that were to follow the World Trade Center explosion (see p. 146).

Shortly after Yousef's arrival in America, the conspirators rented locker number 4344 in the so-called Space Station, a complex of four prefab storage buildings in Jersey City, New Jersey, across the Hudson River from the World Trade Center. From the end of November 1992, regular deliveries of chemicals and other materials were made to the locker. Urea, nitric acid and sulfuric acid, ingredients for high powered explosives, piled up.

Their laboratory was a few miles away in a garage that had been converted into two apartments on Pamrapo Avenue. More than a thousand pounds of explosives were collected, along with nitroglycerin, which would be the igniter of the explosives. Through the month of January 1993, they worked to convert 1,200 pounds of urea into an explosive, mixing the nitroglycerin and freezing it for easy transport.

At the end of the month they drove deep into Pennsylvania to a heavily wooded retreat and tested the bomb. It detonated, but it needed more energy. They went in search of compressed hydrogen and after many failed attempts finally found a company that would sell it to them.

What was interesting and almost as terrible as what was to come at the World Trade Center was the fact that an FBI informer, Emad Salem, a former Egyptian army colonel who was posing as Sheik Abdel-Rahman's security adviser, had gotten wind of the plot and had told the FBI about it. But the FBI was skeptical of his news and failed to follow up on it. Had they acted, the bomb never would have been exploded.

On February 24, 1993, Salameh and Ayyad stopped at a car dealership on Kennedy Boulevard in Jersey City that also rented Ryder trucks. Salameh chose a yellow van and drove off in it.

Over the next few days scouting trips were made to the World Trade Center garage. In between, heavy tanks of hydrogen, the last material left in locker 4344, were loaded into the van.

The plot moved rapidly forward. Salameh reported the van stolen to the Jersey City police and the rental company. It was a weekend. Neither did anything about it.

The van went to the Pamrapo Avenue laboratory, where the bomb itself was loaded into it. Before dawn on February 26, it and two other cars—a red Corsair and a blue Lincoln—moved out. All night the terrorists had been gathering, unloading and reloading the van.

Four containers of nitro-glycerine were placed into four cardboard boxes. Extended fuses, threaded through surgical tubing, were rigged to blasting caps and boxes of gunpowder inserted into each container of nitroglycerin. The hydrogen tanks were propped upright behind the boxes next to the van's rear doors.

The conspirators headed into the Holland Tunnel, inching toward what they felt would be a Holy War. Destroying even one World Trade Center building would kill thousands.

At a little after noon the yellow Ryder truck entered the ramp of the B-2 level of the parking garage beneath Number One World Trade Center. Ten minutes later it pulled into an illegal parking space vacated by a yellow Port Authority truck. The color of the van kept it from being noticed. And because it was in an illegal space, it had not passed through the garage's toll gates.

The two men in the van—Salameh and Eyad Ismoil—took out a disposable cigarette lighter and lit the fuses. They left the van and climbed into the waiting red Corsair, which drove off. The 107-story tower would, they were confident, collapse against its twin in a few moments.

At 12:17 the bomb exploded. It was a monumental eruption, an underground clap of extreme thunder. The walls of the van were torn asunder. The concussion wave created a wind equal to that of a tornado, which ripped through the underground garage. It hit the north wall of the tower with a force of 1,500 pounds of pressure per square centimeter, turning the cinder block wall into dust. A 14,000 pound steel beam was ripped from its welds. Concrete walls collapsed. Eleven inches of concrete above the van disappeared into dust, and the force of the explosion traveled another story, where it picked up a woman who was sitting at an airline ticket counter three stories above and flung her 30 feet.

The blast continued more deeply into the core of the building, ripping down masonry blocks around an elevator shaft. The cars in the garage were turned into fiery skeletons of twisted steel, the paint blasted off their sides, their tires melted into indefinable masses. The concussion sped through the garage at a thousand miles a second. Fires roared in the freight elevators, and the smoke from the car fires in the basement rose a quarter mile through the elevator shafts, then spread along the floors of the building. The floor of the Vista Hotel, above the garage, was a gaping, smoking hole.

And yet the building stood. It absorbed all of this and remained upright; although severely damaged. All of the offices and the lunchroom on level B2 were reduced to fine-grained rubble, and the workers within them were killed.

The streets around the World Trade Center almost immediately became jammed with firetrucks, police cars and rescue vehicles. The smoke became suffocating on practically all of the building's 107 floors—it had shot up the ventilation, heating and cooling vents as these systems were crushed and demolished in the basement. The elevator shafts were other vertical avenues of choking fumes.

It would be late afternoon before the last of the building's several thousand occupants were evacuated and the statistics set down. Six were killed and more than 1,000 were injured by the blast, most by falling walls, exploding glass or airborne debris.

As the maelstrom of activity whirled through the downtown streets of Manhattan, Ramzi Yousef, his mission accomplished, boarded a plane under the fictitious name of Abdul Basit. He had the foresight and planning to escape a set of security details at all the New York airports. His destination was Jordan, and he would remain out of sight of investigators for two years. Later, Mahmud Abouhalima would escape the same security details and fly to Egypt.

The next day politicians of all levels arrived at the site of the blast, along with a dozen agencies, including the FBI, the Secret Service, the NYC bomb squad, the ATF (Bureau of Alcohol, Tobacco and Firearms) and the Port Authority. The previous night New York police precincts had been bombarded by calls from a multitude of organizations taking credit for the bombing. Colombian drug cartels, Iraq, Yugoslavia, Chinese militants, the Black Liberation Front, and Serbia all were mentioned in the calls, sometimes several times.

Four days later the *New York Times* received a letter from the Liberation Army. It read, in part:

> The following letter from the LIBERATION ARMY regarding the operation conducted against the W. T. C.
>
> We, the fifth battalion in the LIBERATION ARMY, declare our responsibility for the explosion on the mentioned building. This action was done in response for the American political, economical, and military support to Israel the state of terrorism and to the rest of the dictator countries in the region.

Later, the computer of Nidal Ayyad at Allied Signal, his place of employment, was seized, and in its deleted files a copy of the letter was found.

Meanwhile, the serious task of assessing damage and looking for clues began. The ATF's bomb experts descended into the darkness of the crater created by the blast. Its sides were as uncertain as those of a mountain before an avalanche. Gnarled metal and teetering piles of concrete met them. The B-2 level was a cliff hovering over a 60-foot hole.

By March 4 a piece of the Ryder van with the vehicle identification number intact was discovered. It was a matter of hours before it was traced to the company to which Mahmud Salameh had gone to rent the yellow van. For days, this not very smart bomber had been pestering the agency for his $400 deposit, which he felt he was owed since the van had been "stolen." On March 4, 1993, FBI agents surrounded the agency while one of their members posed as an employee of the firm. Mahmud Salameh was the first conspirator to be arrested.

That same day an inquisitive clerk at the Space Station in Jersey City had noticed a great deal of activity around locker 4434. He investigated, found the evidence of bomb making material and contacted the FBI. There was great pressure from the media and governmental officials to solve the case quickly, and with the help of Emed Salem, the informant they had ignored earlier, they made a series of swift arrests of the remainder of the conspirators.

In Egypt Egyptian police captured Abouhalima, questioned him, tortured him and delivered him in chains to American agents. He arrived in New York on March 25.

Undercover informant Salem was outfitted with a wire and engaged Sheik Rahman in a conversation that undeniably implicated him. "I learned," Salem said to the sheik over dinner, "they originally planned to bomb the Big House [the UN] and then it was changed to the World Trade Center."

"Yes, exactly," the sheik answered.

"Why the World Trade Center?" asked Salem.

"The operation is to make them lose millions," the sheik answered, "and that is what happened. This is a message. We want to tell them that you are not far from us. We can get you anytime."

The sheik was arrested not long after, and Salem went into the witness protection program and became the government's star witness in the extended trial that followed.

The jury was faced with the testimony of 206 witnesses and more than 10,000 pages of testimony when it was finally sequestered to produce a decision. All of the terrorists were found guilty. The judge calculated how many years each of the victims could have expected to live and sentenced each bomber to a year in prison for each year of life they deprived their victims.

Two and a half years after the bombing, in 1995, Ramzi Yousef was arrested by the FBI in Pakistan and returned to New York. While there, he had concocted a plot in 1994 to use five people from Manila to plant liquid bombs on a dozen U.S. airliners flying from Asian cities, each of them set to explode in a span of 48 hours. A Pakistani court found Yousef and two codefendants guilty, and after his conviction turned him over to the FBI.

Yousef was tried and convicted in New York along with a latecomer to the conspiracy, Eyad Ismoil, a Palestinian who had been in Dallas since 1989 studying engineering, had known Yousef in Kuwait and flew to New York five days before the bombing in order to be part of it. He was the second man along with Salameh in the van as it parked on level B-2 of the World Trade Center. His sentence was 240 years in prison plus a $10 million fine.

The World Trade center bombing, along with the intersecting plans to blow up much of Manhattan, were the most serious threats to Americans in America yet. But terrorism, the guerilla war against innocents, continued in the rest of the world, seemingly a world away in 1993. The words of Sheik Abdel-Rahman echoed prophetically: "We want to tell them you are not far from us. We can get you any time."

UNITED STATES
NEW YORK
NEW YORK
June 23, 1993

A gigantic plan to blow up bridges, tunnels, buildings and the Statue of Liberty in New York City was prevented by an FBI raid on June 23, 1993.

Emad Salem was an undercover agent for the FBI. A former Egyptian army colonel, he had become the security adviser for Sheik Omar Abdel Rahman, the accused head of a terrorist network in America. Sheik Rahman, a blind cleric, had waged an underground war against Egyptian President Anwar Sadat, and was implicated in, tried and acquitted for the assassination of President Sadat.

However, in 1990 he left Egypt and, despite gaining a reputation as an international terrorist, was granted a visa and entered the United States. He became active in storefront mosques in Jersey City and Brooklyn, where he stirred up trouble between factions competing for funding for terrorist activities.

Six months after his arrival in America, Rabbi Meir Kahane was killed in New York City. The assassin was identified as El Sayyid Nosair, a top aide of militant cleric Rahman. Nosair was acquitted of the murder but imprisoned on related weapons and assault charges.

The FBI gained the services of Emad Salem. One of the first tips he brought to the bureau seemed so outlandish, it was ignored. Salem reported in March and April of 1993 on a conception so vast it could only have come from the mind of Sheik Omar, who in one of his sermons in Brooklyn urged his parishioners to "hit hard and kill the enemies of God in every spot, to rid it of the descendants of apes and pigs fed at the tables of Zionism, communism and imperialism." It would be radical Islam's Pearl Harbor attack.

A series of bombs would be planted in various targets across the city of New York that would disable transportation links, destroy crucial government office buildings and high profile landmarks, cripple major portions of New York's infrastructure and kill hundreds of thousands.

One bomb would be planted on the George Washington Bridge, over the Hudson River between New Jersey and Manhattan. It would destroy the bridge and send cars, pavement and victims plunging 100 feet into the river. Two other bombs would be planted and would destroy the Lincoln and Holland Tunnels beneath the Hudson River by opening cracks in the tunnels' sides and drowning civilians in a tidal wave tainted by smoke, flames and toxic fumes. Next, the bombers would blow up the United Nations and FBI headquarters in Federal Plaza during business hours. At the same time, another bomb would destroy the Statue of Liberty. A final bomb would explode and wipe out forever the Diamond District in Midtown New York, run and populated mainly by Hasidic Jews.

"Boom! Broken glass, dead Jews in the street," bragged one of the bombers.

It would be apocalyptic—too much so for the credibility of the FBI, and they ignored the warnings. Had they listened to their informer, the bombing in February of 1993 (see p. 144) might have been avoided, because the cat's cradle of terrorists in America intertwined between those of the World Trade Center plot and the other, more extensive plot targeting American citizens and spirit.

In light of the accuracy of Salem's reporting in the aftermath of the World Trade Center bombing, of Sheik Omar Abdel-Rahman's implicit involvement in that event and the identities of its bombers, the FBI now focused on the grander plan, which was in the process of being implemented. In the middle of May 1993, the team of doomsday bombers, under the guidance of Siddig Ali, leased a large building, somewhere between the size of a large garage and a small warehouse, at 131–09 90th Avenue in Jamaica, Queens. Rather than use the dangerous mix of chemicals employed in the World Trade Center bombing, these conspirators decided upon a simple mixture of fuel oil and fertilizer dubbed ANFO and used extensively by terrorists worldwide. Ignited by a blasting cap, the ingredients would make bombs less powerful than the World Trade Center bomb but easier to put together and easier to assemble from ordinary ingredients. All of this would make their activities less suspicious.

By June the FBI had enough information, gained through Salem, to halt the bombmaking. On the morning of June 23, 1993, sharpshooters arranged themselves on the roofs of houses surrounding the bomb factory and behind cars on the street. Agents with bulletproof vests entered the premises.

The bombers, caught in the midst of mixing their ingredients, offered no resistance. It was one of the bureau's smoothest operations, and it prevented one of the worst potential holocausts in the entire history of terrorism.

As the interrogation of these bombers and the bombers of the World Trade Center progressed, the sheik's involvement became clear, and on July 2 police surrounded his mosque on Foster Avenue in Brooklyn. Negotiations went on within the mosque, a dummy sheik for the media was released through the front entrance and the real sheik was hustled out a back entrance and taken to jail. He was tried and convicted of sedition and sentenced to life plus 65 years in a prison hospital at the Federal Medical Center in Rochester, Minnesota. Blind and suffering from worsening diabetes and heart disease, he was isolated from other prisoners in the hospital. Rumors continue to circulate regarding his condition and treatment. At one point he was reported dead, but this was refuted. The latest report was of a confirmed assault on the cleric on August 8, 1999, by a prison guard, who was dismissed. The rest of the conspirators also received life sentences, but not in prison hospitals.

UNITED STATES
NEW YORK
WASHINGTON, D.C.
September 11, 2001

More than 3,000 innocent people were killed in two suicide terror attacks in New York City and Washington, D.C., on September 11, 2001. Approximately 2,800 died in the collapse of the two World Trade Towers, 265 were killed aboard four hijacked commercial airliners that were used as missiles and 115 died in the Pentagon after one of the planes crashed into its west wall. It was the deadliest terrorist attack of modern times.

It was a September morning to wish for, a morning bursting with a seamlessly blue and brilliant sky, the sharp outlines of the chiseled, right-angled facades of the buildings of New York City leaping into bright focus after the last softening of summer's humidity. The spirit-lifting, manageable warmth of early autumn made it impossible not to feel lighter than usual, and thousands of people headed, with varying degrees of cheerfulness, to their jobs in the twin towers of New York City's World Trade Center, not for a moment suspecting that close to 3,000 of these workers were headed to their deaths.

At 8:45 A.M., as some workers sat at their desks or grabbed one more cup of coffee from the office coffeemaker, others waited for elevators and still others read their morning papers on the subways converging on the financial area of the city, two commercial jetliners, their tanks filled to capacity for their projected transcontinental flights, were changing course.

United Airlines flight 175, a Boeing 767, had taken off from Boston's Logan Airport at 7:58 A.M. for Los Angeles with 56 passengers and nine crew members aboard. One minute later, American Airlines flight 11, another Boeing 767, left the same airport headed for Los Angeles with 81 passengers and 11 crew members aboard.

At the same time, workers were arriving at the Pentagon in Washington, D.C., and two minutes after the American Airlines flight to Los Angeles left Logan Airport, at 8:01 A.M., United Airlines flight 93, a Boeing 757, left Newark Airport bound for San Francisco with 38 passengers and seven crew members aboard. Nine minutes after that American Airlines flight 77, a Boeing 757, took off from Washington Dulles Airport for Los Angeles with 58 passengers and six crew members aboard.

None of these four planes would reach its scheduled destination, nor would anyone on any of them live beyond that morning. Each flight, at a prearranged moment, would be hijacked by terrorists, its pilot disabled and replaced by a suicidal fanatic bent upon bringing Jihad and terror, once and for all and with undeniable certainty, to the mainland of America.

American Airlines flight 11 rose to its cruising altitude over Massachusetts, then headed west over upstate New York. As the plane flew north over Albany, five hijackers rose from their seats armed with knives and box cutters. Four of the five could fly a commercial jet. Mohammed Atta, sitting in seat 8D, took over the controls, while his accomplices, Waleed Alshehri, Abdulaziz Alomari, Wail Alshehri and Satam Al Suqami, herded the passengers and remaining crew to the back of the plane.

Flight 11 took an abrupt change of course and headed straight south, nearly colliding at one point with United Airlines flight 175, which had now been commandeered by Marwan al-Shehhi, Fayez Rashid, Ahmed Alghamdi, Hamza Alghamdi and Mohand Alshehri. It, too, was flying south, toward New York City.

The passengers were allowed to make final phone calls on their cell phones. The flight attendants frantically called for help. But it was all happening very fast, faster than the training for such an emergency could anticipate. At 8:43 a flight attendant aboard flight 11, who had informed the ground of the hijacking and was asked if she knew where they were, said, "There's water, and there are buildings, and—My God! My God!"

The horrendous energy of United Flight 175's collision with the South Tower of the World Trade Center sprays flaming and atomized debris from its upper floors onto downtown Manhattan. (AP/Wide World Photo)

It was the last communication from flight 11 before, at exactly 8:48 A.M., it slammed into the top floors of One World Trade Center and exploded in a thunderous, cataclysmic convergence of steel, concrete and ignited jet fuel, which melted the steel, pulverized the concrete, and instantly incinerated all on the plane and in the ten floors immediately impacted by it. Fire and sickening dense smoke erupted from the building and smeared the absolute blue of the autumn sky. Immediately, fire companies clambered aboard firetrucks and headed for the burning tower to rescue whoever they could and to extinguish the blaze, which at this point was consuming the upper quarter of the tower.

But the horror was only beginning. At 9:06 A.M. United flight 175 appeared out of nowhere, it seemed. It banked sharply, curving around tower number one, and crashed into the south World Trade Center tower, entering the west side and emerging on the south side before it, too, erupted into an enormous fireball that ripped the building apart.

Rescuers, ambulances and police battalions arrived, desperately trying to reverse or ameliorate the holocaust that was burning with increasing ferocity above them. Workers, some of them terribly burned, began to emerge from the towers as firefighters rushed in and upward toward the upper floors of both towers.

Those on the ground were pelted with debris. And more. "I felt this debris falling all around me," one of the firemen recalled later. "And I looked up, and it wasn't debris. It was people. People were jumping out of the windows, and falling."

And now the unthinkable, the unimaginable, except in the minds of the terrorists, happened. At 10:00 A.M., the south tower collapsed. From the top down, it peeled its facade, its floors collapsed upon the floors beneath them, its girders melted, its concrete was atomized and it, along with the thousands of workers, policemen and firemen within it, dissolved into nothingness. Twenty-nine minutes later, the north tower also collapsed, erasing a portion of the mightiest skyline in the world.

But the terrorists were not finished yet. American Airlines flight 77, which had flown west as far as Columbus, Ohio, suddenly reversed its path, now, piloted by Hani Hanjour while Khalid Almihdhar, Nawaf Alhazmi, Salem Alhazmi and Majed Moqed held the passengers and crew in the rear of the airplane. Aboard was Barbara Olson, the columnist and television talkshow personality who was also the wife of Theodore Olson, special counsel to President George W. Bush. She called her husband on her cell phone, agitatedly asking him what to do. He seemed to be at a loss for words and direction.

The plane, uncertainly piloted, headed for the Pentagon in Washington, D.C. David Marra, on his way to work, saw it approaching. "It was 50 feet off the deck when he came in," he told reporters later. "It sounded like the pilot had the throttle completely floored. The plane rolled left and then rolled right. Then he caught an edge of his wing on the ground." The plane cartwheeled and slammed into

United Airlines Flight 175, hijacked and piloted by a suicide terrorist, heads for the South Tower of the World Trade Center just before smashing into it with cataclysmic, flaming force. (AP/Wide World Photo)

the west wall of the Pentagon at 9:40 A.M., tearing a huge hole in the outer perimeter of the building before both exploded in a monstrous fireball of flaming jet fuel.

A different scenario was occurring on United Airlines flight 93. It, too, had changed course after flying as far west as Cleveland, Ohio. It was reversing its path now, transversing Ohio, headed into Pennsylvania and from there, it is assumed, to Washington, D.C., and either the White House or the Capitol. But a small band of passengers who, by now, had been informed of the World Trade Center calamity, determined that although they would probably die in a crash, they could at least prevent the deaths of thousands more, and possibly they might be able to land the plane if they overpowered the hijackers. The last communication from them was a whispered direction from passenger Mark Bingham, "Let's roll!"

They apparently wrested the controls from the terrorist/pilot, and the plane descended over the cattle pastures, farmhouses and shops of Shanksville, Pennsylvania. It was headed straight for the Shanksville-Stonycreek school, filled that morning with 501 kindergarten through 12th-grade students. But two miles short of the school, in a reclaimed section of an old coal strip mine, the plane plowed into the earth and exploded.

Final cellphone calls to home had been made. "Hey, Jules," had said Brian Sweeney to his wife from United flight 175. "I'm on a plane that's been hijacked. It doesn't look good. I just want to tell you how much I love you." Peter Hanson, on the same flight, had called his parents. "I think we're going down," he had said, "but don't worry. It's going to be quick."

President George W. Bush was in Sarasota, Florida, in a second-grade classroom when the news of the World Trade Center calamity occurred. As he was boarding Air Force 1, the news of the Pentagon attack reached him. Air Force 1, accompanied by jet fighters, took off and headed to Louisiana, then Nebraska. Vice President Cheney was lifted bodily from his office in the White House and taken to a safe subterranean shelter. House Speaker Dennis Hastert and other congressmen in line for the presidency in an emergency were hustled off to safehouses.

Finally, that afternoon Air Force 1 returned to Washington, and with it the president, who delivered an Oval Office speech of reassurance to the country. Members of the Congress, Republicans and Democrats, appeared for television cameras on the Capitol steps, shaking hands and singing "God Bless America."

If one of the purposes of the assault was to sow fear and chaos, the mission had succeeded. Both were apparent through the day and night of September 11. In the 35 minutes between the time the second plane hit the World Trade Tower and the third slammed into the Pentagon, Department of Defense leaders did nothing about evacuating the Pentagon. No order was given, and employees were on their own.

Two F-16 fighter jets were scrambled immediately after the World Trade Center collisions took place, but instead of taking off from Andrews Air Force Base, a mere 15 minutes from the Pentagon, they left from Virginia's Langley Air Force Base, 130 miles away. Twenty minutes before they arrived over Washington, the west wing of the Pentagon was ablaze. What they could have done was questionable. Commercial jets could be attacked only by direct order of the president—a regulation that, in a few days, would be severely modified.

Senior Pentagon aides, informed of the New York City attack, rushed to the command center within the complex, but they remained unaware that another hijacked plane was headed in their direction. Even after the plane hit, they did not realize what caused the jolt that jarred papers loose from their planning table. Only Defense Secretary Donald Rumsfeld seemed to be aware of what was going on, and he refused to be evacuated. He stayed at the Pentagon all afternoon assisting in rescue efforts. An Air Force general, interviewed later, admitted, "This was something we had never seen before, something we had never even thought of."

Chaos gave way to heroic rescue attempts by New York City firemen and policemen, who worked around the clock hoping against hope that there would be survivors in the thousands of tons of rubble. All that was found were a few bodies, many more pieces of bodies, and for thousands, nothing. Another building near the towers collapsed. Others seemed on the verge. The entire area had turned into a disorderly, flaming, overwhelming tomb.

Gradually, fear turned to anger, then to resolve, then to a uniting of the country as it had not been united since World War II, the last time U.S. territory had been attacked in any manner near the extent of this attack. And that had been in Hawaii. This was the first and worst terrorist attack of wartime dimensions ever on the continental United States.

President Bush, heretofore gaining a reputation as a somewhat diminished, frequently inarticulate inheritor of

One person clings to the outside of the North Tower of the World Trade Center and another flings himself from the fire engulfing the building's upper floors, shortly before the building's collapse. (AP/Wide World Photo)

his role, took on a new dimension and stature. In a ringing, forceful speech to the country and the Congress, he articulated the feelings of the country and gave it a purpose. If there was no declared war in the sense that war had been known up to this point, there was a new sort of war enunciated, and an enemy. The enemy in a general sense was terrorism, and the man behind it, as the investigation of the events of September 11 would begin to unfold, was Osama bin Laden, the head of al-Qaeda, a far-reaching terrorist group working with and composed of Islamic extremists. Its goal: to overthrow all non-Islamic governments and expel Westerners and non-Muslims from Muslim countries. Bin Laden was linked closely to the bombing of the U.S. embassies in Kenya and Tanzania (see p. 123) and the bombing of the USS *Cole* in Yemen. In 1998 he had issued an edict to his followers stating that it was the duty of all Muslims to kill U.S. citizens and their allies everywhere.

Only one major U.S. figure seemed not to comprehend this. Reverend Jerry Falwell, in a burst of execrable tastelessness, posited that God was punishing the United States for allowing within its borders abortion supporters, the American Civil Liberties Union and People for the American Way. But his voice was dimmed to silence in the national purpose of combatting worldwide terrorism.

With great skill and care, Secretary of State Colin Powell set about forming an international coalition of more than 40 countries, including several Arab countries with overwhelmingly Muslim populations. Bin Laden, present in Afghanistan as a "guest" of the Taliban rulers had already isolated that poverty- and war-racked country, and now it was isolated still further. Saudi Arabia and Pakistan were the only two countries in the world who recognized the Taliban as Afghanistan's government. Saudi Arabia withdrew its recognition, and only Pakistan remained, for a practical reason. Pakistan welcomed the international coalition onto its territory but kept diplomatic lines open to the Taliban, who defiantly rejected America's demand to hand over Osama bin Laden and close down his terrorist camps, where his followers learned their trade.

The son of a rich Saudi contractor, bin Laden had inherited $80 million as a very young man. In 1979 the resistance against the Soviet invaders of Afghanistan drew him to that country. Raising money and Islamic recruits, he aided in the defeat of the Soviets.

Returning to Saudi Arabia, his extremism proved to be a problem as he agitated against the American troops that remained in the country after the Gulf War. He called them infidels and raised the cry that they should be expelled from the holy ground of Saudi Arabia. Saudi Arabia exiled him for organizing the terror network al-Qaeda. Back in Afghanistan, he grew richer and more powerful as he intensified his war against America.

War preparations gained momentum in the coalition, which included the nations of NATO. The United Nations Security Council added its condemnation of the attack and pledged support to the coalition. President Bush confirmed what was becoming more and more apparent. In one explosive instant, America and much of the Western world had left a time of peace and prosperity for one of war and uncertainty. Terrorism had undergone a change, too. From the bombing of buses and embassies, in which hundreds died, it had escalated to wartime casualties of more than 3,000 innocent people murdered in one day's work.

The streets of lower Manhattan are clogged with people fleeing the advancing cloud of perilous fire and debris as the South Tower of the World Trade Center disintegrates. (AP/Wide World Photo)

It was war, indeed, and President Bush prepared the American public for a war like no other before it. Waged on diplomatic, economic and military fronts simultaneously, it would be against an enemy who struck unexpectedly and seemingly without reason. Mr. Bush repeatedly assured the world that the fight was against terrorism, not against Muslims. But pro–bin Laden demonstrations nevertheless erupted in Pakistan, Iraq, Egypt and particularly Indonesia, which contains the world's largest Muslim population. As armament and military personnel arrayed themselves around Afghanistan's borders, the Taliban continued its defiance, and bin Laden burrowed further into Afghanistan's distinctly unfriendly landscape.

All the while, the determination in America quickened as its economy began to suffer from the effects of the World Trade Center bombing. Travel and leisure suffered most, and unemployment rose alarmingly to nearly 500,000 in three weeks. The questions raised themselves: How could this have happened? Why were there those in the world who hated America so much? What sort of terror was next?

The way that this, the most dreadful terrorist act yet in the world, happened was through careful, long-term planning, much as had, the first bombing of the World Trade Center in 1993 (see p. 144). However, these bombers were radically different from the not very bright group who planned the 1993 attack. They were, as are most terrorists, poor, uneducated, hopeless, deeply religious and convinced that their religion bid them to kill in its name. But they were not suicidal.

The 19 suicide bombers who piloted the 2001 holocaust to its conclusion were educated, bright and inventive, in addition to being fanatically dedicated to the Jihad against America and all things Western. They were unobtrusive personalities who faded easily into the fabric of American society. They drank, smoked, spent time weight lifting in gyms and took flying lessons. The initial planning was believed to have begun months and perhaps years before with Osama bin Laden, possibly after the first, mostly failed attempt at the World Trade Center in 1993 and the following aborted plan to blow up New York City's bridges, tunnels, the UN building and the Statue of Liberty (see p. 146). As more and more evidence piled up in the investigation, closer and closer links to bin Laden began to emerge.

Firefighters and inspectors continue to look over the wreckage of the Pentagon which was rammed by hijacked American Airlines Flight 77 the day before. (AP/Wide World Photo)

Evidence gathered after the catastrophe indicated that much of the later planning took place in Hamburg, Germany. Mohammed Atta, the hijacker pilot of the first plane to strike the World Trade Center, had been a student for several years at the Technical University of Hamburg-Harburg and shared an apartment with al-Shehhi. There was a great deal of traffic in and out of the apartment and frequent gatherings of as many as 20 rather noisy Islamic visitors.

In April of 2001 Ziad Jarrah told his girlfriend in Germany that he was going to Afghanistan. He went, instead, to Fort Lauderdale, Florida. The others in his cell—Ahmed Alghamdi, Nawaf Alghamdi and Saeed Alghamdi—joined him later.

At the same time, all the flight 175 hijackers and some of those who would commandeer flight 93 moved to Delray and Deerfield Beach, Florida, where they took flying lessons at Huffman Aviation, between Tampa and Fort Myers. Simultaneously, Mohammed Atta and Marwan Al-Shehhi circulated throughout Florida, remaining in touch with the others and taking flying lessons in Coral Springs. Waleed Alshehri, Satam al-Suqami and Wail Alshehri stayed in Boynton Beach, Florida.

From 1990 onward Hani Hanjour lived in Arizona and took flight lessons at CRM Airline Training Center in Scottsdale in 1996 and 1997. Nawaf Alhazmi and Khalid Almihdhar lived in San Diego, California, and joined Hani Hanjour in Arizona in 2001. Four days before the strike, all 19 moved closer to their targets—some to Laurel, Maryland, some to Fort Lee, New Jersey, some to Newark, New Jersey.

They communicated with one another over the Internet using an al-Qaeda code that was discovered and translated during the investigation. In the days preceding September 11, the hijackers, apparently prepared and determined to die on their mission, drank heavily in one Florida bar and purchased lap dances in another.

And then they converged upon their appointed airports, although not necessarily by direct routes. Mohammed Atta drove to Portland, Maine, and boarded a US Airways flight to Boston, leaving behind a rented car. His fellow conspirators went directly to Boston's Logan Airport. Atta transferred to flight 11 in Boston, but his luggage did not, and in it, as in the abandoned rented cars, investigators found airline schedules, schematics of airliners, instruction booklets and a five-page handwritten letter in Arabic, which turned up in the wreckage of United flight 93. It was part manifesto, part instruction manual in the preparation for a martyr's death.

In part, it read: "Purify your heart and forget something called life, for the time of play is gone and the time of truth has come. . . . God will absolve you of your sins, and be assured that it will be only moments and then you will attain the ultimate and greatest reward.

"Check your bag, clothes, knives, tools, ID, passport, all your papers. Inspect all your weapons before departure. Let each find his blade for the prey to be slaughtered.

"As soon as you put your feet in and before you enter [the plane] start praying and realize that this is a battle for the sake of God, and when you sit in your seat say these prayers that we had mentioned before. When the plane starts moving, then you are traveling toward God and what a blessing that travel is."

Slaughter, then, became the key to the door to paradise, and it was toward paradise these suicide hijackers were pointed on September 11. This leads to at least some of the answers to the second question that was on the minds of many Americans following the disaster: Why do they hate Americans so much?

The answer is partly religious, insofar as Islamic fundamentalists, from Iran to Iraq to Egypt to Syria to Indonesia, painted America as the Great Satan, to be conquered and eliminated from the world. But there is more to it than that. Osama bin Laden's particular, stated reasons for hating America are its support of Israel and the stationing of U.S. troops on the sacred soil of Saudi Arabia, but his power over his adherents is based upon more far-reaching and sometimes historical foundations.

To the Arab world, the United States had propped up objectionable local despots for selfish interests, primarily the protection of its access to Middle East oil. Besides this, the United States had supported other repressive regimes in the area.

There was also the subject of sanctions. The United States was envied for its wealth by millions of Islamic fundamentalists and hated for its withholding of this wealth to punish others. Although few extremists in the Middle East admired Saddam Hussein, it was difficult to understand why, 10 years after the United States had driven the Iraqis out of Kuwait, sanctions that were starving the Iraqi people and accounted for the deaths of 5,000 Iraqi children every month from malnutrition and disease were justified. Sanctions against Syria, Libya, Iran and Sudan—all Muslim countries (and all, it must be stated, considered sponsors of terrorism)—were unacceptable to Islamic fundamentalists. In addition to this, there was a feeling that the Western powers had tolerated for too long the ethnic cleansing by Christian Serbs of Bosnian Muslims and the later killings by Serbs of ethnic Albanian Muslims in Kosovo.

This led to historic grievances, including the excesses of the Christian Crusades and the loss of historic dignity of the Arab world, which it has never regained. Layer upon this abject poverty and hopelessness, two roots of crime, which, compounded and left to fester, had grown into justifications for terrorism fed by fanaticism. Hatred, over a long history, had grown into rage, and rage against anything Western, which was regarded by the most fundamentalist groups within Islam as a morally destructive culture. All of this became a rich growing ground for individuals like Osama bin Laden with their own agendas, and thus for the multiplicity of terrorist groups that grew into a worldwide network.

And this, in turn, led to the final question: What's to come? Preparations were made for a unique confrontation that went beyond any neat or even acceptable definition of war. Events unfolded by the day publicly and, it was understandable to believe, by the hour in the private places of power and intelligence.

The link between Osama bin Laden and the events of September 11 grew more certain every day. Mohamed Atta, who emerged as one of the leaders of the September 11 attack, attended al-Qaeda training camps. At the beginning of October 2001, U.S. and British intelligence officials identified Mohammed Atef, a former Egyptian policeman and one of bin Laden's closest aids of not only commanding the 1998 embassy bombings in Kenya and Tanzania but of masterminding September 11.

A worldwide investigation and freezing of assets proved fruitful and informative. Thousands of dollars had been transferred from Mustafa Mohammed Ahmed, al-Qaeda's chief financial lieutenant, to Mohammed Atta days before the New York and Washington attacks. And all evidence seemed to indicate that, contrary to the first World Trade Center bombing, which was financed at only $20,000, this one received major financing. Ayub ali Khan and Mohammed Jaweed Azmuth, two men who were detained on a train in Texas the day after the hijackings, were found to have $5,000 in cash, box-cutter knives and hair dye in their luggage. The hijackers were seen to peel $100 bills off huge bankrolls to pay for goods in Florida. Although they lived frugally, they spent lavishly on their final plane tickets. Of the tickets purchased by the five hijackers who boarded United Airlines flight 175, the least expensive cost $1,600. Two of the terrorists on that plane spent $4,500 each for business-class seats.

Bin Laden, even before he formed al-Qaeda, was noted as a financial wizard and boasted of his worldwide financial network. Thus, one of the more important aspects of the efforts of the coalition against terrorism has been to find, explore and cut off money supplies to terrorist groups. Although al-Qaeda was the first target, the leaders of the coalition also realized that its high-profile work had emboldened such organizations as Hezbollah in Lebanon and Hamas in the West Bank and Gaza, as well as fundamentalist movements in Saudi Arabia, Pakistan, Egypt, Indonesia and Algeria.

It was, according to President Bush, going to be a long war without a particularly certain conclusion. The country rallied behind the president, though some thought it odd that the only sacrifice people were asked to make was to travel more and spend more.

On October 7, 2001, U.S. missiles were launched against terrorist camps and Taliban installations and infrastructures in Afghanistan. The battle was a joint effort; the war, largely fought with American technical assistance, air power and bombs, coupled with ground action by fighters of the Afghan Northern Alliance, reached a climax, if not a definite conclusion after only two and a half months. The Taliban, except for small pockets of resistance, was effectively defeated, and al-Qaeda was seriously routed.

An interim government was established with Hamid Karzai at its head, and a United Nations peacekeeping force, with British troops playing a major role, was dispatched to the region. Shipments of food supplies, which had ceased to arrive because of the fighting, resumed, but a return to lawlessness and tribal conflict emerged with the departure of the Taliban. The peacekeepers' task seemed to be cut out for them.

The first phase of the war against terrorism was at an end, but the second, more difficult one now began. Although the al-Qaeda training camps and headquarters had been demolished, not all of their soldiers were killed or captured. A large amount of the lower-level fighters had melted into the populace. Hundreds of other non-Afghans who had responded at first to Osama bin Laden's plea to join in the fight against the infidels, were casualties or prisoners of war or refugees who had fled across the porous Pakistan border.

As for Taliban spiritual leader Mullah Mohammad Ossam and al-Qaeda mastermind Osama bin Laden, both, as of this writing, are still at large, rumored to be first here, then there, killed, then alive, in Afghanistan, then escaped to Pakistan. Bounties on their heads and the determination to mark the conclusion of the first phase of the war on terrorism by their capture or deaths seemed to forebode an end to their flight.

Meanwhile, on the home front, some of the goals of the terrorists were playing out. As Dr. Craig Smith, an infectious disease expert at Phoebe Putney Memorial Hospital in Albany, Georgia, put it, "Warfare seeks to conquer territories and capture cities; terrorism seeks to hurt a few people and to scare a lot of people in order to make a point."

Although 3,000 is hardly "a few people," terrorism's other goal did seem to be translating to action, in varying degrees. Airlines reported a huge drop in ridership. The airline lobby made this clear to Congress, and within days Congress appropriated a $15-billion bailout for the airline industry. In addition, when passengers returned to airports, after a period of a little over a week, they found considerable delays as security measures were heightened dramatically. Passengers accepted this readily, however, preferring the inconvenience of heightened security to the insecurity that convenience in boarding might give them.

In some cases, nervousness and overconcentration on security caused innocent men, merely because they wore turbans or were Middle Eastern–looking, to be denied access to airplanes. In extreme situations, anger at Middle Easterners erupted in acts of murder and the desecration of mosques in American cities. Racial profiling, a negative practice before September 11, gained a certain emergency justification in the minds of some citizens and security enforcers.

Attorney General John Ashcroft announced a series of abridgements of civil liberties necessary in wartime, including allowing law enforcement agencies to listen in on conversations between lawyers and their clients if the clients were accused of terrorism, searches and seizures without warrants and the extended detention of suspected illegal aliens or terrorists without charges. Acting upon these emergency measures, the Justice Department seized and held, often for months in undisclosed locations, nearly 1,000 legal and illegal aliens from Middle Eastern countries, while hiding their whereabouts from their families and lawyers. In addition, the attorney general requested local law-enforcement agencies to interrogate 5,000 citizens and noncitizens from the Middle East.

Shortly after this, President Bush announced the establishment of military tribunals to try, sentence and possibly execute captured terrorists who were not U.S. citizens. The powers of the tribunals were to be sweeping, allowing them to set their own rules, allow hearsay evidence, deprive the prisoner of picking his or her legal representation, deny the right of appeal and prescribe a verdict of capital punishment to be imposed by only two-thirds of a jury. Attorney General Ashcroft added the weight of the Justice Department to the establishment of these tribunals, stating unequivocally, at a press conference that "Foreign terrorists who commit war crimes against the United States, in my judgment, are not entitled to and do not deserve the protection of the American Constitution."

Civil rights activists were appalled at most of the abridgments to individual liberties, issued without consultating Congress, but the public at large agreed, by an 85 percent majority, that they were necessary in a time of crisis. Only the military tribunal issue seemed to bother the public, and as time passed, modifications were made in the mandates and structure of the tribunals, bringing them into line with the regulations governing courts-martial. Furthermore, Zacarias Moussaoui, the so-called 20th hijacker, and the first member of the terrorist conspiracy to be jailed, was scheduled to be tried in a federal court rather than a tribunal.

And then there was the anthrax scare. Late in September, Secretary of Health and Human Services Tommy Thompson told television viewers in a reassuring statement that the government was prepared to deal with any kind of bioterrorism attack. However, on October 5, Robert Stevens, a photo editor at American Media, a publisher of supermarket tabloids in Boca Raton, Florida, died from inhaled anthrax poisoning. It was the first case of a death from inhaled anthrax in the United States since 1978, and until this time, anthrax had been found only in two places in the United States: in military laboratories, and in wild animals and livestock, who left spores from the disease in soil.

In a statement designed to calm the public panic that followed the news of the Florida death, Secretary Thompson suggested at a press conference that Stevens might have contracted anthrax from drinking water out of a stream. This was disproven almost immediately, when a sweep of the office in which the victim worked revealed anthrax spores on his computer keyboard, in the company mailroom and ultimately throughout the building. This was no contamination from nature. It was obviously a deliberate act of terrorism.

A series of anthrax-contaminated letters began to appear within days. One was discovered in newsanchor

Tom Brokaw's office at NBC in New York, one at the *New York Post,* one at the offices of Senator Tom Daschle and one at the office of Senator Patrick Leahy. All had similar messages and handwriting, and all had traveled through a major mail-sorting facility in suburban Hamilton Township, New Jersey. Following this, an assistant to Dan Rather at CBS turned up with cutaneous anthrax, contracted from touching contaminated mail.

Further investigation discovered anthrax spores in post offices elsewhere in New Jersey, New York City, the Washington, D.C., area, and the Middle West. High-speed mail-sorting equipment had puffed spores into the air, sickening eight postal workers. It was assumed that tens of thousands of letters might be contaminated.

On October 22, two postal workers died from inhaling anthrax spores. Both worked in the Brentwood Road Northeast postal center, which handled all mail for Washington, D.C., including Congress. The Capitol was swept for evidence of anthrax spores. The House of Representatives recessed early and was closed for one day. The Senate remained in session, but the Hart Senate office building, which housed Senator Daschle's office, was considered unsafe to inhabit and was closed. Several attempts to decontaminate it only partially succeeded, and as of this writing, the building remained shuttered.

Two more deaths from anthrax followed: Kathy T. Nguyen, a 61-year-old New York City hospital worker died on October 31, 2001, and a few days later, so did Ottilie Lundgren, a 94-year-old widow from rural Oxford, Connecticut. Both apparently died from inhaling spores that had worked their way through the postal system and ended up on mail delivered to them.

Now, thousands of terrified people were given Cipro, the only antibiotic approved by the Federal Drug and Alchohol (FDA) Commission for treatment of anthrax, a disease so rare that many doctors could not recognize it. Education came quickly to them and to the public at large. Drug companies were urged by the government to increase the supply of Cipro available to the general public. Furthermore, vaccination against smallpox was urged by some officials, in case this became the next bioterrorist attack.

No further anthrax-laced letters were discovered as 2001 ended and 2002 began, and the source of the attack remained a mystery. Weaknesses in the nation's public health system, which found itself woefully ill equipped to handle any bioterrorism attack, were discovered, but little else.

Analysis of the antrhax in the letters sent to Senators Daschle and Leahy and to the media revealed that they were all of the Ames strain, suggesting that they were the work of one individual or group of individuals. Furthermore, the small concentration of the attack convinced most investigators that it was not the work of al-Qaeda, though Osama bin Laden had been in search of biological and nuclear weapons. Nor did it seem to come from Iraq. The strain differed from that used by Sadam Hussein in his development of biological weapons.

This left two possible sources for the anthrax attack: a well-equipped amateur laboratory or a government laboratory. The presence of additives to keep the anthrax spores from clumping into coarser grains suggested that the terrorist or terrorists either worked in or knew someone who had worked in one of the government laboratories, such as the Dugway Proving Ground, an army facility in Utah, which as late as 1998 was producing small amounts of weapons-grade—that is, non-clumping—anthrax.

And so, it seemed that this was the work of a domestic terrorist, possibly a far right-wing fanatic, since the key targets were the media and two liberal senators. But no one was certain, and as of this writing, the FBI was still without a tangible clue regarding the origin of this compounded terrorism that caused an increase in national nervousness. It was a terrorist's dream come true.

In this context, the war against terrorism being waged by the United States and its allies, moves forward, as all wars must. On September 11, 2001, more than 3,000 people were killed in the two suicide terror attacks in New York City and Washington D.C. It was almost the largest death toll in any single incident in any war in which the United States has ever participated, nearly eclipsing the 4,700 soldiers who died in the Battle of Antietam during the Civil War and exceeding the 2,390 service members who died at Pearl Harbor in the Japanese attack on December 7, 1941. In both these cases, however, only military personnel were involved. The casualties on September 11 were civilians, pursuing peaceful lives, and fire and police officers engaged in rescue missions.

The words of the Japanese general who expressed uneasiness after his military had attacked Pearl Harbor in 1941 arose in many memories. "We may have awakened a sleeping tiger," he had said. So, too, may the followers of Osama bin Laden's worldwide terrorist network in 2001 have awakened the same slumbering fighter.

UNITED STATES
OKLAHOMA
OKLAHOMA CITY
April 19, 1995

One hundred sixty-eight people were killed and hundreds were injured in the bombing of the Alfred P. Murrah Federal Building in Oklahoma City on April 19, 1995.

For two years Timothy McVeigh had fumed over the Waco, Texas, standoff and its fiery conclusion. He was a loner all his life who grew up in upstate New York, dropped out of Niagara County Community College in 1986 and enlisted in the army in 1988. He saw action in the Persian Gulf War and received a Bronze Star for valor there.

Discharged in 1993, he drifted, indulged his obsession with guns, lived in cheap motels, drove old cars and occasionally stayed with two army buddies, Michael Fortier in Kingman, Arizona, and Terry Nichols in Decker, Michigan.

Nichols drifted, too, but with a Filipina wife. He worked a farm in Decker and was sued by a local bank for accumulating and not paying a $35,000 debt on his credit card. He mounted his own defense, lost and countersued the bank, trying to pay off his debt with a bogus check from a right-wing organization. He surrendered his voting rights shortly thereafter, just about the time Timothy McVeigh moved in with him on the farm.

By this time McVeigh was wearing camouflage pants and army underwear, carrying a .45 pistol, going to gun shows and trading guns. He had just come from Waco (see p. 157), and his bitterness over the events that had transpired there was palpable.

Before long, McVeigh and Nichols were setting off explosives on the farm and attending meetings of the Michigan Militia Corps, a right-wing extremist group that trained in the Michigan woods. McVeigh and Nichols soon left the militia because it seemed to them too peace loving. They formed their own cell of a local paramilitary organization called the "Patriots."

In 1994 Nichols moved to Las Vegas, then to Kansas, where he worked as a farm hand. In August McVeigh appeared again, Nichols quit his job to sell surplus Army items at gun shows, and, most important, on the day he left his job, using the name "Mike Havens," bought 40 50-pound bags of ammonium nitrate from a local farm co-op. The day after, 299 sticks of dynamite and 544 blasting caps were stolen from a quarry nearby. During the next two months Nichols rented two storage sheds and bought 40 more bags of ammonium nitrate under the same alias.

He and Timothy McVeigh were building a bomb large enough and strong enough to blast apart a government building in revenge for Waco. Nichols had the expertise; McVeigh had the purpose. It was he who rented the Ryder truck that would carry the bomb.

The previous December, McVeigh had approached his other army buddy, Michael Fortier, who was living in Kingman, Arizona. Both took a scouting trip to the Alfred P. Murrah Federal Building in Oklahoma City. Located in the middle of a midwestern city, it possessed a wide plaza into which a bomb-laden truck could be driven, parked and ignited. McVeigh reportedly urged Fortier to join in the plot, but, according to later testimony, Fortier declined.

He did, however, participate with McVeigh and Nichols in the robbery of an Arkansas gun dealer in December of 1994. His job was to receive the guns from McVeigh and Nichols and sell them. He later confessed to selling 25 of the guns in Arizona and giving the money from the sales to Nichols.

On the morning of April 19, the anniversary of the fiery end of the Waco standoff, Timothy McVeigh slipped into the driver's seat of a rented Ryder truck packed with a 4,800-pound bomb made from a mixture of fertilizer and fuel oil, drove it to Oklahoma City and parked it in front of the main doors of the Alfred P. Murrah Federal Building. He lit the fuse, leaped out of the rental truck, jumped into Terry Nichols's pickup truck, and the two sped from the scene.

At 9:02 A.M. the bomb exploded, pulverizing the van, ripping the north side from the nine-story building, tearing down, as if with a huge hand, half of the floors full of offices and people and, perhaps most horribly, burying a day care center on the second floor. With no support, the upper floors accordioned into the lower ones. Fire broke out, and a sunny morning suddenly shattered into unspeakable chaos.

At the moment of explosion, a wave of superhot gas moving at 7,000 mph hit the building; half a second later the gas dissipated and was replaced by an equally violent vacuum, which sucked back debris that had been blown out. The shock waves tore up the building's inner structure, pulverized windows and weakened buildings blocks away.

"At about 8:58 A.M.," said Clark C. Peterson, who had an office on the fourth floor of the building, "I sat three feet from the north windows as my supervisor gave me final instructions for a project. I returned to my desk, which was about 20 feet south of the windows, and began to type.

"At about 9:02, an electric spark appeared by my computer and everything turned black. Propelled objects raced throughout the darkness amid the sound of moaning metal.

"I caught a glimpse of a terrified girl with her arms straight up in the air. We were apparently falling, but I didn't realize what had happened until a minute or two later. The sight of her was so brief and faint that I couldn't identify her. She yelled, 'Ah!' as if there weren't enough time to inhale air.

". . . As the blackness and dust began to clear, I discovered that the armchair I was seated in had been replaced. I remained in a sitting position, but on a flat, ceiling-like material, which was on top of a three-story pile of rubble.

"I saw the north half of the Murrah Building was gone, except for the east and west sides. I was 10 feet in front of the remaining structure, about level with the third floor. 'This had to be a bomb!' I thought, and a 20-foot crater below confirmed that it was. I couldn't see anybody else in the building or the rubble."

Sergeant Jerry Flowers, one of the first policemen on the scene, described the interior of the wreckage. "It was very dark as we crawled over large pillars of concrete just to get to an area where we could hear people yelling for help," he recalled. "We made our way into a room where large slabs of concrete were lying from one floor down to the next. There was a large hole you could look up through and see the nine broken floors hanging above our heads. Large ropes of steel rebar hung down and others protruded from the floor where we were trying to walk. . . . Firefighters swarmed into the room. A generator was started and portable lights lit up the area. . . . Looking down into what appeared to be a well, we saw a lady trapped on her back by concrete. The floor around her was filling up with water. She kept yelling, 'Don't let me drown!' Rescuers attacked the well to free the woman. And they were successful."

Bloodied survivors stumbled from the building; trapped ones pleaded for rescue. Broken glass and shattered walls covered the ground as if there had been an aerial bombing. Ambulances and firetrucks raced to the scene, and within half an hour St. Anthony's hospital, located eight blocks away, had set up a trauma center. One hundred and fifty-three victims went through this center in the first day. Other hospitals in Oklahoma City received 120 torn and bleeding survivors the same day.

Doctors and nurses fully expected a second wave of survivors on the second day, but there were none. The final survivor, 15-year-old Brandi Liggons, was freed on the night of the day of the explosion. Unwilling to believe this, rescuers, including members of the FEMA rescue team, continued to search for survivors using dogs, fiber-optic cameras and acoustic listening devices. At the same time, rubble was meticulously removed so as not to crush any victims underneath it.

For 16 days the search went on. Even on that day, it was believed that dead victims still lay buried under the base of one of the remaining columns of the building. But work shifted the column ominously, and finally, on May 4 at 11:50 P.M., Oklahoma City special operations chief Mike Shannon gave the ultimate order: "I'm calling the game."

One hundred and sixty-eight people, including 16 children aged four and under, died in the blast. The nation joined in the overwhelming grief of the survivors and the families of the dead. President and Mrs. Clinton joined the national service of prayer at the State Fair arena on the Sunday morning following the disaster.

Even as rescue efforts were barely beginning, the search for the bombers started. Bomb squad sergeant Mike McPherson of the Oklahoma City police force arrived on the scene moments after the explosion. He recognized the crater as the work of a truck bomb. "You almost had to have been brought up here in a truck," he said later. "It had to be a pretty good size truck to do that much damage."

He searched through the rubble for an identifying piece of the vehicle. He found more: A block away, he discovered its rear axle. "The rear axle was sitting down here. It had hit a car, and I had seen it earlier," he recalled. The vehicle serial number was visible under the grime and dust, and within two hours McPherson had traced the number to a Ryder rental truck. After further hunting, he and the FBI discovered that it had been rented in Kansas by Timothy McVeigh.

Only 90 minutes after the explosion, an Oklahoma Highway Patrol officer pulled McVeigh over for driving without a license plate. Nichols, in his rage against authority and all things governmental, had pulled the license plate off.

McVeigh was held in a county jail overnight. Shortly before he was to be released on April 21, word came to the local sheriff that his prisoner was wanted as a suspect in the Oklahoma City bombing. It was not long before Terry Nichols turned himself in to police in Herrington, Kansas. Michael Fortier, who first denied knowledge of the bombers' intent and activity, soon turned state's evidence and became the prosecution's star witness in the trials of McVeigh and Nichols.

In June 1997 McVeigh was found guilty of 11 counts of murder and sentenced to death. On December 23, 1997, Nichols was convicted of involuntary manslaughter and of conspiring with McVeigh. His sentence, delivered six months later, was life in prison.

A week before McVeigh was scheduled to be executed on May 16, 2001, thousands of FBI documents that had been withheld from defense lawyers during his trial were discovered. McVeigh, who had previously refused to appeal his death sentence, was convinced by his lawyers to appeal it. Terry Nichols also appealed his sentence.

A stay of McVeigh's execution until June 11 was granted by Attorney General John Ashcroft. No apparent evidence that would clear him was found, and McVeigh dropped his last appeal and prepared for his execution, the first federal execution in 38 years and possibly the one that attracted the most international attention ever. Terry Nichols continued his appeal until the summer of 2001, when C. Wesley Lane II, the Oklahoma County district attorney who had recently replaced Robert H. Macy, announced his determination to try Nichols on 160 state charges of murder and to seek the death penalty.

Nichols withdrew his appeal and was willing to agree never to renew it, but the district attorney remained adamant, even though the families of some of those who died in the Oklahoma City tragedy made public statements disavowing revenge and despite the negative history of an attempt to supersede the federal sentencing of Mr. Nichols. It had cost Mr. Macy his job, and there was little hope that an unbiased jury could be found in Oklahoma to sit in such a trial, but Mr. Lane vowed to find one.

Victoria Cummock, who had lost her husband on Pan Am Flight 103 over Lockerbie, Scotland, in 1988 (see p. 128), was flown to Oklahoma City to serve as a counselor to survivors after she wrote a letter to President Clinton. In the letter she encapsulated the feelings of many survivors of every senseless tragedy caused by terrorism: "The anger you feel is valid," she wrote, "but you must not allow yourselves to be consumed by it. The hurt you feel must not be allowed to turn into hate, but instead into the search for justice. The loss you feel must not paralyze your own lives. Instead, you must try to pay tribute to your loved ones by continuing to do all the things they left undone, thus ensuring they did not die in vain."

UNITED STATES
TEXAS
WACO
February 28–April 19, 1993

Eighty-five members of the Branch Davidian cult died in the fiery climax to a 51-day standoff in Waco, Texas, from February 28 to April 19, 1993. The standoff was between FBI and Bureau of Alcohol, Tobacco and Firearms agents and the Davidians, who were accused of killing four ATF agents in a raid on the Davidians on February 28.

The Branch Davidians are a religious cult that had its origins in 1934, when Victor Houtoff, a disgruntled Seventh-Day Adventist, argued over interpretation of certain

Biblical teachings, left the main body of the religion and formed his own sect.

Over the years various leaders rose to the top of the sect as it continued to divide. One last division occurred in 1978, as a follower named Vernon Howell led the Branch Davidians away from a group of Davidians led by Lois Roden, who had died and wished her son to lead.

Vernon Howell later changed his name to David Koresh and convinced his followers that he was Jesus Christ. A rivalry between Koresh and George Roden led to an accumulation of arms, violence and attempted murder charges against Koresh and seven others. All were arrested and tried. The seven were acquitted. Koresh's trial ended in a mistrial.

Within the Branch Davidians, Koresh formed a harem. The children produced from the harem would, he told his followers, rule the earth after he and his male followers slayed the unbelievers.

On Sunday morning, February 28, 1993, more than 100 agents of the federal Bureau of Alcohol, Tobacco and Firearms left their hiding places under tarpaulins on two cattle trucks and began to enter the fortified 77-acre compound of the Branch Davidians, 10 miles outside of Waco, Texas. In an unbelievably bungled raid, they advanced on the compound. Their objective: to serve David Koresh with a summons for illegal firearms possession. They were aware that the Davidians in the compound were heavily armed, had a huge cache of munitions and ammunition and that surprise was vital to the success of the raid. But for some reason, their plans and their schedule were well known by local law enforcement officers and a local television station, which had sent a camera crew to the site. Even before the agents could get near the main part of the compound, three helicopters with senior agents aboard were fired upon from the ground.

Even before that, David Koresh had been warned in a telephone call that a raid was about to take place. He told his followers to prepare for it, and an undercover agent fled the compound, dashed up a road and ran into a house that the cult members had already surmised contained federal agents.

Still, the raid went forward. Some of those taking part had requested and been denied heavier firepower; others later claimed that they had not been warned of the firepower within the compound.

The agents divided into two groups. One group headed for the front door, the other for the part of the compound that housed its weapons room. A fusillade of fire erupted. Agents fell at the front door of the main building. A half dozen other agents using two steel ladders climbed onto the roof alongside a window near the weapons room. One agent broke the window with a steel bar and three crawled through, heading down a hallway to the weapons. In an instant they were met with a spray of bullets. The agent left on the roof was hit by shots fired through the wall, but he managed to make his way to a ladder and slide down to safety.

In 45 minutes it was over. Four agents were dead or dying, 16 had been wounded and an undetermined number of cult members had been killed or wounded. A cease-fire was negotiated, and the agents dragged off their dead and wounded.

Now a long standoff began. More than 400 FBI and ATF members drew up battle positions around the compound while within the cult members dug in for a battle. Before the initial standoff they had stockpiled huge supplies of food; the compound contained its own water tower, and members had contacted their relatives, asking them to buy guns with their credit cards and deliver the firearms to them. Koresh had told his followers that they were to prepare for his apocalyptic vision of a bloody conflict that would amount to his crucifixion and would allow his devotees to go with him to heaven.

As the days drifted by, the FBI released reports that they were willing to wait "as long as it takes" to bring the standoff to a conclusion. Over the length of the siege, 21 children and some women would be released from the compound, raising both hopes and fears.

As the first week ended, tempers began to flare not only between the sides, but between the FBI and the ATF members, and between those who were working toward a negotiated settlement and those who were there to do battle.

On March 24 the federal agents began to play music—chants by Tibetan Monks, Christmas carols, "Sing-along with Mitch," Nancy Sinatra recordings and so on—at earsplitting levels aimed at the compound. At dawn a recording of reveille was played over and over and over. Word began to leak from government sources that the initial raid had proceeded despite the fact that the supervisors of it knew that the element of surprise had been lost.

On April 12 coils of barbed wire were strung by the FBI and ATF around the perimeter of the compound. At the same time debris, including vegetation and vehicles, was cleared away from a space around this perimeter. Arguments between the negotiators and the agents determined to fight escalated.

Finally, at 5:55 A.M. on April 18, federal agents notified the cultists by telephone that they intended to begin injecting tear gas into the buildings of the compound. The telephone was slammed down, ripped from its wiring and flung through the front door. Fifteen minutes later an armored vehicle rammed the building near the door, tearing a large hole in the wall. Tear gas was fired through a long boom protruding from the front of the tanklike vehicle. Simultaneously, 75 to 80 shots were fired from within the building. Loudspeakers blared demands to the cult members to surrender.

For the rest of the morning, armored vehicles punctured holes in the side of the building, doing the most damage on the southwest corner and the south side. Each time they would inject tear gas into it, and each time they were met with a fusillade of fire from within.

At 12:05 a wisp of smoke billowed from several windows in the building, and orange flames were seen inside. Four minutes later a huge wall of flame and great gouts of smoke poured from the entire compound. By 12:15 it was entirely engulfed in flame. Much of the main building, including the lookout tower, collapsed.

"My God," a Fort Worth reporter, kept at a distant checkpoint with other reporters, said. "It's horrific. Horrific. A holocaust. Could anyone live through that?" The Waco fire department did not arrive until a full half hour after the fire started, although they were informed of it at 12:09. The FBI kept them back, as they later explained, "for their own safety."

Inside the progressively damaged compound, chaos reigned. Nine people escaped. One woman, her clothes in flames, ran out of the compound, then turned and ran back into the fire. Eighty-five were killed, including 17 children.

The firefighters arrived to a grisly scene of utter devastation. Charred remains, skeletons, burned out pieces of wall and furniture were the only indications that human beings had inhabited the building until noon of that day. "There wasn't nothing left but bone," one of the firefighters described it.

"It's the worst I've ever seen," another remarked. "It was sad. I couldn't believe this."

The 51-day siege was at an end, but not the recriminations. Those who favored negotiation said bitterly that law enforcement officials treated the entire standoff as a hostage situation, one that could only have been resolved ultimately by force. U.S. Attorney General Janet Reno, who gave the final order to insert tear gas into the compound, took full responsibility for the outcome but acknowledged that the tear gas was a mistake.

Debate raged from Waco to Washington. Representatives of the far and militant right accused the FBI of setting the fire with the tear gas. The FBI countered that the tear gas was not flammable and that there was eye witness evidence that the fire had been set in several places within the compound.

The survivors were imprisoned, and their lawyers maintained that the fire began when a battering ram knocked a lantern across bales of hay. They and the survivors maintained matching stories that there was never talk of suicide.

Two investigations were made, one by the Treasury Department, one by the Justice Department. Shortly before the release of the Treasury Department investigation, Stephen E. Higgins, the senior Washington official who supervised the raid, tried to resign, but his resignation was rejected. Two other officials announced their retirements. When the report was released, Mr. Higgins was fired from his job. The investigation found that he had made misleading statements to the public and Congress about what had occurred, concealing fundamental errors in planning, execution and follow-up of the operation. The concealment chiefly dealt with withholding the news that the element of surprise was known to have been breached.

The Justice Department investigation cleared top level federal agents but blamed mid- and lower- level agents who recommended that negotiations be abandoned in favor of a tear gas assault. In fact, the report revealed that at one point the tactical agents turned off the electricity to the compound just as the negotiators were trying to figure out how to reward the cult for releasing hostages.

Ms. Reno was not found to be at fault. In fact, she had once vetoed an FBI proposal to carry out an assault but then changed her mind after being pressed by senior FBI officials.

The report did not silence the far right. Calls for Ms. Reno's resignation continued from her Republican enemies, and law suits were instituted throughout the 1990s to prove that the FBI actually murdered American citizens. None of them succeeded.

But the questions did their work. A young malcontented war veteran named Timothy McVeigh believed they were right and later stated that Waco was one of the major reasons for his bombing of the Alfred P. Murrah Federal Building in Oklahoma City (see p. 155). Waco would become a symbol for the militant right, one that seemed to feed its fervor and justify its existence.

UNITED STATES
WASHINGTON, D.C.
March 9–11, 1977

The assassination by Black Muslims of the family of Hanafi Muslim Hamaas Abdul Khaalis led to a three-day occupation of three Washington, D.C., sites by Khaalis and his followers from March 9 to 11, 1977. One person died; 19 were wounded; 134 were taken hostage.

In the violent 1970s, thousands of blacks took Islamic names, converted to Islam and joined one or another of several Islamic sects. The largest and best known of these was the Chicago-based Black Muslims, who numbered—depending on who was counting—10,000 to 70,000 and claimed among their members boxer Muhammad Ali.

The Black Muslims, however, were considered blasphemous by more devout Muslims, who in the United States numbered some two million. Hamaas Abdul Khaalis, born Ernest Timothy McGhee in Gary, Indiana, was a member of the strict Hanafi sect, a subgroup of the orthodox Sunni Muslims.

In the early 1970s, fierce factional fighting broke out between various Muslim groups, and more than 29 Muslims were murdered in these clashes.

The Washington-based Hanafi group claimed among its 1,000 members basketball star Kareem Abdul-Jabbar, who in 1970 bought the group a mansion. Soon after they took up residence, Khaalis began a verbal war against the Black Muslims. He blamed them and their leader, Elijah Muhammad, for the murder of Malcolm X. At the end of 1972, Khaalis wrote letters to 58 Black Muslim ministers, calling their leader "a lying deceiver."

A few days later, when Khaalis was absent from the mansion, a squad of Black Muslims invaded it and killed two adults and six children in gruesome ways. All of the children were related to Khaalis. His 10-day-old son was drowned in a sink before his mother's eyes. Three other children—two his, one a grandchild—were drowned in a bath, and two sons were shot. One of his wives and a daughter were seriously wounded.

Five Black Muslims were arrested and sentenced to life imprisonment, but it was not enough for Khaalis. At 11 A.M. on Wednesday, March 9, 1977, he and 11 followers began a 39-hour siege of three locations in Washington, D.C. Before they were through, 134 hostages would be taken, one man would be dead and 19 hostages would be seriously wounded, one seriously enough to be paralyzed.

The first building to be seized was the eight-story headquarters of B'Nai Brith, the Jewish service organization. Khaalis and six of his commandos, dressed in jeans and work shirts with long knives strapped to heavy steel chains hung on their hips and carrying guitar cases holding an assortment of rifles, shotguns and a crossbow, leaped from a van and invaded the building's lobby. "They killed my babies and shot my women," screamed Khaalis at terrified passengers from an elevator that had just arrived at the lobby. "Now they will listen to us—or heads will roll."

The invaders set up a command quarters and a concentration camp on the eighth floor of the building. Those who protested were slashed with machetes. The men and women were separated and then were bound and taunted.

At 12:30 P.M. another group of commandos took over the Islamic Center on the edge of Rock Creek Park. Khaalis had been chiding the center's director, Dr. Muhammad Abdul Rauf, for months for supporting Elijah Muhammad, the leader of the Black Muslims, and for being an Egyptian. "Your country is seeking peace with the Jews," was Khaalis's charge.

Meanwhile, Israeli prime minister Yitzhak Rabin was accepting an honorary degree from American University. He was hustled directly from the ceremonies to Andrews Air Force Base, where he caught a plane for New York.

At 2:15 P.M. the third attack, carried out by two gunmen dressed in black and carrying a shotgun and a .22-caliber handgun, took over an office on the fifth floor of Washington's city hall. It was here that the murder took place.

Two elevators, one carrying city councilman (later Washington mayor) Marion Barry and councilman Robert Pierce and the other Maurice Williams, a reporter for radio station WHUR, and Steven Colter of the *Washington Afro-American,* arrived simultaneously at the fifth floor. The gunmen opened fire. Williams received the full force of a shotgun blast and was killed on the spot. Pierce was paralyzed, the bullet that hit Barry stopped less than an inch from his heart and the bullet that grazed Colter's skull miraculously missed his brain.

Police snipers and FBI agents had by now surrounded all three locations, and Khaalis phoned his demands to Max Robinson, a reporter for WTOP-TV. The demands were many and hysterical. First, he wanted a current motion picture, *Mohammad, Messenger of God,* starring Anthony Quinn to be withdrawn from all movie houses and shipped out of the country. "It's a joke. It's misrepresenting the Muslim faith," said Khaalis.

"Next thing I want the killers of my babies," continued Khaalis. "I say we want them right there. I want to see how tough they are. I want the one who killed Malcolm [X] too." The leader also wanted the police to reimburse him for the $750 fine he had received for contempt of court during the 1973 trial of the five Black Muslims who had massacred his family. Finally, he wanted Secretary of State Cyrus Vance to be contacted; he gave as his reason: "We are going to kill foreign Muslims at the Islamic Center [and] create an international incident."

"They were a bunch of crazies," said Andrew Hoffman, one of a number of hostages released early in the siege. The city hall commandos seemed to be the most brutal, tying their victims hand and foot, making them lie face down on the floor and poking them continually with shotguns. In the B'Nai Brith building, the terrorists painted over windows to block out visibility for police snipers.

By 6:16 P.M. on Wednesday, the 9th, negotiations had begun with Egyptian ambassador Ashraf Ghorbal and Pakistani ambassador Sahabzada Yaqub-Khan conferring with Khaalis. The White House remained removed. That night Iran's Ardeshir Zahedi, who had flown in from Paris on the Concorde, joined the negotiations. The offending film was pulled from movie houses in New York and Washington.

The first breakthrough occurred at 5:30 P.M. on Thursday, the 10th, when Khaalis agreed to meet face to face with Yaqub-Khan in the lobby of the B'Nai Brith building. The talks lasted three hours. At the end of that time Yaqub-Khan asked for the release of 30 hostages as a gesture of good faith. Khaalis proposed to let all hostages go, provided he was released without bail.

It would be nearly 2 A.M. on the 11th before the legal tangle of doing this was unraveled, and from then until 5:10 A.M., hostages were released and commandos gave themselves up to police. Later that day bail was set at

The three leaders of the Hanafi Muslim siege of Washington, D.C., from March 9 to 11, 1977 are transported to jail in a Red Cross ambulance. (American Red Cross)

$50,000 each for two terrorists and $75,000 for six others. The three gunmen at the Islamic Center, where no hostages were harmed, were let go on their own recognizance along with Khaalis.

Khaalis would recoup his $750 and succeed in having a movie shut down for a few days. But that would be his only accomplishment, aside from the satisfaction of terrorizing most of America for 39 hours.

EXPLOSIONS

THE WORST RECORDED EXPLOSIONS

* Detailed in text

Afghanistan
- Salang Tunnel
  - * Truck collision (1892)

Austria
- Skoda
  - Arsenal explosion (1916)

Belgium
- Hamont Station
  - Ammunition train (1918)

Canada
- Alberta
  - Lethbridge
    - Mine explosion (1914)
- British Columbia
  - Ferme
    - Mine explosion (1902)
  - Vancouver Island
    - Mine explosion (1887)
- Nova Scotia
  - Halifax
    - * Ammunition ship (1917)

China
- Beijing
  - Arsenal explosion (1925)
- Changxing
  - Mine explosion (1998)
- Fanlin
  - Fireworks factory explosion (2001)
- Lanzhou
  - * Arsenal explosion (1935)
- Manchuria
  - * Mine explosion (1931)
- Tsingtsing
  - Mine explosion (1940)

Colombia
- Cali
  - * Dynamite truck convoy (1956)

Cuba
- Havana
  - * Munitions ship *La Coubre* (1960)

El Salvador
- La Libertad
  - Explosives warehouse (1934)

France
- Courrières
  - * Mine explosion (1906)
- Mons
  - Mine explosion (1875)
  - Mine explosion (1892)
- St.-Etienne
  - Mine explosion (1889)

Germany
- Alsdorf
  - Mine explosion (1930)
- Camphausen
  - Mine explosion (1885)
- Johanngeorgendstadt
  - * Mine explosion (1949)
- Oppau (See NUCLEAR AND INDUSTRIAL ACCIDENTS)
- Volklingen
  - * Mine explosions (1962)
- Westphalia
  - Mine explosions (1908)

Great Britain
- England
  - Barnsley
    - * Mine explosion (1866)
  - Durham
    - Mine explosion (1909)
  - Haydock
    - Mine explosion (1878)
  - Hulton
    - * Mine explosion (1910)
  - Lancashire
    - Mine explosion (1885)
  - Northumberland
    - Mine explosion (1862)
  - Podmore Hall
    - Mine explosion (1918)
  - Sunderland
    - * Mine explosion (1880)
  - Yorkshire
    - Mine explosion (1857)
- Scotland
  - Lanarkshire
    - Mine explosion (1877)
- Wales
  - Abercane
    - * Mine explosion (1878)
  - Clyfnydd
    - Mine explosion (1894)
  - Glamorgan
    - Mine explosion (1856)
  - Monmouthshire
    - Mine explosion (1890)
  - Pontypridd
    - Mine explosion (1867)
  - Risca
    - Mine explosion (1860)
    - Mine explosion (1880)
  - Sengenhydd
    - * Mine explosion (1913)
  - Wrexham
    - Mine explosion (1934)

India
- Asansol
  - * Mine explosion (1958)
- Bombay
  - * Steamship explosion (1944)
- Chasnala
  - Mine explosion (1975)
- Dharbad
  - * Mine explosion (1965)

Japan
- Fukuoka
  - * Mine explosion (1965)
- Nagasaki
  - Mine explosion (1906)
- Otaru
  - Harbor explosion (1924)
- Sapporo
  - Mine explosion (1920)
- Shimonoseki
  - Mine explosion (1915)

Mexico
- Barroteán
  - Mine explosion (1969)
- Guadalajara
  - * Sewer explosion (1992)
- Mexico City
  - Gas storage area (1984)

Prussia
- Rhenish
  - Mine explosion (1907)
- Upper Silesia
  - Mine explosion (1895)

Rhodesia
- Wankie
  - * Mine explosion (1972)

Russia
Jusovka
Mine explosion (1908)
Senegal
Dakar
* Ammonia truck explosion (1992)
Spain
Cadiz
Naval mine and torpedo factory (1947)
Switzerland
Bern
Mine explosion (1921)
Turkey
Kharput
Munitions plant (1925)
Kozlu
* Mine explosion (1992)
Ukraine
Luhansk
* Mine explosion (2000)
United States
California
Port Chicago
* Harbor explosion (1944)
Illinois
Cherry
* Mine explosion (1909)
New Mexico
Dawson
* Mine explosion (1913)
Pennsylvania
Cheswick
Mine explosion (1904)
Jacob's Creek
* Mine explosion (1907)
Mather
* Mine explosion (1928)
Plymouth
* Mine explosion (1869)
Tennessee
Coal Creek
Mine explosion (1869)
Memphis
* Steamship *Sultana* (1865)
Texas
New London
* Gas explosion (1937)
Texas City
* Liner *Grandcamp* (1947)
Utah
Castle Gate
* Mine explosion (1924)
Scofield
* Mine explosion (1900)
Virginia
Pocahontas
* Mine explosion (1884)
West Virginia
Eccles
* Mine explosion (1914)
Monongah
* Mine explosion (1907)
Wyoming
Hanna
* Mine explosion (1903)
USSR
Ufa
* Gas pipeline, passenger trains (1989)
Yugoslavia
Kakanj
* Mine explosion (1965)

CHRONOLOGY

* Detailed in text

1856
July 15
Glamorgan, Wales; mine explosion
1857
February 19
Yorkshire, England; mine explosion
1860
December 1
Risca, Wales; mine explosion
1862
January 16
Northumberland, England; mine explosion
1865
April 27
* Memphis, Tennessee; steamship *Sultana*
1866
December 12–13
* Barnsley, England; mine explosion
1867
November 8
Pontypridd, Wales; mine explosion
1869
September 6
* Plymouth, Pennsylvania; mine explosion
1875
December 14
Mons, France; mine explosion
1877
October 22
Lanarkshire, Scotland; mine explosion
1878
June 7
Haydock, England; mine explosion
September 11
* Abercane, Wales; mine explosion
1880
July 15
Risca, Wales; mine explosion
August 17
* Sunderland, England; mine explosion
1884
March 13
* Pocahontas, Virginia; mine explosion
1885
March 17
Camphausen, Germany; mine explosion
June 18
Lancashire, England; mine explosion
1887
May 4
* Vancouver Island, British Columbia; mine explosion
1889
July 3
St.-Etienne, France; mine explosion
1890
February 6
Monmouthshire, Wales; mine explosion
1892
September 7
Mons, France; mine explosion
1894
June 23
Clyfydd, Wales; mine explosion

1895
June 10
Upper Silesia, Prussia; mine explosion
1900
May 1
Scofield, Utah; mine explosion
1902
May 19
Coal Creek, Tennessee; mine explosion
May 23
Ferme, British Columbia; mine explosion
1903
June 30
* Hanna, Wyoming; mine explosion
1904
January 24
Cheswick, Pennsylvania; mine explosion
1906
March 10
* Courrières, France; mine explosion
March 29
Nagasaki, Japan; mine explosion
1907
January 28
Rhenish, Prussia; mine explosion
December 6
* Monongah, West Virginia; mine explosion
December 19
* Jacob's Creek, Pennsylvania; mine explosion
1908
July 1
Jusovka, Russia; mine explosion
November 11
Westphalia, Germany; mine explosion
1909
February 16
Durham, England; mine explosion
November 13
* Cherry, Illinois; mine explosion
1910
December 21
* Hulton, England; mine explosion
1913
October 14
* Sengenhydd, Wales; mine explosion
October 22
* Dawson, New Mexico; mine explosion
1914
April 18
* Eccles, West Virginia; mine explosion
June 19
Lethbridge, Alberta; mine explosion
1915
April 13
Shimonoseki, Japan; mine explosion
1916
February 6
Skoda, Austria; arsenal explosion
1917
December 6
Halifax, Nova Scotia; ammunition ship
1918
January 12
Podmore Hall, England; mine explosion
August 3
Hamont Station, Belgium; ammunition train
1920
July 16
Sapporo, Japan; mine explosion
1921
June 20
Bern, Switzerland; mine explosion
September 21
Oppau, Germany (See NUCLEAR AND INDUSTRIAL ACCIDENTS)
1924
March 8
* Castle Gate, Utah; mine explosion
December 27
* Otaru, Japan; harbor explosion
1925
March 1
Kharput, Turkey, munitions plant
May 25
Beijing, China; arsenal explosion
1928
May 19
* Mather, Pennsylvania; mine explosion
1930
October 21
Alsdorf, Germany; mine explosion
1931
February 12
* China (Manchuria); mine explosion
1934
March 14
La Libertad, El Salvador; explosives warehouse
September 22
Wrexham, Wales; mine explosion
1935
October 26
* Lanzhou, China; arsenal explosion
1937
March 18
* New London, Texas; gas explosion: school
1940
March 29
Tsingtsing, China; mine explosion
1944
April 14
* Bombay, India; steamship explosion
1947
April 16–18
* Texas City, Texas; liner *Grandcamp*
August 18
Cadiz, Spain; naval mine and torpedo factory
1949
November 29
* Johanngeorgendstadt, Germany; mine explosion
1956
August 7
Cali, Colombia; dynamite truck convoy
1958
February 19
* Asansol, India; mine explosion
1960
March 4
* Havana, Cuba; munitions ship *La Coubre*
1962
February 7
* Volklingen, Germany; mine explosion
1965
May 28
* Dharbad, India; mine explosion
June 1
* Fukuoka, Japan; mine explosion

June 7
* Kakanj, Yugoslavia; mine explosion

1969
March 31
Barroteán, Mexico; mine explosion

1972
June 6
* Wankie, Rhodesia; mine explosion

1975
December 27
Chasnala, India; mine explosion

1982
November 2
* Salang Tunnel, Afghanistan; truck collision

1984
February 25
Cubatao, Brazil; oil pipeline
November 19
Mexico City, Mexico; gas storage area

1989
June 3
Ufa, USSR; gas pipeline; passenger trains

1992
March 3
* Kozlu, Turkey; mine explosion
March 25
* Dakar, Senegal; ammonia truck explosion
April 22
* Guadalajara, Mexico; sewer explosion

1998
December 14
Changxing, China; mine explosion

2000
March 11
* Luhansk, Ukraine; mine explosion

2001
March 7
Fanlin, China; fireworks factory explosion

EXPLOSIONS

Explosions are the most spectacular and dramatic of man-made disasters. Like volcanic eruptions, they occur instantaneously, usually without warning and always with great disturbance to the atmosphere. And their casualty counts are high. Those who die as a result of the initial explosions die cruelly and quickly. Of the secondary catastrophes set off by the blast, fire is the most obvious and pervasive. Thus, those who are not blown to bits are frequently burned to death. And in the case of mine explosions, those who escape either of these fates often expire by asphyxiation.

Historically, the worst and most widespread explosions have occurred in coal mines. Ever since its beginnings in Shropshire, England, near the end of the 17th century, coal mining has been one of the most hazardous of all human occupations. Later methods of strip and open-pit mining were relatively safe, but underground coal mining, in which coal is extracted from the earth through long, interconnecting tunnels, has carried its hazards with it from its inception. Black lung disease and blindness are only the beginning. The added hazards of unsafe working conditions, cave-ins and barely accessible mine shafts—as well as low wages—have made the lot of the miner a rough one. And from the very beginning, the threat of explosions in these mines, which sometimes burrow two miles beneath the earth's surface, has haunted miners and their families.

The number of coal mines in the world and particularly in the United States has shrunk as the demand for coal has diminished. From the beginning of the Industrial Revolution until the middle of the 20th century, the burning of coal heated most homes in the United States, powered the world's railroads and factories and was transformed into the coke essential to the steelmaking process. In addition, the gaseous by-products of coke ovens became the raw materials from which chemicals and other products—from aspirin to nylons—were made.

But as other means of heating and creating power evolved during the 20th century, the role of coal decreased. At the beginning of the 21st century, coal was used predominantly in the United States and the more advanced industrialized nations to produce steam to drive the turbines of power plant generators. Secondarily, it was still used in a steel industry that has increasingly been supplanted by a growing plastics commerce. In addition, coal-fired furnaces still exist in the production of chemicals, cement, stone clay and glass.

The recognition, at the end of the 20th century, that the burning of coal was one of the most significant reasons for an increase in greenhouse gases in the atmosphere also contributed significantly to the decline of coal as a fuel in countries that began to enforce stricter environmental standards and controls on polluting emissions. The combustion of coal is a major source worldwide of carbon dioxide emissions, which constitute a pollutant and a greenhouse gas. Moreover, methane, a major killer of miners in the mines, is also released into the atmosphere.

Admittedly, the major pollution from methane occurs under the surface of the earth, where both methane and coal are formed together during coalification, a process in which plant biomass is converted by biological and geological forces into coal. The methane is then stored in coal seams and the surrounding strata and released not only into horizontal mine shafts but upward through connecting vertical shafts. In addition, small amounts of methane are released during the processing, transport and storage of coal.

The most dramatic release of methane and carbon dioxide from coal mining into the atmosphere in the present is through long-burning fires in coal deposits. Not only are they contributing to global warming, but they are damaging land as well. Ground fractures caused by coal seams reaching the surface are sources of some coal fires, which destabilize the ground and cause it to collapse. In China, as of 2002, coal fires are frequent and ravaging, releasing carbon dioxide equivalent to that produced by all the cars and small trucks in the United States.

In some countries—Turkey, Ukraine and China in particular—where industry is developing or remaining stagnant because of economic pressures, coal mining still abounds. And in these places, the danger that always stalks the miners of coal continues unabated. At the turn of the 21st century, the deaths per million tons of coal mined were a mere 0.1 per year in Australia and the United States, but 119 in Turkey, more than 200 in Ukraine and 1,000 in China, which reported 5,300 deaths in 2000 alone.

There are multiple arguments, then, to cease the production of coal altogether, but the coal-mining industry is a powerful lobbying force in the United States, constantly pressing for the relaxation of provisions in the Clean Air Act, since coal is by nature a dirty burning fuel. Energy interests in the United States also lobby Washington for the same rollbacks in environmental regulations, because burning coal in their plants to produce the steam necessary to turn their generators is the cheapest way to create electrical energy.

Thus, in the United States, energy interests have resisted even newer methods of producing the heat to produce electricity, a method currently in use in Great Britain, which employs the oxidization of coal mine ventilation air, which in turn generates heat that can be used directly on site or to produce electricity.

The combination of these forces in an advanced industrialized nation like the United States and in less-advanced countries that must use their coal resources elongates a dangerous mining process that has changed only slightly for the better in 300 years.

Coal is generally mined in two ways. The first, a laboriously slow method, is to chip away at the walls of shafts, breaking up the coal into sizable chunks that are then transported by cart to the surface.

The second method, more dangerous and widespread, is to drill holes in the face of the wall of coal, pack the holes with explosives, detonate the explosives and blast the coal into manageable chunks. The trick is to blast only the coal, not the mine, and the failure to maintain that delicate balance has resulted in a multitude of tragedies. As in railroading, human error has played the largest role in explosion disasters, both within mines and in other locations.

The hazard of overexploding is further compounded by the usual presence of toxic and ignitable gases. While the need for constant ventilation in the mines has always been apparent, it has not always successfully or diligently been assured. Early on, furnaces were kept burning beneath the surface of the earth to keep the air circulating, and the dangers of these are obvious. Later, manual and then electric fans circulated the air.

But even the most modern methods of ventilation sometimes fail to remove pockets of noxious and ignitable gases, which are called, in the parlance of mining, "damps." These damps come in different varieties, each of them dangerous. A damp, derived from the German *dampf,* which means fog or vapor, results from the decomposition of coal itself. Released by the process of mining, it is an inevitable by-product.

The various categories are:

Firedamp, which consists of methane and other flammable gases, often mixed with air. Explosive mixtures of firedamp with air usually contain from 1 percent to 14 percent methane.

Afterdamp, which is the mixture of gasses remaining after an explosion of firedamp. It consists chiefly of carbon dioxide and nitrogen.

Chokedamp, which is the general name given to any mixture of oxygen-deficient gases that cause suffocation.

And that last description is important, for it is perhaps the most pervasive cause of fatalities in mine explosions. An overload of blasting powder, a careless ignition process or an accident with a miner's lamp can ignite a pocket of firedamp. The explosion kills through the force of the blast and the resultant collapse of the tunnel. Fire spreads through the gases and the ever-present coal dust. This in turn produces afterdamp, which results in chokedamp. These secondary damps remove the oxygen from the air and thus suffocate miners who have survived both the blast and the fire.

The unavoidable presence of damps has always plagued both the operators and workers in mines, and various methods for detecting them have evolved over the ages. Keeping canaries in the depths of the first mines was one way of detecting the presence of damps. The birds have a low tolerance for noxious gases, and their deaths warned miners that damps were present. The Davey safety lamp was one of the first detection devices developed that did not require the deaths of birds. The color and height of the lamp flame indicated the amount of firedamp present. If the flame was extinguished, it was a sign of chokedamp. In modern mines, colorimetric detectors and methanometers are used to detect firedamp.

But no amount or sophistication of detection equipment can overcome human failure, and a quick survey of explosions, both within mines and without, points to human error, miscalculation or carelessness as causes. Too much blasting powder, too little care with a match or a miner's lamp, the storing of old and unstable ammunition in a ship in a crowded harbor (see Bombay, India, p. 180), a school board's decision to save money by channeling unsafe waste gas into a school (see New London, Texas, p. 188), governmental cover-up that resulted in the overcrowding of a steamboat (see Memphis, Tennessee, p. 187) and the misreading of signals by a ship's pilot (see Halifax, Nova Scotia, p. 171) are all human errors and responsible for the explosions that followed them.

No matter what the cause, the effects of explosions are catastrophic, and the magnitude of the heroism that follows has been correspondingly impressive.

The criterion for inclusion in this section is based on the number of fatalities. A general low figure of 100 deaths was used as a cutoff point, and even that seems generous, considering that the high point of fatalities has reached into the thousands.

AFGHANISTAN
SALANG TUNNEL
November 2, 1982

The collision of a Soviet army vehicle with a fuel truck in the Salang Tunnel near Kabul, Afghanistan, on November 2, 1982, caused a massive explosion and fire. Three thousand motorists and soldiers trapped in the tunnel died from either the explosion, fire or fumes; hundreds more were injured.

The 1.7-mile-long Salang Tunnel is a gateway between Kabul, Afghanistan, and the border of the former USSR. Built by the Soviets in the 1970s, it is located 11,100 feet high in the rugged Hindu Kush range, a region in which, in late 1982, there was considerable activity between the Soviet army and Afghan rebels.

On November 2, 1982, a long Soviet army convoy entered the tunnel, which was crowded with civilian buses, cars and trucks. The convoy was traveling from Hairotum, on the Amu Darya, the border stream separating

Afghanistan from the Soviet Union. Midway through the tunnel, which is 17 feet wide by 25 feet high, one of the Soviet army vehicles collided with a fuel tanker. With a gigantic roar that echoed from one end of the tunnel to the other, the tanker exploded, sending gouts of flame outward in all directions. Thirty army vehicles containing Soviet soldiers were consumed instantly, their occupants burned to death on the spot.

Flames rocketed along the narrow passage of the tunnel, setting fire to buses and civilian vehicles. Panic spread as quickly as the flames, and those in cars at either end of the tunnel tried to escape. But the Soviet army, thinking the explosion was the beginning of a rebel attack, blocked both ends of the tunnel with tanks, thus killing hundreds more from asphyxiation.

The nightmare was increased by two other factors: It was bitterly cold, and those motorists who were unable to see the cause of the tie-up assumed that it was just another traffic jam and remained in their cars with the engines running. This increased the carbon monoxide level in the tunnel, killing more unwary occupants. And to further complicate the situation, the tunnel's ventilation system had broken down days before and was not operating.

All of these factors combined to kill thousands of trapped and innocent civilians and soldiers alike, either from the blast, fire or asphyxiation. It would take days to retrieve the dead from the tunnel, and reports that inched their way slowly out of Afghanistan (no foreign reporters were allowed into the country in 1982) stated that there was hardly a person in the capital city of Kabul who did not have either a relative or a friend who had died in the disaster. The Soviet dead were taken to Kabul; the Afghan dead and injured had to be transported 70 miles east of Kabul to Jalalabad in Nangathar Province.

The exact number of dead would never be known. Estimates ranged between 2,000 and 3,000, with credence given by eyewitnesses to the higher figure.

CANADA
NOVA SCOTIA
HALIFAX
December 6, 1917

Eight million tons of TNT ignited and set off the worst accidental explosion in the history of the world when the munitions ship Mont Blanc *collided with the* Imo *in Halifax Harbor on December 6, 1917. Twelve hundred were killed; 8,000 were injured.*

The worst accidental explosion in the history of the world—that of eight million tons of TNT—occurred in Halifax Harbor on the morning of December 6, 1917, at the height of World War I. Commenting to the *Times* of London just after the calamity, Lieutenant Colonel Good of Fredericton opined that he had not seen that much carnage on the battlefields of France. "All that could be seen for a great circumference," he said, "were burning buildings, great mounds of iron and brick in the streets, and dead bodies."

Halifax Harbor was, during World War I, a gathering point for transatlantic convoys. Six miles long with a breadth of about one mile, it provided secure deep-water anchorage at times of both high and low tide. There was not a ship afloat that could not be comfortably accommodated at Halifax, and so, on December 7, 1917, a number of ships had gathered there, to be led by the British cruiser HMS *High Flyer* across the U-boat–infested Atlantic to Europe.

A few days earlier, the French freighter *Mont Blanc* had picked up a lethal cargo in New York. The 3,121-ton ship was loaded to the capacity of its hold with TNT, picric acid, gun cotton and barrels of benzene.

The morning of Thursday, December 6, was fog laden in its early hours, making visibility difficult, except for the experienced pilots aboard ships such as the *Mont Blanc*, who were used to the harbor. The *Mont Blanc* arrived at 8:40, and by then the sun had burned off most of the fog. All that was necessary was for the pilot to navigate "The Narrows," a portion of the harbor that slimmed down to a half-mile-wide channel. On the south shore lay the Richmond section of Halifax; on the north, the town of Dartmouth. Slightly beyond it was the berth into which the *Mont Blanc* was to ease, temporarily.

Suddenly, from around a bend in the channel, the Belgian relief ship *Imo* appeared, heading out to sea. Its course was carrying it directly toward the *Mont Blanc*. The captain of the *Mont Blanc* described what happened then to the London *Times:*

> [Responding to a blast of the *Mont Blanc's* whistle by pilot Frank Mackie,] the Imo signaled that she was coming to port which would bring her to the same side with us. We were keeping to starboard and could not understand what the Imo meant, but kept our course, hoping that she would come down as she should on the starboard side, which would keep her on the Halifax side of the harbour.
>
> . . . Then we put the rudder hard aport to try to pass the Imo before she should come to us. At the same time the Imo reversed engines. As she was light, without cargo, the reverse brought her around slightly to port, her bow towards our starboard. As a collision was then inevitable, we held so that she would be struck forward of the hold where the picric acid substance, which would not explode, was stored, rather than have her strike where the TNT was stored.

It was a correct and safe plan, if it had worked. But it did not. The *Imo* slammed into the *Mont Blanc*, gashing a huge hole in her side and setting fire to the benzene. The fire spread alarmingly. Once it reached the TNT, there would be a cataclysmic explosion. The captain knew this and immediately issued an abandon-ship order.

At pier eight, where the *Mont Blanc* was to dock, two simultaneous activities took place: A fire alarm was set, and Halifax's fire brigade rushed to the scene. At the same time, the captain of the British ammunition ship *Pictou*, moored at pier eight, realized the imminent mortal danger and ordered his ship abandoned too.

The two crews reached shore and began to scramble for the woods, which were only a short distance away, nestled against high cliffs that secured the harbor against winds. Workers in the dockside factories, seeing the running, shouting sailors, swarmed out of their factories and offices and, joining them, scrambled up the cliff toward the Citadel, Halifax's ancient fortress.

Only the captain of the cruiser *High Flyer* thought of trying to contain the blaze, and his heroic decision proved to be foolhardy and fatal. He ordered 23 men to man a launch and try to sink the *Mont Blanc* before she exploded.

The men had scarcely boarded the ship when, with a roar like a concentrated bombardment, it exploded, sending pieces of metal, balls of fire and white-hot explosive incendiaries sky high. A huge wall of water was forced outward from the explosion. It doubled back upon itself in a tidal wave that ripped huge ships from their moorings and tossed them up on the shore.

In Halifax William Barton, eating his breakfast at the Halifax hotel, described it: "In ten seconds it was all over. A low rumbling, an earthquake shock, with everything vibrating, then an indescribable noise, followed by the fall of plaster, and the smashing of glass. A cry went up: 'A German bomb.'"

Richmond, on the other side of the harbor, was hit by pressure waves roaring through the trough of hills with the speed and force of a hurricane. An area two and a half miles in circumference was totally flattened by the blast and its aftermath. The explosion was felt up to 125 miles away; in the immediate vicinity it laid waste to everything. The Intercolonial Railway Station, a brick and stone structure in downtown Halifax, was flattened, crushing crowds of people waiting within. A hundred workers were killed in a sugar refining plant on the docks. Children were just gathering in the area schools to begin their school day. And sadly, every one of the schools would be torn asunder. Of the 550 school children in the Halifax area, only seven would survive.

Now fire began to spread, but every fireman in Halifax lay dead in the midst of the wreckage of all of Halifax's fire equipment. Twenty-five thousand people would ultimately be rendered homeless by either the explosion or its resultant fire.

It could have been considerably worse, but for the tidal wave, an act of heroism and a change in the weather.

The tidal wave caused by the explosion washed over the naval ammunition works at The Narrows, preventing it from catching fire and exploding.

Marine superintendent J. W. Harrison climbed aboard the abandoned British ammunition ship *Pictou*, tied up at pier eight, opened the sea valves and set the vessel adrift. Within minutes, the *Pictou* sank, along with its lethal cargo.

And finally, an hour after the explosion, the weather suddenly turned cold, and it began to snow furiously. The storm extinguished the fires that had been burning out of control and laying waste to large areas of Halifax.

Despite these modifications of a truly cataclysmic disaster, a large part of the city lay in ruins. By afternoon the city militia would take charge. Trains with supplies from New York and Boston began to arrive, and public buildings were opened for the homeless and injured. Emergency hospitals set up by the Red Cross bulged.

Estimates of the dead ranged from 1,200 to 4,000. The official tally was 1,200, with over 8,000 injured. Ironically, and in contrast, only 12 soldiers from Halifax would lose their lives on the battlefields of Europe during the entire war.

CHINA
LANZHOO
October 26, 1935

Sabotage was suspected but never proved in the explosion of an arsenal in the middle of Lanzhoo, China, on October 26, 1935. Two thousand were killed; thousands were injured.

In late 1935 Lanzhoo, in western China on the Yellow River, was, as it is today, an important industrial hub. Linked by rail to Beijing and Mongolia, and on the highway to Tibet, it was, in 1935, important strategically to both the Kuomintang army of Jiang Kaishek (Chiang Kai-shek) and the increasingly powerful Communists. In two short months, Lanzhoo would be the scene of the largest and most effective Communist uprising of the 1930s, and the terrible arsenal explosion in downtown Lanzhoo, on October 26, 1935, may very well have been a grim prelude to that historic happening.

The ancient arsenal, containing thousands of pounds of powder and ammunition belonging to the Kuomintang, exploded with a mighty roar. The force was equal to that of an earthquake or a tornado, and for thousands of square yards around the scene of the explosion, houses were flattened as if they had been visited by a string of tornadoes.

Families sitting down for the midday meal were buried alive and later discovered perfectly in place, but dead. The explosion was so enormous that some residents were literally blown to bits. The portion of the city near the arsenal was totally devastated. Two thousand bodies were recovered, but the total casualty figures were thought to be higher.

CHINA
MANCHURIA
February 12, 1931

A paucity of information available in Manchuria on February 12, 1931, obscured or erased the reason for the gigantic mine explosion of that date. Three thousand died, and an unknown number were injured.

The difficulty of gathering news in Manchuria in 1931, when tensions between the Japanese and Chinese were at their highest, was dramatically represented by the monumental explosion of the Fushun mines, 50 miles east of

Shenyang, on February 12, 1931. Owned by the Fenchihu Company, the mines produced about seven million tons of coal a year and employed thousands of miners.

On February 12, 1931, 3,000 men were working in the mines when the colliery exploded with a thunderous impact. First reports emerging from the scene stated that all 3,000 had been entombed by the blast, which had collapsed not only the shafts but all entrances to them. On the next day Japanese sources denied the report, stating that all 3,000 had been rescued.

On the following day, Chinese papers contradicted the Japanese version, reinstating the report that all 3,000 miners had perished in the explosion. No further details were ever released, and in September of that year, the "Manchurian Incident"—the bombing of the Japanese railway near Shenyang—began the long slide into the Sino-Japanese War of 1937–45.

COLOMBIA
CALI
August 7, 1956

No cause has ever been found for the explosion of seven dynamite trucks parked in the middle of Cali, Colombia, on the night of August 7, 1956. Twelve hundred died, and thousands were injured in the blast.

It certainly seemed as if it was the work of organized terrorists, and General Gustavo Rojas Pinilla, the president of Colombia, stated that it was something planned and executed by "treacherous and criminal conspirators." But the mystery of who caused seven trucks loaded with dynamite to explode in the middle of the city of Cali, Colombia, on the night of August 7, 1956, was never solved.

They were part of a military convoy of 20 trucks loaded with ammunition and dynamite, and they had disembarked at the Pacific port of Buenaventura, on the way to Cali and Bogotá. On August 6 the convoy reached Cali. Thirteen trucks peeled off and headed toward Bogotá. Seven remained behind, parked in front of the main railroad terminal, which was also very near the Codazzi army barracks and the heart of the downtown district of this city of 900,000 located in the western part of Colombia.

At a few minutes after midnight on August 7, all seven trucks detonated with an ear-splitting roar. Every window within a three-mile radius was shattered; the blast scooped out a huge crater in the street in which the atomized trucks had once been parked. Eight city blocks were obliterated, including the Codazzi barracks, in which 500 sleeping soldiers were killed. Even the bronze doors of Cali's enormous and imposing St. Peter's Cathedral, located a full 13 blocks from the blast site, were blown off.

Twelve hundred people were killed in the explosion, and thousands were injured. And the true cause or the persons who set the blast would never be found.

CUBA
HAVANA
March 4, 1960

A broken hoist cable allowed a net full of grenades to plummet to the deck of the Belgian ammunition ship La Coubre *in Havana Harbor on March 4, 1960. One hundred were killed in the explosion; scores more were injured.*

A broken cable on a hoist, a pregnant pause, and a net full of grenades plummeted to the deck of the Belgian munitions ship *La Coubre,* tied up in Havana Harbor on the afternoon of March 4, 1960. With a roar that shook buildings on the waterfront and sent shock waves against the hovering helicopter holding Fidel Castro, the ship blew up, scattering pieces of it for blocks and shooting off ammunition in all directions.

Longshoremen, soldiers and crew members aboard the ship were ripped apart by the force of the explosion and the resulting release of live bullets or burned to death by the enormous fire that burst forth after the blast. G. Delgado, a fireman who was called to the dock to fight the fire, was momentarily trapped in the ship's stern when it rolled over, preparatory to sinking. "It looked like a scene from Dante's *Inferno,*" Delgado told reporters. "Bodies and pieces of bodies were all over. God knows how I escaped. Bullets and shrapnel were flying all around me." One hundred men died in the blast, most of them from the direct explosion of the grenades when they detonated on the deck.

Castro would later blame the explosion on CIA sabotage and even staged a media event in which he ordered two cases of grenades to be dropped 400 feet from a helicopter onto a ball field. The grenades in these cases did not explode on contact, but six soldiers were killed as grenades detonated when they attempted to clean up the debris.

FRANCE
COURRIÈRES
March 10, 1906

The worst mine explosion in French history, in the Courrières Colliery in northern France on March 10, 1906, was caused by a combination of a smoldering fire in the pit and trapped gases. One thousand sixty miners died; hundreds were injured.

The Courrières Colliery in northern France was part of an enormous complex of coal mines that employed 2,000 men and boys. Located in the mountainous region of Pas-de-Calais, the mines formed a series of subterranean tunnels whose multiple outlets were spaced over several towns. Six of the outlets were near Lens; the rest emerged at Courrières, Verdun, and other tinier hamlets populated by the families of coal miners. The output of the mines was a par-

Rescued miners wait to be lifted to the surface after the gigantic mine explosion in the Courrières Colliery, in which 1,060 miners died. (London Illustrated News)

ticularly combustible coal that was largely used in the manufacture of gas and in smelting.

At 3:00 in the afternoon of the day before France's worst mine disaster, a small, smoldering fire began in the Cecil pit of the Courrières mine at a depth of 270 meters, in a location in which masonry work was being done. Engineers tried to cope with the small blaze but were unable, through the night and early morning, to extinguish it. They opted to starve it of air and closed the outlet.

Apparently, fissures in the walls allowed combustible gases to creep into the closed-off portion of the tunnel. At 7:00 A.M., when 1,795 men and boys were working in the mine, the pit exploded with a thunderous roar, spitting cages and debris from the mouth of the shaft and killing several men and horses who were above ground near the shaft opening. The roof of the mine office was blown cleanly off.

Rescuers began their work immediately, drawing up the injured, most of whom were terribly burned. Fires poured from every outlet, driving off rescuers and driving down the hopes of the thousands of family members who also rushed to the mine, pushing against the cordons of police that were hastily formed.

Leon Cerf, a survivor, recounted the scene below at the time of the explosion to the *New York Times:*

> I was working with a gang when the explosion occurred. The foreman immediately shouted for us to follow him, and, dashing into a recess in the gallery, we were followed by a blast of poisonous gases, which, however, rushed by without affecting us. We remained there for eight hours, when, feeling that suffocation was gradually coming upon us, we attempted to escape. We crawled in single file toward the shaft, but several of the men dropped dead on the way, including my son and the foreman. I carried my nephew on my back for forty minutes, and succeeded in saving him. It took us four hours to reach the shaft.

He was one of the fortunate. Others, trapped farther below, could not be reached by rescuers because of collapsed tunnels and noxious afterdamp and chokedamp.

Rescuers inside the Courrières Colliery. (London Illustrated News)

One group of rescuers, descending in a cage, distinctly heard a tapping on water pipes, indicating that there were imprisoned miners nearby. But engineers summoned to the location, listening to the same tapping, dashed hopes by estimating that it would take eight days to dislodge the debris in that part of the shaft. By then, the engineers reasoned, the miners would be dead of either starvation or asphyxiation. During the next few hours of rescue work, the tapping diminished and then ceased.

Rescue workers continued to toil through the night. One party of 40 disappeared into a shaft that collapsed behind them, burying them alive.

A huge mortuary camp was set up, and 400 soldiers were called in to buttress the police presence and control hysterical relatives, crying to see the bodies. Some rescuers descended scores of times into the pits until they themselves collapsed. One rescuer, brought to the surface unconscious, was packed into a closed carriage to be driven home. A group of miners, suspecting that the carriage contained bodies, broke its windows, further injuring the exhausted rescuer.

When hope and rescue work were finally abandoned a week later, the death toll was set at 1,060, making it the greatest single mine calamity in not only France but the European continent to that date.

GERMANY
JOHANNGEORGENDSTADT
November 29, 1949

Soviet security prevented the rest of the world from knowing the cause and details of the explosion in the uranium mine in Johanngeorgendstadt, East Germany, on November 29, 1949. Three thousand seven hundred were reportedly killed and an unknown number injured.

Soviet authorities clamped an airtight lid on details of the horrendous explosion in the Soviet uranium mine at Johanngeorgendstadt, in Saxony, East Germany, on November 29, 1949. The only report that found its way into Western newspapers came from the chief officer of a fire brigade in Leipzig. The officer escaped from the Soviet zone in December and brought with him some data on the explosion, which had been large enough to register on seismographs in Europe.

Four thousand workers had been in the mine when the explosion occurred, according to the officer, and only 200 to 300 of them had been rescued, which meant that at least 3,700 had perished in this, one of the worst mine disasters of all time. Precise details would never be known. Soviet security police cordoned off the mine immediately after the explosion. Eighty members of the Johanngeorgendstadt fire brigade fought the blazes that occurred as a result of the blast. When they finished their work, according to the escaped officer, all 80 were arrested by the Soviet security police and shot to death.

GERMANY
OPPAU
September 21, 1921

See NUCLEAR AND INDUSTRIAL ACCIDENTS, p. 302.

GERMANY
VOLKLINGEN
February 7, 1962

Methane gas exploded in the Luisenthal pit in Volklingen, West Germany, on February 7, 1962. Two hundred ninety-eight miners were killed; more than 200 were injured.

By 1962 it was thought that the danger from the various damps (firedamp, afterdamp, chokedamp) had been solved. Colorimetric detectors and the methanometer had replaced earlier, more primitive means of detection, and miners could toil in relative safety.

That security was blasted into oblivion on February 7, 1962, in the Luisenthal pit in Volklingen, West Germany. One of the most modern collieries, it employed 480 men per shift in well-supported shafts on several levels. At precisely 8:00 A.M. that morning, methane gas exploded in the second-level tunnel, igniting a huge fire that licked along this shaft until it found the main one, where it branched upward and downward.

A second explosion followed swiftly on the heels of the first, when flames found more methane at the 1,800-foot level. Chokedamp filled every inch of the tunnel, and multiple explosions followed in quick succession.

The men trapped in the tunnels had almost no chance of survival. They suffocated immediately or were crushed to death when the shaft ceilings collapsed from the multiple explosions. "The injured looked terrible," said one survivor, George Kneip. "Some looked completely black. Many cried in agony. One body was headless."

Two hundred ninety-eight miners were killed and more than 200 injured in this, one of the worst mine disasters in German history.

GREAT BRITAIN
ENGLAND
BARNSLEY
December 12–13, 1866

An overabundance of blasting powder caused the explosion in the Oaks Colliery, in Barnsley, England, on December 12, 1866. Three hundred forty died; the number of injured was unreported.

The Oaks Colliery, located at Barnsley, a short distance north of London, was one of the largest producers of coal in the South Yorkshire district of England in the 19th century.

The giant explosion in the Oaks Colliery in Barnsley, England, on December 12, 1866, is graphically portrayed in a contemporary lithograph. Three hundred forty men and boys died in the tragedy. (Illustrated London News)

More than 430 miners toiled in its depths, whose principal shaft was known as the "dip" and along which ran a broad roadway. Adjacent to this was the so-called engine plane, a passage that ran for two miles. Underground, the colliery resembled a small city, with horses that pulled the carts of coal upward housed in underground stalls and an air circulation system facilitated by a large furnace that burned night and day.

Boys worked alongside men in the 19th century in England. There were no child labor laws to speak of, and the day shift of December 2, 1866, consisted of one-third boys and two-thirds men. Although the mine was worked around the clock, the day shift was the most active, and it was only during this time that coal was removed from the mine.

Three hundred thirty men and boys entered the mine at 6:00 A.M. on Wednesday morning, December 12. At 1:20 P.M., the ground shook, and a dull, heavy explosion erupted from the mouth of the main shaft of the mine. Dense columns of smoke and dust shot into the air from each of the shafts, and in a few seconds the pit bank was enveloped in a thick black cloud.

The explosion had collapsed all of the air shafts, locking noxious fumes in the tunnels in which the miners worked. Rescuers arrived on the scene almost immediately, and, led by a Mr. T. Diamond, the managing partner in the mine, and Superintendent Greenhalgh, a rescue party immediately lowered itself as far as it could into the main shaft. Eighteen badly injured men were discovered and brought immediately to the surface. That would be the last good news of the day.

The longer they searched and the more deeply they dug, rescuers found only the bodies of miners caught by the afterdamp or crushed under falling timbers. The stables were flattened, and 20 horses lay dead there. Many of the miners were frozen in attitudes of prayer; one group of 20 was clustered together in a last gesture of communal protection.

Rescuers worked all day and all night of the 12th, digging out tunnels and transporting an increasing number of bodies to the surface. By 8:00 A.M. on the morning of the 13th, nearly 800 yards of temporary airways had been constructed, and the searchers had penetrated some of the farthest reaches of the mine's tunnels. Only one level contained fire, and it was minor.

And then, shortly after 8:00 A.M., some of the 37 rescuers in the mine noticed that the air was being rapidly drawn from them. A miner in the rescue party recognized the signs: Another explosion was in the making. Sixteen of the 37 scrambled to the surface and warned another party that was about to descend that there was danger of another explosion. Incredibly, most of this search party refused to believe the escaped rescuers and descended into the mine anyway.

This second party had just reached the bottom of the shaft, at approximately 9:00 A.M. on the morning of Thursday, December 13, when a second explosion tore through the tunnels of the Oaks Colliery. Debris and pieces of the rescue party shot out of the newly made openings as if they were part of an artillery barrage.

Forty minutes later, shortly before 10:00 A.M., a third, lesser explosion rocked the works and made rescuers think twice before resuming their work. By nightfall the mine had apparently quieted, and search parties resumed their digging. Some miraculous rescues occurred; individual miners, nearly dead from chokedamp, managed to climb through the debris and meet the rescue parties, who then raised them to the surface by means of a wooden tub suspended from a makeshift block-and-tackle system.

Not many lived, however. Three hundred forty men, 28 of them rescuers, died in the three explosions.

The reason for the tragedy had been a long time coming, according to a report in the Sheffield, England, *Independent:* "It seems that there have been for some time complaints of the heat of the atmosphere from the long distance the air had to travel through the workings," the newspaper report noted. The engine plane had been the only passage used to ventilate the tunnels, and so Mr. Diamond, the managing partner, had determined that December to dig another ventilating tunnel.

Apparently an impatient man, Diamond ordered a maximum amount of blasting powder to be used to expedite the task. On Wednesday morning, December 12, miners Richard Hunt and John Clayton began to use a long drill. They decided that a large charge of dynamite was needed to drive it through to its destination and thus wired up a large charge. William Wilson, the man in charge of operations learned of this while he was working at another location; realizing the possible consequences of setting off such a charge at that specific place—which was filled with pockets of methane—he dashed to the site where Hunt and Clayton were by now setting a fuse.

He got there too late. The fuse was set and detonated just as he arrived on the scene, and the dynamite went up in a fearful roar, igniting the gas and causing a cataclysmic explosion. Astonishingly, Wilson lived to tell the story. The other two were killed by the blast they set.

GREAT BRITAIN
ENGLAND
HULTON
December 21, 1910

No cause was determined for the explosion in the Little Hulton Mine in Hulton, England, on December 21, 1910. All 360 miners in the mine were killed.

The Little Hulton Mine, owned by the Hulton Colliery Company and located in the small town of that name some four miles from Bolton, England, was entirely demolished by a calamitous explosion and fire on the morning of December 21, 1910.

At a little after 7:00 A.M., shortly after the morning shift descended into the colliery, an ear-shattering explosion rocked the countryside, wrecked the lift mechanism of the shaft and showered the hills with debris. Almost immediately, an inferno of flames spit out of the head of the shaft, preventing rescuers from entering.

When the fire was finally brought under control, rescuers found their way blocked. The shaft had collapsed entirely below the 400-yard level, burying the entire shift of miners. For a few hours, the fate of 400 other men working in an adjoining mine was in jeopardy when passageways between the two mines collapsed. But all 400 men of that shift were brought to the surface, including several who were injured by the impact of the explosion.

The fate of the workers in the Little Hulton Mine was just the opposite. Rescuers found small groups of bodies of men who were working above the 400-yard level, and that was all. By 9:30 that night, the fate of the entire shift of 360 boys and men was obvious. None could have survived; they were either crushed, blown apart or asphyxiated. Not one person escaped alive from the mine.

GREAT BRITAIN
ENGLAND
SUNDERLAND
August 17, 1880

The ignition of afterdamp was responsible for the explosion in the Seaham Colliery in Sunderland, England, on August 17, 1880. One hundred sixty-one were killed; scores were injured.

Large collieries proliferated throughout Britain and Wales in the 19th century and the first half of the 20th century, and one of the largest of these was the Seaham Colliery near Sunderland, England. Sixteen hundred miners worked three shifts in this hugely productive mine, churning out coal for the world.

On August 17, 1880, the village of Sunderland had scheduled its annual flower show. Heavily attended, it drew a large percentage of the village and surrounding areas, and it, more than anything else, probably accounted for the saving of hundreds of lives in the early morning tragedy that blasted apart the lowest seams of the mine's main tunnel and killed 161 members of a "light shift" of only 246. The death toll would have been considerably heavier and the mine certainly more thickly populated had it not been for the flower show. Many miners had decided to sleep in and go to the show rather than work this shift.

The explosion, touched off by the ignition of afterdamp—the highly combustible gas given off by coal, which, when it accumulates in the shafts, becomes lethal to humans—was set off by the lighted lamp of an unsuspecting miner at 2:30 A.M. It collapsed the walls and ceilings of the shaft, burying many of the miners. Others, near the blast, were ripped apart and found in pieces later that day.

The other, latent danger in any mine explosion is the spreading of the afterdamp, which, now released, filled the other tunnels, killing the survivors as quickly and effectively as an army's poison gas. Ralph Markey, a miner who had survived three previous explosions, was trapped along with 18 other men in a pocket of space caused by the collapse of a shaft wall. Speaking later to the *Illustrated London News,* Markey recalled feeling a rush of wind an instant before the blast.

Sizing up the situation, Markey led the men in a digging-out exercise that brought them into one of the main shafts. They followed this for a quarter of a mile, stepping over the corpses of their fellow miners. "A deputy overman named Wardle," recalled Markey "[was] lying insensible, with his face covered with blood, and here [came] the afterdamp." The group pressed cloths to their faces and tried to crawl under the lethal layer of gas.

Eventually, they reached the elevator shaft, but the cage was useless, jammed halfway between tunnels. At last air filtered down to them, and they posted shouters to continually call up the shaft while the rest of them prepared tea.

It would be two hours before shouted encouragement reached them, assuring them that rescuers were on their way, and another eight hours before a party led by Stratton, the owner of the mine, would reach them. Eighty-five miners would be rescued; 161 would die in the great colliery explosion of 1880.

GREAT BRITAIN
WALES
ABERCANE
September 11, 1878

The cause of the explosion in the Ebbw Vale Steel, Iron and Coal Company's Abercane Colliery in Abercane, Wales, on September 11, 1878, was and remains unknown. Two hundred sixty-eight miners died in the explosion; 12 were injured.

The Abercane Colliery, owned by the Ebbw Vale Steel, Iron and Coal Company, was one of the largest collieries in south Wales during the end of the 19th century. Situated a few hundred yards from the Abercane railway station, on the Western Valley section of the Monmouthshire Railway, it nestled in a picturesque valley in the shadow of the Crumlin Viaduct, one of Wales' most charming and well-visited tourist attractions.

It was not unusual to have a working colliery in the midst of town in the 1800s; the production of bituminous coal was to Wales then what oil wells later became to Okla-

An anxious crowd gathers outside the ruined Universal Colliery in Sengenhydd, Wales, following the worst mine disaster in the history of Great Britain on October 14, 1913. *(Illustrated London News)*

homa and Texas. And the Abercane Colliery in its most productive times produced 1,000 tons of "steam" coal daily.

It was also one of the most up-to-date mines in the world. Its winding, pumping and ventilating machinery was the most modern for its day; its use of safety lamps was rigidly enforced.

Thus, the 373 men and boys who entered the mine at 11:00 A.M. on September 12, 1878, felt protected and secure. If there ever was an explosion-proof mine, it was this one. And thus, the reason that the mine exploded with an ear-shattering roar at 12:10 P.M. that day was and would remain an unexplained mystery.

The first explosion was followed by two others of equal force. Within seconds, a huge tongue of flame flew from the main shaft opening followed by dense clouds of acrid smoke, dust and debris. An enormous fire was obviously burning within the pit, incinerating whoever was there.

The winding gear for the buckets that transported men in and out of the mine was damaged by the explosion and had to be repaired before rescuers could begin their work. When it was back in operation, they brought out 82 men and boys who had been working within a few hundred yards of the shaft opening.

But that would be the extent of the rescue. Descending to the bottom of the 330-yard-deep mine, rescuers found horrible devastation. The underground stables yielded 14 dead horses, 12 terribly burned men and 13 bodies. Most of the ambient tunnels had collapsed, and those that had not were unapproachable. The chokedamp was pervasive and lethal, and rescuers were repeatedly turned back.

Finally, at 2:30 A.M. on September 13, to the despair of hundreds of relatives at the scene, it was decided to flood the mine to prevent further fire and explosions. The bodies of 255 men and boys were buried underground; only the 13 bodies that were discovered near the stables were brought to the surface. It would be one of Wales' worst and most mysterious mine disasters.

GREAT BRITAIN
WALES
SENGENHYDD
October 14, 1913

There is no known cause for the greatest mine disaster in the history of Great Britain, the explosion in the Universal Colliery at Sengenhydd, Wales, on October 14, 1913. Three hundred forty-three died; 12 were injured.

The greatest mine disaster in the history of Great Britain occurred on the morning of October 14, 1913, at the Universal Colliery at Sengenhydd, South Wales, eight miles from Cardiff.

The colliery consisted of two pits, side by side, the Lancaster and York. At 6:00 A.M. on October 14, 935 men descended into the two mine shafts, beginning the first and most active shift of the day.

At 8:12 A.M. the earth around the mine shafts shook, and an enormous explosion ripped through the Lancaster pit, spewing a fountain of debris and dust skyward and ripping out the pithead gear. Within minutes, plumes of orange flames shot up from the pit opening.

Hundreds of rescuers came on the scene immediately, but they were unable to descend into the Lancaster pit because of the intensity of the fire. Not only was the heat tremendous, but the fire was blocking the only air intake to the shaft. It would be an hour before the fire would be controlled enough to allow the pit gear from the undamaged York pit, 50 yards away, to be set in place so that rescuers could be lowered into the Lancaster shaft.

The teams found a holocaust. The force of the explosion had been enormous. Scores of headless and dismembered bodies were found, but fortunately, so were hundreds of survivors. All in all, 498 men were raised in groups of 20 to the surface by miners from nearby collieries. Twelve seriously injured men were taken to nearby hospitals.

But that still left nearly 350 men trapped within the collapsed mine shaft. Rescuers attempted to enter these shafts, but the afterdamp threatened to kill them, too, and by nightfall it was apparent that the only survivors of the explosion were now above ground. By that time, 40,000 people had gathered at the pitheads, hoping for a miracle.

The rescuers dug on into the night, recovering 73 more bodies. Finally, on the morning of October 15, the decision was made to seal the mine. Three hundred forty-three men and boys had lost their lives, and slightly more than 200 of them would be sealed in the mine that morning.

Prince Arthur of Connaught and his bride of only a few weeks sent messages of sympathy to the bereaved families of the dead miners and announced that the royal wedding presents would be exhibited in public for the next month as a means of raising money for relief funds.

INDIA
ASANSOL
February 19, 1958

A gas explosion collapsed a mine in Asansol, India, on February 19, 1958. One hundred eighty-three died; scores were injured.

There is a maze of interlocking tunnels in the mines beneath Asansol, India. Like a mountain with a multitude of subterranean ski trails, separate mines interlock and cross.

On February 19, 1958, gases within one of the tunnels ignited and exploded, setting off a chain reaction of echoing explosions throughout the underground system. Walls caved in; the roofs of shafts collapsed and crushed miners beneath them. To add to the terror, water from underground sources poured into the shafts, filling them and drowning many more men.

It would be six hours before rescuers could even reach the first of a mere 17 men. Of the 200 working that particular shift, 183 died.

INDIA
BOMBAY
April 14, 1944

No cause was ever discovered for the explosion of the ammunition ship Fort Stikine *in Victoria Dock, Bombay, India, on April 14, 1944. One thousand three hundred seventy-six were killed, and more than three thousand were injured.*

Victoria Dock in Bombay was nicknamed the Gateway to India. Built up over years of British rule, it was equipped with all of the niceties necessary to supply the residing colonials with their needs and to link India with the rest of the industrialized world.

In 1944, at the height of World War II, it was an abnormally busy place. Armies of the Allies were in omnipresent evidence; troop ships, ships loaded with supplies, aircraft, artillery and ammunition, docked and left regularly. On April 14, 1944, there were no less than 27 ships docked there.

Two months earlier the Canadian-built cargo ship *Fort Stikine* left Birkinhead, England, with 12 dismantled Spitfires, over 1,390 tons of explosives and a million pounds of gold bricks. It was obviously a potential seaborne volcano, and its captain, Alexander J. Naismith, was instructed to remain on the outskirts of the convoy to which it was attached. A well-placed torpedo could cause the *Fort Stikine* to ignite the entire convoy.

However, the voyage from England to Karachi was uneventful. At that port the Spitfires were unloaded, and a masking cargo of 8,700 bales of raw cotton, several hundred barrels of lubricating oil, a holdful of scrap metal and another holdful of fish manure was added. It was, then, not only a dangerous but also an odoriferous voyage from Karachi to Bombay, and the priority off-load when the ship finally tied up at Victoria Dock was the fish manure.

Five gangs of Indian stevedores began the long task of unloading the masking cargo on April 13, beginning with the fish. The next day they were still unloading when smoke began to rise out of the number two hold, which held explosives. (Later investigation would turn up no clues regarding its origin. It remains a mystery to this day.)

Firemen were called, and they arrived immediately, but within moments it was obvious that they would need reinforcements. Unfortunately, the man sent out to call for them found a dial-less phone, which he abandoned before the operator could come on the line. He then set off an alarm box that only signaled a minimum fire force—far less than was required to halt the spreading blaze, which was now advancing on the small-arms ammunition.

The cotton bales went up in flames, and despite the fact that 32 hoses were trained on the fire, it was apparent to the sailors aboard that the ship was going to blow. Shortly afterward, the small-arms ammunition exploded, and some of the fire fighters left the ship.

It was time for a quick decision on the part of Captain Naismith: take the ship out into the harbor, away from the docks and other ships, or scuttle it? At 3:30 P.M. on April 14, while he was still making up his mind, the first of a series of major explosions erupted, blowing away the entire bow from the bridge forward and killing the captain. Pieces of the ship and shrapnel from the ammunition planed across the water and flew through the air. The 400-foot *Japlanda,* anchored alongside the *Fort Stikine,* was blown 60 feet in the air. It landed on the roof of a shed on the dock, setting the shed on fire.

Everyone in the bow of the ship and everyone on the dock near it was either killed or horribly mangled by the explosion. There was not a single fireman left without at least one arm or leg missing.

Rescuers tried to drag the injured and the dying away from the vessel, but they had no sooner begun their task when a second, bigger explosion destroyed the rest of the ship and the remainder of Bombay Harbor. Every single one of the 27 ships docked there was sunk. Gold bricks soared through the air, landing in the heart of Bombay, killing scores of people as effectively as if the bricks were Japanese or German bombs.

The scrap metal was an even more lethal danger. The enormous cloud caused by the explosion, which rose 3,000 feet in the air, was laced with fragments of metal. Some of the heavier pieces fell on parked cars, flattening them and killing the occupants. Others demolished homes. Over a mile away, Captain Sidney Kielly, walking to his office, was cut in half by a piece of falling metal.

It was a horror beyond description for the unsuspecting residents of Bombay, who were suddenly faced with death raining out of the sky. Every ship, every home, every shed, every building, every square foot of docking in the harbor was blown to bits. It would take days for the Indian fire department, aided by over 7,000 Allied troops, to put out the fire, haul off the injured and count the dead. One thousand three hundred seventy-six would die, and more than 3,000 would be injured in this, one of the worst man-made explosions of all time.

INDIA
DHARBAD
May 28, 1965

A methane gas ignition was responsible for the explosion in the coal mine in Dharbad, India, on May 28, 1965. Three hundred seventy-five were killed; hundreds were injured.

Dharbad, 225 miles northwest of Calcutta, was rocked by an enormous explosion in its coal mine on May 28, 1965. A spark of unknown origin ignited methane in one of its shafts, blasting apart a huge section of the mine and sending coal dust into the air over a four-mile radius.

The force of the explosion was so great that it killed over 100 miners who were working on the surface of the mine; it also demolished the record office, the engine room and several nearby houses. Timbers from the shafts were shot upward and soared like airborne battering rams over the area.

The entire main shaft was consumed by raging flames. It would be several days before rescue workers could even begin to enter it. When they finally did, they would find 375 dead.

JAPAN
FUKUOKA
June 1, 1965

Failure to install safety devices by the management of the Yamano coal mine near Fukuoka, Japan, led to an explosion in the mine on June 1, 1965. Two hundred thirty-six miners were killed; 37 were injured.

The early to mid-1960s seemed to be a time rife with mine explosions. One of these, the explosion in the Yamano coal mine near Fukuoka, Japan, was the result of negligence on the part of management.

In 1959 an explosion in a pocket of methane gas had killed seven miners and injured 24. The management had been cited for safety lapses and had agreed to correct them to avoid a repetition of the small tragedy.

But by June 1, 1965, when 552 miners were at work in the mine, nothing had been done. No methanometers had been installed; no colorimetric detectors had been purchased. The mine was still a time bomb waiting to ignite.

And ignite it did, with a stupendous roar. Shafts collapsed; great, multi-ton boulders sealed off passages. Miners suffocated on the spot as the chokedamp filled the tunnels. Two hundred seventy-nine miners managed to scramble to the elevators, which were fortunately still working, and ascend to safety. Of the 279, 37 were seriously injured and were carried to the surface by their fellow miners.

It would be two days before rescuers could pierce the sealed off chambers that contained the dead. More than 2,000 relatives had kept vigil, day and night, hoping that there would be some survivors. There were none. Two hundred thirty-six miners were discovered, every one of them dead.

The trade and industries minister, Yoshio Sakarauchi, resigned his office the next day, admitting that his office had failed to introduce the obvious safety measures that could have prevented the disaster.

MEXICO
GUADALAJARA
April 22, 1992

In a horrendous series of explosions on April 22, 1992, a huge portion of the sewer system in the Reforma district of Guadalajara, Mexico, exploded, killing 215 and injuring more than 1,500.

For three days, beginning on April 19, 1992, the residents of the Reforma blue-collar district in the southeastern section of Guadalajara, Mexico, complained to the municipal police and fire departments that a "foul smell" that was strong enough to cause nausea and eye and throat stinging was permeating their neighborhood. Gas fumes were drifting upward from sewer gratings, according to the complaints. But the pleas were largely ignored. The only sign of a response was the presence of one firefighter, who removed some manhole covers on the 21st.

The smell persisted; the residents became more and more uneasy, and finally, at 11:30 A.M. on April 22, 1992, a gigantic series of multiple explosions reverberated, one after the other, through the district, ripping the earth open in a jagged trench one mile long, 35 feet deep and 10 feet wide. The trench was contiguous with the sewer system, and manhole covers were flung into the air along the length of the opening in the earth.

The trench ran under more than 20 square blocks of buildings in the crowded neighborhood, and damage from the blast radiated out from the initial source in all directions, much like an earthquake from an epicenter. Ultimately, nearly 1,000 buildings were damaged.

In several locations, craters, some of them 300 feet in diameter, opened up, swallowing people and vehicles. A bus was, in the words of one eye witness, "swallowed up by the hole," and bystanders rescued some of its passengers by pulling them through the windows of the precariously perched vehicle.

The dead and dying lay everywhere in the blood-spattered streets. The wounded and dazed wandered helplessly through the rubble, fire and smoke crying out for help.

For days police, fire, rescue and Red Cross workers labored in the ruins of part of Mexico's second-largest city. It would be weeks before the tally of dead and wounded was published. It would be devastating: 215 people had been killed and more than 1,500 had been wounded. More than 25,000 had been evacuated from the area in anticipation of further explosions.

José Trinidad López Rivas, the Guadalajara fire chief, blamed the explosion on, as he termed it, "thousands of liters of gasoline" in the sewer system prior to the explosions. However, PEMEX, the Mexican state petroleum company, whose pipes ran through the sewer system, maintained that the cause was a leak of hexane from a factory in the area. The gas has a low flash point and a mild gasoline smell, and for a while most authorities accepted this version of the source of the tragedy.

Further examination revealed, however, that there had not been a hexane gas leak. A water pipe installed above a PEMEX gas pipe had rusted the gas pipeline, and gasoline had leaked into the sewer line.

A scandal followed. Mexico's attorney general, Ignacio Morales Lechuga, charged nine government and industry officials with negligence. Because of his failure to order an evacuation of the area when the smell of gasoline was first reported, Guadalajara mayor Enrique Dau Flores was indicted, along with four of his subordinates. In addition, four midlevel employees of PEMEX were arrested.

A suspicion that the Institutional Revolutionary Party of Mexican president Carlos Salinas de Gortari, which held a monopoly at the federal, state and local levels, was trying to insulate Salinas by parceling out the blame to lower-level officials was answered with surprising candor by one federal director. "That's exactly what they're going to do," said the anonymous source. "Identify those responsible, say they were negligent, and separate them from the rest of the government." And this was what was done.

RHODESIA
WANKIE
June 6, 1972

A methane gas ignition caused the explosion that decimated the Wankie Colliery in Rhodesia on June 6, 1972. Four hundred twenty-seven miners were killed; 37 escaped unscathed.

Rhodesia's largest colliery is located in Wankie, in the middle veld of that African republic. The town of Wankie contains 20,000 people, and in one way or another, almost all of them are either employed in the mine or related to those who are working in it.

Thus, when on the afternoon of June 6, 1972, an enormous explosion rocked the earth around the number two shaft of the Wankie Colliery, it immediately emptied every dwelling and filled every person with an understandable terror. A cable car had been spit out of the mouth of the shaft and lay in mangled ruins some yards beyond. Methane gas poured from the opening, and rescue parties were driven back for hours.

Finally, by nightfall, wearing masks and lights, teams of searchers dug into the earth around the shaft and lowered themselves into it. They discovered 37 men who had miraculously been able to squirrel themselves away from the main tunnel and thus avoid the lethal afterdamp.

Not so 427 other miners of that particular shift. Every one of them died in one of the worst mine disasters in the history of southern Africa's coal mining.

SENEGAL
DAKAR
March 25, 1992

A sudden explosion of a truck delivering liquid ammonia to a peanut processing factory in Dakar, Senegal, on March 25, 1992, killed 60 workers and injured 250.

One of Senegal's main exports is peanut oil, and its major export crop is the pulp that results from the process of extracting oil from the peanut. Dakar, Senegal's capital and major port city, contains a multitude of peanut factories, extracting peanut oil, and detoxifying, with liquid ammonia, the pulp, which is used both domestically and abroad in fertilizer and animal feed.

On March 25, 1992, at the height of a busy workday, an ammonia truck unloading its contents in the midst of more than 100 workers suddenly exploded. Shards of metal from the dissembling truck pierced walls and workers alike. The fire burned others. Still more were overcome by the deadly fumes that burst forth from the ruptured tank and lingered in a toxic cloud over the entire premises.

The factory, primitive by modern standards, owned only a few gas masks, and they were scattered in various locations. Thus, workers and rescuers who might have saved the dying were unable to penetrate the gas cloud that hovered over the victims.

Police and firefighters rushed to the scene immediately, but they, too, were handicapped by a lack of gas masks and equipment to fight more than the fire that resulted from the explosion.

Finally, French soldiers stationed in the country, a former French colony, arrived and controlled the situation. Two hundred and fifty workers were injured and were taken to nearby civilian and military hospitals. Sixty died.

TURKEY
KOZLU
March 3, 1992

The worst mine explosion in Turkey's history killed 270 miners in the Incirharmani coal mine at Kozlu, Turkey, on March 3, 1992.

The hills for miles around the Black Sea port of Zonguldak, Turkey, 170 miles northwest of Ankara, are black. The streets of the town are black. The beach is black. A patina of pollution from the Incirharmani coal mine at Kozlu, six miles from Zonguldak, in a steep valley that runs down to the sea, is everywhere and has been for generations. This is the way livings are made in this part of the world. Low-quality coal has been produced from its ground for 150 years, and in the 50 years preceding March 3, 1992, when Turkey's worst mine disaster erupted, 17 calamities had claimed the lives of 525 miners. Still, despite this and despite the abominable safety record (the worst in the world) of Turkey's coal mines, 32,000 men and boys continue to toil for between $500 and $600 a month in these mines and several thousand eke out wretched livings at the railyards and in the small port.

The day shift of the Incirharmani mine finished its work at 4:00 P.M. on the 3rd of March, 1992, and Temel Emral, an electrical engineer, crossed paths with his closest friend on the way to the night shift. "We talked," he told reporters later. "I wished him well with his work."

Four and a half hours later, at 8:30 P.M., a roar that felt like an earthquake to those in their homes on the surface shook walls, shattered plates and sent the residents running to the mine. There had been an explosion of built up methane gas in a tunnel 1,300 feet underground, so sudden and rapidly moving that it had defeated the automatic detectors that might have warned the men working in the tunnel. It erupted through the pocket in the tunnel, pushing flames ahead of it, sucking the air from the tunnel. There was no time for escape. Two hundred and seventy men were killed instantly by the blast.

There were no survivors, and yet for days afterward rescuers tried to search the shafts for any who might have escaped death, while knots of miners, some stoic, some openly weeping, clustered around the shaft opening.

"Perhaps," one miner said to the other on the third day, "God is great and there will be a voice calling for help from somewhere down there."

"No," answered another, who had lost a brother and a brother-in-law in the explosion, "all we find is the bodies."

Ironically, the Incirharmani mine, one of the many state-run mines in the area, had become the subject of controversy a month before the tragedy. An Istanbul businessman, Ishak Alaton, had drawn up a plan to close it. Because of huge government subsidies and high production costs, Mr. Alaton reasoned, the coal produced in the mine and its surrounding area was so expensive that it would be cheaper to keep the 32,000 workers on full pay but to close the mine and import better-quality coal from outside Turkey.

"It keeps people employed but it has degenerated," he said later. "This is a blatant system of exploitation. It is a savage way of exploiting human life and human health." It was a humane and logical plan that would be implemented, but too late for the victims of this, Turkey's worst mine disaster.

UKRAINE
LUHANSK REGION
March 11, 2000

The worst to date of many mine explosions in the coal region of Ukraine claimed 80 lives in the Luhansk region of the new republic on March 11, 2000.

Coal mining has always been a dangerous, filthy—and for a vast number—ill-paying job. For those who survive the dangers in the mines, black lung disease, the consequence of inhaling coal dust for years on end, claims many lives. Yet, in the richest coal mining regions of the world, underpaid and overworked miners descend thousands of feet into the earth each day, and each month some die.

The Russian Ukraine is one of those rich mining regions, and for years it was the pride of the Soviet Union, with 200 working mines employing 400,000 coal workers. When the collapse of the old order came in 1991 and the various Soviet republics became independent entities, the government of Ukraine, suddenly financially bereft, slashed subsidies to the industry. Private mine owners took over, and the low safety standards in place in Soviet times worsened considerably.

The equipment became outdated, and low maintenance on it turned it faulty. Ventilation systems, the safety valves that prevent deadly buildups of methane gas, were often either clogged or broken. Exposed wires were fires waiting to happen. Gas sensors and oxygen tanks did not work, and pit props were broken, increasing the danger of roofs collapsing on miners. Mine elevators sometimes possessed rusty wheels, and there were old and fraying ropes from which the elevators hung. Workers had to trudge through long tunnels full of heat and steam, knowing that the management of their mines was hopelessly irresponsible, and willing, if trouble should erupt, to shift the blame for the accident onto the miners themselves. As for pay, many miners lived on promises made and broken. Paychecks were, as a matter of expected reality, months late.

All of this neglect produced a string of mine explosions in the 1990s. In 1996 nearly 340 miners lost their lives in mine accidents and explosions in the Ukraine; in 1997, 260; in 1998, 360; in 1999, 280 and in 2000, 320. Some mines, such as the Zasiadko mine in Donetsk, had repeated fatal explosions. In that particular mine, methane gas and coal dust explosions claimed 63 in 1999 and 36 in 2001.

The worst of these many explosions in the new republic of Ukraine occurred on March 11, 2000, in the Barakova mine in the Luhansk region in eastern Ukraine. At 1:30 P.M. a full shift of miners was finishing up its work at a depth of 2,000 feet when suddenly an accumulation of methane gas exploded, killing some miners.

The survivors began to run to safety when a second explosion, this time of coal dust, felled more of them. Dmitri Kalitventsev, a union leader, later told reporters, "When the accumulated gas exploded, it immediately killed a lot of miners and injured many others, but if it weren't for the dust explosion they would still have had a chance to survive. The dust explosion killed off the rest."

Two hundred and seventy-seven miners were in the shaft at the time, while another shift was descending to take their place. The relief group was far enough away from the explosion to emerge unharmed, but 80 died in this, the worst explosion of many in the sorry coal mines of the Ukraine.

More promises of increased safety from the government and from mine owners were made following this tragedy, but little or nothing was done, and more explosions followed. After the second explosion in the Zasiadko mine in Donetsk on August 19, 2001, Ukrainian deputy prime minister Volodymyr Semynozhenko declared publicly, "We understand once again that we must re-equip our coal industry both technically and technologically to bring it to a proper level. It is one of the state's priorities." Told of this by a Reuters reporter, a coal miner silently lifted his arms in resigned despair. He knew what to expect.

UNITED STATES
CALIFORNIA
PORT CHICAGO
July 17, 1944

Unstable, outdated ammunition being loaded on the ammunition ships E. A. Bryan *and* Quinault Victory *exploded in Port Chicago, California, on July 17, 1944. Three hundred twenty-one were killed; hundreds were injured.*

Port Chicago, a shipping town on San Francisco Bay, received a shot of life from World War II. Moderately active before the conflict, it became one of the main staging areas for naval supply ships headed for action in the Pacific, and it employed enough of the town's residents and brought in enough more to allow Port Chicago to qualify as a boom town.

But on the night of July 17, 1944, that title took on an ironic and tragic significance. Two supply ships, the *E. A. Bryan* and the *Quinault Victory,* were being loaded, and it would take almost until midnight to complete the job. The cargo was TNT and cordite, some of it left over from World War I, some of it newly manufactured, all of it destined for the troops fighting the Japanese.

The last of the shipment had almost been loaded when suddenly an enormous, ground-splitting explosion ripped through the ships and the wharves, sending 50-foot flames into the air that could be seen 50 miles away. San Francisco, Oakland and Alameda felt an earth tremor that most thought was the beginning of a major earthquake.

The shock waves of the detonation spread for 20 miles in all directions; every building in the town of Port Chicago sustained some damage. Every one of the 1,500 residents was aware that something cataclysmic had happened on what were once the wharves of the town.

When the sun rose it became clear that the waterfront had disappeared. It had been atomized, as had the ships and the 321 men who had been on them or loading them. There were no recognizable traces of docks, ships or men left.

There was talk of sabotage. There always was after a disaster in wartime. But the official conclusion was that the ancient World War I supplies were unstable and that this was probably the reason for the explosion.

UNITED STATES
ILLINOIS
CHERRY
November 13, 1909

The November 13, 1909, explosion in the St. Paul Company mine in Cherry, Illinois, was caused by a miner's torch igniting a pile of hay. Two hundred fifty-nine miners died, and an unknown number were injured.

Until well into the 20th century, mules or horses were used to haul iron carts loaded with coal from the mines of both Europe and America. The animals were fed hay, and this hay was piled, as it was on any farm or factory, in an assigned place.

The choice of this place at the St. Paul Company mine in Cherry, Illinois, was an unwise one, to put it charitably. Feed hay was piled next to the entrance to the main shaft of the mine, and at 1:00 P.M. on November 13, 1909, the piled hay caught fire from a discarded miner's torch.

It was a small, smoldering fire at first—hardly noticeable, and that was the problem. By the time the hay had begun to burn vigorously, it was too late. Despite workers' frantic efforts to shove, pull and wheel the bales away from the mine entrance, some of the flames from the fire shot into the shaft, igniting the gases in it. A gigantic explosion rocked the entrance to the mine, and flames quickly spread down the tunnel's timbers, setting off more explosions and forcing lethal smoke and chokedamp into the bowels of the mine.

There was chaos both above and below ground. The first group of six rescuers, led by mine superintendent John Bundy, died from fumes as soon as the elevator cage they rode hit bottom.

Trying to seal off the fire, another mine superintendent, James Steele, ordered the shaft entrance sealed. It merely intensified the heat below, as a second wave of rescuers with oxygen tanks and masks attested. Lowering themselves in one of the elevators, they soon ascended again, fleeing from the unbearable heat that prevented them from entering any of the three tunnels that snaked off from the main shaft.

Five thousand people quickly gathered at the mine, most of them relatives of the 400 men who had been trapped at the time of the explosion. Some men had died at the mouth of the shaft. Andrew McFadden, ordered by his foreman to stay at the entrance and guard the mules, ignored logic and his fellow fleeing workers and followed orders. He was burned to death, along with the mules he was ordered to protect.

Finally, the shafts cooled enough to permit rescuers to penetrate the various subsidiary shafts and hunt for the living, the dead and the dying. The smoke was blinding, the chokedamp still lethal. Rescuers had to crawl on their bellies, holding on to the rails that had guided the carts to the surface in order to find the tunnels.

Rescuer William Vickers told a reporter for the *New York Times,* "At one point we passed about 65 miners sitting by the roadside, almost in a stupor. I tried to rouse them and encourage them to go on, but they seemed to have given up all hope, and did not stir. I had no time to lose and continued on, expecting to send back relief from the shaft. The sight of my doomed comrades is something that will haunt me until my dying day."

Vickers himself escaped death by minutes. Crawling back to the elevator shaft, he collapsed a few feet from it, and only the alertness of two other rescuers already in the cage saved him. The men dragged him to the elevator just as it began to ascend.

He was one of the last out alive. Thirty-six hours of rescue work turned up 170 men who were saved. Two hundred fifty-nine others died in the fire, explosion and release of toxic gases.

UNITED STATES
NEW MEXICO
DAWSON
October 22, 1913

The blast that destroyed one of the Stag Canyon Fuel Company's coal mines in Dawson, New Mexico, was caused by dynamite charges igniting coal dust in the mine. Two hundred sixty-three miners were killed, and 10 were injured in the explosion.

Dynamite was used frequently in the Stag Canyon Fuel Company's coal mines in Dawson, New Mexico. It was an efficient means of loosening coal from the veins, and it was considered a safe method in this meticulously run mine, which contained all of the most modern safety devices developed by 1913. Sprayers were located at intervals in the shafts to cut down on the presence of ignitable coal dust and chokedamp. Dynamite charges were ignited not by torch, as they were in some mines, but by electrical impulses activated from safe areas above ground. There were electric fans to assure free circulation of air and prevent the building up of pockets of methane.

Yet with all of these precautions, a series of dynamite detonations set off an uncontrolled explosion at 3:00 P.M. on October 22, 1913. Coal dust that had not been damped down by the sprinkler system was ignited; it exploded, and the force of the explosion collapsed the main shaft, trapping 284 miners.

The explosion also disabled the main air fans, and toxic gases freed by the explosions rapidly filled all of the tunnels of the mine. It would be two hours before a rescue team carrying oxygen bottles and equipped with masks could enter. The afterdamp was so thick that despite their gear, they were overcome and had to be rescued by a backup team.

Only 21 men were saved, and it would be 8:00 that night before the lethal gases would abate enough to allow men to enter and pile the dead in carts that had heretofore hauled coal from the mine. Two hundred sixty-three miners, almost the entire shift, perished in the explosion or its aftermath.

UNITED STATES
PENNSYLVANIA
JACOB'S CREEK
December 19, 1907

No cause was ever discovered for the explosion of the Darr Mine of the Pittsburgh Coal Company in Jacob's Creek, Pennsylvania, on December 19, 1907. Two hundred thirty-nine miners were killed, and one was injured.

The Darr Mine of the Pittsburgh Coal Company in Jacob's Creek, Pennsylvania, was located in the side of a mountain separated by a gorge from Jacob's Creek and the nearby Youghiogheny River. From its entrance it plunged nearly two miles into the earth. That entrance was almost inaccessible. The only way miners could get to it was via a so-called sky ferry, which was a wooden bucket on a winch that lifted six men at a time from Jacob's Creek to the mine entrance.

The sheer laboriousness of even getting there made the Darr Mine an unpopular place to work, and so the 240-man shift that was working on the morning of December 19, 1907, was composed mainly of Greek and Italian immigrants. Life was tempestuous and sad in the nearby mining community of Jacob's Creek. Not only were there bad feelings between the Greeks and Italians, which sometimes erupted into confrontations that bayonet-wielding law enforcers had to break up, there was also the pervasive fear that hung over everyone. The mine was rife with pockets of chokedamp.

Mrs. John Campbell, the wife of the mine foreman, later interviewed by local journalists, admitted, "I have for a long time feared an explosion in the mine, for I knew it was gaseous. My husband and I had talked of it, and he often referred to the gas in the mine."

That explosion finally came at 11:30 A.M. on December 19, 1907, when a 240-man shift was at work two miles deep in the mine. "About 11:30 o'clock," continued Mrs. Campbell, "there was a loud report and the dishes in my cupboard and on the table were rattled and knocked out of place, while the glass in the windows was shattered. . . . I knew what had happened."

Families and supervisors rushed to the site, but the entrance had collapsed. A huge cloud of black smoke drifted out of it and clung to the countryside. "My husband was about due for his dinner when the loud report came," said Mrs. Campbell, "and I looked out the back door toward a manway from the mine, through which he always came to his meals. Instead of my husband I saw a great cloud of dust and smoke pouring out of the mouth of the mine through the manway. It floated upward and disappeared across the river."

Joseph Mapleton, who had been near a side entry, was the only miner to survive the blast. Two hundred thirty-nine of his fellow workers were either blown up, crushed or asphyxiated deep beneath the earth in an unsafe coal mine a long way from their birthplaces.

UNITED STATES
PENNSYLVANIA
MATHER
May 19, 1928

Gas ignition from an electric locomotive caused the explosion in the Mather shafts of the Pittsburgh Coal Company on May 19, 1928. One hundred ninety-five died; six were injured.

The practice of keeping or releasing canaries in mine shafts to detect noxious or lethal gases was carried to one of its most extreme limits on the evening of May 19, 1928, in Mather, Pennsylvania. One hundred birds were released as part of a rescue effort after the 4:07 P.M. explosion in the Mather coal mine.

One of the Pittsburgh Coal Company's most prolific collieries, the Mather shafts yielded one million tons of anthracite per year and employed 750 men around the clock, 365 days a year. The explosion, caused by the igniting of methane by an electrical arc from a storage-battery locomotive that was later defined by government investigators as "non-permissible," erupted just as shifts were changing. Approximately 400 men were either entering or leaving the main shaft at the time of the detonation; 209 were actually in the mine and directly affected by it.

And affected they were, instantly: 193 died immediately, either from the blast, the collapsing mine shafts or the afterdamp. Two more men died from the effects of the disaster, thus bringing the total to 195. Only eight were saved in a rescue attempt that went on for three days. Of these eight, six were injured but recovered.

UNITED STATES
PENNSYLVANIA
PLYMOUTH
September 6, 1869

Sparks from an underground furnace ignited support timbers, which caused an explosion that collapsed adjacent tunnels in the Lackawanna and Western Railroad's Avondale coal mine in Plymouth, Pennsylvania, on September 6, 1869. One hundred ten miners were killed; 30 were injured.

A graphic recreation of the first discovery of the victims of the explosion in the Lackawanna and Western Railroad's Avondale coal mine in Plymouth, Pennsylvania, on September 6, 1869. (Frank Leslie's Illustrated Newspaper)

Until the end of the 19th century, when electricity made circulating fans possible, it was a standard but hazardous practice to keep a furnace burning, night and day, in the depths of a mine. The theory was that the heat would keep a steady circulation of air that would move the noxious gases and prevent them from accumulating in ignitable pockets.

It was a two-edged sword, as the disaster of September 6, 1869, in Plymouth, Pennsylvania, proved. The Avondale coal mine, owned and operated by the Lackawanna and Western Railroad, contained just such a furnace, and on that morning some sparks from it ignited the support timbers of the shaft. This fire was enough to cause a small explosion, which in turn collapsed more of the adjacent tunnels.

According to *Harper's Weekly,* "Whatever fresh air there was in the mine went to feed the fierce flame, while the sulfurous gases, having no longer an outlet, were forced back into the chambers and galleries of the colliery."

Rescuers rushed to the scene, first lowering a dog with a lantern hanging from his neck into the shaft to test for an abundance of lethal gases. The dog was hauled back up alive, and so the rescue work proceeded. Eighty men were found alive. One hundred ten died of asphyxiation.

The horrendous explosion of the steamship* Sultana *near Memphis, Tennessee, on April 27, 1865. An appalling 1,547 died in one of the worst tragedies in American history. *(Frank Leslie's Illustrated Newspaper)*

UNITED STATES
TENNESSEE
MEMPHIS
April 27, 1865

One of the worst tragedies in American history, the boiler explosion aboard the steamboat Sultana *in Memphis, Tennessee, on April 27, 1865, was caused by human negligence, overloading and an overstoked boiler. Officially, 1,547 deaths were recorded, but this figure is generally thought by historians to be too low; hundreds were injured.*

The month of April 1865 was an eventful one. President Abraham Lincoln was assassinated on April 14. Vice president Andrew Johnson assumed the Presidency on April 15. General William Sherman accepted the surrender of General Joseph E. Johnston on April 26, thus bringing the armed resistance of the Confederacy to an end. On the same day, John Wilkes Booth was shot to death in a Virginia barn.

Thus, when the steamboat *Sultana* blew up in the Mississippi River just north of Memphis, Tennessee, at approximately 2:00 A.M. on April 27, 1865, the horrendous tragedy, one of the worst in U.S. history, went largely unreported and unrecorded. In fact, it would be another 30 years before Congress would enact legislation designed to prevent the sort of disaster that occurred aboard the *Sultana* that night.

Steamboat travel always involved one major hazard, that of an overworked, exploding boiler. From the beginnings of steamboat travel very early in the 19th century until 1850, there had been 185 steamboat explosions resulting in the deaths of 1,400 people.

In just minutes on that one night, at least 1,500 people would meet their deaths in the spectacular explosion of the *Sultana*'s boilers, and the cause would be a compound of personal and political misjudgment.

In 1865 great numbers of Union prisoners were released from the Confederate prisoner of war camps at Cahaba and Andersonville. Vicksburg, Mississippi, was the loading point for these emaciated survivors, who wanted nothing more than swift passage northward to Cairo, Illinois, the debarkation point from which they could then go home.

Thousands of them boarded upriver steamers in the early spring of 1865 bound for safety and familiar sights. By April, however, there were ugly and probably founded rumors that the government was giving all of this lucrative business to one steamboat company in return for a kickback of one dollar a passenger.

Anxious to scotch the rumors before they reached the public and official investigatory agencies, government officials at Vicksburg welcomed the arrival of the *Sultana,* which belonged to a rival company. Two years old, the *Sultana* was not a particularly impressive or big boat, weighing in at approximately 1,700 tons. It had a legal load limit of 376 passengers and crew members. Anything exceeding this would demand forced firing of its four tubular boilers, an extremely dangerous practice.

But neither the owners nor the government seemed to care about the regulations or the hazards that spring. By 2:00 A.M. on April 26, after it had ceased loading its war prisoners and cargo and repaired a faulty steam line leading from one of its boilers, the *Sultana* pulled away from Vicksburg with 2,300 to 2,500 veterans, 75 to 100 civilian passengers and a crew of 80. (The figures concerning both passengers and fatalities have remained approximate, since no accurate records were kept.)

Low in the water and lumbering against the current, the *Sultana* carried the greatest load any steamboat had ever carried on the Mississippi when it left Vicksburg. It would take an unprecedented 17 hours for it to reach Memphis, where it docked shortly after 7:00 P.M. on the 26th.

The coal bins had been almost emptied by the time the *Sultana* reached Memphis, and once it had unloaded 100 hogsheads of sugar, it was taken to the Arkansas side of the river to pick up another 1,000 bushels of coal. Once the ship had been loaded, stokers were ordered to "pour on the coal, and keep this thing moving." A plausible rumor ran along the river then and afterward that the chief engineer wired the safety valves in place so that every available bit of steam was available to drive the side wheels of the *Sultana*.

Midnight came and went, and the boat beat against the current, heading steadily northward. At 2:00 A.M. on the 27th, while most of its passengers slept, the number three boiler exploded with a cataclysmic roar. Hot metal ripped through the ship like white-hot knives. Within minutes, two more boilers exploded, ripping half the steamer apart, collapsing the various decks and crushing those hapless passengers or crewmen who had not been either scalded to death or ripped apart by the metal pieces of boilers.

Some passengers were blown into the water and survived. Others who got through the initial explosion were trapped by the roaring fire that followed it. A small group huddled on the bow of the boat and were quickly pushed into the water by the advancing flames and the collapse of the *Sultana's* two smokestacks.

Steamboats in the area, hearing the explosion, rushed to the rescue, as did the Union gunboat *Grosbeak*. They found a scene of unbelievable carnage. Having survived the hell of Andersonville, hundreds of homeward bound men were blown apart or burned by a disaster that, in retrospect, was nearly inevitable.

A later investigation failed to either fix the blame for the tragedy or accurately estimate the casualty figures. Trained observers guessed the number of dead to be 1,900, a total generally considered to be too high. A U.S. Army board of review released a figure of 1,238, a total obviously designed to minimize the tragedy. The estimate by customs service officials at Memphis of 1,547 has been the generally agreed on, if inconclusive, one, which makes the death toll of the explosion of the *Sultana* 30 more than that of the much more celebrated sinking of the *Titanic* (see p. 263).

UNITED STATES
TEXAS
NEW LONDON
March 18, 1937

The cataclysmic explosion of natural gas that destroyed the New London, Texas, Consolidated School on March 18, 1937 was caused by "wet" gas, used as an economy measure by the school system and ignited by a spark. Two hundred ninety-seven students and teachers were killed and 437 were injured in the worst accident in the history of the American public school system.

On a blackboard that survived the worst disaster in the history of public education in the United States and the worst explosion in America in terms of deaths since the burning of the *General Slocum* in 1904 (see MARITIME DISASTERS, p. 292) was scrawled, in the writing of an elementary school student in the New London, Texas, school: "Oil and natural gas are East Texas' greatest natural gifts. Without them, this school would not be here and none of us would be learning our lessons."

Ironically, the New London school *was* no longer standing on the night of March 18, 1937, and 297 students and faculty were indeed no longer there. They had been killed by an explosion caused by a combination of the natural gas that was East Texas's greatest natural gift and a parsimonious decision by the local school board.

New London, Texas, was not an underprivileged community, nor were its educational facilities wanting. On a clear day, you could see 10,000 oil derricks through most of the classroom windows of its Consolidated School, 11 of them on the school grounds themselves. The 1,500 students school facility was the most modern available, built at a cost of $1 million in 1937 depression dollars.

But in January of that year the school board made a mysterious and tragically unwise decision. Until that time, Union Gas Company had sold the New London school board a natural gas mixed with a pungent odorant. It was safe; it was a so-called dry gas from which impurities had been removed, and it was also cheap. The fuel bills for the school amounted to only $250 to $350 per month.

But the school board decided to economize even more by tapping into a pipeline that carried waste gas from a plant operated by the Parade Oil Company. This was a common practice in East Texas in towns close to oil fields. Most of the time it caused no problems. But this waste gas was notoriously unstable. "Raw" or "wet" gas had a heating and ignition point that varied widely from hour to hour because of the variety of impurities in it. It was used in many of the homes in New London, and the board members decided it could be used in the school, too.

So it canceled its contract with the Union Gas Company and authorized a school janitor to install a connection that would pipe waste gas through the school's heating system. In testimony given after the tragedy that followed, University of Texas expert E. P. Schoch stated that if just one of the school's main lines was accidentally left flowing for half a day, "the saturation point" would be reached, creating conditions for a potential explosion.

At 3:05 P.M. on Thursday, March 18, 1937, just 10 minutes before dismissal time for 694 high school students and 40 teachers in the Consolidated School, there was a horrific explosion that blew the roof off the school, caved in the walls, shook the ground for 40 miles and buried practically everyone in the building under tons of rubble and steel girders. Several small explosions followed.

Rescue squads, many of them oil workers and many of them parents of children buried in the rubble, rushed to the sickening scene. Men plunged immediately into the debris, and some emerged with terrified but safe children. One rescuer, Don Nelson, came upon a heavy bookcase, tilted against a wall. In the tentlike space formed by it, he found 10 children, frightened beyond belief but safe.

But these were the exceptions. The number of dead was first reported by the Associated Press as 455 children, "crumpled under steel and concrete or squeezed bloodless by the blast." But fortunately this theatrical first report was revised downward to 297 students and teachers. Some were killed so instantaneously that they still had smiles on their faces. Others were mutilated beyond recognition. Their bodies were lined up in the school yard; a consignment of 200 coffins was sent for from Dallas. By the time they arrived, officials were forced to wire for more. Bodies were placed on trucks, and the trucks moved in a steady convoy to improvised morgues.

Meanwhile, the injured—437 of them—were taken to overflowing first-aid stations set up in New London and neighboring Tyler, Overton, Kilgore and Henderson.

Outrage, despair and rage forced immediate hearings into the cause of the accident, which was determined to be an accumulation of wet gas ignited by a spark. The specific spark was never identified, but it was theorized that it could have been from either a light switch or a buildup of static electricity.

A positive result of the New London schoolhouse tragedy was the immediate adoption by oil-producing states of a standard law that prohibited oil companies from allowing tapping of their unprocessed "wet" gas. By law, it must now be burned at site.

UNITED STATES
TEXAS
TEXAS CITY
April 16–18, 1947

Human error caused the worst harbor explosion in American history, the explosion of the French ship Grandcamp *in Texas City Harbor, Texas, on April 16, 1947. Seven hundred fifty-two were killed and 3,000 aboard neighboring ships and onshore were injured.*

The worst harbor explosion in American history stretched itself over a three-day period and destroyed fully one-third of Texas City, Texas, 10 miles across Galveston Bay from Galveston.

"For God's sake, send the Red Cross—thousands are dying!" yelled an operator into one of the few remaining lines linking Texas City with the world after the explosion. She was not exaggerating, and the worst of it is that the bizarre circumstances leading up to the initial blast indicate that human misjudgment might have been the overriding cause.

The French ship *Grandcamp*, a liberty ship used during World War II to move supplies from manufacturing ports to theaters of war, was loaded with, among other cargo, fertilizer destined for French farms. The fertilizer was composed of nitrate and ammonia—two components of, among other products, TNT. The fertilizer was in the number four cargo hold of the *Grandcamp*. Next to it, in number five, investigators were told that there was ammunition, but this was either untrue or a military secret. It was neither confirmed nor denied by the government, and in wartime reporters were not given to pursuing such details.

At 8:00 A.M. on April 16, 1947, a smoldering fire was discovered in some of the sacks of fertilizer by the ship's carpenter, Julian Gueril. He attempted to extinguish it but could not.

Shortly thereafter the ship's captain, Charles de Guillebon, ordered the hatch closed and the steam jets turned on. It was an accepted way, under ordinary circumstances, of starving a fire. Under the circumstances of April 16, 1947, it was an unfortunate decision on the part of the ship's captain. Ammonium nitrate decomposes, often violently, at 350 degrees Fahrenheit. The steam easily raised the temperature in the hold to at least that in a very short period of time.

The fertilizer continued to burn, and this prompted the captain to summon the Texas City fire department, which arrived swiftly—within 10 minutes. The water from its hoses seemed to make little impression on the billows of black smoke that poured continuously from the *Grandcamp*'s hold and attracted hundreds of curious onlookers, who now lined the docks.

And then, at exactly 9:12 A.M., the *Grandcamp* exploded, instantly killing everyone aboard and everyone on the nearby dock. It was an awesome, thundering blast, heard as far away as 160 miles and containing the force of a small atomic bomb. Its devastation would by no means be confined to the ship and the dock area, where 32 sailors and 227 firemen and observers were instantly killed, chopped up into unidentifiable fragments.

An area of the city 20 blocks by 12 was flattened into smoldering desert in an instant. A one-ton piece of the ship's propeller was flung 13,000 feet and embedded itself in the driveway of a private home. In one of the more incredible side effects of the blast, two private planes, flying 1,000 feet above the explosion site, were blasted out of the air as surely as if they had been hit by anti-aircraft fire.

But that was still only part of it. The Monsanto Company maintained a huge complex in Texas City, and its storage tanks, containing styrene products, were a mere 700 yards away. They went up in flames and black smoke. Farther down the dock, the oil storage tanks of the Humble, Stone and Republic oil companies erupted in flames hundreds of feet high.

The entire Monsanto plant began to buckle and crumble, burying those workers under the debris who had not either escaped or been blown through windows and doors at the moment of the explosion. All in all, 3,300 buildings in Texas City would be completely flattened. The water

Heavy clouds of acrid smoke hang over Texas City, Texas, following the cataclysmic explosion of April 16, 1947. (American Red Cross)

system would erupt, preventing the fire department from controlling the subsequent fires. A dockside ghetto simply disappeared. The Texas Terminal Railway building went up in flames. An elementary school with 900 students in attendance lost all of its windows in the blast, and almost every one of the students was slashed by flying glass. Every electric line and all but one telephone line in the city was knocked down.

The streets were littered with chunks of human flesh, and this was the sight that greeted Red Cross units that sped to the scene from San Antonio, Galveston, Port Arthur, Houston, Dallas and Beaumont. Buildings were crushed. The shock waves of the blast even reached residences several miles away, where Mrs. Tena Lide was reported to have been lifted up and tossed out of a second-story window by them.

Rescuers did what they could in the unchecked inferno of fire that engulfed Texas City. The city hall was turned into a hospital. "I carried out pieces of bodies all afternoon," a rescue worker told the *New York Times,* "[but] I don't believe they added up to two people."

Meanwhile, back at the dock a secondary horror was building. The *Highflyer,* docked near the *Grandcamp* and loaded with munitions, had caught fire and was in imminent danger of exploding. Two versions of its last moments survive: One, reported by the Associated Press, had it locked hopelessly with a nearby ship, the *Wilson B. Keene.* In this version it became impossible for tugs to haul it out to sea. In another version, Deputy Mayor John Hill was quoted as saying, "We asked the tugboats to pull the *Highflyer* out to midstream. They refused. They heard she was carrying ammunition."

For whatever reason, the *Highflyer* remained, blazing, at dockside, and at 1:10 P.M. on April 17 it too blew up, killing most of its crew and hundreds more in the city. Texas City burned for the rest of the night. Most of its surviving residents, like the survivors of the Chicago fire, prepared to evacuate.

Through that night and into the next day, rescue teams continued to comb the city for survivors. McGar's Garage, the city's largest repair shop, became its largest morgue. Five hundred fifty-two bodies or body parts were laid out on its floor. Three thousand other residents were injured. Two hundred were reported missing and thought to have been atomized by the blast. The financial loss would amount to $100 million.

UNITED STATES
UTAH
CASTLE GATE
March 8, 1924

No specific source was discovered for the ignition of accumulated gases that caused the explosion in the Utah Fuel Company's coal mine in Castle Gate, Utah, on March 8, 1924. One hundred seventy-three miners were killed; 30 were injured.

National newspapers failed to report the details of the sudden explosion that rocked the Utah Fuel Company's coal mine on the outskirts of Castle Gate, Utah, on March 8, 1924.

A mine that was widely touted as safe, with modern ventilation and anti–gas-lock equipment, it was not thought to be a candidate for a disaster. But that is what occurred. Accumulated gases that the ventilating system apparently failed to disperse ignited—from what source it was never determined—and the mine's main shaft exploded and collapsed. A full shift was at work at the time, and 173 died in the blast.

UNITED STATES
UTAH
SCOFIELD
May 1, 1900

Human misjudgment was responsible for the storage of blasting powder underground in the Scofield, Utah, coal works. The powder ignited on May 1, 1900, causing an explosion that killed 200 miners.

A particularly grisly explosion in the number four shaft of the Scofield, Utah, coal works was caused by neither pockets of gas nor faulty ventilation. This time, it was just plain carelessness.

Thirty kegs of hybrid "blasting powder," as it was colorfully called at the time, were stored in the number four shaft, an act of monumental daring or stupidity, depending on one's point of view. At 10:25 A.M. on May 1, 1900, in the middle of a shift, one of the kegs was ignited. The cause was never determined, for there was not enough of either shaft or men left to be able to piece together a scenario. All 30 kegs detonated with a terrible roar, blowing apart that portion of the number four shaft, atomizing all who were near it and collapsing the walls and ceilings of the tunnel on the remainder of the 140 men at work in that shaft.

Unfortunately, the number one shaft intersected with number four near the keg storage area, and the rolling afterdamp caused by the mass unearthing of bituminous coal floated a lethal cloud into shaft number one, asphyxiating 60 more miners. All in all, 200 men would lose their lives to reckless disregard of safety and sanity on the morning of May 1, 1900, in Scofield, Utah.

UNITED STATES
VIRGINIA
POCAHONTAS
March 13, 1884

A combination of overuse of blasting powder and faulty ventilation caused the explosion in the Laurel Mine of the Southwest Virginia Improvement Company in Pocahontas, Virginia, on March 13, 1884. One hundred twelve miners were killed; two were injured.

The worst-run mines in America in the 19th century were frequently worked by unsuspecting and hardworking immigrants. In Jacob's Creek, Pennsylvania (see p. 185), they were Italians and Greeks. In Pocahontas, Virginia, they were Hungarians and Germans. The Virginia mine carried impressive credentials. It was called the Laurel Mine, and it was owned by the Western Railroad Company and operated by the Southwest Virginia Improvement Company, which was apparently dedicated to improving conditions in Virginia other than the lives of its coal miners.

The mine was a monument to carelessness. There were no sprinkler systems in the shafts to cut down on the coal

A young wife discovers the body of her husband, blown from the mouth of the Laurel Mine of the Southwest Virginia Improvement Company in Pocahontas, Virginia, on March 13, 1884. One hundred twelve were killed and two were injured in the blast. *(Frank Leslie's Illustrated Newspaper)*

dust and its concomitant destroying gases and the incidence of black lung disease. As in the Scofield, Utah, mine (see previous entry), the blasting powder used to dislodge coal from stubborn veins was not stored above ground but deep in the bowels of the mine, where its ignition would cause certain disaster. In addition, the mine had a reputation for being heavy-handed with blasting powder, using more than was necessary and thus endangering safety every time the powder was used. Pockets of lethal gas were everywhere. Ventilation in the mine was accomplished through one fan at the mouth of the shaft—a woefully inadequate proviso that further threatened the well-being of the workers every minute they were underground.

Thus, the accident that had been waiting to happen occurred at 1:00 A.M. on March 13, 1884. The customarily extravagant amount of blasting powder was distributed deep in the main shaft. It was detonated, and suddenly, with an ear-decimating roar, the entire shaft exploded, spewing debris and pieces of miners throughout the Virginia countryside, flattening the shacks built by the miners and their families near the shaft and generally turning the entire site into an inferno.

Like the cork in a champagne bottle, a steam engine and its cars of coal soared out of the shaft's mouth, catapulting its engineer 100 yards to his death and running down two more miners at the shaft's mouth. The shock waves flattened homes and scorched the forest. "The very trees on the mountains, which have withstood the beating storms of ages," wrote the reporter for the *New York Times,* "were shriveled, torn and blasted, their branches scattered in every conceivable direction. Steel, wood and flesh were blown as far as a mile. Even the coal dust was blown over the mountain, and covered the earth on the opposite side to a half of an inch."

As for the miners who were in the mine at the time of the explosion: They were no more. *Frank Leslie's Illustrated Newspaper* later reported that there were "fragments of bodies lodged in tree-tops and on the roofs of houses and sheds."

In every shaft, fire emitting from every exit prevented rescuers from entering the decimated mine. Finally, later that day, Colonel George Dodds, a mining engineer, ordered all of the entrances and exits sealed. Fire hoses were hooked up, and the mine was flooded. Two weeks later, when the mine had cooled enough to allow rescue teams to enter it, squads descended through the acrid and devastated ruins. They found 112 dead miners. Those who had not been killed in the blast were drowned by their rescuers.

UNITED STATES
WEST VIRGINIA
ECCLES
April 28, 1914

A dynamite blast that ignited gases caused the explosion in the New River Colliers Company mine in Eccles, West Virginia, on April 28, 1914. One hundred seventy-nine were killed; 51 were injured.

The New River Colliers Company, owned by the Guggenheim family, maintained a mine in Eccles, West Virginia. At that time the laws of West Virginia allowed children to work in mines, and five of the miners who would die on the afternoon of April 28, 1914, were 15 years of age.

The disaster occurred at the 500-foot level of the mine's number five shaft at exactly 2:10 P.M., while full shifts were working in each of its six tunnels. A dynamite charge was being set in the shaft at that time, and the blasting fuse ignited a pocket of gas that had accumulated from previous controlled explosions.

This one was far from controlled. It blew apart the shaft, sending enormous tongues of flame all the way to the surface. Every one of the 172 men working in the shaft was killed on the spot. Others were so disfigured that they could only be identified by the brass checks that were distributed to each miner as he entered the shaft opening at the beginning of his shift.

Six miners working in the number six shaft, 250 feet above the blast, were rocked by the explosion but initially unhurt. They attempted to escape by placing handkerchiefs over their faces, but the rising afterdamp overtook them before they could get to a safe place, and all six died of asphyxiation.

Rescue efforts began immediately, and 66 men were retrieved from the number six shaft. One of these died later, bringing the death toll to 179.

UNITED STATES
WEST VIRGINIA
MONONGAH
December 6, 1907

A runaway rail car severed electrical cable that in turn ignited gases in the Monongah mine in Monongah, West Virginia, on December 6, 1907. The resultant explosion killed 362 miners and injured four.

Ironically, in both the Eccles (see previous entry) and Monongah mine disasters, a life insurance salesman was present in the mine hawking his wares of continuance for the families of miners when these two explosions occurred.

The Monongah catastrophe occurred at 10:28 A.M. on the morning of December 6, 1907, while a full shift of 366 men and one insurance agent were working in the intersecting number six and eight shafts. Shortly before this, a train of coal cars being guided to the surface by a young miner became uncoupled. The cars behind the broken coupling rolled backward down the steep slope into the number six shaft. The miner ran ahead, hoping to cut the electric current before disaster struck. But the switches

Wreckage in the Monongah Mine in Monongah, West Virginia, on December 6, 1907. Expectant families and rescuers line the hill that encloses the mine. Three hundred sixty-two miners died in an explosion caused by a runaway rail car. (Frank Leslie's Illustrated Newspaper)

were too far away. The cars slammed into a wall, severing electric cables that sent up showers of sparks that in turn ignited trapped methane in the tunnel. The enormous blast set off by the ignited gas blew the young train operator out of the mouth of the tunnel.

He survived, but 362 of the 366 men working below the blast would not be as lucky. Every one of them was killed either by the force of the explosion or the falling timbers and collapsing ceilings and walls of the tunnels in which they were trapped.

The 3,000 residents of the nearby coal miners' shanty village rushed to the mine to wait out the laborious and heartbreaking rescue efforts. Several hours after the blast, four men, bleeding profusely and nearly dead from breathing the afterdamp, crawled through an outcrop at the top of the number eight shaft. They would be the only survivors.

It would be five full days before rescue parties would be able to reach the grisly sight of the blast and its victims. Three hundred sixty-two miners and the salesman were dead in one of the worst mine explosions in U.S. history.

UNITED STATES
WYOMING
HANNA
June 30, 1903

Blasting powder ignited gas that caused the explosion in the Union Pacific Railroad's Hanna, Wyoming, mine on June 30, 1903. One hundred sixty-nine died; 27 were injured.

The Hanna, Wyoming, coal mine was nothing of which the Union Pacific Railroad could be proud. It was, in fact, a carefully kept secret, its grim and dangerous presence known only among American miners, most of whom would not work it.

Thus, as in the Eccles and Monongah mines (see p. 192), the company employed recent immigrants at low wages and under inhuman conditions. In Hannah the vast majority of the immigrant coalminers were Finnish—in fact, the mining village had a distinctly Finnish flavor to it—but there were also groups of Polish and Chinese immigrants and some American blacks.

There were practically no safety precautions taken in the mine. Pockets of methane abounded, and every charge of blasting powder that was ignited was a potential harbinger of calamity.

At 10:30 A.M. on June 30, 1903, blasting powder used to loosen a vein of coal one and a half miles deep in the mine's one shaft ignited a gas pocket and set off an enormous explosion. Two seconds later, in a chain reaction, a second explosion occurred, rippling along the pockets of chokedamp and afterdamp, collapsing huge sections of the shaft and belching forth enormous orange flames that reached above the surface of the mine opening.

All but one of the manways into the mine were clamped shut by the timbers collapsing around them. It was through this manway that 46 men managed to escape. Twenty of these were pulled out by a black miner, William Christian, who repeatedly entered the superheated, choking interior of the mine to haul out the injured until he himself collapsed.

One hundred sixty-nine men were killed in the dual explosions of June 30, but that would not be the end of the story. The mine would continue to operate on half shifts, even though no precautions were made by management following the 1903 tragedy. Just five years later, on March 28, 1908, another explosion roared through the mine, killing fifty-nine miners. Twenty-seven bodies were found and retrieved, but 32 others were sealed forever in the mine when it was finally closed down for good.

USSR
UFA
June 3, 1989

A leak in a liquefied petroleum gas pipeline was ignited by a spark from a passing passenger train near Ufa, USSR, on June 3, 1989. The explosion killed 190 and injured 720. Another 270 were presumed dead.

Early in the morning of June 3, 1989, partway between the two Soviet cities of Asha and Ufa in the Ural Mountains of the USSR, a liquefied petroleum gas pipeline erupted. It was a Sunday morning, and the gas, which was being transferred from oil fields in Nizhnevartovsk to refineries in Ufa, was being monitored, presumably by a skeleton crew. Pressure gauges undoubtedly showed a drop in pressure, an indication of a leak. But for some unexplainable reason, instead of investigating the leak, the pipeline operators on duty simply turned up the pumps, thus feeding a mixture of propane, butane and benzene vapors into a ravine leading to a nearby railroad. By the time the vapors had settled into the valley surrounding the train tracks, they were composed mainly of methane, the highly volatile gas responsible for a multitude of mine explosions.

Shortly after this, two trains traveling in opposite directions between the Siberian city of Novosibirsk and the Black Sea town of Adler passed each other in that ravine. The trains, loaded with vacationers, were not scheduled to pass at that particular point at that particular moment, but one was behind schedule, and as fate would have it the two were parallel when they entered the valley. The heavy aroma of gas, hanging like a fog to the level of the train windows, became sickeningly apparent to the engineers of both trains as they sped through the pass.

Suddenly, a spark from one of the trains ignited the gas, which exploded with a deafening roar and bright orange flashes of flame. Its force—that of 10,000 tons of TNT—felled every tree within a three-mile radius and blew both locomotives and the 38 cars of the two trains completely off the tracks. Pieces of metal, smashed windows and fragments of bodies were blown in several directions. A metal-melting fire followed instantly, incinerating the surviving passengers before they could extricate themselves from the mangled coaches.

Speaking later to Tass, the Soviet news agency, a Soviet army officer noted that he had been standing at an open window when he noticed the acrid, petroleum smell coming from the gas leak. "I sensed that something must be wrong," he said, "but before I could do anything there was a glow and then a thunderous explosion." The officer escaped from the burning car through a broken window.

Rescue squads immediately poured into the region from both Ufa and Asha, and surgeons, burn specialists and medical supplies were airlifted from Moscow throughout the day and night. The final casualty count was appalling: 190 were known dead, at least 270 were missing and presumed dead and 720 were injured seriously enough to be hospitalized.

YUGOSLAVIA
KAKANJ
June 7, 1965

Gas ignited by a blasting fuse set off an explosion in a mine in Kakanj, Yugoslavia, on June 7, 1965. One hundred twenty-eight were killed; 41 were injured.

Yugoslavia's worst mine disaster occurred in the small town of Kakanj, just outside Sarajevo, on June 7, 1965. By the middle of the 20th century, modern safety precautions had, in practically all countries, reduced the risk of mine explosions dramatically. But the mine at Kakanj was rife with violations of the government's safety regulations, and on June 7, 1965, the miners working the day shift at the Kakanj mine paid the price. Methane, ignited by a blasting fuse, exploded with a thunderous roar, collapsing the mine's main shaft and setting fire to several adjacent ones.

Fortunately, modern rescue apparatus allowed the rescue parties to haul out small groups of dazed and injured workers. One hundred twenty-eight died.

Six months after the blast, a Belgrade court sentenced six officials of the Kakanj mine to seven and a half years at hard labor for gross negligence and malfeasance.

FIRES

THE WORST RECORDED FIRES

* Detailed in text

Austria
- Kaprun
  - * Cable car (2000)
- Vienna
  - * Ring Theatre (1881)

Bangladesh
- Shibpur
  - * Garment factory (2000)

Belgium
- Brussels
  - * L'Innovation department store (1967)

Brazil
- Niteroi
  - * Gran Circo Norte-Americano (1961)
- São Paulo
  - * Joelma building (1974)

Canada
- Quebec
  - Montreal
    - * Laurier Palace Theatre (1927)
- New Brunswick
  - Forest fire (1825)
- Ontario
  - Forest fire (1916)

Carthage
- Sacking of Carthage (146 B.C.)

Chile
- Santiago
  - Jesuit church fire (1863)
  - Braden copper mine fire (1945)

China
- Chongqing
  - Burning of the city (1949)
- Chow-t'sun
  - Fire outside city (1924)
- Dandong
  - Movie theater fire (1937)
- Fuxin
  - * Dance hall (1994)
- Guangzhou
  - Theater fire (1845)
- Hankou
  - Fire on the docks (1947)
- Karamay
  - * Movie theater (1994)
- Luoyang
  - * Disco (2000)
- Shenzhen
  - * Toy factory (1993)
- Tangshan
  - * Department store (1993)
- Tuliuchen
  - Theater fire (1936)
- Wuchou
  - Tea District fire (1930)

Colombia
- Bogotá
  - * El Almacén Vida department store (1958)

Egypt
- Cairo
  - Burning of much of the city (1824)

France
- Paris
  - * Opéra Comique (1887)
  - * Paris charity bazaar (1897)
- St. Laurent du Pont
  - * Cinq-Sept Club disco (1970)

France/Italy
- * Mont Blanc Tunnel (1999)

Germany
- Dresden
  - * Firebombing of city (1945)

Great Britain
- England
  - Exeter
    - * Exeter Theatre (1887)
  - London
    - City burned (1212)
    - * Great Fire of London (1666)
    - * Underground (1988)

Guatemala
- Guatemala City
  - * Guatemala City Insane Asylum (1960)

Indonesia
- Continuing forest fires

Iran
- Abadan
  - * Movie theater (1978)

Jamaica
- Kingston
  - * Eventide Nursing Home (1980)

Japan
- Hakodate
  - Much of the city burned (1934)
- Osaka
  - * Playtown Cabaret (1972)
- Tokyo
  - * Gambling club (2001)
- Yokohama
  - Catholic Old Women's Home (1955)

Mesopotamia
- Babylon
  - City burned (538 B.C.)

Mexico
- Acapulco
  - Flores Theater (1909)

Puerto Rico
- San Juan
  - * Dupont Plaza Hotel (1986)

Rome
- * Burning of city by Nero (64)

Russia
- Berditschoft
  - Circus Ferroni (1883)
- Igolkino
  - * Factory fire (1929)
- Moscow
  - Burning of city (1570)

Saudi Arabia
- Mina
  - Tent city burned (1975)

South Korea
- Seoul
  - * Taeyunkak Hotel (1971)

Spain
- Madrid
  - * Novedades Theater (1928)

Syria
- Amude
  - * Movie theater (1960)

Thailand
Nakhon Pathom
* Doll factory (1993)
Turkey
Constantinople
Burning of 12,000 houses (1729)
* Fire originating in Armenian district (1870)
Smyrna
Burning of city (1922)
United States
Connecticut
Hartford
* Ringling Brothers, Barnum & Bailey Circus (1944)
Georgia
Atlanta
* Winecoff Hotel (1946)
Illinois
Chicago
* Great Chicago Fire (1871)
* Iroquois Theater (1903)
* Our Lady of the Angels School (1958)
Massachusetts
Boston
* Cocoanut Grove Night Club (1942)
Minnesota
* Forest fire (1918)
Hinckley
* Forest fire (1894)
Mississippi
Natchez
* Rhythm Night Club (1940)
New Jersey
Coast (See MARITIME DISASTERS)
Hoboken
* Docks (1900)
New York
Bronx
* Happy Land Social Club (1990)
Brooklyn
* Brooklyn Theatre (1876)
New York
* Windsor Hotel (1899)
General Slocum (See MARITIME DISASTERS)
* Triangle Shirtwaist Factory (1911)
North Carolina
Hamlet
Chicken processing plant (1991)
Ohio
Collinwood
* Lakeview School (1908)
Columbus
* Ohio State Penitentiary (1930)
Pennsylvania
Boyertown
Rhoades Theater (1908)
Wisconsin
Peshtigo
* Forest fire (1871)

CHRONOLOGY

* Detailed in text

538 B.C.
Babylon, Mesopotamia; Burning of Babylon

146 B.C.
Carthage; Sacking of Carthage

64 A.D.
July 19
* Rome; Burning of Rome by Nero

1212
London, England; Burning of city

1570
Moscow, Russia; Burning of city

1666
September 2–6
* London, England; Great Fire of London

1729
Constantinople, Turkey; Burning of 12,000 houses

1824
March 22
Cairo, Egypt; Burning of much of the city

1825
November 7
New Brunswick, Canada; Forest fire

1845
May
Guangzhou (Canton), China; Theater fire

1863
January 17
Santiago, Chile; Jesuit church fire

1870
June 5
* Constantinople, Turkey; Fire originating in Armenian district

1871
October 8
* Chicago, Illinois; Great Chicago Fire
* Peshtigo, Wisconsin; Forest fire

1876
December 5
* Brooklyn, New York; Brooklyn Theatre

1881
December 8
* Vienna, Austria; Ring Theatre

1883
January 13
Berditschoft, Russia; Circus Ferroni

1887
May 25
* Paris, France; Opéra Comique
September 4
* Exeter, England; Exeter Theatre

1894
September 1
* Hinckley, Minnesota; Forest fire

1897
May 4
* Paris, France; Charity bazaar

1899
March 17
* New York, New York; Windsor Hotel

1900
June 30
Hoboken, New Jersey; Docks

1903
December 30
* Chicago, Illinois; Iroquois Theatre

1904
June 15
* New York, New York; See MARITIME DISASTERS, *General Slocum*

1908
January 13
Boyertown, Pennsylvania; Rhoades Theater
March 4
* Collinwood, Ohio; Lake View School

1909
February 14
Acapulco, Mexico; Flores Theater

1911
March 25
* New York, New York; Triangle Shirtwaist Factory

1916
July 30
Ontario, Canada; Forest fire

1918
October 12
* Minnesota; Forest fire

1922
September 13
Smyrna, Turkey; Burning of city

1924
March 24
Chow-t'sun, China; Fire outside city

1927
January 9
* Montreal, Canada; Laurier Palace Theatre

1928
September 22
* Madrid, Spain; Novedades Theater

1929
March 12
* Igolkino, Russia; Factory fire

1930
April 21
* Columbus, Ohio; Ohio State Penitentiary
October 19
Wuchou, China; Tea District fire

1934
March 21
Hakodate, Japan; Much of city burned
September 8
* New Jersey Coast; See MARITIME DISASTERS, *Morro Castle*

1936
March 15
Tuliuchen, China; Theater fire

1937
February 13
Dandong, China; Movie theater fire

1940
April 23
* Natchez, Mississippi; Rhythm Night Club

1942
November 28
* Boston, Massachusetts; Cocoanut Grove Night Club

1944
July 6
* Hartford, Connecticut; Ringling Brothers, Barnum & Bailey Circus

1945
February 13
* Dresden, Germany; Fire bombing of city
June 19
Santiago, Chile; Braden copper mine fire

1946
December 7
* Atlanta, Georgia; Winecoff Hotel

1947
December 28
Hankou, China; Fire on docks

1949
September 2
Chongqing, China; Burning of city

1955
February 17
Yokohama, Japan; Catholic Old Women's Home

1958
December 1
* Chicago, Illinois; Our Lady of the Angels School
December 16
* Bogotá, Colombia; El Almacén Vida department store

1960
July 14
* Guatemala City, Guatemala; Guatemala City Insane Asylum
November 13
* Amude, Syria; Movie theater

1961
December 17
* Niteroi, Brazil; Gran Circo Norte-Americano

1962
September 7
Parana, Brazil; Coffee plantation fire

1967
May 22
* Brussels, Belgium; L'Innovation department store

1970
November 1
* St. Laurent du Pont, France; Cinq-Sept Club discotheque

1971
December 25
* Seoul, South Korea; Taeyunkak Hotel

1972
May 13
* Osaka, Japan; Playtown Cabaret

1974
February 1
* São Paulo, Brazil; Joelma building fire

1975
December 12
* Mina, Saudi Arabia; Tent city burned

1978
August 20
* Abadan, Iran; Movie theater

1980
May 20
* Kingston, Jamaica; Eventide Nursing Home

1986
December 31
* San Juan, Puerto Rico; Dupont Plaza Hotel

1988
November 18
* London, Great Britain; King's Cross underground station

1990
March 25
* Bronx, New York; Happy Land Social Club

1991
January 1
Indonesia; Continuing forest fires
September 3
Hamlet, North Carolina; Chicken processing plant

1993
February 15
* Tangshan, China; Linxi department store

May 10
* Nakhom Pathom, Thailand; Kader Industrial Co. doll factory

November 19
* Shenzhen, China; Zhili toy factory

1994

November 29
* Fuxin, China; Dance hall

December 8
* Karamay, China; Friendship Hall Cinema

1999

March 24
* Mont Blanc, between France and Italy; Truck fire in tunnel

2000

November 11
* Kaprun, Austria; Cable car in mountain tunnel

November 25
* Shibpur, Bangladesh; Chowdhury Knitwear and Garment Factory

December 25
* Luoyang, China; Dongdu disco and building

2001

September 1
* Tokyo, Japan; Gambling club

FIRES

Even the staunch civil libertarian and guardian of the Bill of Rights, Justice Oliver Wendell Holmes, agreed that the protection of freedom of speech does not extend to a person yelling "Fire!" in a crowded theater. That particular outcry has caused some of the worst catastrophes in history, as a quick glance through this section will amply prove. In fact, pandemonium during a fire is as responsible for its fatalities as the flames themselves or the smoke that causes asphyxiation.

And yet, how can anyone really be blamed for feeling terror at the very thought of a death by burning? Fire has always been a treacherous friend to humankind. It brings comfort, warmth, a romantic glow when the time and the season are right, and it stimulates a fertile imagination (how much of our youth was spent finding forms in the fire that crackled in the family fireplace?).

To the Greeks, fire, along with earth, air and water, was one of the four basic elements from which all things were composed, and the Greeks attached mythological powers to it. One of the greatest of the Greek myths is that of Prometheus, who stole fire from the gods and gave it to man and then suffered eternal torture for his generosity. Other religions have attributed the same fiery origins to either the entire religion or to aspects of it. Vesta, goddess of the hearth, and her virgins guarded the holy fire in ancient Rome. Fire is the earthly representation of the sun in Zoroastrianism. In Kashmir Shaivism, the fire of faith in the efficacy of spiritual practices burns away the karmas of the past and present.

Consider the wonder with which primitive people must have discovered fire—probably witnessing lightning igniting a forest. What a monumental discovery it must have been when these primitives first discovered the uses for fire; they made it the very center of their civilizations, and this continued for thousands and thousands of years. The connection between the Greek colony and the metropolis was the fire kindled in the colony from a brand brought from the mother city's fire. And think of the Olympic flame. And think of the monumental moment in 1827 when an English druggist named John Walker invented the first match.

Fire has warded off the terrors of the dark and the life-robbing chill of the cold. When we love, we say the object of our love warms our heart, and we kindle the flame of love.

But as much as we love fire, we fear it. Rather than die by fire, human beings, over and over, have flung themselves from the tops of high buildings. Were they crazed at the moment? Perhaps. But possibly not. Those who have miraculously survived these falls have affirmed that they would rather have the swift death at the end of a fall than the horrible, prolonged pain of death by fire. Medieval zealots knew this; execution at the stake was one of the most inhuman and barbarous practices ever conceived by humankind.

And it is true; death by fire is an agonizing death, for fire consumes slowly and relentlessly. Some victims have had their lungs burst because the fires around them have superheated the air (3,000 degrees Fahrenheit is the usual temperature in the middle of a firestorm) or robbed it of its oxygen. Toxic gases unleashed by fire cause asphyxiation.

And fire, being as fickle as it is, can turn from benevolent provider to destroyer in an instant. A turned back, a momentary distraction, an error in judgment, and a small fire can become a conflagration. The friendliest campfire or barbecue, the smallest match struck against the darkness can ignite infernos.

Knossos, in Crete—the greatest metropolis of the world in 1400 B.C.—was destroyed by a fire set by invaders. Carthage, Rome, Ninevah, Babylon, Moscow, London, Constantinople, Smyrna, Copenhagen, Munich, Stockholm, St. Petersburg, Cairo, Chicago, New York—every one of them has either been totally destroyed or severely crippled by fire. It is the most devastating destroyer we know—for no matter what humankind invents, no matter how much radiation its atomic or hydrogen weapons produce, even they assume the secondary, pervasive destruction of fire.

Consider its effect upon the world and its population today. The very atmosphere and the continuum of life are threatened by the wholesale destruction by fire of the world's rain forests.

And so, perhaps it is because of humankind's continuing dependence on fire that it is the most devastating of all destructive forces, an energy so pervasive and dangerous that it has produced at least two human professions: fire insurance and fire prevention.

Fire insurance sprang into being as a direct result of the London Fire of 1666. Since then, it has become as necessary as having a roof for every homeowner and a universal guardian against bankruptcy for the owners of structures, from skyscrapers to cottages, museums to country churches, ancient edifices to neighborhood garages.

The first organized firefighters protected the cities of ancient Greece and Rome, and since then, no major city in the world has been without a fire department, nor has any small town lacked a volunteer fire department. In the early days of the colonization of America, there were no fire brigades, but in 1638, the first fire service in the United States was established in New Amsterdam—later New

York—when Director General Peter Stuyvesant appointed four fire wardens.

Still, the first fire departments in America were all volunteer units, sometimes loosely organized, sometimes more rigidly controlled as true fire companies—as in Boston, New York and Philadelphia in the early 18-century. Benjamin Franklin, George Washington and Thomas Jefferson were some of the more prominent volunteer firefighters of this era, in which fire alarms were either given verbally or by rattles, gongs or bells. The fighters employed bucket brigades, one- or two-story high ladders and hand-pumped engines imported from Europe.

Two of the less savory reputations that volunteer fire departments enjoyed in the mid-19th century, when they were at their zenith, were boisterous rowdyism and a resistance to the technology of the steam engine, already a fact of life in Europe. Before long, pressure from insurance companies, politicians and influential citizens gradually brought about paid fire departments in major cities, all of which employed self-propelled steam engines.

Twentieth-century technology such as the internal combustion engine, radio communication and particularly SCBAs—the self-contained breathing apparatus that has saved uncountable numbers of firefighters from death as the severity and variety of fires have increased—benefited volunteer and paid fire companies alike. Flames in themselves are lethal enemies, but today these flames more often than not are partially or entirely fueled by toxic chemicals, petroleum distillates or radioactive material. The life of a modern firefighter, therefore, requires training and courage in equal amounts, as proved by the heroic actions of the New York City firefighters who plunged into the flaming carnage of the twin towers of the World Trade Center disaster on September 11, 2001 (see CIVIL UNREST AND TERRORISM, p. 147). More than 300 of them perished when the towers collapsed on them, and hundreds who played a central part in searching through the pulverized and still-burning ruins for their companions and civilians, emerged not only with commendations for their bravery but with diminished lung capacity.

Fire has always played a double role of sustenance and destruction in both inhabited and uninhabited parts of the world. Wherever humans have gathered, they apparently have been aware of the duplicity of fire and the necessity to balance dependence with an equal amount of protection. Still, even then, if fire wants to destroy, it will, no matter the efforts to control it.

The criteria for inclusion in this section are far more complex than in any other category. First, the decision had to be made regarding the inclusion of fire disasters in the volume on natural disasters or in *Man-Made Catastrophes.* Certainly, there are forest fires that are begun by lightning and the smoldering fires brought on by long droughts. But these are small in number compared with those caused by human error, carelessness or design. Even the worst forest fire in U.S. history, that of the 1871 burn that destroyed the city of Peshtigo, Wisconsin, and 23 other villages (see p. 251), did not begin from wholly natural causes. Though it is thought to have started spontaneously, its devastation is directly attributable to the mess left behind by loggers who continued to fell trees during a rainless summer and to railroad workers who, at the same time, burned debris in the forest. Even spontaneous combustion in a pile of oil-soaked rags is, ultimately, the responsibility of the human being who piled the rags there in the first place. Thus, fires rightfully belong in this volume.

In some cases fire is often the secondary disaster. Explosions cause fires; earthquakes cause fires. And so whenever it seemed as though the fire damage was specifically caused by another primary source for which there was a category, it was not included in this section—hence, for instance, the omission of the San Francisco Earthquake fire, the 1934 fire aboard the *Morro Castle,* and the 1904 fire aboard the excursion steamboat *General Slocum* in New York Harbor. The first can be found in the earthquake section in the volume on natural disasters; the other two can be found in the maritime disasters section in this volume.

Although war disasters have been omitted from both volumes, one exception was included in this section: the firebombing of Dresden by Allied bombers in 1945. One hundred thirty-five thousand civilians lost their lives, not from bombs but from the firestorm set by incendiaries. This was the worst fire catastrophe in the world during any age and, because of this, demanded to be included.

Finally, human suffering and casualty figures again dictated the inclusion or noninclusion of a particular fire. Generally speaking, a cutoff of 75 deaths was utilized, with one exception: the London Fire of 1666.

This fire, in which the bacteria that caused the bubonic plague were incinerated and out of which modern firefighting equipment and materials and the concept of fire insurance evolved, only claimed eight lives. But its impact was enough to warrant—perhaps demand—its inclusion in any compendium of the world's fires.

AUSTRIA
KAPRUN
November 11, 2000

One hundred and fifty-six skiers were killed in Austria's worst skiing tragedy on a cable car in a tunnel connecting Kaprun with the peak of Kitzsteinhorn mountain on November 11, 2000. Only 11 of 167 survived.

On Saturday morning, November 11, 2000, 167 skiers and snowboarders climbed aboard the cable-drawn car that would take them through a two-and-a-half-mile-long tunnel that pierces Kitzsteinhorn mountain, rising above the picture card Alpine village of Kaprun, Austria, a short distance south of Salsburg. The snow-coated glacier that resides at the mountain's peak is located at the top of an ideal ski run frequented by 100,000 skiers and tourists a season.

The entrance to the tunnel resembled an ultra-contemporary train station, and the ninety-foot-long car, riding on tracks but pulled by a cable, was solid and modern in design. The skiers arranged themselves, and the car began its ascent.

But seconds after the car entered the concrete shaft, passengers in the rear of the car spotted flames. At 9:02 A.M. the automatic disruption indicator halted the train. The conductor called the base and reported a fire, then lost contact.

Meanwhile, flames were racing through the car, and toxic smoke was pouring from it, filling the tunnel and turning it into a superheated chimney. "We tried desperately to get the doors open, but couldn't," Gerhard Hanetseder, one of the 11 survivors, told reporters later. "In the meantime, this little fire had set the whole cabin aflame. Already at that point we began to lose hope. There was no way we could get out."

But one man, described as a "burly Bavarian," had the presence of mind to smash the Plexiglas window at the rear of the car. Survivors crawled through the jagged opening and into an inferno of flames, heat and smoke. "We've got to go down!" a man identified only as Erwin G., a German, yelled, "Everybody down!" And he ran with eleven others down the narrow iron stairway that ran the length of the tunnel and headed back to the opening at the bottom. It was only a 600 yard trip, and they stumbled out into snow and sun and arriving rescuers.

The others—155 of them—suffered terrible, painful fates. Some broke through doors at the front of the car and began to climb up the iron stairway. But the draft of air from the top sucked the flames and the smoke up with them, incinerating some so badly that only DNA evidence would be able to identify their remains.

Another car descended from the top of the tunnel just as the ill-fated fire-consumed car entered at the bottom. The flames and toxic fumes were sucked up at lightning speed toward the car and instantly burned the driver of the descending car to death. At the very top of the tunnel three people, two early morning skiers and an employee, were enveloped by the same rushing juggernaut of flame and poisonous gas and were burned to death where they stood. It would be nightfall before rescuers could enter the tunnel safely, and when they did they encountered a terrible scene, "so horrific," said one rescue worker, "I do not want to describe it for the sake of the families."

Bodies were in, under and outside the car, which was melted, twisted and charred beyond recognition. Some were found as far as 60 yards above, indicating the distance they were able to cover before being overcome. The fire had obviously sped through the car so swiftly that passengers could not use the fire extinguishers within it. The extinguishers were still in their appointed places.

It would be September of 2001 before the results of the investigation into the cause of the tragedy would be released. A defective, illegally installed space heater in the driver's cabin had become blocked, causing hydraulic oil in nearby pipes to overheat. The hot brake fluid had then dripped onto the car's plastic-coated floor and caught fire, apparently even before the car left the base station. It had taken only seconds for the fire to become uncontrollable.

AUSTRIA
VIENNA
December 8, 1881

Human error on the part of a stagehand caused the most tragic theater fire in history at the Ring Theatre in Vienna, Austria, on December 8, 1881. Eight hundred fifty died and hundreds were injured.

The most tragic theater fire in history took place the night after Offenbach's *Les Contes d'Hoffmann* premiered at Vienna's elegant, ornate Ring Theatre. It, like so many human catastrophes, was caused by human carelessness compounded by human error and inaction.

The Ring Theatre, one of the jewels in the most elegant and artistically productive time in the history of this fabled city, had been built by the imperial government of Franz Joseph in 1873. Located off the famous Ringstrasse, which was already festooned with the Burgtheatre, the Opera House, the Kunstlerhause and the Musikverein, it immediately became a popular mecca for the city's lovers of popular entertainment.

Vienna under Franz Joseph was alive with the arts at the end of the 19th century. Brahms, the Strauss family and Mahler had all been drawn to it in the same way that Mozart and Beethoven had at the end of the 18th century and the beginning of the 19th. And while the Opera House was the home of grand opera, the masses flocked to the gilded splendor of the Ring Theatre, where the great Sarah Bernhardt and Signor Salvini's dramatic troupe appeared and the lively and racy operettas of Jacques Offenbach were performed. Royalty and the rich were also drawn to this theater by its glitter and its comfort, but they rarely arrived on time. Offenbach knew enough to write long overtures to fill in the time between the announced curtain time and their bejeweled entrances after 7:00 P.M.

Thus, on the night of December 8, 1881, the night after *Les Contes d'Hoffmann*'s premiere, only the two balconies were full at 6:45 P.M. Eager tradespeople, students, actors and actresses, attracted by critical praise and enthusiastic recommendations of the new Offenbach work, filled these two upper parts of the auditorium, while a few renegade knights and bank directors occupied the few boxes and stalls downstairs.

At that precise moment, a stagehand went about his usual task of lighting the upper row of gas jets above the stage. Possibly he was careless. Possibly the elaborate scenery required for the operetta was too abundantly or negligently hung. For whatever reason, his long-handled igniter set fire to the canvas trappings of several theatrical clouds. Within seconds the flames swept to the stage curtains. The stage doors were open; the air blowing in through them fanned the flames and billowed the curtains outward toward the audience. Huge tongues of flame leaped from canvas to canvas onstage and out into the auditorium.

At this point the iron fire curtain that existed in every completely equipped theater of the time could and should have been lowered. It would have contained the fire onstage, curtailed the draft and snuffed out some of the flames. But, inexplicably, this was not done, nor was the fire brigade summoned, nor was the onstage water hose pressed into service.

Instead, panic spread as quickly as the fire. As the flames shot outward from the stage, crawling up drapes and running in fiery streams across the ceiling, the patrons stood up in their seats, screamed "Fire!" and began to shove at one another. To compound the hysteria, a stagehand shut off the gas, plunging the entire premises into darkness save for the light of the rapidly accumulating fire.

The occupants of the stalls and boxes got out safely, walking rapidly to the lobby doors and out to the square, where gilded carriages containing royalty and the wealthy were just beginning to draw up.

In the balconies the crush of humanity battered its way toward the exits, only to find them blocked by impenetrable walls of fire. Some patrons, pushed or panicked, leaped or fell from the front of the balconies. One woman landed on two other audience members, killing herself and both of them.

Summoned by spectators, the fire brigades arrived, but their ladders were too short to reach even the first balcony. By now patrons were smashing the Gothic windows behind the balconies and leaping hysterically to their deaths in the square below. Firemen frantically ransacked the theater for drapes from which to fashion life nets, but most had been burned to charred threads. They finally found one huge stage drape, and, shouting to those in the balconies to jump into it, they stretched it taut beneath the balcony rails.

The patrons calmed. A commanding, aristocratic man ordered the children to jump first, then the women and finally the men. One hundred twelve children, women and some men thus survived before the walls began to cave in and the flames and smoke became so intense that the rescue attempt had to be abandoned. Those who remained were either incincerated or crushed under the falling walls and pieces of decor.

Members of the royal family—among them Franz Joseph's grand nephews Charles, Albrecht, William, Salvatore and Eugene—arrived at the scene, took one look at the inferno before them and, on the spot, began a collection of relief funds for the victims. Crown Prince Rudolf wept openly at the catastrophe, which claimed 850 victims—the highest number of fatalities that would ever be recorded in a European theater fire. Hundreds more were injured.

BANGLADESH
SHIBPUR
November 25, 2000

Forty-eight women and children died and more than 150 were injured in a stampede in the Chowdhury Knitwear and Garment Factory in Shibpur, Bangladesh, on the night of November 25, 2000.

Child labor is rampant in Bangladesh, where nearly 2,000 garment factories churn out products that constitute 70 percent of this impoverished country's total annual $5 billion export revenue. The 1.5 million workers, overwhelmingly women between the ages of 14 and 24 and some younger work six days a week from 8 A.M. to 8 P.M.—and frequently overtime to 10 P.M.—for $9 a month. Although factory workers signed an agreement in 1995 with the International Labor Organization under the auspices of UNICEF to stop using child labor, 6.3 million child workers under the age of 14 still work in Bangladesh, in which poverty and hunger force parents to send their young children to work rather than school. The working conditions are frequently intolerable, and, as has been tragically customary in factories throughout the world, emergency exits are uniformly padlocked against pilfering by the employees.

This, again, was the situation when a fire, apparently sparked by a short circuit on the third floor of the Chowdhury Knitwear and Garment Factory at Shibpur, 25 miles east of Dhaka, erupted at 7 P.M. on November 25, 2000. There were 900 workers in the building at the time amid piles of flammable towels and knitwear. The crush was particularly apparent on the fourth floor, where workers ironed these fabrics.

Instantly, hundreds of crates full of towels caught fire, sending heavy smoke throughout the factory. The workers, terrified, scrambled for the one staircase that would take them out of the building. It was piled high with bundles of garments, which narrowed the passageway down to a slim path. The workers tumbled forward but met another, insurmountable obstacle: the first floor gate on the single staircase had been locked.

"I was working on the second floor," a 22-year-old worker named Aziza told reporters from her hospital bed later. "When I heard the others shouting 'fire,' I ran to the stairwell, but everyone was pushing each other. I tripped on the stairs. . . . The metal gates at the entrance of each floor of the factory are usually closed during working hours. On Saturday, only the second floor gate was open. The others were padlocked but the factory security guard couldn't find the keys during the fire. . . ."

The rush became a stampede. Workers were packed into immobility in the stairwell, where they were burned alive or suffocated. Some girls smashed windows and flung themselves out of them, only to be impailed on the pointed tops of the iron railings surrounding the factory.

Firefighters and rescuers living nearby arrived and broke the padlock on the exit door, but it was too late for 48 of the workers, who were dead when the other survivors trampled them in the rush from the building.

The management of the factory denied responsibility, claiming that it was neither the locking of the door nor the fire, but the panic of the workers that had killed and injured them. They also claimed that the origin of the fire was not in the building's wiring but in a thinner spray gun. Authorities, however, established that it had been an electrical short circuit.

It was one more sorry chapter, part of a litany of violated or uninforced safety standards in Bangladesh. Investigation revealed that 227 garment factories in Dhaka alone did not have emergency exits, nor did they conduct required monthly evacuation drills, nor were there the required number of fire extinguishers. As a result, there had been 200 reported deaths in factory fires in the country between 1998 and 2000—and three factory fires in the three months before the November 2000 conflagration. None equaled the fire at the Chowdhury Knitwear and Garment Factory, however. It is, as of this writing, the worst factory fire in Bangledesh's history.

The survivors and their families held out little hope for justifiable compensation. After all, following the Globe Knitting Company fire in August of 2000, which had claimed 12 lives, the government had offered each victim's family just $9 and 25 kilograms of rice.

BELGIUM
BRUSSELS
May 22, 1967

Panic and the lack of a sprinkler system combined to cause the tragedy of the L'Innovation department store fire in Brussels, Belgium, on May 22, 1967. Three hundred twenty-two died; scores were injured.

Panic kills as many people in fires as smoke or flames, and panic accounted for many of the deaths in the store fire with the greatest fatality count in history. Three hundred twenty-two people died in the noontime fire at L'Innovation, the five-story department store located in the heart of the old city in Brussels, Belgium.

Spring is a time of innovative sales, and L'Innovation, true to its name, featured a Salute to American Fashion in May 1967. On May 22 the "million dollar showcase" attracted approximately 2,500 customers to the store. L'Innovation prided itself on its service, and another 1,200 clerks—one for every two customers—were in attendance at the height of the shopping day, when office workers on their lunch hour swelled the ranks of shoppers.

It was nearly noon when a fire broke out in three places on the fourth floor of the crowded store. There was some inconclusive evidence of an accelerant being used to begin the fire, though arson was never proved. Whatever its source, the fire spread rapidly and unchecked. The old building was without a sprinkler system. It did have 15 full-time firemen on duty at all times to compensate for its lack of mechanical fire-fighting means, but for some reason only two of the 15 responded to the alarm that day. Their sincere but ineffectual efforts to control the wildly spreading blaze with hand-held fire extinguishers did nothing to stop the gathering holocaust.

All 4,000 people in the store tried to reach the exits and elevators at the same time. Many were trampled to death in this insane rush; others had limbs broken and clothing stripped away. Those who could not reach stairways, elevators or doors fought their way to windows. Some smashed them out with their bare hands and leaped for the street, hoping to land on the forgiving hoods of parked cars. Some did and only suffered broken limbs. Others missed and died.

Firemen, hampered by the narrow, twisting streets of the Old Quarter, took an unconscionable amount of time to arrive on the scene. By the time they finally got there, hundreds of canisters of butane gas destined for summer campers and stored on the store's roof exploded, feeding the inferno still further. Desperate people still clinging to upper stories were turned into human torches as the flames consumed the entire building, destroying it totally and burning to ashes many of the 322 people who perished in the fire.

BRAZIL
NITEROI
December 17, 1961

Either arson or sparks from a passing train were suspected of causing the worst circus fire ever recorded in Niteroi, Brazil, on December 17, 1961. Three hundred twenty-three died; 500 were injured.

This incredible circus fire killed 323 persons—most of them children—and cruelly burned 500 more. And it all happened in a little more than three minutes.

As part of its annual Christmas week celebration in 1961, Brazil featured the Gran Circo Norte-Americano, a Brazilian version of Ringling Brothers. In the town of Niteroi, which is located across the bay from Rio de Janeiro, the circus played out its thrills and fantasies in a blue and white nylon tent large enough to accommodate high-wire acts, animal acts, clowns and 2,500 spectators.

On the afternoon of December 17, 1961, the tent was packed to capacity. Most of the audience was composed of children on holiday from school. They were transfixed by the death-defying high-wire acrobatics of the featured trapeze artist, Antonietta Estavanovich. And it was Ms. Estavanovich who first saw the flames. What must have gone through her mind as she soared through the air toward her partner and saw flames beneath her, in the upper wall of the tent, she never said. But by the time she and her partner had spun into the safety net and headed for the exits, the fire had made its way to the center of the tent and was edging downward along the tent poles.

Within three minutes the entire tent had become one huge flame, and the screaming children were stampeding. Three hundred of them ran toward the center ring. The tent collapsed around them, suffocating them. Some others fell as the mob surged in several directions, and they were trampled underfoot.

Sergio Pfiel Manhaes, a heroic young Boy Scout, pulled out his knife, cut a hole in the side of the tent, hauled his family through it and then went back into the conflagration and led an adult, blinded by smoke, to safety.

João Goulart, the president of Brazil, broke down in tears when he went into the children's ward in Niteroi's Antonio Pedro Hospital to visit the 500 injured. The investigation he ordered turned up no conclusive reason for the fire. Opinion on the cause of the disaster was divided between arson and sparks from a passing train.

BRAZIL
SÃO PAULO
February 1, 1974

An overheated air-conditioning vent ignited plastic construction material piled near it in the Joelma building, which housed the Crefisul Bank, in São Paulo, Brazil, on February 1, 1974. Thrill seekers hampered firemen, and 220 died. Hundreds more were injured.

São Paulo is one of the wealthiest cities in Brazil. Boasting a population of six million—roughly the same as Chicago—it also boasts some of the country's most modern office buildings.

But São Paulo is lacking in elementary safety protection for its populace. The six million residents of Chicago are serviced by 300 fire stations. The six million in São Paulo must rely on a mere 13. Safety codes in most major cities of the world decree that nonflammable materials be used in major office buildings. The interior of the Joelma building, a skyscraper in downtown São Paulo that housed the offices of the Crefisul Bank, was composed almost entirely of highly flammable materials. And these two factors accounted for the deaths of 220 people on February 1, 1974.

It was one of the worst office building disasters in history, and it will be forever memorable for the extremes of human behavior that it revealed. At one end of the spectrum were the acts of touching and staggering heroism on the part of firemen swinging on ropes high over the streets to rescue panicked victims. At the other was the crush of spectators straining to watch flaming people fling themselves to their deaths from upper stories. Over 300,000 cars, abandoned by these morbid, sensation-hungry spectators, clogged streets and prevented rescue equipment from getting through.

The fire began in an overheated air-conditioning vent on the 12th floor of the 25-story building. Plastic material piled near the vent quickly ignited and spread to other plastic constructions built into the building. The first six floors of the structure were occupied by a car park; thus, most of the employees trapped by the flames were in upper stories beyond the reach of firemen's ladders, which only extended to the seventh floor.

Some managed to battle their way to exits and ran from the building. Others rushed to save themselves, trampling some of their fellow employees to death. Thirty-four people locked themselves in a washroom and turned on the watertaps in hope of keeping the flames away. They were discovered the next day, every one of them suffocated to death.

People at some of the windows and on some of the ledges of the building, seeing that the ladders could not reach them, jumped, preferring a quick death to a slow one. One man hit two firemen on a ladder, carrying them with him to their deaths. One woman jumped with a baby in her arms. She died; the baby survived. Twenty-five people tried to leap to the roof of a nearby building. All twenty-five died.

Others heeded the large signs that firemen held up to them reading, "Courage. We are with you. Don't jump."

One heroic fireman, Sergeant José Rufino, swung on a rope secured to a nearby building, grabbing 18 survivors and swinging with them on his back to safety. During one attempt a man leaping from the 16th floor collided with Rufino, peeling the man from the fireman's back and sending him to his death. Rufino managed to hang on to the rope and thus saved himself, but his hands were torn and bleeding when he finally rejoined his fellow firemen.

Helicopters sent to rescue survivors from the roof could not land because of the intensity of the heat and the density of the smoke. At one point the paint began to scale off the doors of one helicopter. Firemen finally dropped cartons of milk to survivors who had made their way to the building's roof. The detoxifying properties of the milk were credited with keeping these near-victims alive until an army helicopter was brought to the roof of the building. The helicopter landed on the slowly buckling roof and lifted off, in a series of staccato landings and takeoffs, 85 people. As the last 10 people were rescued, the roof collapsed.

It would be four hours before firemen could bring the blaze under any sort of control. Almost all of the interior of the building from the 12th floor upward was reduced to charred and sodden rubble.

There was some talk of sabotage, some reports that a telephone operator at the Crefisul Bank had received an anonymous call the day before saying that a bomb would explode on Friday morning. But the report was never considered in the inquiry that followed the fire. The cause was multiple; 220 people were dead; hundreds had been injured; and the municipality had much to do to prevent a recurrence.

CANADA
MONTREAL
January 9, 1927

Employee negligence and political payoffs that allowed fire code violations to exist were the causes of the fire in the Laurier Palace Theatre in Montreal, Canada, on January 9, 1927. Seventy-eight children died; 30 were severely injured.

Fate is often ironic, as anyone who has reached the age of reasoning can attest. But it was never more so than on the afternoon of January 9, 1927, when 800 patrons, almost all of them children, gathered in Montreal's Laurier Palace Theatre to watch a film titled *Get 'Em Young.*

The Laurier Palace was a movie house that had existed well past its prime. It had never entirely conformed to the existing fire code of the city of Montreal and was, in January 1927, operating illegally, without a license. The reason: It had failed to correct hazardous safety conditions, chief among them an absence of unobstructed and well-marked fire exits. That the theater was open at all must have had something to do with local politics, since it was located directly across the street from a police station.

On the afternoon of January 9, the theater was conducting a special children's matinee, and it was packed. Most of the children were well under the age of 16. Approximately 500 filled the orchestra floor; the younger children, aged six and seven, were crammed into the balcony. There were no chaperones present. Only the usual theater staff was there to control the children.

Sometime during the showing of the movie, somebody—an employee, it was charged in the investigation—dropped a lighted cigarette in the middle of the balcony. Within seconds inflammable material in the seats had ignited, and a conflagration was under way.

The children on the first floor, removed from the immediate fire, escaped unharmed through several exits. But the small children, now choked and blinded by smoke and surrounded by flames, understandably panicked. The only egress from the balcony was by a narrow, unlighted stairway that descended a short way, made a sharp right-angle turn at a landing and then dropped another few steps to a door that was only 37 inches wide. The stairwell itself was a scant 10 inches wider than this cramped doorway.

As the flames shot upward and outward from their beginnings, the young children stampeded for the stairwell. Some managed to get down the stairs safely—their small size allowed them plenty of room. But then one child either tripped or was pushed down the stairs. She formed a wedge that blocked off the only escape route for the remaining screaming youngsters. They plummeted down the stairs, piled up on one another and became wedged into the stairwell.

It would take less than two minutes for the entire balcony to go up in flames, and during that time the projectionist worked heroically to save nearly 30 children by shoving them, two at a time, through a window in the projection booth onto the top of the theater marquee.

When the firemen arrived, the pile of bodies at the bottom of the stairwell was eight deep and could only be reached by cutting a hole in a side wall. Only a few of the children trapped there were still alive. Seventy-eight young people died, 52 from smoke and asphyxiation, 25 from being crushed to death in the stairwell and one from the flames. Thirty were severely injured.

CHINA
CHONGQING (CHUNGKING)
September 2, 1949

Arson was suspected in the fire that began in the slum district of Chongqing, China, on September 2, 1949, and destroyed 10,000 homes and left 100,000 homeless. One thousand seven hundred died in the fire; thousands were injured.

The year was one of extreme turmoil in China, the turning point between Nationalist and Communist control of the country. Until 1947 Jiang Kaishek's (Chiang Kai-shek) Nationalists, supported by U.S. supplies and money, had tenaciously held on to the control of the country. But by November 1948, when the Chinese Communists, under Mao Zedong (Mao Tse Tung) captured Mukden and thus the industrial heartland of the country, the standoff between the two factions had all but been won by the Communists. Sweeping inflation, increased police repression and a grinding, endless famine had so eroded public confidence in the Nationalists that a state of civil strife trembling on the brink of civil war existed.

In January 1949 Beijing fell to the Communists. From April to November, other major cities also fell, most without a fight. Nanjing, Hankou, Shanghai, Guangzhou and eventually Chongqing, the Nationalist capital, all surrendered.

In September the tensions in the city were at their highest point. And on September 2, 1949, at 4:00 P.M., a fire of mysterious origin began in Chongqing's slum district. Whether the arsonist was a Communist or whether it was someone directed by the Nationalist government to set the blaze in the hope of turning public opinion against the Communists will probably never be known. In the holocaust's aftermath the Nationalists rounded up suspected Communists and, a week later, executed for arson a man known to be part of the Communist underground.

The human toll was staggering. The fire, once begun, spread unchecked in several directions. It ate into the residential district, consuming nearly 10,000 homes and leaving more than 100,000 people homeless. It devastated the business district and then, fanned by winds, roared toward the waterfronts of the Chang (Yangtze) and Jialong Rivers. Refugees, running ahead of the advancing wall of fire, had come to this part of the city in the hope of escaping in one of the hundreds of boats docked there. Hundreds of people were burned to death both on the docks and in the moored boats as the roaring inferno outraced and enveloped them.

Chongqing would burn for 18 hours that afternoon and night, and when it was over more than 1,700 residents of that embattled city would be dead. One day less than a month later, the Nationalists would be in Taiwan, and the civil strife would be at an end. Chongqing's terrible fire would be one of its last and most dramatic manifestations.

CHINA
FUXIN
November 29, 1994

Two hundred and thirty-three weekend revelers were killed and 16 injured in a fire that swept through a dance hall in Fuxin, China, on November 29, 1994.

Fuxin, in China's northeastern Liaoning province, possessed a one-story, 3,250-square-foot building, one of the rapidly stitched together structures of the new market economy in China. It was used for public gatherings, chiefly as a dance hall, rented by entrepreneurs from a local government work unit.

On Sunday afternoon, November 29, 1994, it was filled with dancing revelers celebrating a day off. The interior of the dance hall was decorated copiously—and inflammably. At a little after 1:30 P.M. a fire broke out in the hall and with amazing speed raced through the entire premises, igniting the decorations and filling the hall with flames and toxic fumes.

The partygoers scattered for the exits and, as in so many factories that burned in China in the 1990s, found them locked. They tried frantically to break through them, but to no avail, and the windows of the building were too high to reach. Sixteen people, including the entrepreneur who had rented the building, did manage to break windows and, though injured, they survived.

The local fire department, located only 1.2 miles away, arrived within five minutes with 14 fire engines and 85 firefighters. It took them only eight minutes to control the blaze from the outside. Once they had forced their way through the blocked entrances, a grisly sight greeted them.

Two hundred and thirty-three men and women, some charred into unrecognizability, others suffocated from the fumes, lay dead in the gutted dance hall. "Perhaps the main reason why the people could not escape was because of the smoke and poisonous fumes," one official posited to reporters later. It was partly that, but the main reason for the 233 deaths rested with the blocked exits, a continuing killer that made this tragedy the most tragic and lethal fire to strike the new China to that date. Soon, that record would be broken (see below).

CHINA
KARAMAY
December 8, 1994

In China's worst fire disaster, 323 people, most of them children, died in a conflagration that destroyed the Friendship Hall Cinema in Karamay on the afternoon of December 8, 1994.

Barely one weeks after the staggering loss of life in the Fuxin, China, dance hall fire of November 29 (see p. 207), a worse tragedy by fire occurred in the remote oil town of Karamay. The town, chiefly a living place for employees of a local oil field that provided employment for 90 percent of its populace, is located on the border of the former Soviet Union 180 miles northwest of Ürümqi, the capital of Xinjiang province.

On the afternoon of December 8, 1994, roughly 800 of these people, including 500 schoolchildren, were assembled in the Friendship Hall Cinema. The occasion was a school cultural performance attended by the children, their teachers, local officials and an education inspection team.

Midway through the performance, an explosion resounded through the hall, startling both performers and children. Within instants a raging fire leaped through the floor, ran up the walls and enclosed all of those within the theater in a cocoon of flame.

Panic followed. With screaming children and adults fighting to find exits, a theater that plunged in an instant into darkness made blacker by suffocating smoke closed in on them. The destruction was monstrous and instantaneous. By the time firefighters arrived, the grim price for shoddy construction of public places in the new China had begun to reveal itself.

More than half of the 800 people in the theater escaped, some with nothing more than a few burns, some terribly burned, some with scorched lungs. It would be days before the fire and rescue teams were able to clear the cruelly charred human wreckage from the ruins of the theater. Three hundred and twenty three people, most of them children, had burned to death in the worst fire yet in China's long history of infernos in public buildings.

CHINA
LUOYANG
December 25, 2000

An engulfing fire destroyed the four-story Dongdu building in Luoyang, China, on the night of December 25, 2000. All of the 311 fatalities occurred in a top floor disco. Fifty other victims were severely injured but survived.

The Dongdu building in the commercial center of Luoyang, the capital of the central Chinese province of Henan, was a four-story conglomeration of shops, a supermarket and, on its top floor, an illegal but popular disco, thumping its way into the early hours of weekend mornings, entertaining a large portion of the young population of the city. The illegality of the disco was the product of repeated safety violations within the building, built in the late 1980s and, by 2000, the center of a warren of street stalls and other attractions to service the throngs of people who shopped or amused themselves within the structure.

The Dongdu building seemed to be in a constant state of renovation, and some illegal construction debris stood immediately outside the fire exits, blocking them and making them useless. The building had no sprinkler system, no fire alarms and no smoke detectors, and although it had failed the last of many fire inspections in the middle of December, the owners and the owners of the various concessions within it were unconcerned. As in the past, inspectors and officials could be bribed to forget the citations. It was a pattern that had earned the Dongdu building the reputation of being ranked among the 40 most dangerous buildings in Henan province.

On Christmas night 2000 the premises were crowded with shoppers, and, on the top floor, a private Christmas party was in full progress. The darkly lit disco reverberated with the vibrations of music and the voices of the partygoers.

At 9:35 P.M. Wang Chengtai, one of a team of welders working on the renovation of one of the basement floors, became careless, and the sparks from his welding torch ignited some rubbish. The fire flared out of control, and the welders, alerted by Chengtai, scurried to safety without warning anyone of the fire in the basement.

The flames exploded upward, and those customers in the lower floors dashed to safety through the large entrance. But those on the upper stories were hindered from escaping by narrow corridors piled high with merchandise and construction material.

Fire companies arrived, but they were prevented from getting to the building by the cat's cradle of stalls and shops that blockaded the streets. By the time they were able to surround the building and begin pouring water on it, some desperate victims had broken windows and jumped, some into cushions the firemen spread on the street below, some to their deaths.

The situation in the disco was, however, catastrophic. Smoke from the fire rose through elevator shafts and stairwells. The music had been so loud that the revelers had been unaware of the fire until smoke began to fill the club. Then panic raced through the crowd, who fought its way to the exits. There was only one that allowed them through. The emergency exit to the roof was locked, and of two other exits, one was concealed behind a bar.

Abruptly, the power went out in the building, plunging the disco into stifling darkness and rendering the elevators useless. More than 300 partygoers clawed and pummeled their way toward the one exit, only to be overcome by smoke, which smothered then suffocated them.

It would be three full hours before the firemen were able to extinguish the conflagration and still longer before they could finally smash their way into the charred remains of the disco. There they found 311 bodies piled near the exit or collapsed in various parts of the disco. No one on the top floor survived; every one of them suffocated.

There were 50 survivors wandering or crumpled in corridors and parts of the shops. Given oxygen, they were led or carried out by rescuers.

Authorities, shocked into action, moved swiftly to investigate and fine the workers and shop owners. By the end of December, arrests were made, and on August 22, 2001, 23 people were jailed, including the welders, nine officials from the local police, construction, commerce and fire departments, the building operators and two Taiwanese investors in the building. After the trial, dissatisfied relatives of those who died took to the streets of Luoyang to protest the short sentences given to those responsible for this cataclysm in a country that had in the preceding few months executed more than a thousand people for crimes including robbery and forgery.

Officials pledged to stop the seemingly ceaseless wave of tragedies that had occurred during the frantic building boom that ushered in private enterprise in China. Collapsed buildings that were erected without architects or construction knowledge, inferior and dangerous materials interwoven in hundreds of public places, bribery and a failure to install proper safety precautions had produced a sorry record.

By the Chinese government's own statistics, more than 300,000 fires occurred between 1993 and 1998, killing 12,628 people and injuring 22,382. In the first three months of 2000, the last for which statistics were available, 36,832 fires or explosions took place and claimed 971 lives. Most of these catastrophes took place in private buildings catering to an unwary public.

CHINA
SHENZHEN
November 19, 1993

A tragic fire that was nearly identical to the Triangle Shirtwaist Factory Fire of 1911 and the Thailand doll factory conflagration of May 1993 erupted in a toy factory in Shenzhen, China, on November 19, 1993. Eighty-seven died and 47 were injured.

Once again, history, given human manipulation, repeated itself. Only eight months after the horrendous Kader doll factory fire in Thailand (see p. 226), the Zhili Toy Factory in Shenzhen, China, repeated the enormous waste of life in the 1911 Triangle Shirtwaist factory fire (see p. 246) and for almost exactly the same reasons.

In the same way that the Triangle Company employed migrant women laborers, mostly from Italy, Eastern Europe and Germany in New York in 1911, and the Thailand company employed immigrant girls, the Hong Kong–based Zhili Toy factory, a subcontractor for the Italian multinational Aretsana S.P.A./Chicco, was staffed by female migrant workers from the Chinese interior provinces of Sichuan and Henan.

There was a startling similarity, also, in the attitude of management toward its workers in the two factories. In an almost eerie imitation of the 1911 factory owners, those in Shenzhen suspected their employees of stealing from them, and so not only were the exits and fire doors locked from the outside, but bars were placed over the windows.

The Zhili factory was, in addition, a so-called three in one structure endemic to the frantic building boom of the 1990s in China (see Linxi department store fire, p. 210). A three-story building, it contained a production floor, storage space, and a dormitory for the workers under the same roof.

On November 19, 1993, fire broke out in the storage space in the factory and spread with astonishing rapidity through the building. Management, located in unlocked offices, escaped immediately, but the trapped workers who tried to flee from the flames were locked in. One fire exit was smashed open by brute force, but the rest remained solidly barred against the frantic women. They dashed from door to door as the fire consumed the entire building.

Firefighters arrived and tore through some of the doors. They were able to rescue 47 of the workers, but 87 were incinerated, their charred bodies piled up against the locked fire exits.

The suffering, however, did not end with the extinguishing of the fire. Huang Guoguang and Lao Zhaoquan, the factory owners, were arrested and charged with criminal negligence. Not only had they barred all the exits, turning their premises into a crematorium, but they had ignored safety regulations and bribed inspectors to overlook the obvious safety problems on their premises.

They were given short prison sentences, and within five years were back in business, running another toy factory in Dongguan, a short distance from Shenzhen.

After four years of legal wrangling, Aretsana S.P.A./Chicco agreed to establish a fund to compensate the 130 victims and/or their families. Three hundred million lire was agreed upon, which would amount to less than $1,300 for each family.

But in 1997 this money still remained with Aretsana and none had reached the survivors, some of whom, like Chen Yuying, who was 17 years old when she was trapped in the fire, suffered burns over 75 per cent of their bodies, had limbs amputated and remained crippled.

Xiao Chun, who was hailed as a hero for saving the lives of four women, running back into the flames, suffering burns to half of his body and losing his right hand, was left helpless and unable to provide for himself after the tragedy. Yet he, too, had received no compensation.

CHINA
TANGSHAN
February 15, 1993

A fire destroyed the Linxi department store in Tangshan, China, on February 15, 1993, killing 79 and injuring 51.

China's economic growth in the early 1990s produced a nationwide construction boom. New buildings and factories were built with lightning rapidity, while existing stores and restaurants were remodeled to accommodate an emerging consumer society. In the process shoddy techniques were frequently employed to meet deadlines. Construction accidents were numerous and widespread. And worst of all, safety precautions were often trimmed or ignored in the feverish race for completion.

The new clusters of department stores were among the most dangerously erected symbols of a new society. Often packed with customers navigating among counters piled with flammable items, and possessing fewer exits than common sense would dictate they were disasters in disguise.

On Sunday, February 15, 1993, the Linxi department store in Tangshan, a city 100 miles east of Beijing, was crowded with shoppers becoming accustomed to the unaccustomed feeling of shopping in an indoor marketplace. While business was being conducted in the store, workmen were working with welding equipment, finishing up the construction of the new building. They, as well as the authorities, were lax in following safety precautions, and a flame from a welding torch started a fire that spread abruptly to the entire store at its peak midafternoon shopping hour. It leaped with ferocious speed through the entire three stories of the crowded building, and within minutes the store was turned into a consuming inferno, trapping hundreds of screaming, stampeding victims inside. Some smashed windows and leaped to safety; others battered their way through the few doors.

Firefighters arrived speedily and hacked holes in the store's walls to let choking and burned people out. But they could do little to stop the fire. The entire three stories were gutted and blackened; the exterior walls were as smeared with smoke and fire damage as the interior. Seventy-nine people died in the blaze and 51 were taken to hospitals with severe injuries.

COLOMBIA
BOGOTÁ
December 16, 1958

A light bulb ignited a creche in El Almacén Vida, one of Bogotá, Colombia's, largest department stores, on December 16, 1958. Eighty-four died in the resultant fire; scores were injured.

One of Bogotá, Colombia's, biggest department stores is called El Almacén Vida—the Life Department Store. It became anything but that on December 16, 1958. Early in the day the store was jammed with holiday shoppers shoving and pushing their way past displays of new merchandise and elaborate Christmas decorations.

One of the most elaborate and effective of these was the creche that was set up in the toy department. Realistic, artistically arranged and highly detailed, it had as its centerpiece Christ lying on a bed of straw. Brightly colored lights outlined the roof of the creche, blending with the general joyous atmosphere of the store.

Sometime during the day one or more of these brightly burning lights became dislodged and fell into the straw surrounding the figure of Christ. Unnoticed, it set fire to the straw, and this in turn spread to a pile of plastic toys next to the decorations. Within minutes the entire toy department was engulfed in flames, and the bitter smoke of burning plastic billowed to the ceiling.

Panic spread as hysterical patrons tried to run for the front and largest exits of the store. But the toy department was located near these exits, and a solid wall of flames blocked them.

Some male customers smashed sizable holes in the glass of the store's display windows and scores of survivors gingerly made their way through these openings. But 84 people trapped by the flames and the smoke in the rear of the store were burned to death.

FRANCE
PARIS
May 25, 1887

A gaslight igniting scenery followed by human error—the failure to lower a fire curtain—caused the fire in Paris's Opéra-Comique on May 25, 1887. Two hundred died.

The venerable Opéra-Comique, on Paris's Place Boieldieu, was packed on the evening of May 25, 1887, for a performance of Massenet. The Opéra-Comique, like the Ring Theatre in Vienna (see p. 203), was a livelier place than the Opera, and like the Ring Theatre, it would be the scene of a gigantic tragedy that could have been averted if only the iron fire curtain had been lowered.

As in most 19th-century theater fires, the conflagration at the Opéra-Comique started onstage when a gaslight positioned in the fly space over the stage ignited a piece of canvas scenery. Witnessing this, two singers in the company, Taskin and Soulacroix, stepped to the footlights and tried to calm the immediately restive audience. Had one stagehand had his wits about him and lowered the iron curtain at that moment, the fire would have been contained, and the tragedy that followed would not have occurred.

But this did not happen, and the flames soared out over the audience, sending streamers of black and acrid smoke ahead of them. In other theater fires (see pp. 216, 235 and 244) aristocratic, high-paying customers had no difficulty escaping the conflagration from their stalls and boxes. But in the case of the Opéra-Comique, tragedy was more democratic; it spared no one. Obstructions that a uniform fire code would never have allowed blocked everyone's way to exits, and the well-to-do found themselves joining the hysterical working-class patrons of the galleries trying to climb to upper windows and the roof, away from the flames.

Outside the arriving *pompiers* threw up ladders, but few of them could reach the people trapped on ledges, and none of the ladders came close to reaching the roof. In despair, people began to hurl themselves to their deaths from windows and ledges. Firemen rescued whomever they could. Two bejeweled women were taken from a window sill and brought into a nearby druggist's shop, where they were laid upon two nearby counters. They died soon after.

By 11:00 P.M. the upper-story dressing rooms and music library had caught fire, and the roof had collapsed, sending scores of trapped people hurtling through the flames to their deaths. The last two people to be rescued by firemen were huddled on a brick cornice. The woman had passed out and was lowered by rope to the street; the man was guided onto a ladder, which he descended to safety.

The fire was finally brought under control at dawn of the next day. Two hundred bodies were discovered in the wreckage. Most of them had died of asphyxiation; some had been crushed in the stampede or by falling pieces of the opera house.

A spectacular fire interrupted a performance of Massenet's Mignon at Paris's Opéra-Comique on the evening of May 25, 1887. Two hundred died, and the opera house was gutted. *(Illustrated London News)*

FRANCE
PARIS
May 4, 1897

A lamp used to illuminate a kinematograph ignited the structure erected at Paris's annual Grand Bazar de Charité on May 4, 1897. One hundred fifty people died in the fire; hundreds were injured.

A uniform code of fire regulations governing public places was put in place in Paris as a result of a shocking tragedy that, on May 4, 1897, burned to death 150 of the wealthiest women of that city.

The annual Grand Bazar de Charité, designed to raise thousands of francs for the destitute, was held in May 1897 in a 220- by 300-foot structure built especially for the occasion and decorated like a street in a medieval French city.

The decor was not new. It had been used a year before at an exposition at the Palais d'Industrie. Built from linen, coated with turpentine and filled between the surfaces with papier mâché, it was clearly a stage set and obviously inflammable.

More than 1,500 socialites packed themselves into the structure on the evening of May 4 to buy semiprecious

A molten roof collapsed on hundreds of society patrons attending the Grand Bazar de Charité at the end of a flash fire on the night of May 4, 1897. Here, the grim job of identifying their remains is conducted. (Illustrated London News)

objects at inflated prices for charity. Midway through the evening a lamp used to illuminate a kinematograph set fire to the scenery. Within seconds the flames shot up to the roof, which was made of tarred felt.

There was only one exit, and all 1,500 patrons and society ladies stampeded toward it while flaming pieces of the roof descended upon them, turning some of the escapees into flaming torches. Within a few moments the molten roof collapsed and the walls caved in, burying and burning to death 150 people and brutally injuring hundreds more.

FRANCE
ST.-LAURENT-DU-PONT
November 1, 1970

A dropped lighted cigarette combined with multiple violations of fire codes caused the fire in the Cinq-Sept Club, a disco in St.-Laurent-du-Pont, France, on November 1, 1970. One hundred forty six patrons died; hundreds were injured.

The Cinq-Sept Club, in the small French village of St.-Laurent-du-Pont, 20 miles south of Grenoble, was a disaster aching to occur. A multiple array of safety violations and a disregard of common sense rendered this huge disco, a gathering place for young people from Grenoble, Aix-les-Bains and Chambéry, a dangerous fire trap.

In clear violation of French fire regulations, one of the two required access doors as well as the main entrance was sealed and the other locked. The two regulation fire exits were unlit; one was hidden by a screen behind the bandstand, and the other was blocked by stacks of chairs. The main entrance itself was an eight-foot-high, spiked turnstile.

The psychedelic decor in this hangar-like club consisted of an arched grotto sculpted from highly flammable polystyrene, another violation of fire regulations. Above the dance floor were tiny alcoves reachable by one spiral staircase. There were no fire extinguishers on the premises. And finally, and most astonishing of all, the club possessed no telephone.

On the night of November 1, 1970, the Cinq-Sept was packed with youngsters in their late teens and early twenties gyrating to the sounds of "Storm," a new group from Paris. Around 1:40 A.M. the group had just begun its last set with the Stones' "Satisfaction."

Upstairs in one of the alcoves someone dropped a lighted cigarette on a cushion. It immediately caught fire, and several patrons tried to beat it out with their hands and jackets. But the fire was stubborn, and in moments it had spread to the plastic arches that separated the alcoves. A vast tongue of flame shot the length of the dance floor as the plastic arches began to melt, dropping molten lumps of plastic on those near them.

At first there was an orderliness about the exodus of the crowd. Some in the alcoves descended the staircase and exited through the one obvious and available, if obstructed, exit. Thirty left this way. But moments later panic took over. The flames and the heat intensified enormously, and heavy, suffocating fumes filled the club, asphyxiating some couples who were later discovered still locked in each other's arms on the dance floor and near the bar.

One barman hurled himself against one of the emergency exits near him, and he and a handful of patrons escaped through it. Simultaneously, one of the owners, 25-year-old Gilbert Bas, saw an emergency light come on in his office. Walking toward the door, he heard the anguished cries of "Fire!" but did not open the door to the club. With no telephone, he was unable to call the fire department. He exited through his office door and drove almost two kilometers to the fire station to report the fire in person.

Meanwhile people in the club were dying. The pandemonium induced by panic had pressed the crowd against the turnstile at the main entrance, jamming it. Later rescuers would discover the body of one young man impaled upon one of the spikes of the turnstile.

The club had become an inferno. The corrugated iron roof turned red hot and collapsed on those inside. One hundred forty-four young people would perish horribly in the flames. Two more would die later of their burns, bringing the mortality total to 146.

The next day morbid curiosity would attract thousands to the tiny village to view the grisly sight. It would take a combined force of 200 policemen and law enforcement officials to move the crowd away from the ruins of the club.

The sheer magnitude of the disaster forced an intense investigation, and the village's mayor, Pierre Perrin, and secretary-general of the prefecture de l'Isere Albert Ulrich were immediately suspended from their jobs. As the investigation progressed, a tangled web of bureaucratic fumbling, backturning and compromise was revealed. There was scarcely a municipal agency that was not involved in some way. The rules were in place, but they had never been enforced.

In June 1971 the mayor and two building contractors were charged with causing injury through negligence. Gilbert Bas, the sole surviving owner (his two partners had died in the blaze), was charged with manslaughter. In November 1971 all were found guilty but received suspended sentences—Bas for two years, Mayor Perrin for 10 months, the three building contractors for 15, 13 and 10 months each. The fire would go down in record books as the worst in the history of France.

FRANCE/ITALY
MONT BLANC
March 24, 1999

Thirty-nine motorists were killed when a fire broke out in a truck in the Mont Blanc tunnel connecting France and Italy on March 24, 1999.

The seven-and-a-half-mile long Mont Blanc tunnel, connecting France and Italy through the gigantic mountain, opened with great fanfare in 1965. A favorite scenic route for tourists, it accommodated as many as 5,000 vehicles and was absolutely free of accidents until March 24, 1999.

At 11:00 A.M. that day a Belgian-registered refrigerator truck carrying 20 tons of flour and margarine began to smoke alarmingly. The truck was at the halfway point in the tunnel, and Gilbert Degrave, the driver, said later that he looked in his rear-view mirror and saw the smoke rising. "Fuel must have leaked onto an exhaust pipe," he continued. "Everything was ablaze in half a minute. I ran for my life. Behind me, all hell broke loose. In a few minutes, the tunnel was like an oven."

Degrave and four other drivers escaped. Thirty-nine others were engulfed in smoke, flames and heat that reached 1,800 degrees. Several other large tractor-trailer trucks ran into the abandoned refrigerator truck and caught fire as well. In the 1,000-foot stretch on either side of the abandoned and burning Belgian truck, eight other trucks driving toward France and 12 trucks and 11 cars heading for Italy became trapped by the flames and toxic fumes.

When firefighters were finally able to enter the tunnel, they were met with a scene of blackened devastation. Their firetrucks stalled for lack of oxygen, and the firemen made their way on foot past cracked tunnel walls and into a melee of cars, some with their engines still running, others with their drivers slumped at the wheel, either burned or suffocated from the fumes. One firefighter died of a heart attack in the rescue attempt and 27 more were injured.

The fire caused an avalanche of protests and recriminations. The tunnel's ventilation system was found to be obsolete and inadequate. Dwellers on either side of the tunnel had long complained of the pollution, rumbling and clogging of the roads from the 800,000 trucks a year that passed through the tunnel. Now they pressed for a change to the Swiss method of transferring the cargo of heavily loaded trucks to trains that used their own tunnels.

The Mont Blanc tunnel remained closed for two years while renovations and repairs were done, and more accusations flew. As of July 2001 12 officials of the French and Italian companies had been put under investigation to determine if criminal negligence had been committed.

GERMANY
DRESDEN
February 13, 1945

The Allied firebombing of Dresden on February 13, 1945, caused a firestorm that destroyed the city, killed 135,000 residents and injured hundreds of thousands more.

Picture fire falling from the sky, carried by the wind, consuming land, buildings, people and the very oxygen in the air. This is a firestorm, and this was the ghastly terror that took the lives of 135,000 people during the firebombing of Dresden, Germany, from February 13 through the 15, 1945.

Before it died Dresden was a beautiful city, the Florence of Germany, a center of art, architecture, sculpture and music from the 14th century onward. From 1500 it was the seat of Saxon princes, who fostered and collected the art that distinguished the city. By the late 17th and early 18th century, under Frederick Augustus I and Frederick Augustus II, Dresden became a showplace of baroque and rococo architecture. Bach, Handel, Telemann, Wagner and Richard Strauss lived and wrote there as it developed from a rococo center to a seat of romantic art and German opera. It was one of the loveliest cities in the world and a repository of thousands of pieces of priceless art.

For historians there were famous collections of watches and chronometers, geometric instruments, arms and armor. In the Zwinger, a former medieval execution field, friezes and statuary from the classic age of Greece were collected.

By the beginning of World War II, Dresden had become the seventh-largest city in Germany, with an *Auldstadt,* or old city, that looked as it had in medieval times, with narrow streets, shops, museums, a zoo and a number of ancient churches; and, beyond this, an old/new city with residences and light industry. It was also a rail center, but it was located far enough off the autobahn, which crossed the River Elbe west of the city, not to be choked with the traffic from this main national artery.

The autobahn bridge, the rail yards and the industrial complexes on the city's outskirts were virtually untouched by the firestorm that totally destroyed the city's center in 1945. The bombs and the fire were meant for civilians, and that is who they killed.

It is important to note that the firebombing of Dresden occurred at the end of World War II. In a last-ditch attempt at terrorizing the populace of Great Britain, the Nazis had unleashed V-1 and V-2 rockets. In the five years of the war, England had suffered a total of 60,595 casualties.

There were several tactical reasons for the launching of Operation Thunderclap, a firebomb raid on a leading German city planned for late January or early February 1945 with a combined force of RAF Lancaster bombers

Survivors prepare a mass grave for the dead following the firebombing of Dresden, Germany, on February 13, 1945. Some 135,000 died, hundreds of thousands were injured and the so-called Florence of Germany was leveled. (Library of Congress)

and USAF B-17s: (1) It was needed, it was argued, as a retaliation for the V-2s that had been hitting Britain in order to lift civilian moral; (2) It was needed to further lower German morale; and (3) It was needed as a bargaining chip for the upcoming Yalta Conference between Churchill, Roosevelt and Stalin. As an internal RAF memo stated it: "The intentions of the attack are to hit the enemy where he will feel it most, behind an already partially collapsed front, to prevent the use of the city in the way of further advance, and incidentally to show the Russians when they arrive what Bomber Command can do." There were several cities that were considered for the raid, among them Dresden and Berlin.

By now refugees had swollen the population of Dresden from 600,000 to one million. They lived in huts on the exterior of the city along with some Allied POWs, some of whom were kept in an abandoned slaughterhouse. Among these was writer Kurt Vonnegut, who would later write, devastatingly, of the Dresden firebombing in his novel *Slaughterhouse 5.*

The Yalta Conference, postponed, opened on February 4, 1945, and concluded on February 11. On February 3, acting on its own, the Eighth U.S. Air Force launched a twin bombing raid on Berlin and Magdeburg. One thousand Flying Fortresses bombed Berlin; 400 Liberators bombed Magdeburg. In the words of author Alexander McKee, in his *Dresden, 1945, The Devil's Tinderbox,* "the basic reasons for making such a raid on Berlin one day before the Yalta Conference opened were political and diplomatic: to make clear to the Russians that, despite some setbacks recently in the Ardennes, the United States of America was a super-power capable of wielding overwhelmingly destructive forces."

Thus, the job had already been done in Berlin. Why Thunderclap still went forward, two days after the end of the Yalta Conference, remains a mystery to this day.

A three-wave attack was planned and became reality on the night of February 13th and the afternoon of the 14th. Two waves of 1,299 Lancasters, carrying a bomb weight of 3906.9 tons, left England on the afternoon of the 13th.

It was Shrove Tuesday, and the streets of Dresden were thronged with people in holiday finery shopping or merely celebrating. At 10 P.M. the first "Christmas Trees," as the green flares released prior to a bombing were called, appeared over Dresden. There were no anti-aircraft defenses; only the air-raid sirens signaled the beginning of the raid.

The populace headed for bomb shelters and cellars and remained there while the RAF laid down a "carpet" of incendiary bombs. It was over in 15 minutes, and the populace came out to find flames everywhere. The Historical Museum in the old city center was afire, but a great many of its paintings had been removed, along with an immense china collection, to be transported out of the city before the Russians arrived. It was in a lorry parked outside the museum, and although it survived the first foray, it and the lorry would eventually be consumed.

Jets of flames swept up from the charred wreckage of ceilingless buildings and, driven by a west wind, began to travel horizontally across the city.

Meanwhile, the second, much greater force of bombers was on its way. At 1:22 A.M. the sirens sounded again. By now the streets were full of refugees from the first attack. The avenues, the parks, every open space was crowded with people trying to settle down for the night away from buildings.

Charlotte Mann, interviewed by Alexander McKee for his book, recalled, "It was as if fire was poured from the sky. Where there was darkness at one moment, we could suddenly see flames lick up . . . as I looked back to the center, I noted that it was just one single sea of flames. Now everyone started to make a run for the outskirts in order to reach some open space."

Margret Fryer, who had been questioned by the Gestapo that day and had escaped the concentration camps, now found herself in an inferno. "Because of flying sparks and the fire-storm I couldn't see anything at first," she recalled.

> A witches' cauldron was waiting for me out there: no street, only rubble nearly a meter high, glass, girders, stones, craters. I tried to get rid of the sparks by constantly patting them off my coat. It was useless. I stopped doing it, stumbled, and someone behind me called out: "Take your coat off, it's started to burn." In the pervading extreme heat, I hadn't even noticed. I took off the coat and dropped it.
>
> Suddenly, I saw people again, right in front of me. They scream and gesticulate with their hands, and then—to my utter horror and amazement—I see how one after the other they simply seem to let themselves drop to the ground. I had a feeling that they were being shot, but my mind could not understand what was really happening. Today I know that these unfortunate people were the victims of lack of oxygen. They fainted and then burnt to cinders. . . .
>
> It's dreadfully hot . . . I'm standing up, but there's something wrong, everything seems so far away, and I can't hear or see properly any more. I was suffering from lack of oxygen [too]. I must have stumbled forward roughly ten paces when I all at once inhaled fresh air.

What she had experienced, and what saved her, was the cool winter air rushing in to replace the boiling hot air of the fires and blowing through the inferno of flames that had heated the center of the old town to a temperature of 3,000 degrees Fahrenheit (sandstone begins to alter its form at 1,200 degrees). The cool air was felt by some as a sort of suction that actually collapsed more fragile buildings.

Margret Fryer stumbled on, climbed into a car to rest, was directed out of it and informed that it, too, had caught fire.

"Dead, dead dead everywhere," she continued.

> Some completely black like charcoal. Others completely untouched, lying as if they were asleep. Women in aprons, women with children sitting in the trams as if they had just nodded off. Many women, many young

> girls, many small children, soldiers who were only identifiable as such by the metal buckles on their belts, almost all of them naked. Some clinging to each other in groups as if they were clawing at each other.
>
> From some of the debris poked arms, heads, legs, shattered skulls . . . Most people looked as if they had been inflated, with large yellow and brown stains on their bodies. People whose clothes were still glowing . . . my face was a mass of blisters and so were my hands. My eyes were narrow slits and puffed up, my whole body was covered in little black, pitted marks.

The fire was so fierce it collapsed stone structures. Almost all of the houses lost their roofs, and the walled enclosures then acted like stoves, belching up huge balls of flame and hot smoke. Superheated air rose miles wide and miles high above the infernally flaming city. At 20,000 feet, the crews of the bombers saw the sky as a roseate bowl above them, and the turbulence caused by the rising hot air currents buffeted them about.

Under the collapsed buildings thousands of people were crushed or suffocated or were simply burned alive. Nearly every household in the city was swelled by refugees, and so the casualties were multiplied over and over.

By now the flames of the burning city were visible on the ground from 50 miles away. The wind was blowing from the west, which drove those who survived either toward the open flood meadows of the River Elbe to the southeast, where the Grosser Garten, a designated refugee space, lay, or south through the narrow streets of the old city.

Daylight was darkened in the city by thick clouds from the still-burning buildings. At dawn the third wave of 1,300 U.S. bombers and 900 Thunderbolts and Mustangs took off from King's Cliffe in Northamptonshire. The people of Dresden were surveying the horrible aftermath.

Annemarie Waehmann, who had survived the bombing of the Friedrichstadt hospital complex, approached what had once been the central station. She smelled "thick smoke everywhere." She continued:

> As we climbed with great effort over large pieces of walls and roofs which had collapsed and fallen into the street, we could hear behind us, beside us, and in front of us, burnt ruins collapsing with dull crashes. The nearer we came to the town center, the worse it became. It looked like a crater landscape, and then we saw the dead. Charred or carbonized corpses, shrunk to half size. Oh dear God! At the Freiburger Platz we saw an ambulance, with the male nurses just about to put a stretcher into it. A number of people were sitting on the ground. But why didn't they move? As we came nearer, we saw it all. They were all dead. Their lungs had been burst by the blast.

When the third noon raid arrived, there was no warning. All of the air-raid sirens had been silenced by fire. No incendiaries were dropped. The idea was to kill the escaping refugees.

Another attack—the fourth—occurred that evening on the outskirts of the city. But the rail yards, the autobahn bridge and the industrial complex were again untouched.

The following day, as it does after every firestorm, rain fell, turning the ghost of a city into a sea of mud, wreckage and half-hidden corpses. The discovered diary of a family named Daniels recorded some of the aftermath:

> First they brought [the dead] in wagons to the outskirts of Dresden for burial. Then they burnt them in the Altmarkt. The recoveries and burials took weeks, and there was the danger of epidemics breaking out. It was a miracle we survived. What misery existed. There were children dead whose parents were still alive, and parents dead whose children were left behind. For them, life is hardly worth living any more. We are very grateful that we are still alive and together. When the war is over, all we have to do is build everything up again.

And Dresden was built again, but like Rotterdam, which had been gutted and leveled by the Nazis, it was not, nor would it ever again be, one of the most beautiful cities in the world. That city had died, along with 135,000 people, in the most catastrophic fire ever set or experienced by any humans anywhere on earth in any age.

GREAT BRITAIN
ENGLAND
EXETER
September 4, 1887

A gaslight ignited scenery onstage, causing the fire in the Exeter Theatre in Exeter, England, on September 4, 1887. Two hundred were killed; hundreds were injured.

"The bodies were lying so thick [at the bottom of the gallery stairs] that they quite occupied the entire width of the staircase," said Harry Foot to the *Illustrated London News* on September 5, 1887; "in some cases they were four and five rows deep. At the bottom of the stairs they lay thicker than at the top, almost as if shot down a shoot. In the majority of cases the arms were outstretched beyond the head, as if they had struggled to the last to drag themselves forward; but their legs were rendered immovable by the bodies of those who had followed and partly fallen on them."

Foot and nearly 1,400 other playgoers had attended a performance at the stately Exeter Theatre, one of the prides of the city of Exeter, England, on the night of September 4, 1887. The performance had hardly begun when the nemesis of safety in 19th-century theaters (see pp. 203, 211 and 244), an onstage gaslight, ignited some canvas scenery. As in the Ring Theatre disaster in Vienna six years before (see p. 203), the initial ignition occurred in the flies

The Exeter Theatre fire in progress on the night of September 4, 1887. The theater, one of the prides of Exeter, England, was totally destroyed; 200 died and hundreds were injured. (Illustrated London News)

above the stage and slightly behind the top of the proscenium. Overhead gas lamps set fire to the uppermost reaches of a tall piece of scenery, just behind the act drop, and spread rapidly to the act curtain, then to the drapes in front of the proscenium. From there the flames shot out in lethal sheets into the audience.

The actors onstage and the wealthy in the stalls and boxes were able to file out without injury. Some 900 of them emerged unscathed.

It was, as usual, a different scene entirely in the gallery. There, pandemonium and hysteria took an early toll. Men, women and children fled toward the one stairwell that might allow them to escape the huge, billowing clouds of smoke that were now blotting out whatever light had been left in the theater.

By the time many of them reached the stairwell, it had become a fatal flue, collecting the smoke from other parts of the structure, containing it and shooting it upward. Some gallery patrons were trampled underfoot; others who managed to reach the stairwell suffocated from the smoke, fell in place and blockaded the exit, trapping others behind.

George Cooper, a soldier, William Hunt, a sailor, and the aforementioned patron Harry Foot were among the heroes of the day. Ignoring their own safety, they dashed into the theater, plucking survivors from the steadily accumulating piles of the dead and dying and dragging or carrying them from the flaming theater.

Eventually the flames reached the lead roof and heated it to the melting point. Flames descended the stairwells, followed by drops of molten lead. It was only at this point that the rescuers abandoned their efforts. "It would have been suicidal to have continued our work," Foot confessed to reporters later.

Two hundred patrons died in this fire, most of them in one stairway.

GREAT BRITAIN
ENGLAND
LONDON
September 2–6, 1666

A fire in the chimney of a bake shop, coupled with a long drought, caused the Great Fire of London on September 2, 1666. Thirteen thousand houses and 87 churches were destroyed; only eight people died.

The Great Fire of London was monumental in many respects. It destroyed 13,000 houses and 87 churches—including Saint Paul's—on 400 streets, laid waste to the Royal Exchange and Guild Hall and reduced a score of other public buildings to charred ashes. It burned for five days and was stopped only by a change in the direction and velocity of the wind. It reduced to rubble a large portion of the largest city in the world at that time and exposed to the public the woeful inadequacy of the fire-fighting apparatus and techniques of the age.

Yet there were positive aspects to this holocaust. If records are at all accurate, it claimed only eight lives. The bubonic plague, which had raged through Europe for decades, disappeared in England, apparently burned out of existence by the Great London Fire of 1666. Scientists theorize that the intense heat incinerated the plague bacillus, thus freeing the British Isles from what would continue to roam the continent of Europe for another 150 years.

The methods of forcing water through hoses by compressed air had been known and feasible since the invention, in 1590 by Cyprian Lucar, of the "portable squirt"—a brass tank powered by three men and used to some effect in fighting the 1666 fire. Decaus's "rare and necessary engine," developed in 1615 and outfitted with a swivel joint, and Hans Hautch's engine at Nuremberg, built in 1655, which was designed to force, by air, a steady stream of water at a fire, were also in limited use. But only the least effective of these, the portable squirt, was brought into play during the Great Fire, and even then on a very limited basis. Firemen simply pulled down flaming houses by grappling their walls with iron hooks on poles.

Afterward, however, modern methods of climbing ladders, extinguishing flames and carrying people to safety were begun. After the fire, when a new engine designed to fight fires was invented or introduced, it was not ignored, as it had been before 1666, but tested and, if found effective, adopted.

Furthermore, as a result of the extent of this calamity, the concept of fire insurance was developed, and less than a year later the world's first fire insurance policy was written by Dr. Nicholas Barton, who had built houses in the burned-out districts of London following the fire. His policies guaranteed to replace a house if it was destroyed by fire, and he did an immediate, brisk business, which eventually developed into Phoenix Fire Insurance, a firm that is still functioning today.

A long, pervasive drought preceded the London fire, very much like the long dry spell that occurred before the famous Chicago fire of 1871 (see p. 230). That fire began early in the evening. The London conflagration started at 2:00 in the morning in the chimney of the King's Baker's Shop on Pudding Lane, near London Bridge. From there, borne on a brisk wind, it ignited house after house and worked its way to the Thames wharves, where piles of flammable goods were stored.

Two of the most important men in the fighting of the Great Fire were Samuel Pepys and William Penn. Pepys, the son of a London tailor, was then secretary of the admiralty, an accomplished musician, a critic of painting, architecture and drama, a charming host and a connoisseur of beautiful women. He also kept a meticulous diary, and it is in the pages of this diary that the most vivid and precise record of the fire was set down:

> September 2, 1666. Some of the maids sitting up late last night to get things ready against our feast today . . . called us up about three in the morning to tell us of a great fire they saw in the City. So I rose, and slipped on my nightgown, and went to [the] window; and thought it to be on the back side of Mark Lane at the farthest, and so went to bed again and to sleep.

The next morning, Pepys's wife Jane informed him that 300 houses had been burned down and that all of Fish Street by London Bridge had been consumed. Pepys went to the Thames:

> I . . . got a boat, and through the bridge, and there saw a lamentable fire, everybody endeavoring to remove their goods, and flinging into the river, or bringing them into lighters that lay off; poor people staying in their houses till the very fire touched them, and then running into boats or clambering from one pair of stairs by the water—one side to another . . . Having stayed and in an hour's time seen the fire rage every way, and nobody, to my sight, endeavoring to quench it, but to remove their goods and leave all to the fire . . . and the wind mighty high and driving into the city; and everything, after so long a drought, proving combustible, even the very stones of churches, I to White Hall.

The Great 1666 Fire of London is depicted in a period painting. Thirteen thousand homes and 87 churches were destroyed; eight people died, and the black plague was thought to be incinerated in the flames. (New York Public Library)

At White Hall Pepys informed the king and the duke of York of the horrendous state of London and got from them an order to pull down every house that might carry the fire forward.

"At last," he goes on, "met my Lord Mayor in Canning Street, like a man spent, with a handkercher about his neck, to the King's message, he cried, like a fainting woman:

> 'Lord! What can I do? I am spent. People will not obey me. I have been pulling down houses; but the fire overtakes us faster than we can do it . . .' The houses so very thick thereabouts, and full of matter for burning, as pitch and tar in Thames Street, and warehouses of oil and wines and brandy and other things.

As the days and nights ached forward, the fire seemed to increase. Pepys went back and forth between the lord mayor, the king and the duke of York, bearing one repeated order: "Pull down the houses." He laments:

> and to the fire up and down, it still increasing, and the wind great. So near the fire as we could for smoke; and all over the Thames, with one's faces in the wind you were almost burned with a shower of firedrops . . . and, as it grew darker, appeared more and more; and in corners and upon steeples, and between churches and houses, as far as we could see up the hill of the City, in a most horrid, malicious, bloody flame, not like the fine flame of an ordinary fire . . . We saw the fire as only one entire arch of fire from this to the other side of the bridge, and in a bow up the hill for an arch of above a mile long. It made me weep to see it. The churches, houses, and all on fire and flaming at once; and a horrid noise the flames made, and the cracking of houses at their ruin.

By the morning of the fifth, William Penn, an important enough personage to command attention both in court and in the city, had taken a hand, and instead of simply pulling down houses, fire brigadiers were now, under his direction, blowing them up.

Pepys looked on:

> I up to the top of Barking steeple, and there saw the saddest sight of desolation that I ever saw; everywhere great fires, oil cellars and brimstone and other things burning. . . .
>
> I walked into the town, and find Fenchurch Street, Gracious Street, and Lombard Street all in dust. The exchange a sad sight, nothing standing there of all the statues or pillars but Sir Thomas Gresham's picture in the corner. Into Moorfield's our feet ready to burn walking through the town among hot coals and flint that full of people and poor wretches carrying their goods there

. . . Thence homeward, having passed through Cheapside and Newgate market, all burned . . . and took up, which I keep by me, a piece of glass of the Mercers' Chapel in the street, where much more was, so melted and buckled with the heat of the fire like parchment.

The wind changed; the fire abated. Less than a third of the walled city remained after this cataclysmic fire, which consumed most of London and taught many much about survival and prevention. One of those to learn from it was William Penn's Quaker son, who came to America 16 years later and founded Philadelphia. When he mapped out the city, he made sure it was filled with wide streets that would never become conduits of the sort of fire that he had witnessed as a child in London.

GREAT BRITAIN
ENGLAND
LONDON
November 18, 1988

Thirty-one people died and 80 were injured in the worst fire in the history of the London Underground at the King's Cross station on November 18, 1988. A lighted match dropped on grease and debris under a wooden escalator started the blaze, which trapped its victims in a wall of flame.

The King's Cross Underground station is London's busiest, a place where no fewer than five lines converge and British Railways maintains two terminals, the Kings Cross and St. Pancras lines. The Piccadilly, Northern, Metropolitan, Victoria and Circle underground routes all disgorge thousands of commuters on weekday afternoons and before holidays, and this was the situation on the late afternoon of Wednesday, November 18, 1988.

The Piccadilly line connects to the main ticketing concourse with a long, wooden escalator. Sometime during that afternoon, near the bottom of the staircase, someone dropped a lighted match into a gap beside the moving treads of the Picadilly line escalator. It was a thoughtless move or perhaps a prank; no one knows precisely which, but its effect was gradual and lethal.

As the afternoon turned to early evening and more and more commuters poured into the station and rode the escalator upward to the ticket concourse, the match ignited grease and debris that had accumulated beneath the escalator. The flames built steadily, and at 7:50 P.M., just as the last wave of commuters was emptying onto the upper level of the station, clouds of smoke began to emerge from the escalator. Suddenly an immense fireball erupted out of the regions below the moving stairs, shot up the railing in a white hot wall of flame and exploded into the turnstile area.

Horrified people began to panic as smoke obscured the station, and flames enveloped some who were in the path of the fireball. Attendants directed Picadilly passengers to the Victoria escalator, but it, too, emptied into the turnstile area, which was now a roaring inferno. "We followed their instructions and got on to the other escalator," survivor Andrew Lea told reporters. "About halfway up, a sheet of flame shot across the top of that escalator and very soon the ceiling was on fire and debris started falling down. . . ."

Chaos consumed the station. Attendants used no fire extinguishers, and no one seemed to know where to direct the panicked passengers. Finally 150 members of the London Fire Brigade arrived and began to gather the dead and dying, lead the choking survivors from the station and extinguish the blaze.

Thirty-one people died; 80 were injured, 21 of them seriously enough to require hospitalization. It was the worst fire in the history of the London Underground, surpassing the one in 1985 in which one person died and 47 were injured.

A year-long inquiry resulted in charges of negligence and incompetence on the part of attendants and safety and maintenance personnel in the London Underground. Sir Keith Bright, chairman of London Regional Transport, and Dr. Tony M. Ridley, chairman and chief executive of the London Underground, resigned. The next week the installation of new metal escalators and $465 million in safety improvements were announced in Parliament, and the repairs were carried out.

GUATEMALA
GUATEMALA CITY
July 14, 1960

Either faulty electrical wiring or a candle collapsing onto flammable material in front of a religious statue ignited the fire in the Guatemala City Insane Asylum on July 14, 1960. Two hundred twenty-five died; 300 were injured.

The Guatemala City Insane Asylum was madness personified. A structure built in 1890, its facilities, its design and its safety had all outlived their capacity by the summer of 1960. Sometime during the early hours of July 14 of that year, a fire began in the asylum, started either by faulty electrical wiring or a candle collapsing onto flammable material in front of a religious statue. Within minutes the ancient structure was ablaze, and 600 of its 1,500 inmates and attendants were trapped behind nonfunctioning doors.

The children who were housed in the asylum were the first to be evacuated, and every one of them survived. But there were still hundreds of adults who were incapable of saving themselves, and they were driven to wild hysteria by a fire that resisted every effort of the Guatemala City fire department to extinguish it.

Finally, realizing that most of the exits were blocked and that those that were open were not being used by the patients to free themselves, the fire department, led by Guatemalan president Miguel Ydigoras Fuentes, brought a bulldozer onto the premises and knocked down a wall.

Hundreds fled the building to safety through the hole in the wall, but others still had to be led, fighting and screaming, from the flames by rescuers. Thirty-one maximum-security patients, each considered dangerous, were never freed from their cells and burned to death in them.

The fire was brought under control by early morning of the 14th, and by that evening 27,000 pounds of relief supplies had arrived from the United States. Two hundred twenty-five patients, most of them women, perished in the fire, and 300 were severely injured.

In the grim aftermath murder and arson in the city increased after the fire. They were attributed to 48 criminally insane inmates who escaped from the burning asylum that night and were never recaptured.

IRAN
ABADAN
August 20, 1978

Arson caused a fire in a movie theater in Abadan, Iran, on August 20, 1978. Four hundred twenty-two died in the conflagration.

In 1978 Shah Mohammad Reza Pahlavi was trying to Westernize Iran. He ran head on into Muslim extremists, who announced that, in his efforts to emancipate women and redistribute clerical holdings, the shah was violating the teachings of the Koran. Further, the general atmosphere engendered by his "Westernizing" process had resulted in a general laxity on the part of segments of the public in observing the strict dictates of the Muslim holy month of Ramadan. One of the consequences of this that angered the extremists was the showing of movies during Ramadan.

Saturday night, August 20, 1978, was a hot and muggy night in the oil-refining city of Abadan, at the northern tip of the Persian Gulf. A crowd had gathered at the Rex Theatre to see the Persian-language film *The Deer* and escape the heat.

Partway through the evening, a group of Muslim terrorists aided by two employees of the Rex who were sympathetic to their cause approached the theater from the outside. They carried several cans of gasoline, which they proceeded to splash on every outside wall. Then, as several of them ignited the gasoline, others, aided by the employees, opened the only exit door and doused the interior section of the theater near it with flaming gasoline. They then slammed the door and locked it.

Inside the terrified audience went berserk. Flames roared through the building, consuming its interior and the people within. An enormous hill of grappling human beings piled up at the barricaded exit. Those who arrived there first were crushed under the pile; others were overcome by smoke; those at the top were incinerated.

The heat rose to inhuman levels. The entire building was consumed by flames. Some managed to smash windows; still others discovered a roof exit that had been overlooked by the arsonists and escaped. But they were the lucky few. By the time firemen arrived on the scene, smashed windows and broke in the barricaded door, the screaming from within the inferno that was once a theater had stopped.

Four hundred twenty-two people were burned to death, died of suffocation or were trampled to death. Entire families from the working-class neighborhood in which the theater was located were wiped out.

Ten arrests were made the following Monday, and the theater's manager was arrested and charged with negligence for ignoring police orders to hire more employees and guards. It would be one of the last futile gestures of opposition to the Muslim extremists, who would soon command the country.

JAMAICA
KINGSTON
May 20, 1980

Overcrowding combined with a short circuit in the wall of Eventide Home, a nursing home in Kingston, Jamaica, caused a fire on May 20, 1980. One hundred fifty-seven perished in the blaze.

On the night of May 20, 1980, there were 204 elderly, indigent women asleep in a 110-year-old, two-story building in a three-building complex called Eventide Home, located in Kingston, Jamaica. The other two buildings housed elderly men and handicapped children, and the entire complex was city owned and city run.

The structure sheltering the women was particularly decrepit and dangerous. Built of highly inflammable pitch pine wood, it had been branded a "tinder box" by Kingston fire chief Allen Ridgeway several times, but the city had ignored his warnings and had packed the building, which had a legal capacity of 180, with 204 elderly women.

"It was a place of indigent people," the fire chief explained to reporters after the fire that destroyed it. "The ratio of indigent rose and the capacity of the complex couldn't be expanded. The normal statutes just couldn't be kept."

And this breaking of its own statutes by the city only intensified the tragedy when the inevitable finally happened. At 1:00 A.M. on Wednesday, May 20, 1980, a short circuit in the building's electrical system started a fire in one of the walls. By the time anyone had even smelled the smoke, the flames had begun to consume the building. Screaming women, some unable to leave their beds, remained helpless before the onslaught of the flames, which raced with lightning speed through the entire building, collapsing walls and floors and igniting everything burnable within moments.

Some women managed to reach windows, but the fire spread so rapidly that it had become a hopeless situation long before the fire department even arrived. There were neither ladders nor safety nets for them, and the women who jumped from the upper level of the two stories injured themselves seriously.

The fire department arrived on the scene within five minutes. Four minutes later the entire building collapsed upon itself with a sickening roar that mixed with the piercing screams of the women still trapped within it. A huge funeral pyre, it instantly silenced the last frantic efforts of any remaining survivors. Of the 204 women who had just 20 minutes before been sleeping peacefully within the shelter, only 47 would escape. One hundred fifty-seven died in the flames, most of them burned alive. It would be almost impossible to identify most of them the next day.

The best the fire department could do was to evacuate the children from their nearby shelter. The men, at a far corner of the complex, were not disturbed, and many slept through the entire holocaust.

A political campaign was warming up in Jamaica, and both sides in the contest irresponsibly accused the other of sending arsonists to start the fire. Prime Minister Michael N. Manley informed local radio stations that night that arsonists began the blaze, and a Kingston police spokesman perpetuated the rumor that telephone wires to the complex had been cut shortly before the blaze started. But Fire Chief Ridgeway steadfastly refused to blame arsonists, and his investigation proved that the immediate cause of the fire had been an electrical short circuit. The resultant tragedy was caused by housing helpless people in an overcrowded fire trap.

JAPAN
OSAKA
May 13, 1972

A short circuit in a room containing oil-soaked rags plus obscured fire exits combined to turn the fire in the Playtown Cabaret in Osaka, Japan, on May 13, 1972, into a fatal inferno. One hundred eighteen died; 38 were injured.

The sealing or blocking of fire exits is the major cause of fatalities in fires that occur in public places. This was tragically true in the Cinq-Sept fire in France (see p. 212), and it was also true in the Playtown Cabaret fire in Osaka, Japan, on May 13, 1972.

The cabaret, a club frequented by businessmen and young couples and overseen by an army of partially clothed hostesses, occupied the top story of the Sennichi department store in Osaka. When the department store closed, the fun began in the cabaret.

On the evening of May 13, 1972, a lone workman in the store, an electrician named Keiji Kewashima, was making some electrical repairs on the third floor. Wires apparently shorted in a room that contained oil-soaked rags, and Kewashima soon found himself surrounded by flames. He ran from the room, shouting a warning that could not possibly be heard three stories above him in the Playtown Cabaret.

Thus the fire spread unopposed into the elevator shafts, up the walls and through the ductwork. When it finally reached the cabaret, practically all escape routes had been rendered useless. The cables in the elevator shafts had been burned apart. Flames licked at the outsides of windows. Hallways were filled by blinding, suffocating smoke.

The fire exits that might have provided some safe escape were hidden behind drapes. Later rescuers would find piles of charred corpses, their hands reaching out as if searching along the superheated walls for the fire exits. An emergency fire chute *was* discovered by some, and 20 people scrambled down it. But halfway to the street the chute collapsed, sending all 20 to their deaths.

Hysterical patrons, finding themselves trapped, smashed windows with tables and chairs. Some leaped from the windows, killing themselves in the fall. Nineteen tried to jump to the next building. None made it; all died. Others climbed to the roof, but by then the entire building was a roaring, consuming torch, and they were either burned alive or forced to leap to their deaths.

When the firemen finally arrived with their extension ladders, frantic patrons were falling past them to the earth. One hundred eighteen people died from suffocation, burning or falls, and 38 were injured. Only 48 were rescued.

JAPAN
TOKYO
September 1, 2001

Forty-four young men died and nearly a hundred were injured in a consuming fire in a building housing a gambling club in the Shinjuku Ward of Tokyo, Japan, in the early morning hours of September 1, 2001.

"I heard a sound and thought maybe some luggage had fallen out of a window," said a passerby in the Shinjuku Ward of Tokyo on September 2, 2001. "When I turned to look, I saw that it was a man who had fallen from the building."

"I heard glass breaking and saw a man falling," said another. "There was another man holding his head in his hands. I didn't think much of it at first but then I heard a loud bang and saw smoke billowing out."

They were recreating for the press the event of the night before: the sudden and catastrophic fire that had ripped through a narrow building in the Kabukichoco section of the Shinjuku Ward of Tokyo—a seamy part of the city that was devoted to the sort of entertainment that attracted men after a long day at the office: peep shows, sex parlors, "soapland parlors" where naked attendants applied sudsy massages to patrons, bars with nude waitresses and gambling parlors. A neighborhood of narrow, dimly lit alleys, neon-bright islands of pleasure and slim, multistoried buildings with a floor devoted to each attraction, it was crowded as always shortly after midnight on September 1, 2001.

The third floor of one of these four-story buildings housed a mah-jongg club, where gamblers gathered nightly to test their skill and patience at the club's 17 tables while being served drinks by waitresses dressed as high school girls. Shortly before 1:00 A.M. an explosion occurred somewhere in the slender building, and heavy smoke began to fill the third and fourth floors. "I saw black smoke coming from the elevator hall and then flames," an employee of the mah-jongg parlor later told police. "So I jumped from a kitchen window."

Panic erupted with the flames. Patrons shoved and scrambled for the one narrow passage that served as a hallway, exit and entranceway for the entire structure. The hall was made even narrower by arrays of lockers consuming half of its space. The building was also almost windowless, but some survivors managed to smash the windows that existed and fling themselves from them, injuring themselves as they landed. Others climbed to the roof. Three men either fell, jumped or were pushed from the roof to their deaths; others remained there and were rescued by firemen.

The fire, police and rescue teams found a large hole in the side of the building caused by the explosion and thick, black smoke emitting from every orifice. As they entered the building, frantically searching for survivors, they were met with blasts of enormous heat funneled through the one stairwell and confined by the lack of windows and fire doors.

"The heat and the smoke were just too great," one rescuer said later. That in itself could have been relieved if the building's meager safety measures had worked. But the fire doors did not shut soundly. "If the emergency fire hatches had operated properly, it would not have been this bad," a fire official told reporters at a news conference.

It was a lapse in a country that prided itself in the strict enforcement of its stringent fire codes. Devastating fires from major earthquakes had taught Japan the lesson of vigilance against spreading fires.

But this was an exception, and even as the survivors were being taken to 22 different hospitals, an investigation into the causes of the fire was launched. Many of the survivors died from smoke inhalation or consuming burns. The death toll was 44, mostly patrons in the bar and gambling parlor; the injured numbered nearly 100.

At first an open gas pipe was suspected, but inspectors from Tokyo Gas Company found no cracked pipes or faults with the gas meter. Later inspection uncovered the remains of a stairway that had been clogged with garbage, hand towels and old newspapers placed there by the owner of the mah-jongg parlor. Because many of the attractions in the ward were run by Japanese crime syndicates, arson was suspected. As of this writing the investigation is ongoing.

PUERTO RICO
SAN JUAN
December 31, 1986

Labor troubles led to arson that caused the catastrophic fire on New Year's Eve, December 31, 1986, in the Dupont Plaza Hotel in San Juan, Puerto Rico. Ninety-six died; hundreds were injured.

For nearly two weeks preceding New Year's Eve of 1986, there were bad feelings between the management of the 22-story luxury Dupont Plaza Hotel, located in the Candado Beach area of San Juan, Puerto Rico, and Local 901 of the International Brotherhood of Teamsters, which represented the hotel's employees. During those 10 days three small, smoky fires had broken out and been extinguished in various parts of the hotel. When guests—a great number of them from the United States—called the front desk and mentioned rumors that there had been bomb threats and fires in the hotel, they were told that the rumors were "false and groundless."

The hotel was crowded with holiday celebrants, many of them gambling in its casino, on the afternoon of December 31, 1986. Some of them had received disquieting telephone calls in their rooms that morning warning that they would be "burned out." Again, when they reported these calls to desk clerks they were told that they were probably the work of holiday pranksters.

At 1:41 P.M. that afternoon a phone call was received at a police station near the hotel from a man who identified himself as "Santiago." The message was that a bomb had been planted on the premises of the hotel. Two policemen were dispatched to the Dupont Plaza. There is no record that they made a thorough investigation of the premises, but according to José L. López, a senior spokesman for the commonwealth's police department, they did speak with members of the hotel staff, who told them that "there were no problems, everything was normal." The officers, Mr. Lopez said, "went back to the station and made a report."

Meanwhile, members of the hotel staff who belonged to the Teamsters' local and union officials were holding a stormy meeting in the hotel's ballroom. There was a vote on management's latest contract proposal, and it was turned down. A strike was called for midnight.

Among the employees present at the meeting was a 35-year-old maintenance worker named Hector Escudero Aponte. He was particularly frustrated and angered by the attitude and the offer of the hotel's management. He felt, as he later told investigators, that something had to be done to prove that the union meant business. He knew, he thought, how to intimidate management into listening to the union.

The previous day a shipment of new furniture wrapped in plastic had been delivered to the hotel. It was stacked in the ballroom in piles six feet high.

Aponte stopped by the kitchen and picked up a can of Sterno-type cooking fuel. He entered the ballroom a little after 3:00 P.M., placed the canister next to one of the piles of furniture and lit it. It would, he believed, produce a nice, contained but smoky fire. He would later confess that he only meant to start "a small fire that would damage the personal property of the hotel."

But that is not what happened at all. The six-foot towers of plastic-wrapped furniture became incendiaries. The plastic exploded, sending huge gouts of black smoke and orange flames skyward through the walls and into the casino.

There were 70 people there concentrating on winning money to spend at New Year's celebrations and beyond. According to Kevin W. Condon of Ansonia, Connecticut, "Somebody [at the blackjack table] said there was smoke. But nobody paid any attention, and we continued playing. Then there was a big burst of smoke and we went running toward one of the exits. When someone opened the door, we saw that the whole hallway was covered with black smoke. We slammed the door, went running toward the other exit and that was filled with black smoke. Then the panic began."

"My supervisor shouted there was fire and for everyone to run," said Susana González Pérez, another survivor and a hotel employee, "but there was no time. Fire coming like a ball shot across the ceiling. It hit my supervisor and killed him in his tracks."

The casino was on the mezzanine level, and according to later accounts all but one of its exits were closed and locked for security reasons. Hysterical patrons, now fully cognizant of the fire that was creeping into the casino, smashed the floor-to-ceiling tinted windows that overlooked the ocean.

From here they jumped several feet to the poolside patio below. "People were hysterical," said Larry Roberts of Manhattan. "Everyone at the pool took off like crazy, jumping over barbed wire and knocking over older people who couldn't get over it."

"People were running out of the pool area bleeding from cuts from glass that showered the area," said Alexander Leighton, the owner of the Casablanca guest house across the street. "Two croupiers came running across the pool area and jumped into the pool with their clothing smoldering."

Meanwhile, terrified hotel guests were blinded by the thick smoke that filled the hotel corridors. Puerto Rican law did not specify that sprinkler systems were mandatory in public buildings, and the Dupont Plaza did not have a sprinkler system.

Dominick Pannunzio and his wife, on the 15th floor, opened the door of their room and plunged into a cloud of dark gray smoke. They groped their way to an exit stairwell with a panicking crowd of screaming and shoving people.

"We ran down to the eighth floor," Pannunzio recalled, "and we ran into a solid wall of people, yelling, 'Go up! Go up! You can't get through here!' You couldn't see. Everyone was in panic.

"At the 17th or 18th floor, a bunch of people were coming down yelling, 'Down! Down! You can't go up!' We tried to get into one of the halls but all the doors were locked. People were gagging and falling down."

Finally a maid let Pannunzio and 20 others into a room that had a balcony. They soaked towels and linens with water from the bathroom sink, held them to their faces and huddled on the room's balcony for four and a half hours before firemen finally rescued them.

The smoke was as dangerous as the flames. The plastic wrapping around the piled furniture in the ballroom was capable of producing noxious fumes, and the furniture itself, as well as other furniture throughout the hotel, was made of a fake leather composed of polyvinyl chloride. When burned this substance produces hydrochloric acid, which scorches the lungs, the nostrils, the eyes—and can cause death immediately. Most of the dead were discovered near the pool area, the ballroom and the lobby—some of them sitting peacefully in chairs.

A great many guests ascended to the roof of the building in the hope that helicopters would rescue them. A police helicopter hovered near, but according to radio transmissions received by private pilots its pilot felt that he could not land because the roof was not flat.

At the same moment another helicopter pilot, Pat Walker, a 41-year-old charter pilot based in St. Thomas, was unloading a flight of four people at San Juan International Airport. The four had flown from the Virgin Islands to a New Year's party in San Juan.

Seeing the thick column of smoke, he radioed authorities, who said he was needed and gave him clearance to refuel at Isla Grande and proceed to the Dupont Plaza.

Fortunately, Walker had no timidity about landing on the roof. The closest he could get, however, was to put one landing skid down on its edge. He had no hoist aboard and had to maneuver his helicopter perilously close to the roof. Blinded at times by the column of heat and smoke spiraling upward from the hotel, he coaxed four hysterical women aboard the craft and brought them safely to the ground.

Taking a police officer with him who helped calm the guests and organize the evacuation, Walker returned over and over again during the next 45 minutes, plucking 21 people from the roof of the hotel in his five-passenger Ranger and depositing them safely on the beach below before larger military helicopters arrived on the scene.

It would be 7:00 that night before the fire was brought under control and days before the dead, the dying and survivors were removed from the wrecked hotel. Some were found in their rooms asphyxiated from the smoke that spread through the air-conditioning system. There were bloody towels on the adjoining beach, but a few yards away at the pool, palm trees were pristinely untouched. Paperback books and bottles of suntan lotion were strewn around, dropped by fleeing vacationers.

In the end, 96 died and hundreds were injured. Hector Escudero Aponte confessed to arson and was sentenced to the maximum 99 years.

ROME
July 19, A.D. 64

Imperial arson ordered by Nero caused the fire that consumed three of Rome's 14 districts and damaged seven more on July 19, A.D. 64. No fatality or injury figures survive.

One of the more despicable pictures of ancient times is that of Nero fiddling while Rome burns. But the picture is not entirely accurate. Actually, the mad emperor fingered the lyre while he sang verses from *The Fall of Troy* and watched the conflagration from a safe hilltop.

There is unanimous consent among contemporary historians that Nero ordered this terrible fire set—possibly so that he could expand his already grandiose palace, which occupied two of Rome's seven hills, possibly because he had tired of the drabness of Rome's ancient buildings, possibly because he liked fires.

In any case, Nero departed on a short trip to Actium on July 17, A.D. 64, and on July 19 a mysterious blaze began in the vicinity of the Circus Maximus, at the bottom of the Palatine Hill. From the Circus Maximus it spread swiftly, helped by a strong wind and the narrow streets of the quarter. It moved on without mercy, consuming buildings that had stood since the time of Romulus, the founder of the city 800 years before. Romulus's temple dedicated to the god Jupiter was one of the many venerable and irreplaceable buildings that burned during the six days and seven nights of the conflagration.

Tacitus, the historian of Rome, described the sorry scene:

> Terrified, shrieking women, helpless old and young . . . fugitives and lingerers alike—all heightened the confusion. When people looked back, menacing flames sprang up before them or outflanked them. When they escaped to a neighboring quarter, the fire followed—even districts believed remote proved to be involved. Finally, with no idea where or what to flee, they crowded onto the country roads, or lay in the fields. Some who had lost everything—even their food for the day—could have escaped but preferred to die. So did others, who had failed to rescue their loved ones.

That the fire was officially set was supported by Tacitus, too. "Nobody dared fight the flames," he wrote. "Attempts to do so were prevented by menacing gangs. Torches, too, were openly thrown in, by men crying that they acted under orders."

Finally, before the entire city was destroyed, fire brigades demolished buildings in the fire's path, and it ended, but not before consuming three of the city's 14 districts entirely and severely damaging seven more. Nero forbade homeowners from returning to salvage what they could from the ruins of their homes. The reason? Tacitus answers: "to collect as much loot as possible for himself."

Rumor based on fact spread through the city as fast as the fire; Nero had ordered it. To stop the rumor, Nero publicly speculated that the Christians in Rome, among them Saint Peter, were behind the arson that had wreaked such havoc. He ordered mass arrests and public crucifixions. Christians were set afire in Nero's gardens, and others were forced to enter the Circus dressed as animals, where killer dogs tore them to pieces.

According to the historians, even this failed to hide Nero's guilt. He rebuilt the city after reconstructing his own palace on a hitherto unprecedented scale of opulence. There was a 120-foot-high statue of himself in the entrance hall, a pillared arcade a mile long and gardens containing lakes and complete forests. In the city rebuilt public buildings were restricted in height, built of nonflammable stone, and porches were dictated as part of their approved design so that fire fighters could have easy access in case of future fires.

RUSSIA
IGOLKINO
March 12, 1929

Drunken negligence on the part of a projectionist compounded by the overcrowding of a room with inadequate exits above a factory in Igolkino, Russia, caused the March 12, 1929, fire in that city. One hundred twenty died in the blaze.

March 12, 1929, was the 12th anniversary of the abdication of Czar Nicholas II. In the tiny Russian village of Igolkino, 250 miles northeast of Moscow, a group of drunken workers and their families decided to celebrate by viewing Victor Seastrom's classic film *The Wind*. Igolkino possessed no movie theater, but this did not dissuade the celebrants. They commandeered a 24-by-24-foot room above a factory. The factory manager had protested vehemently against the use of the room. First, according to *New York Times* reporter Walter Duranty, "[he] feared the peasants would steal tools stored in the room." But more importantly and perhaps a bit more believably, he knew that 30 gallons of gasoline had been accidentally spilled on the floor of the room the day before, that there was only one exit from the room and that the windows were too small to accommodate people trying to flee from a fire.

His protestations fell on deaf ears. The village Soviet warned him that he would be arrested if he tried to prevent the workers from using the room. The factory manager acquiesced, and workers led by Bazarnof, a drunken projectionist who carried the projector and film in one hand and a bottle of vodka in the other, crammed themselves into the fetid room.

Most of the revelers could not have cared less about the motion picture. They in fact shouted for music. Bazarnof complied, turning the running of the film over to an unskilled and equally drunk friend. Lighting up a cigarette and strapping on an accordian, Bazarnof squatted in the doorway of the only exit and began to play Russian folk songs. The substitute projectionist allowed the film to run off the take-up reel and accumulate in a pile on the floor.

Unconcerned, Bazarnof continued to play the accordian. When his cigarette had burned down to a butt, he flipped it. The still-glowing cigarette landed in the middle of the nitrate-treated film and instantly ignited it. The flames rushed to the gasoline-soaked floor, and within seconds the entire room and its occupants were ablaze. Bazarnof leaped up and ran. He did not stop until he reached a nearby village, where, a day later, he was arrested.

Meanwhile, people choked and suffocated on the thick black smoke generated by the ball of fire that had now consumed the room. Some were trampled to death underfoot; others were burned alive.

In the midst of this someone discovered a trapdoor that opened onto the factory below. One hundred thirty people managed to squeeze through either the trapdoor or the one exit, but 120 died in that 24-by-24-foot cauldron.

One more victim would be claimed in a ghoulish and grisly charade. Furious and distraught over the mayhem and death, the village's peasants vented their rage not on the absent projectionist who had caused the fire, but on the factory manager who had tried to warn their dead comrades away from the firetrap. A mob of workers cornered him, stoned him, beat him unconscious and flung him into the still-raging fire, where he burned to death.

SOUTH KOREA
SEOUL
December 25, 1971

Human negligence in failing to contain a small fire caused by a propane tank explosion in a coffee shop led to the huge fire in the Taeyunkak Hotel in Seoul, South Korea, on Christmas Day, December 25, 1971. One hundred sixty-three died in the fire; 50 were injured.

Holiday times seem to be particularly vulnerable to tragedies resulting from human carelessness. Eight workers and executives of the luxurious 21-story Taeyunkak Hotel in the center of Seoul, South Korea, were arrested and charged with negligence after the December 25, 1971, fire that raked the hotel with roaring flames and caused the death of 163 persons.

The fire began at 10:00 A.M. when a propane tank used for cooking in a second-floor coffee shop exploded and burst into flames. Under ordinary circumstances this manageable fire should have been extinguished or at least contained within the confines of the coffee shop. But it was not, and the flames soared up through conduits and elevator shafts, climbing 20 stories to the hotel's roof within minutes.

Fortunately, because of the Christmas holiday, the offices in the building were unoccupied. Still, 317 people—187 guests and 130 hotel employees—were in the building when the fire began. Again, as in so many fires in public buildings, there were too few fire escapes, and those that existed were blocked by fire, smoke or debris.

Firemen arrived quickly, but an incredible situation developed as soon as they came upon the scene. Amazingly, in a city of skyscrapers, they had ladders that reached only to the fourth floor. Their hoses only drove water as high as the ninth floor, and the flames were shooting out of the building all the way to the 22nd story and beyond.

Panicked, hysterical people began to fling themselves from windows. Even when 13 helicopters arrived, the mayhem and dying scarcely ceased. The roof was consumed in flames; there was no landing space for the helicopters, and so their pilots and crews attempted to rescue survivors by ladder from window ledges. It was a risky exercise for professionals under ideal circumstances. It proved disastrous in this situation. Only a small number of people managed to clamber up the swinging ladders to safety, and two who were rescued from the flames lost their grip and fell to their deaths. Everything was tried, even the pieced-together poles of circus acrobats, but little could be done for those trapped on the upper floors of this flaming modern hotel, and it was considered fortunate that only 163 people died and approximately 50 were injured.

SPAIN
MADRID
September 22, 1928

A short circuit set fire to scenery in the Novedades Theater in Madrid on September 22, 1928. The resultant fire killed 110 people; 350 were injured, many seriously.

Every major city in Europe seems to have its favorite theater, and in Madrid in the 1920s it was the Novedades Theater. A venerable, ornate wooden structure built in 1860 and converted to electricity soon thereafter, it was in need of further refurbishment in 1928. But the nearly 3,200 spectators who jammed it on the night of September 22, as they did practically every night it was open, were unaware of the old structure's aging innards.

The consequences of time burst into flame during the intermission that evening. A short circuit in the ceiling of a room used to store unused scenery set fire first to the ceiling, then the scenery, then adjacent rooms and finally the auditorium of the theater itself.

The performers, who saw the flames first, were able to escape through their stage exits and entrances. The wealthier patrons on the orchestra floor were, for the most part, either outside or in the outer lobby and thus escaped unharmed.

But as usual in theater fires, the less affluent customers, the working class who could only afford balcony seats, were trapped, hemmed in by the height of the balcony on one side and inadequate, cramped, smoke-filled staircases on the other. Panic struck immediately, and scores were trampled in the rush toward the stairways. Some patrons, shoved by other hysterical audience members, lost their footing and plunged over the balcony rail and into the auditorium.

"Many persons," reported the *New York Times,* "mad with terror tried to fight their way out stabbing with knives right and left or biting, scratching or shoving aside weaker persons in their way."

For each of these despicable acts there were equal numbers of heroic ones. Men carried children on their shoulders through walls of flames and then reentered the inferno to rescue more. A woman usher stood her ground holding a flashlight on a dark exit so that people could be guided to it. She was discovered dead the next day, still clutching the flashlight.

The two staircases soon became piles of wedged bodies caught between flaming wooden walls. Firemen could do nothing to save the old building. Their ladders could not be raised in the narrow streets bordering the Novedades, and so they watched helplessly as people threw themselves to the street and their deaths from upper stories of the theater.

One hundred ten died; 350 were injured, some of them for life.

SYRIA
AMUDE
November 13, 1960

An unexplained explosion in the projection booth of a movie theater in Amude, Syria, on November 13, 1960, caused a fire that gutted the building. One hundred fifty-two children attending a special program were killed; 23 were injured.

As in Montreal (see p. 206), a children's movie program in the tiny town of Amude, Syria, near the Turkish border, turned into a fiery nightmare. The Montreal holocaust took place during a special matinee. The Amude disaster occurred at a special evening program, in which 175 children were gathered in the small local movie theater to see a special film.

The children, who accounted for a large part of the village's population under the age of 15, had just begun to settle back and become involved in the film when suddenly jets of flame shot out of the projection booth accompanied by pieces of flaming film. In the inquiry that followed, some witnesses said an explosion had occurred in the projection booth; some refuted this. The projectionist was severely burned and thus could neither verify nor dispute it.

In any case, the theater was consumed by raging flames in minutes. There simply was neither enough time nor exits to save most of the children. One hundred fifty-two died from suffocation, trampling or fire. Twenty-three escaped, but every one was badly injured.

THAILAND
NAKHON PATHOM
May 10, 1993

Two hundred six workers were killed and 500 were injured in the worst factory fire in history at a doll factory in Nakhon Pathom, Thailand, on May 10, 1993. As in the Triangle Shirtwaist fire (see p. 246), substandard working conditions, blocked exits and a lack of safety precautions caused the appalling loss of life.

The parallels were eerie and disturbing:

- The Triangle Shirtwaist Factory in New York City employed young, immigrant girls in 1911 and paid them starvation wages of $18 a week.
- The Kader Industrial Company doll factory in Nakhom Pathom, Thailand, employed immigrant girls in 1993 and paid them starvation wages of between $120 and $160 a month.
- Working conditions at the Triangle Shirtwaist Factory were crowded and abominable, and the doors were bolted shut in order to prevent the girls from stealing the stock.
- Working conditions at the Kader Industrial Company were substandard, and the doors were bolted shut to prevent the girls from stealing the dolls and novelty items they assembled.
- Fire escapes and safety precautions at the Triangle Shirtwaist Factory were either inoperative, decaying or nonexistent, and when, on March 25, 1911, a fire began in a rag bin and rapidly consumed the factory, hundreds died needlessly, piled up against the blocked doors or in headlong leaps from upper stories when the fire escapes failed.
- Fire escapes and safety precautions at the Kader Industrial Company were nonexistent, and when, on May 10, 1993, a fire began in the cloth-cutting area and rapidly consumed the factory, hundreds of helpless girls died needlessly, piled up against the locked doors or in headlong leaps from the factory's roof when they found no fire escapes.
- 145 girls were burned to death and scores more were injured at the Triangle Shirtwaist Factory.
- 206 girls were burned to death and 500 were injured at the Kader Industrial Company, making this the worst factory fire in the history of the world.

There were four buildings in the complex of the Kader Industrial Company, Ltd., a Southeast Asian–owned manufacturer for export of dolls and novelty items. On May 10, 1993, 800 employees were working in the four-building complex when a short circuit apparently caused a fire in the ground floor cloth cutting room of one of the buildings.

Within minutes the flames spread to the top floors of this building, then leaped to two other structures. Workers clawed their way to staircases leading to the exits, but the doors were bolted shut, and as fireballs of flame pursued them, more and more frantic workers piled onto the staircases, which collapsed under their weight.

Scores were incinerated as they flung themselves in clusters against the doors or were crushed by the collapsing stairs. Meanwhile, those who escaped this fate dashed to windows, from which fire escapes should have taken them to the ground. But there were no fire escapes, and so they did what they had to do and leaped six stories to the earth below. Most suffered broken bones or concussions.

"It's not our fault," one guard in charge of a locked door told reporters. "The company told us to lock the doors so people would not sneak out or steal."

"It's not our fault," echoed company executives to television reporters. "We simply complied with government regulations."

The next day, as Thai soldiers began the grim task of searching through the smoking rubble of the factory, in which human limbs poked up through the ashes like charred tree trunks after a forest fire, the government launched an investigation into the numerous safety violations in the factory. Not a single fire alarm was discovered, and charges of gross negligence were filed against the company's executives.

TURKEY
CONSTANTINOPLE
June 5, 1870

Hot charcoals spilled from a brazier onto the wooden steps of a home in the Armenian section of Constantinople, Turkey, on June 5, 1870, and fanned by high winds led to a conflagration that destroyed 3,000 homes and set fire to the entire city. Nine hundred residents died.

There seems to be some dispute about some of the details of the great fire that swept through Constantinople, Turkey, on Sunday, June 5, 1870. Several versions indicate that it was a balmy spring day, and a large portion of the population was out of the city enjoying picnics and the country. But these same reports also indicate that a gale-force wind was blowing, and this wind was responsible for the wildfire nature of the disaster. Considering the enormous number of casualties—900 persons burned to death, more than 3,000 buildings destroyed—it would seem that the population was at home, not out battling the winds on open picnic grounds.

One detail runs consistently through the chronicles of that terrible day, however: the origins of the fire. An Armenian family in the Valide Tchesme district was definitely at home at dinnertime, and the mother of the household instructed her young daughter to go upstairs, fill an iron pan with burning charcoal and bring it downstairs to the cooking quarters. The daughter obeyed, but on the way back she dropped some of the glowing charcoal on the steps. The gale, blowing through an open window on the staircase, scattered the sparks onto the roof of an adjoining home, and the blaze was under way.

Flames leaped from home to home, leveling both the Armenian and Christian quarters in a matter of hours, and then roared to the docks on the Bosporus and up Feridje, the grand street that contained churches, shops, hospitals, legations and consulates.

The churches, hospitals and diplomatic missions were surrounded by stone walls and sustained little damage. Sir Henry Elliott, the British consul, suffered only a singed silk dressing gown, which he wore while directing fire prevention within his compound. But the damage to the remainder of the city was devastating: 900 dead, 3,000 buildings in ruins and more than a square mile of Constantinople reduced to rubble.

A fire brigade rushes down a Constantinople street during the consuming fire of June 5, 1870. *(Illustrated London News)*

UNITED STATES
CONNECTICUT
HARTFORD
July 6, 1944

The most tragic circus fire in history, the Ringling Brothers, Barnum & Bailey fire in Hartford, Connecticut, on July 6, 1944, was caused by a combination of arson and a shortage of fireproof materials because of World War II. One hundred sixty-eight died; more than 480 were injured.

The greatest circus tragedy in history and one of America's worst fires resulted from two conspiring causes: World War II and arson.

The 6,000 patrons who half filled the main tent of Ringling Brothers, Barnum & Bailey Circus on Thursday afternoon, July 6, 1944, in Hartford, Connecticut, were almost entirely children (two-thirds of the audience was under the age of 12), mothers and grandparents. There were very few young fathers in the audience; most young American men in 1944 were fighting in World War II.

The war had also commandeered something besides young men: Safe fireproofing material of the sort that circuses needed and had until now used to keep the highly inflammable canvas of the big top from burning had been redirected to the war effort. Thus, Ringling Brothers, Barnum & Bailey had treated their new, $60,000 big top with a stopgap mixture of paraffin and gasoline. It was a fatal mistake. Rather than stopping the fire, it accelerated it. Seeing the extreme effects of its refusal to make fire-proofing material available to civilians, the government reversed itself 24 hours after the Hartford circus tragedy.

One more fateful piece fit into the structure of tragedy that afternoon. The management of Ringling Brothers, Barnum & Bailey employed several "firewatchers," who were stationed at strategic points within the tent. The firewatcher stationed at the main entrance became apprehensive about the safety of the crowd near the animal runways. If the roustabouts disassembling the runways should hit some of the huge jacks supporting the stands, they might collapse, reasoned the firewatcher, and he left his post to make sure this did not occur.

The fire was deliberately set. It would be six years before the man who set it, Robert Dale Segree, would come forth and confess, but confess he did, in great detail. He had been only 14 years old when he set the fire, but before that he had killed a nine-year-old girl with a rock, had strangled three other people and had set fire to a store, a boat pier, a Salvation Army center, a schoolhouse and various other buildings. He claimed that he was driven to perform such acts by a rider on a fiery red horse who came to him in his dreams. The night before the Hartford holocaust, this rider had appeared, and so at 2:30 P.M. on July 6, 1944, just as the Flying Wallendas were climbing to their high-wire perch at the top of the tent and as the wild lions, tigers, jaguars and leopards of Alfred Court's wild animal act were being prodded into the wire runways that would direct them out of the main arena and into their outside cages, Robert Dale Segree touched a lighted cigarette to the canvas of the big top, near its main entrance. And at that moment the fire started.

It spread swiftly, powered by the paraffin and gasoline coating and helped by a sudden wind that whipped through the main entrance. Three ushers, Paul Runyon, Mike Dare and Kenneth Grinnell, saw the flame when it was still no bigger than a bouquet of roses and, grabbing three buckets of water, dashed toward it. But the heat of the fire was already so great that it scorched their clothes and drove them back before they could empty their buckets.

The crowd did not panic at first, even though there were isolated cries of "Fire!" Most apparently felt that it would be put out, and they filed in orderly fashion toward the multiple exits. Merle Evans instructed his band to keep on playing, and they did in an attempt to maintain calm.

But none reckoned with the speed of acceleration of the fire. Suddenly, it roared up the wall of the tent and then toward its peak. Large chunks of flaming canvas started to fall onto the exiting audience. And pandemonium began. Evans, sensing that the fire was going out of control, launched into "The Stars and Stripes Forever," the musical equivalent of "hey rube!"; and roustabouts from all over the circus grounds grabbed water buckets and hoses and dashed for the main tent.

Only five minutes had elapsed, and the tent was a collapsing inferno. The three animal runways cut off escape for the reserved-seat section, and these patrons trampled one another in an attempt to get to the main arena. Older people and children, unable to crawl over the runways, fell before them and were buried by others trying to thrash their way past. People, then bodies piled up at the end of each of the runways. Flaming chunks of canvas fell on them, creating ghastly funeral pyres. A hundred people at a time were set afire by these cometlike hunks of flaming canvas.

Thomas E. Murphy, a reporter for the *Hartford Courant,* described the scene:

> I saw one woman fail to make it over the runway. She slid back and slumped to the ground. A man tried to fend off the crowd but the pressure was too great. I was slammed against the steel barrier and my knee caught momentarily between the bars. Then, taking my five-year-old son in my hands, I tossed him over the barrier to the ground beyond. The flames at this point were nearly overhead and the heat was becoming unbearable.

At this point the six gigantic support poles that supported the big top began to tumble, thundering down like falling redwoods, crushing people who were in their path. Merle Evans witnessed this. "[The fire] just kept coming," he said, "and as it raced, the center poles, burned from their grommets, fell one by one." Seeing the hopelessness of the situation, Evans ordered his 29 musicians, their uniforms scorched, to evacuate.

The Flying Wallendas had descended from their perches and had crawled out over the tops of the runway cages and escaped. The Wallendas, the roustabouts, policemen and clowns Emmett Kelly and Felix Adler dashed back into the inferno, rescuing whoever they could. They carried out bodies piled up in the entranceways, hoping to clear it for survivors. But by now the tent had collapsed, sealing the fates of those within.

Outside the big top, mayhem took over as mothers tried to run back into the flames in search of their children. One policeman held a distraught woman who screamed, "Let me go! Let me go! For Christ's sake, my kids are in there!"

Bandmaster Evans, sitting on a bench, shook his head. "I have been through storms and blowdowns and circus wrecks," he said, "but never anything like this. I hope to God I never see a thing like this again."

Clown Emmett Kelly, tears running down his cheeks, comforted a sobbing child. "Listen, honey," reporter Murphy heard him say, "listen to the old man. You go way over there . . . and wait for your mommy. She'll be along soon."

In 10 minutes—some said less—168 people died, and more than 480 were injured, many of them seriously. Two-thirds of those killed were children, and almost all of the rest were women.

Help arrived swiftly and in many forms. Fifteen hundred volunteer workers, 1,000 nurse's aides and staff assistants from the Hartford Chapter of the American Red Cross; Connecticut State Police; Hartford Police; civil defense units; soldiers from nearby Camp Bradley; nurses and doctors and fire apparatus from Hartford, East Hartford, West Hartford and Bloomfield. But the fire had spread so rapidly that they could do nothing to save anyone. Thus, more than 100 ambulances took the injured to the municipal hospital; other vehicles took the unidentified dead to the state armory.

An investigation was launched immediately, and five circus officials were indicted on technical charges of manslaughter. It would be 1950 before Segree would come forward to confess his guilt. He would be sentenced on November 4, 1950, on two counts and receive a comparatively mild—when one considers the enormity of his crime—two to 20 years on each count.

Ringling Brothers, Barnum & Bailey would spend the next seven years paying off its debts. There would be 676 suits by relatives of those who died in the fire, and every one of them would be paid without a court fight. The final total of claims paid was more than $4 million, and since the circus carried only $500,000 worth of liability insurance, the money would come out of the next 10 years' profits. The arbitration agreements have been recorded as one of the most forthright and honest settlements in modern legal history.

Terrified audience members flee the Barnum & Bailey Circus fire in Hartford, Connecticut, on July 6, 1944. (American Red Cross)

UNITED STATES
GEORGIA
ATLANTA
December 7, 1946

The worst hotel fire in U.S. history took place in the "fireproof" Winecoff Hotel in Atlanta, Georgia, on December 7, 1946. Caused by a smoldering mattress that burst into flames in a momentarily unattended corridor, it killed 119 and injured 100.

For 33 years W. Frank Winecoff's "fireproof" signature hotel existed safely on Peachtree Street in the heart of downtown Atlanta, Georgia. And then, in the worst hotel fire in the history of the United States, it burned out of control with no fire escapes, no sprinkler system, no alarm system and therefore no hope of survival for its 285 guests. That 66 of them escaped unharmed was a miracle. One hundred nineteen died, and more than 100 were injured in one night of terror.

The Winecoff was a substantial, boxy 15-story hotel containing 210 rooms. It was built, true to its claim, of entirely noncombustible materials. The walls and floors were made of steel, reinforced concrete, face brick, marble and terra-cotta. The dividing walls consisted of hollow tile. With this sort of construction, city officials failed to enforce fire codes, and thus the hotel operated in 1946 without outside fire escapes, a sprinkler system or an automatic fire alarm.

At 3:30 A.M. on December 7, 1946, a smoldering mattress, left in the third-floor corridor, burst into flame. It would go undetected for only 10 minutes. The night bellhop, Bill Mobley, delivering ice and ginger ale to room 510 and forced to wait in the hall while that room's inhabitant finished taking a shower, smelled smoke. So did the elevator girl, who informed Comer Rowan, the night clerk. He told her to go to the fifth floor and inform the hotel's engineer, who was making his nightly inspection rounds.

Meanwhile, Rowan himself sprinted to the stairwell in the lobby and looked up. Flames were licking at the walls a few floors over his head.

He dashed back to his desk and called the fire department. The first engine would be on the scene in slightly over a minute; the entire 60-man force would be there in 10. But the fate of everyone in the hotel had already been sealed the moment the flames from the mattress had ignited new paint on the corridor walls, raced down the corridor and entered the center stairwell. Fed by an updraft, the fire became a conflagration racing for the upper stories. With both elevators now immobilized, the stairway impassable and no fire escapes, there was no safe way out.

Unaware of this, Rowan began to call rooms, warning the guests of the fire. He was able to complete only a few calls before the phone went dead.

The firemen arriving on the scene had two choices: trying to control the fire or trying to rescue the trapped guests, who were by now fashioning makeshift ropes from knotted bedsheets. Some were merely perched on window ledges. The firemen opted for the second choice and began to run up their ladders, which reached the 10th floor, to try to rescue some of the terrified guests.

Within a few moments panic erupted, and as the fire and smoke began to reach their rooms, some people began to fling themselves to the street rather than be burned alive. One man trying to lower himself from a bedsheet toward one of the fire ladders was struck by two people who had jumped from a higher floor. All three fell to their deaths. A woman leaped from an upper floor and struck a fireman carrying a woman down a ladder. All three were flung to the sidewalk. The two women were killed; the fireman was seriously injured.

Safety nets were frantically spread, and some of the people survived by jumping into them. Others missed. On the 13th floor a group of people pooled their bedsheets to make an extra long rope but then defeated themselves fatally when two tried to descend simultaneously. The sheets parted, and they fell. A four-year-old boy, flung from an upper window, would have been killed if a man on the street had not caught him.

Individual acts of bravery abounded. Major Jake Cahill and his wife were rescued by fire ladder. Once they had reached the ground, Major Cahill entered an adjacent building via the alleyway behind the hotel, stripped off his coat, found a wooden plank and, placing it on two adjacent window sills, crawled across it and rescued his mother from her room by guiding her back across the wooden plank. While he was rescuing his mother, someone stole his fountain pen and traveler's checks from his abandoned coat.

Eventually firemen turned to the fire itself. Entering the lobby, they made their way up the stairwell a floor at a time. They battered their way into rooms hoping to find survivors and found very few. Most were burned beyond recognition or asphyxiated. In one 11th-floor room they found a mother kneeling in the bathroom holding three small children in her arms. The fire had fused their bodies together.

Hospitals in Atlanta, at Fort McPherson and the Atlanta Naval Air Station were crammed with the injured, the dying and the dead. Georgia governor Ellis Arnold and Atlanta mayor William B. Hartsfield launched an immediate investigation and discovered that the Winecoff was not the only hotel in Atlanta without fire escapes, sprinkler systems or automatic alarms. The presence of these plus proper metal fire doors in the corridors would have prevented this tragedy.

The hotel's lessees were indicted for involuntary manslaughter, but the charges were dropped six months later. Ironically, 70-year-old W. Frank Winecoff, who had built his fireproof hotel in 1913 and had been given a 10th-floor suite in perpetuity, also perished that night, a victim of America's worst hotel fire.

UNITED STATES
ILLINOIS
CHICAGO
October 8, 1871

A combination of a long drought, wooden construction and the overturning of a kerosene lantern in the O'Leary barn *on DeKoven Street led to the Great Chicago Fire of October 8, 1871. Some 250 to 300 died and 90,000 were made homeless. There was $196 million in property damage.*

Chicago had grown swiftly from a village of 4,000 people in 1840 to a thriving city of 300,000 in 1871. There were 60,000 buildings spread over 36 square miles by that year, and almost all of them were built of wood. Even the stone and brick buildings possessed either tarred or wood-shingle roofs.

The poor of Chicago were legion, and they lived in squalid two- or three-room shacks, flimsy shanties with winter wood piled against their outer walls and ramshackle barns and sheds behind them. Weather-beaten tenements sheltered thousands of people in minuscule cubicles. And all of this was tied together by miles and miles of wooden sidewalks and fences—which would become superhighways for the flames of the Great Chicago Fire, the worst fire disaster ever to occur in the United States.

The condition of rapid growth in Chicago from 1840 to 1871 certainly contributed to the magnitude of the fire. But the weather of the summer preceding it was also a major factor. From July 3 to the beginning of October, Chicago had received only two and a half inches of rain. Every stick of wood in the city was tinder dry; even the leaves on the trees had been scorched and dehydrated by the summer heat and lack of water, and they lay in piles along the sides of the streets.

The Chicago Fire Department was led by the wise, dedicated and usually frustrated Robert Williams. His manpower and engine power were not exactly immense: 200 men, 17 steam-fired, horsedrawn engines, three hook and ladder trucks and six hose wagons. The 43-year-old Williams, who bore a striking resemblance to Robert E. Lee, had repeatedly asked the city council for more ammunition and troops with which to fight fires, but he had been repeatedly turned down. The council had even refused a fireboat for the Chicago River—a horrendous oversight considering that the river flowed directly through the city, was flanked by warehouses and wooden docks and was intersected by 24 wooden bridges.

Furthermore, during the first week of October there had been 35 major fires in Chicago, the worst of them at 10:00 P.M. on October 7, one night before the big one. A planing mill on the West Side had caught fire, and before it had been extinguished some $750,000 in damage had been done.

That very night author George Francis Train gave a lecture in Farwell Hall in downtown Chicago. "This is the last public address that will be delivered within these walls," he intoned. "A terrible calamity is impending over the city of Chicago. More I cannot say; more I dare not utter."

The beginnings of the Great Chicago Fire, according to accepted belief. This contemporary lithograph captures the moment, on the evening of October 8, 1871, when Mrs. O'Leary's cow kicked over a lantern. (Lithography Collection, Smithsonian Institution)

The wind-whipped flames of the Great Chicago Fire advance on the fleeing refugees in this contemporary painting of the event. (Library of Congress)

Whether he had some inside information, was psychic, or was guessing nobody ever knew. But he was terribly, absolutely right.

At 8:30 on Sunday night, October 8, 1871, something happened at 137 DeKoven Street on Chicago's West Side. Myth and fancy and the rewriting of history by everyone from local newpaper reporters to screenwriters to Katie O'Leary herself have probably buried forever the true facts. But all narrators agree on one: The Great Fire of Chicago began in the O'Leary barn. All also agree that she kept five cows, had a milk route, had five children and had a neighbor named Patrick McLaughlin.

Then the stories begin to part company. One account, by historian Hal Butler, says that one of the O'Learys' cows was sick, that Katie O'Leary went out to the barn at 8:30 that night to examine the sick animal, examined her, decided to go back into the house to get some salt as a remedy, left a kerosene lamp on the straw-strewn floor and the restless cow kicked it over, igniting the straw. Katie first told this story and then later denied it.

Another version has her neighbor Patrick McLaughlin knocking on her door, waking her up at 8:30 and asking her for some more milk for a party the McLaughlins were having. In this version Katie goes to the barn and attempts to milk her cow for the second time that day; the cow resists and kicks over the kerosene lamp.

Another has Patrick McLaughlin sneaking into the barn to milk the cow, which kicks over the lantern.

"Big Jim" O'Leary, Katie's politician-gambler son, played by Tyrone Power in the movie *In Old Chicago,* claimed afterward that some small boys in the neighborhood sneaked into the barn for a smoke and started the blaze.

And still another version has elements formed in the soil throughout the Midwest by a comet many thousands of years ago making the ground—particularly the ground under Katie O'Leary's barn—combustible.

Whatever the ignition, at approximately 8:45 on October 8, 1871, the O'Leary barn went up in flames. Simultaneously, the heat that had gathered under the roof blew a hole through it and puffed out a billow of bright, yellow-red fire surrounded by dense white smoke.

The watchman of Little Giant Company Number 6, whose members had had only four hours sleep after battling the Saturday night blaze at the planing company, was the first to see it, and he routed out the company, which was situated only five blocks away from DeKoven Street and the O'Learys.

As the Little Giant Company started for the fire, a southwest wind kicked up, flinging sparks from the barn toward neighboring structures. Before the engine company arrived on the scene, a second barn, a paint shop and a flimsy wooden home were also on fire. Flames were flinging themselves 50 to 60 feet in the air. The company pulled up at DeKoven and Jefferson, laid a line down through DeKoven and then through the passageway to the O'Leary barn and soon had a solid stream of water directed on the blaze.

Bruno Goll, the owner of a nearby drugstore, ran to a new fire alarm box and turned in the alarm. But the wiring of the box was apparently faulty, and the message never reached the central fire control at the courthouse. This, it turned out, was a key failure, for if reinforcements from other fire companies had arrived within the next half hour, it is generally conceded that the fire could have been contained and extinguished. Instead, it started to spread northward.

Even without the faulty wiring, however, the fire might still have been brought under control if a series of events had not occurred in the courthouse tower. The fire watcher there, Mathias Schafer, spotted the blaze and called down to fire operator William Brown to mark the spot at call box 342, at the corner of Halsted and Canalport, more than a mile from the fire's actual location. Every other engine company in the city, alerted by telegraph, sped toward that spot.

Taking a second look, Schafer realized his error and signaled to Brown to send out another alarm. The fire was not at call box 342, but at call box 319. Astoundingly, Brown refused to send out the second signal, asserting that it would confuse the fire companies. Schafer argued with him, but Brown remained adamant, thus preventing the two most powerful fire engines in the city, housed only a few blocks from the fire, from arriving in time to put it out.

By the time the other companies reached the real scene of the fire, two square blocks of houses and barns were blazing, and the fire was rapidly growing beyond the control of anyone. One company's hose burst; another unit's hose burned up; a third could not find enough pressure to pump water.

Six blocks north of the O'Leary home was Saint Paul's Catholic Church, bordered by Bateman's Lumber Mill and two furniture factories. The wind-driven blaze now picked up a burning timber and hurled it six blocks through the air. The firebrand hit the steeple of Saint Paul's, setting it and the church ablaze. Within minutes the fire had spread to the lumberyard. A half million board feet of lumber, 1,000 cords of kindling and almost a million wooden shingles ignited.

An hour passed, and three major fires now blazed in the city. The first fire had split into two columns rushing north at a terrifying speed. The third was in the nearby Bateman Lumber Mill. Smoke, powered by the terrific heat beneath it, shot up into the wind. Sparks, flaming shingles and pieces of clapboard exploded upward through the smoke like volcanic debris.

The draft caused by the flames, like a suction and very much like the situation that would exist in Dresden 74 years later (see p. 214), plus the southwest wind became

A view of the Great Chicago Fire of 1871 from the West Side, captured in a contemporary lithograph. (New York Public Library)

strong enough to blow people down in the streets. Sparks rained lethally.

Frightened residents began to evacuate their homes. The wind increased, sending flaming ash and debris north toward the downtown district.

Now the three fires merged. Near midnight the wind hurled pieces of burning wood across the Chicago River, where they set fire to the Parmalee Omnibus and Stage Company, and this fire now spread to the Chicago Gas Works, where two enormous gas storage tanks stood in the direct path of the flames. Tom Burtis, the night superintendent, prevented an unholy explosion from happening by releasing the gas into the North Side Reserve and the Chicago sewers.

But his heroism had mixed results. The gashouse did not explode, but the gas fumes emitting from sewer manholes throughout the South Division erupted in thunderous bursts of fire, which burned uncontrollably and set fire to nearby buildings.

From the gasworks the fire leaped to a nearby armory where stored ammunition exploded, adding a sense of Armageddon to the already abundant terror of the night.

Hysteria turned the streets into bedlam. Conley's Patch, Chicago's 24-hour red-light district, was soon ablaze and emptying its brothels, saloons and rooming houses of their inhabitants. Drunks were everywhere as the saloons became free houses. Roaming gangs of looters began to strip battered businesses of their stock. Jacob Klein, a resident of the West Side, was killed by several toughs as he was carrying two bolts of cloth from his building. The toughs tried to wrest it from him; he resisted; they crushed his head with a shovel and made off with the cloth.

As the flames advanced on the courthouse, Mayor Roswell Mason telegraphed other cities for help. Fire equipment was immediately loaded onto railroad cars in Aurora, Illinois, as well as St. Louis, Cincinnati and Milwaukee. New York City dispatched a special train loaded with men and supplies.

Ex-alderman James Hildreth broke into an armory and had 3,000 pounds of explosives removed. Collaring the mayor, he talked him into authorizing the blowing up of buildings to form firebreaks. Aided by two policemen, Hildreth went out into the city dynamiting buildings to absolutely no avail. The fire merely leaped across the open spaces where the buildings once stood.

At 1 A.M. a blazing piece of timber struck the tower of the courthouse, setting it on fire. The huge five-and-a-half-ton bell within it was set in motion, to toll continually for an hour until it would fall through the fire-pitted roof to the ground below. Before that happened the building was evacuated. One hundred sixty prisoners housed in its basement jail were set free to roam and pillage the city. Five murderers were, however, not freed but were handcuffed together and placed in the custody of the police.

The entire business district was engulfed in flames. Two plush new hotels, the Bigelow and the Grand Pacific, had not even opened their doors for business. They never would. The Palmer House (which would later be rebuilt), the Sherman House and the Tremont House were all consumed by flames. The courthouse, the chamber of commerce, the armory, the opera house, Marshall Field's store, the YMCA, four railroad depots and all of the dockage, every theater, every bank, every newspaper and every hotel in Chicago were among the 3,650 buildings in the south district that would be destroyed during the 30 hours the fire storm blew back and forth across Chicago.

By 3:00 A.M. stone decomposed or exploded, girders melted, streetcar tracks were twisted and the iron wheels of streetcars were transformed into shapeless globs as the firestorm heated to—as one clerk in a paint store attested—3,000 degrees Fahrenheit. The night turned saffron. It was said that a man 20 miles away read a newspaper by the light of the flames.

People living in second-story apartments threw boxes and trunks out of windows in an effort to save their possessions. Many people were struck by these objects, and some were killed. Unscrupulous draymen, capitalizing on hysteria and misery, charged $100 to haul property to safety, drove a few blocks and demanded more. If the property owner could not pay, the goods and the owner were dumped into the street.

Joe Medill, editor of the *Chicago Tribune,* desperately tried to get out an issue of his paper, even as the flames roared around the new, supposedly fireproof *Tribune* building. Although the heat inside was almost unbearable, a few reporters tried to write their stories. But the stories would never be printed. The heat melted the press rollers, and shortly after that the building burst into flames, sending the reporters out into the streets to join the thousands of hysterical refugees roaming, running, brawling and looting.

And now the ultimate irony occurred. An enormous firebrand, a huge 12-foot-long plank lifted by the galelike wind, soared through the air and landed on the roof of the new waterworks building. Smashing through the slates of this fireproof stone building located in the middle of a huge park, it set fire to the wooden support timbers and the walls and completely immobilized the pumping machinery. There was no longer any water for the fire engines, no way to keep the fire from spreading.

At 7:00 in the morning on Monday, October 9, the Galena Elevator and McCormick's Harvester Works caught fire, the flames wrapping around the brick buildings and crushing them. The walls of the great structures folded in and sent out immense gusts of heat and tons of sparks and embers, which hastened the destruction of everything north of the river.

All the bridges over the south branch of the Chicago River were burned away. Fire engines drew water from the river now and made ineffectual forays into the inferno. By doing this they managed to keep the fire from doubling back to the south.

One by one the North Side mansions of the wealthy began to go up in flames. Their owners and their servants managed to save some valuables. Davis Fales, a lawyer, dug a huge hole in his backyard and buried his favorite piano in it. One man was seen burying his family up to their necks in the mud by the river. He then scooped up water and kept them wet, saving them until tugs began to rescue people from the scalding beach.

Samuel Stone, the assistant librarian of the Chicago Historical Society, located at the corner of Dearborn and Ontario Streets, did what he could to move valuable documents to the society's basement. One of the most precious of these was the original copy of Abraham Lincoln's Emancipation Proclamation. Stone attempted to smash the glass that protected it but could not. "At this moment," he later told reporters, "again the wind and fire filled the whole heavens, dashing firebrands against the reception room windows." The ceiling began to give way, and Stone escaped, but the Emancipation Proclamation was burned to a cinder.

Finally, on Monday afternoon, the fire began to show signs of burning itself out. Flames continued to spread only in the northern sector. On Monday night the rain that always follows a firestorm began to fall. It was a soaking rain, and it snuffed out all that remained of the fire.

On Tuesday morning one could stand in the ashes of Katie O'Leary's barn and look out across 2,200 acres of ash-strewn wasteland. From DeKoven Street to the lake and northward to the prairies, nearly four miles away, there was hardly a board or a brick left standing. Fifteen thousand dwellings, 80 office buildings, 170 factories, 39 churches, 28 hotels, 39 banks, six railroad terminals, nine theaters, 21 public buildings, 1,600 shops and stores, grain elevators, coalyards and lumberyards, breweries and distilleries, warehouses, bridges, wharves and shipping were all gone. Ninety-thousand people were homeless. The property damage was fixed at $196 million, or one-third of the city's wealth.

Fifty-four American fire insurance companies were ruined by the conflagration, some paying as little as three cents on the dollar. The city of Chicago carried no fire insurance, and the $470,000 courthouse, the police and fire stations, schools, bridges and other public property that had been destroyed were a total loss. Interestingly, the biggest item of damage suffered by the city was 122 miles of wooden sidewalks, valued at more than $941,000.

The exact number of dead will never be known. Most authorities, including the Chicago Historical Society, place it at 250. Others say 300. It was amazingly small considering the extent of the fire.

The famous Civil War general Philip Sheridan, in command of an army post near Chicago, joined forces with Allan Pinkerton's detective agency to enforce the martial law that Mayor Mason imposed on the city.

Relief began to flow into the city overnight. Food, clothing and money poured in from all over the country. Jim Fisk drove around New York in his coach and four picking up relief bundles. Commodore Vanderbilt ran special trains into Chicago to speed the relief effort. The most generous contribution—$550,000—came from the people of Boston, a city whose main section would be gutted by fire a year later.

The city began to rebuild immediately, although 50,000 of its inhabitants would leave during the first month of its laborious recovery. Still, it recovered, fueled by the sort of spirit represented by real estate agent William D. Kerfoot, who hammered together a wooden shanty from scraps of wood that had not been burned too badly and then opened for business with a sign that read: ALL GONE EXCEPT WIFE, CHILDREN AND ENERGY.

UNITED STATES
ILLINOIS
CHICAGO
December 30, 1903

The worst theater fire in U.S. history, that of the Iroquois Theatre in Chicago on December 30, 1903, was caused by a combination of negligence in design, blocked fire exits, a snagged fire curtain, an absent stage manager and a calcium light igniting scenery. Five hundred ninety-one died and scores were injured in the resulting inferno.

"Does freedom of speech extend to yelling 'Fire!' in a crowded theater?" Oliver Wendell Holmes asked, and the consequences of this, the instances of ensuing, fatal panic, replay themselves over and over in the worst theater fires in the world. As in Vienna (see p. 203), Brooklyn (see p. 244) and Montreal (see p. 206), this was true in Chicago's Iroquois Theatre, a mere 24 years after the Great Fire (see previous entry). It would be the scene of the worst theater fire in U.S. history.

The Iroquois Theatre was, frankly, fabulous. Designed by 29-year-old architect and wunderkind Benjamin H. Marshall in French Renaissance style, it was glamorous, dazzling, plush and fireproof. At least this is what Marshall and co-owners and theatrical entrepreneurs Will J. Davis and Harry J. Powers both believed and advertised. George Williams, building commissioner of the city of Chicago, also deemed it safe after inspecting the theater just 39 days before its cataclysmic final performance.

That it was grand nobody would dispute. It consisted of an ornate foyer with two magnificent staircases, a plushly draped auditorium, an elaborately designed proscenium arch, a mammoth stage, dressing rooms galore and the latest in electrical equipment. It would seat 1,724, but there was standing room for 300 more. It had more exits than any theater in the country—30 in all, 27 of them double-door fire exits—and each floor—orchestra, balcony and gallery—was equipped with emergency exits feeding into the foyer.

But fireproof? William Clendenin, the editor of a magazine called *Fireproof*, toured the premises of the theater before it was finished, and his assessment was far less sanguine that that of its builders and owners. There was no sprinkler system over the stage, he pointed out. There was no ventilating flue above the stage to carry flames up and away from the audience. The skylight above the stage was nailed shut. There was heavy use of wood trim throughout the theater. There was no direct fire alarm connection with the fire department. And on the night of December 30, 1903, all but three of the 30 exits would be locked, some with iron gratings securing them.

Clendenin was dismissed by both the theater owners and the office of the Chicago building inspector. He wrote a scathing editorial about this in *Fireproof,* but it was an exercise in futility; hardly anyone read his magazine.

On November 22 Ed Laughlin, an inspector from the building commissioner's office, inspected the building "from dressing rooms to capstone" and pronounced it "fireproof beyond all doubt." Two weeks later, Joseph Daugherty, a stagehand, sounded a preliminary alarm, which was also ignored. He informed owner Will J. Davis that there had been a small trash fire backstage, which he had put out. However, when he had tried as a safety measure to lower the asbestos curtain, it had snagged on a reflector some 20 feet above the stage floor. Daugherty suggested that although the actors wanted the reflector there it should be moved. Davis agreed to take it under advisement but then did nothing about it.

The December 30 matinee performance of a Klaw and Erlanger extravaganza titled *Mr. Bluebeard,* starring comedian Eddie Foy, was more than sold out. There were 2,000 people in the audience that day, 1,724 sitting and the rest standing. The first act passed without incident. Everything was going so well, in fact, that stage manager Bill Carlton left his post backstage and strolled out to the foyer to watch the second act from the front. Several stagehands slipped out for a drink in a nearby saloon. More importantly, Edward Cummings, a stage carpenter in charge of the electrical mechanism controlling the asbestos curtain, also slipped out to go to the local hardware store.

Firemen at work controlling the worst theater fire in U.S. history, Chicago's Iroquois Theatre fire on December 30, 1903. (New York Public Library)

The second act began with a double octet of eight men and eight women singing a song titled "In the Pale Moonlight." Above them, in one of the wings, spotlight operator William McMullin noticed that his calcium light was dangerously close to a flimsy tormentor. And as he watched, the tormentor caught fire. He tried to snuff out the tiny flame with his hand, but it was just beyond his reach.

The flame twisted upward, and theater fireman Bill Sallers, summoned by McMullin, grabbed an extinguisher and aimed it at the small blaze. The stream fell short.

By now more and more pieces of scenery were on fire. The stage was filled with 40,000 cubic feet of scenery, wooden sticks, frames, paint and canvas, and 180 drop scenes hung with 75,000 feet of new, oily and highly inflammable manila rope.

The fly space above the stage was a mass of flame, and sparks began to rain down on the double octet. The dance routine wavered and then began to break down. One of the women fainted, and the rest of the dancers panicked and sprinted for the wings. The crowd, not quite sure of the danger, became slightly restive.

Eddie Foy, who smelled the smoke in his dressing room, walked briskly onstage and down to the footlights. "Stay seated," he said to the audience. "It is nothing. It will be out in a minute."

And then he turned to the conductor in the pit and hissed, "For God's sake, play, play and keep playing!" Like the band on the *Titanic,* the pit orchestra began to play, but by now several audience members had cried out "Fire!" and the audience had begun to stampede. Foy tried to calm them, pleading with them to walk slowly to the exits. Then, realizing that this was useless, he shouted to the stagehands to lower the asbestos curtain.

The man in charge was gone; fireman Sallers fumbled with the mechanism and finally got it working. But as it had two weeks ago, the right side of the curtain caught on the reflector 20 feet above the stage floor. The left side descended to 12 feet above the floor and stopped. And the flames licked out from beneath the curtain into the auditorium.

Foy left the stage just before the entire catwalk and loft rigging thundered to its floor, spouting sparks like a fireworks display. Some of the mechanism landed on the central lightboard, and every light in the theater went out, terrifying the fleeing audience.

Even at this point the fire was not enormous, and the horror that followed would not have occurred if the dancing girls from the double octet had not run for the stage door. Two stagehands obligingly opened it for them, and the huge draft that swept in through the open door exploded the onstage fire into an enormous fireball. It roared and rolled through the entire theater, spanning the 50-foot space to the balcony in one huge leap and then splitting in two. The lower part swirled under the balcony and out into the foyer; the other half roared upward to engulf the people in both the balcony and the gallery. Everything in its path burned.

Bedlam ensued. People clawed at one another. Some were stripped naked. Some fell or were pushed over the balcony rail, setting fire to or crushing many of those below. Vicious fights broke out. Men, women and children were trampled.

The 16-year-old ushers on the first floor went berserk, actually holding the doors against the crowd. Abandoning this, they fled for their lives without opening the emergency exits. Some calmer audience members managed to find some of these doors and opened them only to find that there was a four-foot-high ledge between the door sill and the sidewalk. People dashing out fell, breaking legs and arms and becoming rugs for other terrified escapees.

Almost all of the 800 patrons on the orchestra floor escaped. But in the balcony more than half of the 1,100 spectators died. There were no exit signs; bewildered and panicked, these hapless humans dashed about trying to find a door. Some did, only to discover that the fire escapes had no ladders to take them to the ground.

The crowd surged like a tidal wave for the 60-foot-high gilt and marble foyer. But all of the exits from the balcony, the gallery and the orchestra, met here, and the crush became horrendous. People flung themselves over and on one another, falling over balustrades, landing on top of the swirling mass of humanity trapped at the middle of the foyer. And then the flames reached this pile of humanity and fused it into a hill of burning corpses.

Behind this, in the balcony, firemen later found bodies piled six deep in the aisles. Many of the people were nude; others were horribly mangled and mutilated. A few people were still alive, buried under the charred bodies.

It had all taken 15 minutes. That was all, and 591 people were dead. The stage, boxes, main floor, balcony, gallery, all of the wood trim and draperies and even the asbestos curtain—which was discovered to have been made of nothing but paper—were destroyed.

Twelve people were indicted, including the theater owners, building commissioner Williams and Mayor Harrison. The coroner's inquest was a national sensation for three weeks.

The case went to a grand jury, which exonerated Mayor Harrison and several others but upheld the manslaughter charges against Davis and two minor theater officials and charged Williams with misfeasance.

Not one of them served a jail sentence. All got off on technicalities. None of the relatives of the dead collected on any of the hundreds of damage suits. The only person to serve a jail sentence was the owner of a nearby saloon. His property had been used as a temporary morgue, and he was convicted of robbing the dead.

UNITED STATES
ILLINOIS
CHICAGO
December 1, 1958

A lack of fire drill regulations led to tragedy in the fire, begun in a pile of trash in the basement, in Our Lady of the Angel's grade school in Chicago on December 1, 1958. Ninety-three perished in the blaze.

Chicago's Our Lady of the Angels grade school, run by the Sisters of Charity, was an old building lacking in many amenities and necessities. It did not have a sprinkler system, and it is generally acknowledged that the presence of one might very well have averted the tragedy that occurred on December 1, 1958. However, it was the general breakdown of certain elementary fire drill regulations that probably caused most of the deaths that day.

Ordinarily, school concluded at 3:00 P.M. Just 30 minutes before this on December 1, 1958, a fire started in a heap of trash in the basement. Two teachers who had classes on the first floor noticed the smell of smoke first. They led their classes out of the building to safety but, inexplicably, did not inform other classes or make any effort to ring the fire alarm.

At 2:42 a janitor found the fire, and he did shout for the parish housekeeper to turn in the alarm. Even then, teachers and students on the second floor were not informed of the steadily increasing blaze. And it was on the second floor that all of the fatalities would occur. By the time the firemen arrived five minutes later, all of the first-floor classrooms had been safely evacuated.

On the second floor many students panicked, running through the smoke filled halls, coming up against walls of furiously burning flames in the stairwells, retreating to the classrooms, climbing on windowsills and leaping to their deaths on the sidewalk below.

Horrified firemen and passersby tried to calm the hysterical children and attempted to talk them back into the rooms so that firemen on ladders could safely take them from the windows. But their pleas were drowned out by the screams of the children plummeting to the ground around the fire ladders. Some firemen caught falling youngsters by their legs, their arms, their hair. As quickly as they could, they raced up and down ladders carrying hysterical children, some of them on the brink of asphyxiation.

There were some clear heads in the midst of this madness. One nun instructed her students to crawl under the smoke in the hallway and then roll down the stairs. All did; all survived. Another teacher barricaded the door of her second-floor classroom against the smoke and instructed her charges to pray. They did; firemen appeared at the windows of the classroom and rescued the class.

Priests, passersby and firemen rushed into the rapidly disintegrating building and carried out other children. Of the 1,515 students, 1,425 survived. Ninety students and three nuns died in the fire.

UNITED STATES
MASSACHUSETTS
BOSTON
November 28, 1942

A smoldering match carelessly tossed onto an artificial palm caused the tragic fire in Boston's Cocoanut Grove Night Club on November 28, 1942. Four hundred ninety-one died, and hundreds were injured.

The interior of the Cocoanut Grove Night Club after the horrendous fire of November 28, 1942, which claimed 491 lives. (American Red Cross)

The fad of nightclubs was at its zenith during the years of World War II. Elaborate, multileveled, gaudily decorated and dimly lit, these places of entertainment offered several rooms for several moods. There were small, intimate bars with piano players and an occasional singer, slightly larger cabarets with jazz combos and huge dining rooms with dance floors, big bands, and opulent shows with singers, dancers, acrobats and production numbers.

Boston's Cocoanut Grove Night Club, owned by Barnet Wilansky and located in a one-story building between Piedmont Street and Shawmut Avenue in Boston's midtown theater district, was typical of this genre. The main entrance through a revolving door opened into a foyer, which connected with the main dining room, complete with a rolling stage. There was a double-door exit from the dining room to Shawmut Avenue and a hallway exit that led to a service door. A new cocktail lounge affixed to the dining room opened in early November 1942 to accommodate the increased crowds of servicemen and war workers.

The Piedmont Street side of the building contained two bars. One, which was part of the main dining area, was called the Caricature Bar; the other several steps down from street level was dubbed the Melody Lounge. It was the most darkly lit, intimate part of the club, where dates could sit in booths in semidarkness while a pianist played. It was reached by a narrow stairway from the main level. Throughout, the club was decorated with a forest of artificial palm trees.

On Saturday night, November 28, 1942, the biggest crowd in its history—more than 800 patrons—jammed the Cocoanut Grove. In addition to the soldiers, sailors, marines, coastguardsmen and warplant workers, there was a large contingent of football fans celebrating Holy Cross's unexpected and thorough thrashing of Sugar Bowl–destined Boston College. In the main dining room movie cowboy Buck Jones entertained a large table of motion picture executives and flunkies.

At 10 P.M. the floor show, headlined by singer Billy Payne, the dance team of Pierce and Roland, acrobatic dancer Miriam Johnson, violinist Helen Fay and a chorus of dancers backed by Mickey Alpert's band was about to begin.

In the Melody Lounge, which was, like every room in the club, vastly overcrowded, more than 130 people talked, drank or listened to a ragtime piano player. In a corner booth a soldier unscrewed a light bulb hanging above him.

John Bradley, one of the five bartenders on duty, told Stanley Tomaszewski, a 16-year-old bar boy, to screw the bulb back in. Tomaszewski grabbed a bar stool, informed the unhappy soldier of his mission, climbed up on the stool and attempted to find the light bulb. Failing on his first try, he lit a match, located the bulb and tightened it. The smoldering match, discarded carelessly, landed on an artificial palm. Within seconds the tree incandesced in a rush of flame. By the time the terrified boy had descended from the bar stool, the fire had leaped from tree to tree, and the ceiling of the Melody Lounge was ablaze. The bartenders rushed over, tearing down drapes and trees, trying to smother the flames, but it was hopeless. The conflagration had begun.

The lounge's occupants panicked and bolted for the narrow flight of stairs leading to the street-level foyer. It was a hopeless horror. Flaming pieces of palm trees rained down on the kicking, clawing mass of people who piled up in the stairwell, turning it into a grim funeral pyre.

In the main dining room, just as the show was about to begin, a young woman, her hair on fire, suddenly dashed across the dance floor, screaming. Panic spread here, too, and customers, overturning tables and battering at one another, ran for the only exit they knew—the revolving door to the street. But before they got there they ran into the few, fire-seared refugees from the Melody Lounge who had managed to crawl over the mass of bodies on the stairs.

A huge battering ram of humanity slammed against the revolving door, instantly jamming it. Next to the door was an auxiliary exit equipped with a panic lock designed to open easily if pressure was applied from within. But the door had been bolted, and the first wave of refugees were slammed so tightly against it that they could not locate the bolt.

At the same time customers in the new cocktail lounge clawed their way to the customer exit on Broadway. But the door opened *inward,* and before the first terrified customers there could pull it toward them, the crush of those behind smashed them up against the now immovable door. More than 100 bodies were later found piled in a terrible heap against the door.

The lights went out, making the scene more terrifying. Toxic fumes, fed by burning paint and blazing decorative material, killed many before they had a chance to leave their tables in the dining room. Buck Jones was thought to have died this way.

Fifty patrons made their way to the basement. Many of them, unable to find an exit, were asphyxiated. Others did manage to find windows and, breaking them, made their escape. One veteran waiter, Henry W. Bimler, had gone to the basement kitchen immediately. But he found the employee exit locked. He asked the dishwasher for the key, and the dishwasher stentorially refused, saying that only the boss could give him permission to surrender it. Bimler went back upstairs, where he encountered several young women who had become separated from their escorts and now tearfully asked him for help. He guided them back to the kitchen and the nightclub's walk-in refrigerator. He led them into it and walked in himself; firemen later rescued the group—slightly chilled but safe—from the ruins of the club.

Providentially, the Boston Fire Department had been summoned to a minor automobile fire just around the corner from the Cocoanut Grove, and when the alarm went out and someone dashed down the sidewalk yelling that the club was on fire, they only had half a block to travel.

They smashed windows and pried some survivors loose from the piles of humanity at the exits. And several of the entertainers managed to rescue not only their fellow players but a score of patrons as well.

The fire lasted only a few minutes. It spread with awesome speed, consuming the entire premises and then exploding through the roof. The firemen could do little more than hose down the fire and carry out the dead.

Hundreds of morbid onlookers rushed to the scene, hampering rescue efforts. Several naval offices and their men locked arms and formed a human chain to hold back the crowds. Servicemen helped firemen stretch their hoses. When it became apparent that there were not enough ambulances available to carry the injured to hospitals, nearly 100 taxicabs were commandeered. Trucks carried off the dead.

The Boston chapter of the Red Cross sent 500 workers to the scene; the New York Red Cross sent disaster relief workers; civil defense units tended to the injured and dying. Sulfa supplies dipped at hospitals; a special plane reached Boston with an adequate supply an hour later from New York.

By 2 A.M. martial law had been declared, enforced by both civilian and military police. The final toll was sickening: 491 died and hundreds more were injured.

In the aftermath of the tragedy, fire safety rules that had fallen into place after the tragic Iroquois Theatre blaze in Chicago in 1903 (see p. 235) were extended to nightclubs, which, until the Cocoanut Grove fire, had been without fire regulations. Capacity limits, sprinkler systems and plainly marked exits became the rule as a result of the terrible and avoidable holocaust of November 28, 1942.

UNITED STATES
MINNESOTA
October 12, 1918

Ordinary spring and summer smolderings were fanned into Minnesota's worst forest fire by 60-mile-per-hour winds on October 12, 1918. Eight hundred people died.

Minnesota's worst and most extensive forest fire to date began as a six-month-long expected series of spring and summer brush fires. The peat terrain in northern Minnesota frequently experienced fires in dry seasons, settlers in the area were told, and so the fires glowed silently on with neither incident nor cure.

But at 1:00 P.M. on October 12, 1918, a 60-mile-per-hour wind suddenly sprang up fanning the glowing ashes into a roaring conflagration with flames as high as the towering trees they consumed. Fed and nudged by the wind, the firestorm spread wildly, first in a 175-mile stretch of land that ranged from Bemidji to Two Harbors, then from the Mesaba Range in the north to a midway point between Duluth and Minneapolis.

There were scores of settlements, logging towns and summer resorts in the affected area, and the inhabitants scarcely had time to prepare for the onslaught of the fast-advancing forest fire. The tiny village of Brokston, west of Duluth, was the first to be consumed, but a special train evacuated every one of its inhabitants. Cloquet, a logging town of 10,000, was similarly evacuated to nearby Carlton, but the fire pursued its inhabitants there, and, adding the population of Carlton to the passengership of the train, they sped 20 miles farther to Superior, where the entire population of the two villages remained in safety.

Resort dwellers fared much less well and much more tragically. Moose Lake was surrounded by a solid wall of fire that closed in on the resort like pincers and with such force that the fire funneled down a 30-foot well, burning to death a family that had taken refuge there. All 400 people in this town died.

The lake area north of Duluth suffered similar damage. The flames, powered by the relentless gale, licked across large bodies of water, burning to death people who dove into the waters in the hope of escaping the fire. Boatloads of frantic refugees overturned, drowning their occupants. Another 400 perished in the lake region.

The wall of fire advanced on Duluth in two sections, one from the west and one from the north. But just as it was entering the suburbs, the winds shifted, and the city was spared. Nevertheless, 800 people died in this horrendous and wide-reaching forest fire, one of the worst in U.S. history.

UNITED STATES
MINNESOTA
HINCKLEY
September 1, 1894

Smoldering ashes on a forest floor were fanned into flames by a sudden burst of wind on September 1, 1894, near Hinckley, Minnesota. The resultant forest fire killed 413 people—one-third of the population of the town—and destroyed every building.

A raging forest fire that in many ways equaled the more famous one in nearby Peshtigo, Wisconsin (see p. 251), completely destroyed the town of Hinckley, Minnesota, and killed one-third of its population on the afternoon of September 1, 1894. A smoldering beginning (see previous entry) lulled the populace of the small town, located 75 miles from St. Paul, into acceptance of these small fires as the consequence of a summer without rain.

Terrified escapees from the devastating forest fire that destroyed one-third of the town of Hinckley, Minnesota, on September 1, 1894, run from a burning evacuation train and fling themselves into a swamp. Some were burned to death by the hot mud, but most survived. *(Frank Leslie's Illustrated Newspaper)*

In August of that year, in fact, much of the northwestern United States was beset by forest fires, which gutted 25 other towns in Minnesota, Michigan and Wisconsin. Small settlements in the wilderness, built mainly of wood and existing along railroad lines, these towns fell, one by one, as shifting winds fanned glowing ashes into fulminating fires.

On September 1, 1894, a sudden gust of wind did just this on the outskirts of Hinckley, and a three-story-high wall of flame began to advance on the town. Terrified residents ran for the few bodies of water in the town. But the long drought of the summer had nearly emptied most of the ponds and had lowered the level of the Grindstone River to a knee-high level. The hundreds who threw themselves into it could not submerge deeply enough to escape the long fingers of flames that rushed across the water, and so they burned alive.

Another hazard in this rural community turned out to be stampeding animals, who ran over fleeing townspeople, crushing them to death.

Two trains took hundreds of people from the town, and one of these escapes was miraculously and touchingly dramatic. The Limited, on the St. Paul and Duluth Line, paused at the Hinckley station to receive refugees who had not been able to leave on an earlier Eastern Minnesota Line train. By the time the Limited drew into town, the fire had cut off escape on all sides, and 105 people died trying to reach the train.

Hundreds of others clambered aboard, and the engineer threw the train into forward, heading directly into the wall of fire. Flames poured into his cab, burning him horribly, but he stayed at the controls and physically restrained his fireman from jumping. Meanwhile, passengers lay down on the floor of the wooden coaches, which were now afire.

The engineer got the train to the side of a lake, waved the hundreds of men, women and children who were aboard the train toward the lake and then fell dead of the burns he had just received. The passengers staggered to the lake and flung themselves in. Many had already been cruelly burned, and the mud they spread on their bodies instantly baked on to their skin. Their clothes were in blackened tatters, but most of them survived thanks to the engineer.

Four hundred thirteen of their fellow townspeople, one-third of the population of Hinckley, did not, however, and there was not a building left standing in the town when the fire that had consumed it roared on.

UNITED STATES
MISSISSIPPI
NATCHEZ
April 23, 1940

The ignoring of fire regulations by both owners and officials and a carelessly thrown match were the causes of the fire that devastated the Rhythm Night Club in Natchez, Mississippi, on April 23, 1940. One hundred ninety-eight died; 40 were injured.

In the segregated world of Natchez, Mississippi, in 1940, there were some small joys for blacks. One of them was the decrepit old St. Catherine Street church, transmogrified into a nightclub that featured nationally famous black jazz musicians. The best, passing through, played here, and every corner of the Rhythm Night Club, as it was called, was packed for these occasions.

The white city government of Natchez neither cared about nor enforced any sort of fire regulations in the club. First, fire surveillance was extremely loose regarding nightclubs in the United States before the Cocoanut Grove fire in 1942 (see p. 237), and second, the club was operating illegally, anyway, and local law enforcement authorities let these kinds of operations function pretty much on their own.

On the night of April 23, 1940, the Rhythm Night Club was full and a little more. More than 250 patrons funneled through its one combination entrance and exit that night and danced, drank and listened to the music under the strung Spanish moss that made the church into a sort of swinging grotto. Its windows had long since been boarded up and decorated on the inside.

Somewhere, sometime during the evening, a carelessly thrown match ignited the Spanish moss. The spread of the fire was immediate and terrible. Stifling smoke and raging flames whipped around the space as if it were the interior of an infernal tornado. Crazed patrons, trying to escape, pounded on the boarded up windows with no success. A battle raged for the front door, and 40 people, most of them men, managed to fight their way through it. All were badly burned, but they at least survived.

The others—198 of them—were forced toward the opposite side of the building by the whirling conflagration. There they died in enormous piles of fused bodies. The club burned to the ground. Not one person there escaped injury.

UNITED STATES
NEW JERSEY
COAST
September 8, 1934

See MARITIME DISASTERS, *Morro Castle* (p. 262).

UNITED STATES
NEW JERSEY
HOBOKEN
June 30, 1900

A smoldering fire of unknown origin in cotton bales piled on the docks in Hoboken, New Jersey, on June 30, 1900, suddenly burst into flames, igniting the entire docks and four German Lloyd ships loaded with Sunday sightseers. Three hundred twenty-six died in the blaze; 250 were injured.

The Hoboken piers were busy places in 1900, and their busiest sector housed the ships of the German Lloyd Line. On Saturday, June 30, 1900, four of German Lloyd's most impressive and modern passenger liners were docked at the pier: the magnificent 20,000-ton *Kaiser Wilhelm der Grosse,* the 5,267-ton *Saale,* the 10,000-ton *Bremen* and the 6,398-ton *Main.* The ships were large and famous. The *Main,* commissioned a mere month before, was the modern pride of the German commercial fleet; the leviathan-like *Kaiser Wilhelm,* 648 feet long, held the eastward Atlantic record from New York to Southampton of five days, 17 hours and eight minutes.

A large percentage of the crews of these ships were ashore on leave; a skeleton crew, carpenters and longshoremen were loading and preparing them for departure within the next few days. Roiling the waters around the four vessels were 18 canal boats laden with oil, coal, cotton and gasoline.

It was the practice in 1900 to allow sightseers on board ships in port, and hundreds of them were aboard all four ships that sunny afternoon. Hundreds more lined the Manhattan shoreline, across the North River from Hoboken.

Pier number three was a particularly crowded place. The *Saale* and the *Bremen* were docked there, and stacks of cotton bales sentineled the docks. Next to the cotton were 100 barrels of whiskey, also ready to be loaded aboard.

At approximately 3:55 that afternoon a smoldering fire of unknown origin in one of the bales of cotton suddenly burst into flames. Within an instant the fire had spread to the whiskey barrels, which went up with a roar, sending orange and red flames licking skyward toward the storage sheds on the dock. The wood of the pier ignited next, and the flames shot like racing animals toward 200 long-shoremen. The longshoremen did not stand on ceremony or curiosity. They ran, with the flames pursuing them. Forty of them were not fast enough and perished in the fire that overtook them.

At the same time the conflagration was now shooting up the wooden gangplanks to each of the four ships tied up at the German Lloyd docks. Hundred-foot-high flames leaped from pier to pier, igniting warehouses and the wooden decks of the four steamers.

The sightseers were trapped aboard as flames descended into the living quarters and innards of each ship. The *Kaiser Wilhelm* was farthest from the source of the fire, and a dozen tugs raced to it and pulled it away from the dock and into midstream. With its captain standing at the bridge with two pistols at the ready, the entire crew set to putting out the fire. They succeeded, beating back the fires systematically. No one was killed, though many sailors suffered extreme burns.

Chaos consumed the other three liners. They burned out of control as tugs pulled the *Bremen* and the *Saale* away from the pier and into the waters of the North River. Simultaneously, these tugs took crew members and sightseers off the ships. One hundred four men were rescued from the *Bremen* by the tug *Nettie Tice;* 40 were removed from the *Saale* by the *Westchester.*

But that was a small effort. Most of the personnel and visitors on the ships were below decks when the fire began. With the blaze beginning on the wooden decks, their way to escape was completely cut off. Portholes on all ships were only 11 inches in diameter—far too small to allow an adult to escape.

Sailors on the tugs looked on in horror and helplessness as faces appeared at the portholes with flames reaching up behind them. The *Main,* its steel hull turning red hot and then white hot, was irremovable. Snugged against the flaming pier, it took the full brunt of the fire. A stewardess appeared at one of its portholes frantically pleading for help. Rescuers tried to pour water through an adjacent porthole to put out the flames around her but could not.

Finally, according to *Munsey's Magazine,* the woman, realizing that it was hopeless, said, "Now listen! Listen! Tell my mother—she lives in Bremen—tell her my last thought was of her—tell her all my money is in the bank—tell her she can have it all—tell her—." And the flames closed around the stewardess.

It would be three hours before all of the frantic people—crew members, visitors and sailors—would die, while rescuers looked on in frustration and fury. Only 15 men, holed up in the bowels of the *Main* and able to signal via an oil lamp of their whereabouts to the tug crews, were saved on that ship. Rescuers, spotting the lamp at 11:00

On Sunday, June 30, 1900, a horrible fire engulfed the Hoboken docks and four German Lloyd ships. Three hundred twenty-six sightseers, passengers, firemen and crew members died in the tragic conflagration. *(Frank Leslie's Illustrated Newspaper)*

P.M., pulled up alongside the ship and began to cut through the red-hot metal. They reached the 15 men, delirious, stripped naked and nearly suffocated. But all survived without aftereffects except for one elderly sailor who went blind in the hellish heat.

It would be the next day before the bodies could be removed from the ships. A conservative estimate put the number of dead at 326, the injured at 250. But most who were there agreed that there were far more dead than the official estimates admitted.

UNITED STATES
NEW YORK
BRONX
March 25, 1990

In the worst mass murder in U.S. history and the worst fire in New York City since the Triangle Shirtwaist tragedy of 1911, Julio Gonzalez set fire to the Happy Land Social Club in the East Tremont section of the Bronx, New York, on March 25, 1990. Eighty-seven people perished in the blaze.

Exactly 79 years to the day after the Triangle Shirtwaist fire, which killed 149 young women in New York City (see p. 246), the Happy Land Club, an illegal social club crowded with Honduran immigrants, was the scene of an infernal catastrophe that took the lives of 87 people at 3:30 A.M. on Sunday, March 25, 1990.

In March 1990 there were 177 illegal social clubs dotted throughout the five boroughs of New York City. Usually open only on weekends and catering to groups of people united by ethnicity, nationality, geography or shared interests, they sold liquor illegally, allowed dancing without benefit of cabaret licenses and generally provided congenial, if illegitimate, neighborhood nights out.

The Happy Land Social Club was one of these illegal oases. Located on the west side of Southern Boulevard in the East Tremont section of the Bronx in the heart of a Honduran community, it sported a stone face painted red and adorned by a large sign that read:

HAPPY LAND SOCIAL CLUB INC.
LITTLE LEAGUE—PONY
FOR HIRE HALL
ALL SOCIAL EVENTS

A smiling face beamed down from the space between "Happy" and "Land."

The club was 22 feet wide by 58 feet deep and was on two stories. On the left side was the entrance door, which led to a coat-check and admission area where patrons paid a $5 cover charge. There was a bar at the rear of the downstairs room, and one in the same position upstairs, whose windowless room most people favored, since it contained a disc jockey and a dance floor. Celebrants paid $3 a drink to talk, sing and dance in the small, low-ceilinged room. A narrow, steep front staircase and a back set of stairs connected the two parts of the club.

"It was like a headquarters for Hondurans," said Steven McGregor, who lived near the club and was interviewed by the *New York Times* after the fire. "They threw parties every weekend."

The club had been pronounced a firetrap by the city fire department, and its landlord had issued an eviction notice 10 months before for nonpayment of rent. On November 21, 1988, the city had ordered the building vacated because it lacked a second exit, a fire alarm or a sprinkler system. In the meantime there had been two arrests made for selling liquor without a license. On November 1, 1989, the police visited the club but found it padlocked. It always was during the week, neighbors said, and it was on a weekday that the police made their visit.

But March 24, 1990, was a Saturday, and the place was packed with noisy, friendly partygoers, except for one disgruntled customer. Early in the morning of Sunday the 25th, 36-year-old Julio Gonzalez came to the club, had two beers and had an argument with his former girlfriend, Lydia Feliciano, who worked there as a coat-check attendant.

Gonzalez was well known at the club and in the neighborhood. In 1980 he had deserted the Cuban Army and had fabricated a record of drug trafficking in Cuba in order to win expulsion to the United States in the 1980 Mariel boatlift. In the intervening 10 years between his arrival in New York City and the fateful night of March 25, 1990, he had worked at and been fired from various jobs. His latest job as a warehouse worker at a lamp company in Long Island City had terminated just six weeks earlier.

Most of the time he lived as a street person who hustled money washing cars or peddling. He and Ms. Feliciano had lived together for eight years and had recently broken up, and for the past few days he had tried to convince her to come back to him. That night, in his entreaties and demands, he became loud, boisterous and profane. A bouncer, noting the escalating argument, kicked him out.

Gonzalez became furious and shouted a vow that he would shut down the club. He picked up a plastic jug that was sitting near its entrance, walked to a nearby Amoco station and bought $1 worth of gasoline. Returning to the Happy Land, he entered its one street door and encountered a lone man, who was exiting. He pretended to make a phone call on the pay phone in the hallway until the other man was out of sight and then poured a trail of gasoline from the street through the entrance and into the inside hall. He drew out a match, tossed it into the gasoline and watched it ignite. Satisfied that it was burning, he turned and walked home.

Upstairs, the disc jockey, Ruben Valladares, began his favorite Jamaican reggae song, "Young Lover," by Coco Tea.

"The floor got so full that a lot of people couldn't dance," Felipe Figuero, one of the only three people on the second story of the club who would survive, later told a *Times* reporter. "Then you heard it coming up the stairs."

What he heard were shouts from one of the club's two doormen. (The other was on the dance floor with his girlfriend.) He was yelling "Fire! Fire!" and shoving his way through the mass of humanity.

"Right away, people went crazy," said Figuero. "Some ran this way and that way. Some people didn't make a big deal out of it. I could see a little smoke coming upstairs."

The doorman shoved through the crowd looking for his girlfriend, and Figuero took up the cry, in Spanish. "Fuego!" he yelled, while his friend, disc jockey Ruben Valladares, turned up the lights and shouted "Fuego!" into a microphone.

"People were already desperate," continued Figuero. "Everybody was running around."

Figuero headed for the stairs. "Everyone saw me go for it," he said. "I yelled 'Down, let's go!' There were a lot of people around those stairs, but nobody followed me. I could hear all the cries, lots of people saying 'Mama!' I heard something explode, like a light."

It was the fire, roaring up and onto the second story. Figuero plunged down the stairs into the smoke, threw himself against the crash bar of the back entrance to the club and found himself on the sidewalk with Ms. Feliciano and several patrons from the downstairs room.

A minute later Ruben Valladares exploded through the door, his clothes in bright flames. "He let out some screams that I remember too much," Figuero said. "I didn't even know who he was—he was so burned." Valladares would survive with burns over 50% of his body.

The fire had been set at 3:30 A.M. The alarm was turned in at 3:41 A.M. According to Albert Scardino, New York City Mayor David Dinkins' press secretary, fire equipment was there within three minutes. But it was already too late for 87 people.

Nineteen died from burns or smoke inhalation on the stairs or in the ground-floor room. But the greater number perished in a mountain of humanity that resembled the mass murders in the gas chambers of Auschwitz. The upstairs room, so low-ceilinged that the 5-foot, 6-inch tall Felipe Figuero could stand flat-footed and touch the mirrored, revolving globe suspended above its dance floor, contained precious little air under ordinary circumstances. When the flames reached it they abruptly sucked all of the oxygen out of the room, and all of the second-floor victims died in seconds of suffocation and smoke inhalation.

"It was shocking," First Deputy Mayor Norman Steisel told reporters afterward. "None of the bodies I saw showed signs of burns. They looked waxen." "Some looked like they were crying," one of the firefighters from Ladder Company 58 told reporters. "Some were horrified. Some looked like they were in shock. There were some people holding hands. There were some people who looked like they were trying to commiserate and hug each other. Some people had torn their clothing in their panic to get out."

Later, emergency crews would break a hole through the wall of the upstairs room and into a construction office next door. They would drag the bodies through it and then take them to nearby Public School 67, which was turned into a temporary morgue. All day Sunday its corridors were choked with grieving relatives and friends, who identified their loved ones through Polaroid pictures supplied by police officials.

Detectives, acting on information supplied by eyewitnesses, arrested Julio Gonzalez at his home that day. He offered no resistance and was brought before Bronx Criminal Court Judge Alexander W. Hunter, Jr., at 2 A.M. on March 26. There, Bronx District Attorney Robert T. Johnson charged Gonzalez with 87 counts of arson felony-murder and 87 counts of murder by depraved indifference to human life.

He went to trial in July 1991, pleading not guilty by reason of insanity. Justice Burton B. Roberts of the state supreme court denied a motion by the defense to suppress Gonzalez's admissions to detectives that he had set the fire as well as physical evidence that included his sneakers containing residue from the gasoline that had been used to fuel the fatal blaze.

On August 19, 1991, the jury found Julio Gonzalez guilty on all charges, and he was sentenced to 25 years to life in prison, the maximum penalty under the law.

UNITED STATES
NEW YORK
BROOKLYN
December 5, 1876

A gaslight ignited scenery in the fly space of the Brooklyn Theatre in Brooklyn, New York, on December 5, 1876. Two hundred ninety-five died; hundreds were injured.

Actor Harry S. Murdock may never be remembered for the memorable roles he played. But in theater annals his performance in the face of disaster on the evening of December 5, 1876, at the Brooklyn Theatre will probably never be equaled. In fact, the entire cast of *The Two Orphans,* which was playing to a capacity audience of 900 that night, behaved in the best "show must go on" tradition.

Like practically all theater fires of the 19th and early 20th centuries (see pp. 203, 211 and 235), this one began in the fly space above the stage. It was noticed by stagehands as it began partway through the performance and rapidly spread to scenery and drapery not visible to the audience. The stagehands did what they could with their coats and hands, but there was no fire hose backstage, and the waterbuckets that were supposed to be filled every day for just such an emergency were empty.

One of the stagehands whispered to the actors through an upstage curtain, "Fire," and simultaneously, a patron, seeing a wisp of smoke near the top of the proscenium arch, yelled "Fire."

Murdock, who was playing the part of a cripple, immediately gained the use of both legs, walked calmly to the stage apron and said, "Now, now. None of that."

Another actor, J. N. Studley, added his assurances. "There is a small flame," he said, "but it will be put out. Please stay calm and keep your seats."

Harry S. Murdock, Kate Claxton and two unidentified actors plead vainly with panicking audience members to stay calm during the Brooklyn Theatre fire on December 5, 1876. (Frank Leslie's Illustrated Newspaper)

And that is what a few in the first rows of the orchestra did. But behind them in the gallery and the balcony, where most fatalities occur in theater fires, there was bedlam. "The dress circle and galleries," commented *Frank Leslie's Illustrated Newspaper* after the blaze, "seemed from the stage to be filled with raving lunatics."

In one last desperate attempt to calm the increasingly panicky audience, the play's star, Kate Claxton, entreated, "We are between you and the fire. Sit still, for God's sake, sit still."

But it was too late for that. The patrons on the orchestra floor and the actors did remain calm, and most of them escaped through the ground-floor exits. But not all. By the time the last of the orchestra patrons had reached the entrance foyer, flames and smoke had bellowed out from the stage, igniting the drapery in the auditorium, ringing the balcony with flame and then entering it. The balcony patrons stampeded down a narrow staircase to the foyer floor and poured into the crowd of exiting ground-floor audience members.

A huge pileup ensued, and children, women and men who fell were trampled underfoot. Some women fainted, and their inert bodies were carried over the heads of the crowd by desperate husbands and escorts. Hundreds were trapped and began clawing their way down the staircase to the lower floor.

Meanwhile, keeping his aplomb, Harry Murdock went back to his dressing room, calmly removed his makeup and changed into evening street clothes, complete with cape and top hat. The other actors, sensing the danger, had long since left by emergency stage exits. When Murdock finally emerged from his dressing room and attempted to descend the stairs to the street, he was met with a wall of fire.

Again, calmly, according to witnesses, he shrugged and turned back. The fire department had arrived by now, and firemen were rescuing frantic patrons who had smashed windows and were struggling toward fireladders. Murdock flung open the window of his dressing room and began to squirm through it. Halfway there, the window closed on his midsection, wedging him tight. Wordlessly, he turned and began to free himself, shoving the window upward while firemen hosed down his face and upper torso. He got the window loose, but then, abruptly, he disappeared inside the building. The fire had burned away the floor beneath him, and Murdock, along with scores of other hapless, trapped patrons, plunged to the basement of the theater. There, hours later, his charred corpse would be found fused into a pile of other victims.

The wooden theater was a total loss. It burned into the night, and the next day was spent transferring the remains of hundreds of dead to coffins and, when they ran out, to bags. Two hundred ninety-five people were buried in a cylindrical ditch in Greenwood Cemetery. Hundreds more were injured, and an unknown number were incinerated beyond discovery.

UNITED STATES
NEW YORK
NEW YORK
March 17, 1899

New York's Windsor Hotel was totally destroyed by a fire on March 17, 1899, caused by a combination of a carelessly thrown match igniting dining room curtains, a stubborn policeman and St. Patrick's Day crowds. Ninety-two died; there is no record of the number of injuries.

The seven-story Windsor Hotel, located at 46th Street and Fifth Avenue in New York City, was an unprepossessing, boxlike structure that nevertheless housed its 250 guests comfortably and safely—that is, until St. Patrick's Day of 1899. And it very well might have continued to do so for a goodly number of years if it had not been for the carelessness of one of the 300 extra spectators added to the registered roster of guests and the unforgivable stubborness of a New York City policeman.

It was the practice of the management of the Windsor in the late 19th century to open the doors of the hotel to a certain number of unregistered guests to allow them an unimpeded and somewhat more luxurious view of the annual St. Patrick's Day Parade on Fifth Avenue than they might have had were they on the jammed sidewalk. This was the case on March 17, 1899. Unregistered guests outnumbered registered ones by 50, and the combined total of 550 festively crowded the hotel.

Partway through the afternoon a careless spectator lit a cigar and nonchalantly tossed the still-burning match out of a window facing 46th Street. The wind blew the match into a lower-story window, where it ignited a set of curtains in the dining room. An alert headwaiter saw the fire start and attempted to beat it out. Unable to extinguish it, he summoned a porter to tell the manager that there was a fire in the dining room. Wasting no time, the headwaiter dashed through the lobby and out to the sidewalk, which was seven deep in spectators. The alarm box was on the opposite side of Fifth Avenue, and the frantic headwaiter shoved his way through the crowd and reached the curb in seconds. There he met one of New York's finest, on duty to maintain crowd control.

The waiter explained his mission. He had to cross the street, through the marchers, and turn in a fire alarm. The Windsor Hotel was on fire. The policeman looked at him askance. He'd seen this sort of con job before. He shoved the waiter back into the crowd. Frantically the headwaiter tried to explain again. Once more the policeman shoved him back. The harried man tried to circumvent the policeman, but the upholder of the law grabbed him by the arm, threatening to arrest him.

Firemen try desperately to bring the Windsor Hotel fire on March 17, 1899, under control. The holiday crowds, a stubborn policeman and a carelessly thrown match combined to totally destroy the famous New York City hotel. (New York Public Library [Brown Brothers])

At that moment, after precious minutes of time had been wasted, someone in the sidewalk crowd yelled, "Fire! The Windsor Hotel is on fire!" The policeman looked up, Flames and smoke were pouring from the hotel. Now he himself ran across Fifth Avenue, through the paraders, smashed the window on the alarm box and pulled its lever.

It would be another half hour, at least, before the fire department would be able to maneuver itself through the throngs of parade watchers. By that time the Windsor fire would be totally out of control. Fourteen frantic, trapped people flung themselves from upper-story windows. Another 78 died from asphyxiation or the flames. The hotel would burn completely to the ground, a $1 million loss, which might have been prevented if one obstinate policeman had unbent for a moment.

UNITED STATES
NEW YORK
NEW YORK
June 15, 1904

See MARITIME DISASTERS, *General Slocum* p. 292.

UNITED STATES
NEW YORK
NEW YORK
March 25, 1911

One of the most tragic fires of all time, the Triangle Shirtwaist Factory fire in New York City on March 25, 1911, began unexplainably in a rag bin but was compounded by overcrowding, inadequate, decaying or bolted fire exits and fire escapes and a wholesale ignoring of safety regulations. One hundred forty-five died; scores were injured.

One of the grisliest and saddest fires ever to take place lasted a mere 18 minutes. If it had begun only 30 minutes later, it might have consumed only three floors of the 10-story Asch Building, located on the northwest corner of Greene Street and Washington Place in New York City. Instead, it took the lives of 145 young and trusting immigrant employees of the Triangle Shirtwaist Factory.

The factory, a sweatshop in the worst sense of the word, was located on the top three floors of the Asch Building and was owned and operated by Max Blanck and Isaac Harris. There were almost 800 such factories in New York in 1911. Because of a shortage of appropriate factory space, the top floors of existing buildings were commandeered as loft factories.

Working conditions were unbearable by today's standards—standards that began to be established as a result of the Triangle Shirtwaist Factory fire. On the eighth and ninth floors of the Triangle Shirtwaist loft, young women worked elbow to elbow at sewing machines that were arranged in long lines. The backs of the chairs on one line touched the backs of those on the next, making it difficult to move about. A few men, called cutters, worked at long tables nearby.

Most of the young workers were between 13 and 20 years old, and practically all were Italian, Russian, Hungarian and German immigrants who could speak little or no English. Most had worked up from messenger status, for which they had been paid $4.50 a week. They moved up to sewing on buttons for $6 a week and from there to the position of machine operator at $12. By working overtime—13 hours a day, seven days a week—a few facile girls could and did earn as much as $18 a week.

There were exits from the building, but they were criminally inadequate. There were four elevators, but only one operated efficiently. Access to the elevator was down a long, narrow corridor that was made narrower by piled remnants, so that the girls would have to pass, single file, by inspectors, who examined their purses to make sure they did not steal anything.

There were two stairways leading to ground level, but the Washington Place doors were bolted shut and could not be opened from the inside—again, another device to keep the employees from stealing. The other door opened inward.

The only other exit was a decrepit fire escape a foot and a half wide and rotting. After the blaze it was estimated that it would have taken three hours for those working on the top three floors to descend it.

The Triangle Shirtwaist Factory had several fire buckets full of water lined up at its walls; there was a fire hose, but it had rotted to pieces long ago; there was a No Smoking sign, but it was regularly ignored and unenforced.

For years safety regulators were aware of the dangerous conditions that existed in the brick buildings from Canal Street north to Eighth Street. A fire in one such building had taken the life of Assistant Fire Chief Charles W. Krueger. But the efforts of Fire Chief Edward Croker were frustrated by Wall Street interests, factory owners and the apathy of the city government.

The Triangle Shirtwaist owners were particularly arrogant in their defiance. They, after all, had been responsible for destroying the shirtwaist strike in 1910. It had started there and spread until 40,000 workers were out in the industry. Triangle refused to sign a contract and was credited with finally breaking the strike.

At 4:40 on Saturday afternoon, March 25, 1911, roughly 600 workers were working an overtime shift to make up for a backlog in orders. The narrow aisles were stacked with baskets of cut goods of lace and silk. On the cutting tables layers of linen and cotton fabric were piled high. Huge bins were filled with scrap and waste material; the floor was littered with remnants. On overhead lines, finished shirtwaists were hung.

The shift was almost over; some employees had already begun to draw on their coats. And then, for a reason that has never been established, the fire started in a rag bin on the Greene Street side of the eighth floor. It was tiny, and nobody noticed it until it had gotten an impressive, fatal start.

The bodies of young women burned to death in the Triangle Shirtwaist fire of March 25, 1911, in New York City are lined up in the morgue for identification. (New York Public Library [Brown Brothers])

One of the women workers spotted it then and screamed "Fire!" Factory manager Samuel Bernstein and foreman-tailor Max Rother, who were on the Washington Place side of the eighth floor, heard the cry and, grabbing fire buckets, raced to the fire. But it had gained too much headway to be snuffed out by two buckets of water. Other men rolled out the hose, which rotted away in their hands. The valve was rusted shut.

Now the flames leaped upward, igniting the shirtwaists that were hung overhead. Women shoved at one another, knocking over chairs as they tried to squeeze toward the exit from the room. The fire vaulted to a cutting table, setting a blaze there. The narrow hallway to the elevators became jammed with crying, terrorized women. The elevator held only 12 people, and its operator, Giuseppe Zito, could only make four or five round-trips before his car was rendered immovable by burned cables and the weight of the bodies of those who had flung themselves down the elevator shaft.

Those who got down the narrow staircase to the Greene Street exit ran into the inward-opening door. Scores of bodies piled up against it before some men bodily wrested the door open and shepherded a few frantic workers through.

On the Washington Place side of the building, workers piled up against the unyielding, bolted door. Tearing at one another's clothing, most of them died there as flames roared down the stairwell, burning them alive.

The fire had spread to the ninth floor now, where 300 more workers stampeded to escape the flames. On the 10th floor, in the executive offices, Blanck and Harris, along with Blanck's children and governess, who had come to visit him in his office, got to the roof and made their way to an adjacent building via a fire ladder.

Meanwhile, the eighth floor had become a raging inferno. Those who were trapped at the Washington Place exit were already blackened corpses. Some workers frantically tried to leave via the eighth-floor fire escape. Too flimsy to stand either heat or the weight thrown upon it, the iron ladder warped and then gave way. Those on it fell to the ground.

The fire roared on to the ninth floor. The heat was intense enough to curl sheet-iron shutters on a building 20 feet away.

The fire alarm had been turned in, and in moments Engine Company Number 18, led by Foreman Howard Ruch, Company Number 72 and Hook and Ladder Company Number 20 were all there. But even though they arrived no more than eight minutes after the blaze began, they were too late. The most horrible phase of the drama had begun before they could get there.

Foreman Bernstein later told a UPI reporter that a shopgirl named Clotilda Terdanova was the first to jump from the building to her death. "She tore her hair and ran from window to window," he said, "until finally, before anyone could stop her, she jumped out. She was young and very pretty. She was to leave us next Saturday to be married three weeks later."

More and more women, some of them clinging together, jumped. It was literally raining bodies. The firemen could not unravel their hoses because of the smashed corpses piling on the sidewalks. Some finally did, and Battalion Chief Edward Worth used his first two lines to cool the building over the heads of the hysterical girls clinging to ledges and standing in windows. Then, a gust of wind sucked flames out of windows and onto the clothes of the trapped girls.

A cry of "Raise your ladders! Raise your ladders!" came from the spectators who had begun to accumulate. A girl on a ninth-floor ledge waved a handkerchief, directing one of the ladders, which ascended toward her. But Chief Croker had warned long ago that their ladders would only reach to the seventh floor of any blazing building, and that is where the ladders stopped that afternoon. The girl, her skirt ablaze, leaped for the ladder 30 feet below her and missed. Her flaming body hit the sidewalk.

All of the fire hoses were now crushed by falling bodies, and no water was reaching the blaze. Company Number 18 spread the first life net, a new one 14 feet long. Three girls dove simultaneously from the ninth floor. When they landed, they ripped the net to shreds and pulled a dozen firemen inward on top of their bodies.

Company Number 20 set up a 20-foot Browder net. Bodies rained on it so rapidly that the tube steel frame buckled and gave way.

Two policemen improvised a blanket and caught one girl dead center. The blanket held for a minute and then gave way, and she crashed against the grating of a sky-light.

Another net received a girl who landed safely. Battalion Chief Worth pulled her upright. "She blinked and said nothing," Chief Worth later told reporters; "I told her to 'go right across the street.' She walked ten feet—and dropped. She died in one minute.

"Life nets?" continued Worth. "What good were they? The little ones went through life nets, pavement, and all. I thought they would come down one at a time. I didn't know they would come with arms entwined—three and even four together."

Bill Shepherd, the UPI reporter, wrote of the floods of water from the firemen's hoses that ran into the gutter. "[They] were actually red with blood. I looked upon the heap of dead bodies, and I remembered these girls were shirtwaist makers. I remembered their great strike of last year, in which these same girls had demanded more sanitary conditions and more safety precautions in the shops. These dead bodies were the answer."

In 18 minutes it was over. The doors on the ninth floor were chopped down, and the firemen quickly extinguished the flames. All of the damage had been done in the first 10 minutes. Firemen found 49 burned or suffocated bodies on the ninth floor. Thirty-six more were found at the bottom of the elevator shaft. Fifty-eight lay on the sidewalk. Two more would die of their injuries. All in all, 145 innocent, exploited young immigrants would perish in those terrible 10 minutes.

The reaction to the tragedy was immediate and immense. The Waistmakers Union organized a mass funeral for the victims, and 10,000 mourners attended. New York's East Side, from which most of the dead had come, seethed with anger. On April 5 more than 80,000 people marched up Fifth Avenue following an empty hearse pulled by six horses draped in black.

The testimony at the trial of Harris and Blanck brought forth horrifying admissions. Safety expert H. F. Porter, who had pleaded with the Triangle owners to institute fire drills, told the *New York Times,* "One man whom I advised to install a fire drill replied to me: 'Let 'em burn. They're a lot of cattle, anyway.'"

In December a grand jury exonerated the Triangle owners of manslaughter charges, claiming that the bolted door might have been locked by an employee. The *New York Times* erupted in an editorial. "The monstrous conclusion of the law is that the slaughter was no one's fault," it thundered, "that it couldn't be helped, or perhaps even that, in the fine legal phrase which is big enough to cover a multitude of defects of justice, it was 'an act of God!' This conclusion is revolting to the moral sense of the community."

But the country was shocked, and labor gained much. The International Ladies Garment Workers Union was formed as a direct result of the tragedy. The day of the sweatshop was nearing an end. Uniform fire and factory codes, led by New York's Sullivan-Hoey Fire Prevention Law of October 1911, were instituted all over the country. Fire Chief Edward Croker turned in his badge in order to lead a crusade for safety. Nothing would be quite the same in either fire prevention or factory working conditions ever again, and it had taken a tragedy of incredible proportions to bring this about.

UNITED STATES
OHIO
COLLINWOOD
March 4, 1908

The fire that began unexplainably in the boiler room of the Lake View School in Collinwood, Ohio, on March 4, 1908, turned tragic because of blocked and faulty exits. One hundred seventy-six died; scores were injured.

The Lake View School in the Cleveland suburb of Collinwood was a daytime home for 325 students. On March 4, 1908, 176 of them died by burning or asphyxiation, trapped behind a front door that opened inward and a rear door that was bolted shut.

The fire's origins were never established. It began, apparently, in the boiler room, which was located directly under the steps of the front staircase of the three-story school. Two young girls discovered the smoke first and reported this to a janitor who turned in the fire alarm and then ran for his life, apparently locking the rear door of the school behind him and thus condemning to death by burning scores of children.

The fire spread swiftly; the building was constructed entirely of wood and other inflammable materials. The first-floor, first-grade classroom of teacher Grace Fiske was, because it was located over the boiler room, the first to be invaded by smoke and flames. Fire burst through the closed classroom door, and Ms. Fiske, assessing the situation, directed her hysterical class to the windows and the fire escape beyond them. When some did not respond immediately, she gathered them up and shoved them toward the windows. Others, who ran berserk around the room, she picked up in her arms and carried to the windows. Most of the children in the class escaped, but some recalcitrant ones—and their teacher, who went back into the flames to try to rescue them—were burned to death.

Meanwhile the fire was spreading rapidly throughout the building, racing through the narrow hallways and igniting drapes in the auditorium. (One small boy would later try to escape by climbing, hand over hand, along the top of the stage curtain. He would fall to his death halfway across.) Fire drills seemed to have meant little to the children. They and some of their teachers ran helter-skelter into the halls and piled up in massive, fatal drifts against both the front and rear exits. A Miss Golmar, finding the staircases blocked, deserted her class; she climbed over struggling children and through a rear window to safety.

Eventually the back door was battered in by Henry Ellis, a real estate broker, and I. E. Cross, a train superintendent. They encountered a huge pile of charred bodies, with more children climbing on every minute. "Flames reached out from the walls to catch first one and then another child," said Ellis later to UPI. Both men hauled dozens of still-living people from the pile, including Pearl Lynn, a teacher, who, leading her children to the door, stumbled and fell and was buried beneath the hill of bodies and still survived.

Mrs. John Phillips, whose daughter went to the school, dashed to it when told of the fire. The front door had been pried partially open by the time she arrived, and she saw her daughter struggling near the top of the mass of children jammed against it on the inside. She reached for the child's hands and pulled but could not free her. She stroked her head, "trying," as she said later, "to keep the fire from burning her hair. I stayed there and pulled at her," she went on, "and tried to keep the fire from her till a heavy piece of glass fell on me, cutting my hand nearly off. Then I fell back and my girl died before my face."

The fire department arrived, but there was little they could do, and that little was diminished by their ancient equipment. Neither their ladders nor the water from their hoses would reach above the first floor, and the roaring inferno of the schoolhouse was now a heartrending sight, with bedlam inside and the faces of children at the windows being overcome, one by one, by flames or smoke. Eight thousand spectators—practically the entire population of Collinwood—eventually crowded into the schoolyard, some of them jeering and kicking the firemen as they put forth their futile efforts.

Charles G. McIlrath, the chief of Collinwood's police force, arrived just as his eight-year-old son Hugh appeared on the fire escape, leading a group of younger children to safety. The fire escape ladder ended eight feet from the ground, and Chief McIlrath reached up and lowered those who were afraid to jump to the ground. His son, meanwhile, ran back into the school to rescue more, while his father pleaded with him to jump to safety. The boy's body was later found on a stairway. His arms were around two smaller, dead children.

It would be three hours before the fire burned down enough for rescue teams to enter the building. What they found was appalling. All night, bodies and some few injured children were carried from the wreckage of the school. A makeshift morgue for the 176 killed was fashioned at the Lake Shore Depot.

A coroner's inquest was held, and the coroner asked for prosecution of the builders of the school for erecting it with narrow hallways and escape doors that opened inward. The builders were never brought to trial or account.

UNITED STATES
OHIO
COLUMBUS
April 21, 1930

Arson and overcrowding conspired to intensify the fire in the Ohio State Penitentiary on April 21, 1930. Three hundred twenty-one died; 130 were injured in the resultant blaze.

Most of the prisoners had been locked in their cells at the Ohio State Penitentiary on the evening of April 21, 1930. The prison, like most penitentiaries in the first years of the

Great Depression, was severely overcrowded. Forty-three hundred inmates were imprisoned in accommodations built for 1,500. The Ohio State Penitentiary, one of the largest prisons in the country, was also one that had been under fire for 12 years by prison authorities for not only its overcrowding but its substandard conditions.

Thus, partially to accommodate its critics, the prison was undergoing a token expansion. A minuscule construction project designed to increase its capacity had begun in the west cell block. A cat's cradle of scaffolding stood outside the older building, and it was here that a fire began. A year later two convicts would admit having set the blaze as a protest against being forced to work on the scaffolding. They had poured oil on a pile of garments stored beneath the structure and ignited it with a candle stolen from the chapel.

The flames, fanned by a stiff breeze, quickly spread to the six-tiered cell block that housed approximately 800 prisoners. Running up the scaffolding, the flames leaped to the roof, eating their way through the tar paper and timber and showering sparks on the prisoners in the upper layer of cells. Bedding and mattresses began to ignite, and prisoners, trapped in their cells, screamed in fear, pleading to be let out of their fiery cubicles. The guards, following the orders of head guard of the upper tier Thomas Watkinson, refused to unlock the cells.

Some of the prisoners in the lower rows of cells had not yet been locked in for the night. Hearing the yells of their fellow prisoners, smelling the smoke and seeing the raining cinders, some of these prisoners refused to be locked up. The guards, again mindlessly following orders, tried to force them into their cells.

A riot resulted. Prisoners wrestled guards to the floor, slammed them up against walls and generally managed to drive them back from the first-tier cell block. The smoke grew more dense, obscuring the fighting and choking the combatants, but the prisoners finally forced the guards back to the door leading to the safety of the prison yard. It was a logical, humane solution that some of the lock-stepped guards still resisted.

One obstinately tried to bar the only other door to the yard. Two convicts, John Sherman and Charley Simms, grabbed him and ripped his keys from him. Dashing to the second and third tiers, they began to unlock locked cells. They managed to free 68 men and direct them to the yard before, almost overcome by the increasing smoke, they had to abandon the effort and run for the yard themselves.

The sixth tier was a horror. Finally realizing the murderousness of their stubbornness, two guards, Thomas Little and George Baldwin, pleaded with chief guard Watkinson to unlock the cells. The roof had begun to crack, and each cell was now an oven. Some men had already died, turned into human torches when pieces of the roof caved in on them in their confined spaces. Watkinson still refused, citing his orders.

Finally the two guards wrestled Watkinson to the floor and took his keys from him. But it was too late. The bars on the cells and their doors had become red hot, and the locks had fused together. The keys would not work.

Now, with a sickening roar, the roof collapsed entirely into the cell block, incinerating every single prisoner there. One hundred sixty-eight men burned alive while Little and Baldwin watched helplessly. Not one man survived. "I saw their faces," recalled one guard for reporters, "wreathed in smoke that poured from their cells. With others I tried to get them out, but we could not move the bars. Soon flames broke into the cellrooms and the convicts dropped before our eyes. They were literally burned alive."

On other tiers more rational behavior took place as prisoners and guards used sledges and crowbars to pry open cell doors. There were heroic rescues. Frank Ward, an ex-policeman, now a prisoner, alone released 136 men; "Big Jim" Morton, a bank robber, at liberty in the yard when the fire began, rushed into the cell block time and again to drag out convicts who were half dead from the smoke. Morton himself was eventually overcome and had to be carried to the yard.

Despite the frantic rescue efforts, men continued to die. A fireman described the scene to reporters afterward: "While we were trying to cut through the steel, the trapped prisoners climbed up the bars of the cells pleading with us to save them. We could hardly see through the smoke. We were driven back and these men died before our eyes. They were overcome toward the end and did not scream, so I think they were unconscious by the time the fire reached them."

Firemen faced dual danger when they arrived at the prison. Four thousand convicts, enraged at the fate of their fellow prisoners who had been locked into their iron deathtraps, barred the firemen from fighting the fire. A standoff ensued while the fire grew to horrendous proportions. Finally, National Guardsmen and federal troops, summoned by prison officials, moved in with bayonets.

Firefighters began to put out the blaze, but now prisoners began to pelt the firemen with rocks. The firemen turned their hoses on the prisoners, forcing them back. But the riot began to take on insane proportions. Some convicts tried to set fire to a fire truck. Others succeeded in igniting the Catholic chapel and a woolen mill. Firemen managed to put out both fires before they caused extensive damage. When an entering ambulance was rushed by 20 convicts, the warden issued a "shoot to kill" order, and the riot quieted.

Firemen finally extinguished the conflagration, and they and prison officials entered the smoldering building. A horrendous sight greeted them. There were the crushed and incinerated corpses of the 168 men of the sixth tier, and in isolated other cells, the agony-twisted and charred remains of hundreds more. Of the 800 prisoners housed in the cell block, 321 were dead, and 130 were severely injured.

The shocking details of the early minutes of the fire were made public. Watkinson, accused of signing the death warrants of 168 men, denied responsibility, testifying that he was merely carrying out the orders of his superior, Captain John Hall. Hall denied ever giving such an order, and Watkinson was suspended from his job.

It was revealed that the fire alarm was turned in from a box outside the prison wall long after the fire had gained a firm upper hand. The main cage door leading to the cells, which was usually unlocked, was found to have been locked all during the conflagration. Thus, there was a strong indication that someone had locked it after the fire had started.

The Reverend Albert O'Brien, the prison's Catholic chaplain, issued a written condemnation: "The disaster was a crime on the part of the state—a greater crime than any of those dead boys ever committed against the state," he wrote.

The *Columbus Evening Dispatch* noted the horrendous overcrowding and editorialized: "For many years successive legislatures have dawdled over the prison problem while defenseless human lives remained in jeopardy."

The *Cleveland Plain Dealer* echoed this. "The State must abandon a policy of neglect and indifference," it stated. "The cries of men behind steel bars, held in a vise for creeping flames to destroy, are ringing in Ohio ears. The State is more cruel than we believe if the cries are unanswered."

Ironically, with all of its loss of life, the fire caused a mere $11,000 in damage to the west cell block of the Ohio State Penitentiary.

UNITED STATES
WISCONSIN
PESHTIGO
October 8, 1871

The worst forest fire in U.S. history destroyed the entire town of Peshtigo, Wisconsin on October 8, 1871. No specific cause was recorded, but it was probably a combination of smoldering fires, a drought, indiscriminate logging and the burning of debris by railroads. Two thousand six-hundred and eighty-two people died in Peshtigo and the rest of Wisconsin, and hundreds were injured.

On October 8, 1871, the very day that the Great Chicago Fire began in Mrs. O'Leary's barn, the worst forest fire in the history of the United States started 250 miles to the north of Chicago. It would claim 2,682 lives and level an area of 400 square miles of forest.

Peshtigo, built in the midst of an enormous forest a few miles from Green Bay, was a hugely successful logging town in 1871. It boasted 350 houses, three hotels, two churches, four saloons, a dozen stores, a sawmill and a woodenware factory. The Peshtigo River ran through the heart of town, and a large wooden bridge allowed those on one side of the settlement to get to the other. Most of Peshtigo's population of 2,000 was in the logging business, and they looked forward to the not too distant day when the Chicago and Northwestern Railroad would link their town with Milwaukee and Chicago.

All of the spring and summer preceding the fire was relentlessly dry. Early in July a day of rain held out false hope. From then until September, no more rain fell. Springs began to dry up; rivers fell. During the summer a number of small, fussy fires smoldered in dry peat bogs and in the webwork of roots that dried-up swamps revealed.

In August citizens of Peshtigo carved a fireline out of the forest ringing the city. Accomplishing this, they felt safer, and when, on September 5, rain finally fell, they felt even more secure. But the rain was shortlived and inadequate, turning the forest into a steaming jungle. It was ripe for a forest fire, and the forest dwellers and logging interests only increased the danger by two foolish activities.

First, figuring that the rivers would soon be filled with water from expected rains, the logging interests continued to harvest trees. Limbs and slashings carpeted the forest floor. Second, railroads cutting through the forest south of the city ignored the dryness of the brush around them and burned their debris.

Whether the final fire began from the spontaneous combustion of marsh gases, or the small smoldering fires finally burst into full-blown flames, or the railroaders kindled it with their fires, or a careless forest person began it, no one would ever determine. In fact, the fire itself received little notice in the press, except in the pages of Luther B. Noyes's three-month-old *Marinette and Peshtigo Eagle,* which published a Fire Extra on October 14 and has supplied practically all of the information on this fire for historians. The national Fourth Estate was too busy at the time with the Chicago Fire.

But to the people of Peshtigo, on the chilly Sunday of October 8, 1871, the approaching fire was very real. A pall of brownish smoke hung over them as they went to church to hear sermons warning of Judgment Day.

By seven o'clock that night the wind had picked up, and ashes were beginning to fall on the city in a steady rain, much as if a volcano were erupting nearby. When the townspeople left services an hour later, a steady, full-throated roar could be heard nearby in the forest. Frightened, many of them went home, closed their doors and windows against the smoke and waited.

Meanwhile, farmers in the forest were in dire trouble. The flames of the fire were undeniably visible to them. By seven o'clock huge tongues of flames licked at a score of farms and then rushed forward, consuming them. The network of paths through the woods became impassable, and forest animals began to mix with farm animals as they tried to escape the advancing fire.

One farmer gathered his wife and five children together into a wagon and tried to outrace the flames. They overtook them, killing everyone but the farmer.

By 9:00 P.M., just as the Chicago fire was starting, church bells began to toll in Peshtigo warning of the approaching inferno, which was now brutally apparent. The night was rimmed in the crimson glow of the fire advancing from the south. Cinders began to fall more rapidly on the city, setting small fires. The fire department shuttled between them, putting some out, being beaten back by others.

Now women began to gather up their children and, wrapping them in covers, fled to the streets with them. Sparks ignited trees within the city. Some townspeople began to turn toward the Peshtigo River.

Suddenly, a tornado-like gale whipped around the city, scattering a rain of fire. Sheets of flame and huge firebrands began to fly through it. Balls of burning grass uprooted from the swamps and explosions of methane gas from the marshes rocked the night air.

The fleeing townspeople turned into a mob. Men joined the women and children. Those on one side of the river tried to reach the other side, and those on that side tried to cross the bridge at the identical moment. Two frantic, terrified mobs met each other at midspan, milling, fighting, tearing at each other. The bridge groaned and collapsed, flinging the mob into the river, where many drowned.

In the midst of this, the telegraph operator at the railroad station received some news that caused him to dash out into the street and inform nearby firefighters: "This fire is bigger than we thought. We just got a message on the telegraph from Green Bay. Chicago is burning!" The firefighters figured that it was indeed Judgment Day.

The firestorm erupted in another destroying wind. Flames roared through the city. Houses were bowled over. Roofs exploded from the tops of houses and flew through the superheated air. Burning trees became flaming battering rams. And the wall of fire roared on.

The air was so hot now that a person could burst into flames without being touched by fire at all. People simply became human torches, incinerated on the spot. Vacuum pockets developed in the air, and those who ran into them lost their breath, collapsed and died within a few steps.

Bizarre incidents flared up as easily as the fires set by the firestorm. One husky husband running through the streets toward the river with his wife in his arms collided with someone in the semidarkness brought on by the smoke. They all crashed to the ground. Frantically, he picked up his wife and continued toward the river. Reaching it, he plunged in up to his shoulders and set her down on her feet.

But it was not his wife. She was a total stranger. In the confusion after his fall, he had picked up the wrong woman. His wife was now a blackened corpse by the side of a road.

Another man dragged a heavy bed with his wife, who was suffering from a fever, in it to the river, where he submerged it to a depth that covered his wife's body but left her pillowed head above the surface. The family huddled together there all night and survived.

Still another man, realizing he was too far from the river to make it, decided to take refuge in a horse watering trough. He was boiled alive.

Terrified cattle stampeded through the streets, trampling some people. A hysterical man, rather than face death by fire, killed his wife and children and then slit his own throat. Another fastened a noose around his neck and hanged himself in his well. Seventy people huddled together in the middle of a cornfield were burned to death. And on the banks of the Peshtigo River that night, several pregnant woman gave premature birth.

The fire raged till dawn unchecked. T. J. Teasdale, one of the survivors, gave his story to the *Marinette and Peshtigo Eagle:*

> When the fire struck the town it seemed to swallow up and drown everything . . . a fierce, devouring, pitiless rain of fire and sand, so hot as to ignite everything it touched. . . .
>
> Within three hours of the time the fire struck, Peshtigo was literally a sand desert, dotted over with smoking ruins. Not a hencoop or dry goods box was left . . . Cattle and horses were burned in their stalls. The Peshtigo Company's barn burned with over fifty horses in the stable. A great many men, women and children were burned in the streets, and in places so far away from anything combustible that it would seem impossible they should burn. But they were burned to a crisp. Whole families, heads of families, children were burned, and remnants of families were running hither and thither, wildly calling and looking for their relatives after the fire.

Peshtigo was no more. Monday morning dawned silently. No dogs barked, no cows mooed, no birds sang. Even the fish in the river were dead and floating on its surface.

By afternoon a steamer from nearby Marinette, which had miraculously escaped the fire, arrived with food and clothing and the news that 23 other towns had been destroyed by the same inferno—Casco, DePere, Shite Rock, Ahnepee, Elm Creek, Forestville, Little Sturgeon Bay, Lincoln, Brussels and Rosiere among them.

That night, it began to rain. It was more than the usual rain that follows a firestorm. It would be the beginning of 15 years of the most bountiful rainfall in the history of Wisconsin.

The survivors in Peshtigo set about rebuilding their city and retilling their land. Within three years the town would be rebuilt as a dairy center, which it is today. But the worst forest fire in the history of the nation would claim 1,182 lives in Peshtigo alone, and 1,500 others in the state of Wisconsin—an unenviable, unbroken record.

MARITIME DISASTERS

THE WORST RECORDED MARITIME DISASTERS

* Detailed in text

Atlantic Ocean
- * *Andrea Doria/Stockholm* (1956)
- *Guiding Star* (1855)
- * *Lusitania* (1915)
- *Monarch of the Sea* (1866)
- * *Morro Castle* (1934)
- * *Titanic* (1912)

Baltic Sea
- * *Wilhelm Gustloff* (1945)

Bay of Gibraltar
- * *Utopia* (1891)

Belgium
- Zeebrugge
  - * *Herald of Free Enterprise* (1987)

Canada
- Nova Scotia
  - Halifax
    - * *Atlantic* (1873)
    - *Mont Blanc* (1917) (see EXPLOSIONS)
  - Sable Island
    - * *La Bourgogne* (1898)
  - Wingo Sound
    - * *St. George/Defence* (1811)
- Quebec
  - St. Lawrence River
    - * *Empress of Ireland* (1914)

Cape of Good Hope
- * *St. James* (1586)

Caribbean Sea
- Near Veracruz, Mexico
  - * Grand Fleet (1591)

Chile
- Valparaíso
  - *L'Orriflame* (1770)

China
- Chang (Yangtze) River
  - Chinese troop carrier (1926)
  - *Hsin Hsu-tung* (1928)
- Manchuria
  - Chinese Army vessel (1949)
- Shantou Harbor
  - * *Hong Koh* (1921)
- Wusong
  - * *Kiangya* (1948)
- Zhoushan Islands
  - * *Hsin-Yu* (1916)
  - *Tai Ping* (1949)

Congo
- Brazzaville
  - * *Ferry* (1993)

Egypt
- Safaga
  - * *Salem Express* (1991)

English Channel
- * Spanish Armada (1588)

Finland
- Uto
  - * *Estonia* (1994)

France
- La Rochelle
  - * *Afrique* (1920)

Great Britain
- England
  - Spithead
    - * *Royal George* (1792)
  - Woolwich
    - * *Princess Alice* (1878)
- Scotland
  - North Sea
    - * *Piper Alpha Oil Rig* (1988)
    - * Rockall
      - *Norge* (1904)

Greece
- Paros
  - * *Express Samina* (2000)

Gulf of Finland
- *Leffort* (1857)

Gulf of Mexico
- Flota de Nueva España (1590)

Haiti
- Port-au-Prince
  - * *Neptune* (1993)

Hispaniola
- Spanish fleet (1502)

Holland
- Texel River
  - * *Minotaur* (1810)

Hong Kong
- River steamer (1945)

India
- Indian Ocean
  - *Blenheim* (1807)
- Madras
  - * *Comorta* (1902)
- Manihari Ghat
  - * Passenger ferry (1988)

Indonesia
- Manado
  - * *Cahaya Bahari* (2000)
- Sumatra
  - * *Gurita* (1996)
  - * *Harta Rimba* (1999)

Italy
- Leghorn
  - * *Queen Charlotte* (1800)

Japan
- Hakodate
  - * *Toyo Maru* (1954)
- Hokkaido
  - *Indigirka* (1939)
- Sasebo
  - *Mikasa* (1905)
- South coast
  - * *Ertogrul* (1890)
  - *Kichemaru* (1912)
- Tokyo Bay
  - * *Kawachi* (1918)

Java Sea
- * *Tamponas II* (1981)

Labrador
- Egg Island
  - * English Armada (1711)

Martinique
- French-Dutch convoy (1776)

Mexico
- Veracruz
  - Flota de Nueva España (1600)
  - *Nuestra Señora de la Concepción* (1732)

Myanmar (Burma)
- Gyaing River
  - * Ferry (1990)

New Zealand
Auckland
* *Cospatrick* (1874)

North Sea
York (1803)

Norway
Skagerrak
* *Scandinavian Star* (1990)

Philippines
Tablas Strait
* *Dona Paz/Victor* (1987)

Romania
Galati
* *Mogosoaia* (1989)

Tanzania
Mwanda
* *Bukoba* (1996)

Turkey
Constantinople
Neiri Shevket (1850)

United States
Florida
Coast
* *Capitanas* (1715)
Gulf of Florida
Narvaez expedition (1528)
Keys
Spanish convoys (1622)
Tampa
Spanish expedition (1559)
Hawaii
Honolulu
* *Ehime Maru/ Greeneville* (2001)
Illinois
Chicago
* *Eastland* (1915)
New York
New York
* *General Slocum* (1904)
Tennessee
Memphis
Sultana (1865)
(See EXPLOSIONS)

Venezuela
Aves Island
French fleet (1678)

West Indies
Mona Passage
Sisters (1787)

CHRONOLOGY

1502
July
Hispaniola; Spanish fleet

1528
September 22
Gulf of Florida; Narvaez expedition

1559
Tampa, Florida; Spanish expedition

1586
* Cape of Good Hope; *St. James*

1588
August–October
* English Channel; Spanish Armada

1590
July
Gulf of Mexico; Flota de Nueva España

1591
August 10
* Caribbean Sea near Veracruz; Grand Fleet

1600
September 12
Veracruz; Flota de Nueva España

1622
September 6
Florida Keys; Spanish convoys

1678
May 3
Aves Island, Venezuela; French fleet

1711
August 22
* Egg Island, Labrador; English Armada

1715
July 31
* Florida coast; Capitanas

1732
January
Veracruz; *Nuestra Señora de la Concepción*

1770
Valparaíso, Chile; *L'Orriflame*

1776
September 6
Martinique; French-Dutch convoy

1787
Mona Passage, West Indies; *Sisters*

1792
August 29
* Spithead, Great Britain; *Royal George*

1800
March 17
* Leghorn, Italy; *Queen Charlotte*

1803
North Sea; *York*

1807
February 1
Indian Ocean; *Blenheim*

1810
December 22
* Texel River, Holland; *Minotaur*

1811
December 24
* Wingo Sound, Nova Scotia; *St. George/Defence*

1850
October 23
Constantinople, Turkey; *Neiri Shevket*

1855
January 9
Atlantic Ocean; *Guiding Star*

1857
September 23
Gulf of Finland; *Leffort*

1865
March 26
Memphis, Tennessee; *Sultana* (See EXPLOSIONS)

1866
April 3
Atlantic Ocean; *Monarch of the Sea*

1873
April 1
* Halifax, Nova Scotia; *Atlantic*

1874
November 17
* Auckland, New Zealand; *Cospatrick*

1878
September 3
* Woolwich, Great Britain; *Princess Alice*

1890
September 19
* South coast, Japan; *Ertogrul*

1891
March 17
* Bay of Gibraltar; *Utopia*

1898
July 4
* Sable Island, Nova Scotia; *La Bourgogne*

1902
April
* Madras, India; *Camorta*

1904
June 15
* New York, New York; *General Slocum*

June 28
* Rockall, Scotland; *Norge*

1905
September 10
Sasebo, Japan; *Mikasa*

1912
April 14
* Atlantic Ocean; *Titanic*

September 28
Mikasa, Japan; *Kichemaru*

1914
May 29
* St. Lawrence River, Quebec; *Empress of Ireland*

1915
May 1
* Atlantic Ocean; *Lusitania*

July 24
* Chicago, Illinois; *Eastland*

1916
August 29
* Zhoushan Islands, China; *Hsin Yu*

1917
December 30
Halifax, Nova Scotia; *Mont Blanc* (See EXPLOSIONS)

1918
July 12
* Tokyo Bay, Japan; *Kawachi*

1920
January 12
* La Rochelle, France; *Afrique*

1921
March 18
* Shantou Harbor, China; *Hong Koh*

1926
October 16
Chang (Yangtze) River; Chinese troop carrier

1928
August 15
Zhoushan Islands, China; *Hsin Hsu-Tung*

1934
September 8
* Atlantic Ocean; *Morro Castle*

1939
December 12
Hokkaido Island, Japan; *Indigirka*

1945
January 30
* Baltic Sea; *Wilhelm Gustloff*

November 8
Hong Kong, China; Chinese river steamer

1948
December 3
* Wusong, China; *Kiangya*

1949
January 27
Zhoushan Island, China; *Tai Ping*

1954
September 26
* Hakodate, Japan; *Toyo Maru*

1956
July 25
* Atlantic Ocean; *Andrea Doria/Stockholm*

1981
January 27
* Java Sea; *Tamponas II*

1987
March 6
* Zeebrugge, Belgium; *Herald of Free Enterprise*

December 20
* Tablas Strait, Philippines; *Dona Paz/Victor*

1988
July 5
* North Sea, Scotland; *Piper Alpha* Oil Rig

August 6
* Manihari Ghat, India; Indian passenger ferry

1989
September 10
* Galati, Romania; *Mogosoaia*

1990
April 7
* Gyaing River, Myanmar (Burma); Ferry
* Skagerrak, Norway; *Scandinavian Star*

1991
December 14
* Safaga, Egypt; *Salem Express*

1993
February 17
* Port-au-Prince, Haiti; *Neptune*

March 1
* Brazzaville, Congo; Ferry

1994
September 28
* Uto, Finland; *Estonia*

1996
January 21
* Sumatra, Indonesia; *Gurita*

May 21
* Mwanda, Tanzania; *Bukoba*

1999
February 10
* Sumatra, Indonesia; *Harta Rimba*

2000
June 29
* Manado, Indonesia; *Cahaya Bahari*

September 26
* Paros, Greece; *Express Samina*

2001
February 9
* Honolulu, Hawaii; *Ehime Maru*/USS *Greeneville*

MARITIME DISASTERS

For those who could afford it, there was no more romantic or peaceful way to travel abroad than on a transatlantic ocean liner. Even in the late 1960s, when jet travel was beginning to make ocean travel obsolete, it remained a remnant of a more gracious, less frantic age, when time was not the tyrannical monarch of a person's life and a little bouillon calmed both the stomach and the nerves.

Those leviathans of the North Atlantic and the South Pacific were self-contained cities, and like the great cities of the world, each had its own personality. The Cunard Line, for instance, was definitely British and guarded the doors between first, cabin and tourist class. On the French Line, sybaritic considerations—a shortage of men for the after-dinner revels in first class, for instance—sometimes relaxed those same barriers. The Italian Line was similarly relaxed, as was the round-the-world Moore-McCormick Line.

But what distinguished these liners from the dull world as it was was a sense of opulence married to a sense of fun. Although the present resurgence of cruise liners is a welcome testament to people's need for the sea, they are really pale imitations of the real thing. Floating summer camps, they cannot begin to approximate the dignified feeling of comfort and relaxation unto relief that life aboard the grand transatlantic liner offered its passengers.

Perhaps it was the long tradition of sailing that reaches back to the Phoenicians that made it so comfortable and even insular. There was a feeling of being cared for that was almost familial, and perhaps that was necessary, for the sea has been, and always will be, the master of all it surveys, borders on, or floats.

Aboard one of these posh, floating metropolises, one never really felt threatened by the sea, only occasionally tormented. Storms at sea were only fun for the very stalwart, and there were those to whom even the gentle sway of the grand saloon was too much for their centers of equilibrium. Seasickness is not fun, as anyone who has experienced it will attest. But that was the only drawback to this supremely romantic and restful way of travel.

Still, disasters did occur at sea, and this section details the worst of them. Interestingly enough, only one—the *Andrea Doria/Stockholm* collision (see p. 259)—occurred in the last three decades of transatlantic-transpacific travel.

Unlike air crashes, disasters at sea have rarely killed everyone aboard. Except in cases of wholesale stupidity and criminal neglect, as in the events involving the *Titanic* (see p. 263) or the *General Slocum* (see p. 292), adequate lifeboats and life preservers are provided, and mandated safety precautions and drills assure passengers that there is at least some chance of surviving a disaster at sea.

Storms have, naturally, been the scourge of sailors since the Phoenicians. Some of the earlier maritime disasters were caused by hurricanes. The sinking of the entire Spanish Armada in the English Channel in 1588, with its staggering death toll (see p. 275), and the equally dramatic demise of the English Armada off the coast of Labrador in 1711 (see p. 286) were, at least partially, the result of storms.

But even in these cases, the enormous loss of life would have been avoided had it not been for the misjudgment of commanding officers. In fact, as in all disasters involving the transportation of large numbers of people, be it by rail, air or sea, so-called human error has proven to be the most pervasive culprit. Captains have driven their ships into storms and onto reefs, ignored warnings and sped recklessly through fogs. And their ships have sunk, sometimes with them aboard.

In addition, this incompetency, recklessness or downright stupidity has often been compounded by cowardice. The nameless captain who slammed the Portuguese packet *St. James* onto the rocks of South Africa in 1586 (see p. 272) leaped into the first lifeboat, leaving all of his passengers behind; the crew of the French liner *La Bourgogne,* after colliding with another ship because its captain refused to slow down in a heavy fog, saved themselves by deliberately drowning hundreds of passengers (see p. 270). In still other circumstances, such as the fire aboard the *General Slocum* and the *Dona Paz* (see p. 288), the parsimonious shortcuts taken by profit-motivated steamship companies accounted for needless deaths.

But these are the bizarre extremes. Although the tragedy of the *Titanic* proved conclusively that there is no such thing as an unsinkable ship, most passenger liners, lake steamers and even ferries have the good manners to take their time in sinking, thus giving their passengers a reasonable chance of surviving. And with increasingly precise navigation equipment developed by the middle of the century, sea travel has become one of the safest and most gracious ways of travel yet evolved.

A pity, then, that speed proved to be the one barrier reef upon which all the ocean liners of the world except one would founder. The *Queen Elizabeth II,* at a whopping premium and with an assurance that passengers only need to take their time across the North Atlantic in one direction and can wing themselves back by Concorde, still plies the sea-lanes between New York and Southampton. But that is it, at least for now.

Gone are the riotous midnight sailings of the 1920s, the heart-stopping thrill of immigrants and returning expatriates at seeing the Statue of Liberty rise up in a morning mist as the ship on which they sailed eased in on the early tide to New York Harbor; gone are pulse-slowing, leisurely ways to get from here to there.

And yet, the lure of the sea persists. Though the era of the transatlantic and transpacific voyage has all but ended, shipboard life has not. It survives in a plenitude of cruise lines and their cruise ships, wandering from island to island in the Caribbean and the South Pacific, from port to port in the Mediterranean and up and down the west coast of Europe and Africa and both coasts of South America.

Life aboard many of these ships is considerably more frantic and not necessarily as luxurious as it was on the leisurely, five-day crossings of the Atlantic aboard the great ships of the great lines, but there are similarities. Nonstop eating is possible, at no extra charge, and nonstop entertainment is available at any hour and for almost any taste. Graciousness is still there, too, but fun is the focus.

The newest of these cruise ships are, with their multiple above-the-waterline decks, more floating hotels than seagoing craft. Vestiges of the past still survive here and there: The old SS *France* has become the *Norway,* for instance, and thus the merging of two time periods and two worlds is possible on one of this elegant ship's cruises. Tour companies run smaller ships and boats for river cruising, particularly through Europe in the summer, and these are more personal and time-insensitive ways to satisfy the urge to be on the water, at least temporarily.

The safety record of all of these present-day ships has been admirable, with a few exceptions. Fires have broken out on certain cruise ships, forcing them to abort their cruises and discharge their passengers in unexpected ports. Others have experienced mechanical breakdowns and have drifted helplessly until repairs have been made or rescue ships have arrived. But there have been no major disasters resulting in the sinkings of this new fleet of pleasure ships.

The same cannot be said for contemporary ferries. Though the *Herald of Free Enterprise* (see p. 267) and the *Scandinavian Star* (see p. 286) were the finest of their class, they were not immune to human error, which sank both of them. The bizarre grounding of the *Express Samina* (see p. 279) on a plainly marked rock outcrop while the captain and crew were watching a soccer match on television was yet another example of destructive carelessness.

The situation of ferries in Africa, Haiti and Indonesia is another matter, however. Old, poorly maintained and overloaded, they still carry those who cannot travel from place to place any other way. But regularly, it seems, they capsize and sink, in direct contradiction of the technological progress and safety record of much of the rest of the world.

The criterion for inclusion in this section was a purely mathematical one. Since most passenger ships carry upwards of 3,000 passengers, a cutoff figure of 100 deaths was employed (exceptions are the 1956 sinking of the *Andrea Doria,* and the *Ehime Manu* disaster of 2001, which are included in this book despite a lower number of fatalities). Sadly enough, there were more than enough maritime disasters that fit within that parameter.

As in other categories of man-made catastrophes, wartime sinkings were, with one exception, not recorded. The one exception, the torpedoing of the German transport *Wilhelm Gustloff* by an unidentified Soviet submarine in the waning years of World War II (see p. 266), is so monumental in its all-time record toll of human life that no book that pretends to be comprehensive could ignore it, despite the fact that, to this day, little more than the bare statistics are available.

ATLANTIC OCEAN
July 25, 1956

The Andrea Doria *sank to the bottom of the Atlantic Ocean on July 25, 1956, after the SS* Stockholm *rammed it. Forty-three passengers, all aboard the* Andrea Doria*, died.*

The *Andrea Doria* was the pride of the Italian Line, a 29,083-ton floating marvel. Eleven decks high and 697 feet long, she was outfitted with no less than 31 public rooms, air conditioning and—a first for ocean liners—three outdoor pools. The first luxury ship built by Italy after World War II, she contained not only up-to-date appointments and safety features but also a hand-decorated interior that glowed with good taste and luxury. Apropos of the clientele it hoped to attract, the Italian Line assigned the *Andrea Doria* with her captain, Piero Calamai, to the fashionable Genoa—New York run, which called at Cannes, Naples and Gibraltar. Her maiden voyage, on January 14, 1953, was a triumph, and from then until July 1956, she made 50 uneventful but memorable transatlantic crossings.

On July 17, 1956, the *Andrea Doria* departed from Genoa on her 51st transatlantic trip. By the time she cruised out of Gibraltar, she was carrying 190 first-class passengers, 267 cabin-class passengers, 677 tourist-class passengers and a crew of 572. Her hold contained 410 tons of freight, nine automobiles, 522 pieces of baggage and 1,754 bags of mail.

The crossing was smooth and festive. July at sea was easy, with none of the stormy weather of midwinter crossings. The only difficulty at this time of year was fog, which wrapped itself around the *Andrea Doria* when it approached the U.S. coast on July 25, its last full day before docking in New York in the early morning of the 26th. By 9:00 P.M. that evening the ship was nearing Nantucket Light, and the visibility was virtually zero. The captain

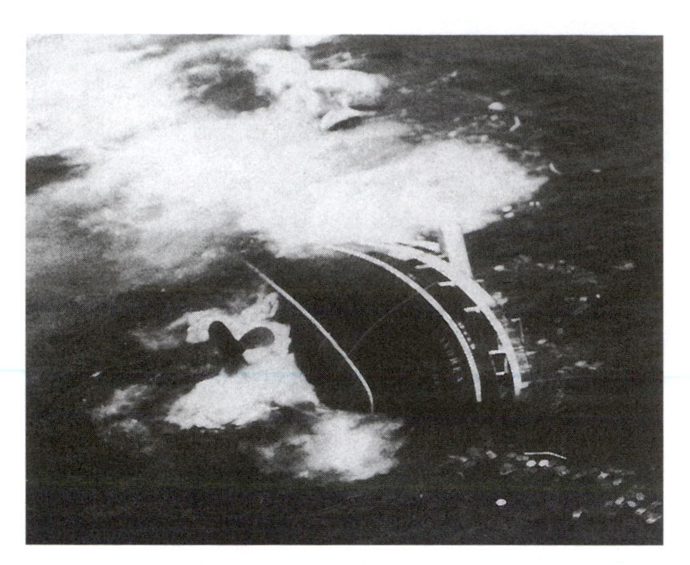

The stern of the Andrea Doria *tips to starboard just before the ocean liner sinks beneath the surface of the Atlantic after its collision with the SS* Stockholm *on July 25, 1956.* (Library of Congress)

ordered the ship's speed lowered from 26 knots to 21. The two radar repeaters on the bridge whirred with green calmness. Anything significant, animate or inanimate, would show up on their screens.

Meanwhile, the small 12,644-ton Swedish-American liner *Stockholm,* which had left New York at 11:30 that morning, was cruising eastward at 18 knots bound for Europe. It was in the same clinging, soupy fog that enveloped the *Andrea Doria* and bound for the same sea-lane.

At 10:40 P.M. the *Stockholm* appeared as a tiny green blip on the *Andrea Doria's* radar screen. It drew steadily closer, headed for the *Andrea Doria*'s starboard side. But the other ship should be outfitted with radar, too, the officers on the bridge of the *Andrea Doria* assumed, and would turn to avoid a collision.

As fate would have it, the *Stockholm* had neither radar nor a senior officer on the bridge. Third Officer Ernst Johannsen-Carstens was manning the controls alone.

Horrified, the captain and officers of the *Andrea Doria* watched their radar screen as the other ship continued to bear down on them. The captain ordered the fog horn to be sounded at 100-second intervals. Still the other ship came on, and at 11:45 P.M. it burst through the wall of fog, its lights blinding the *Andrea Doria*'s wheelhouse crew.

Captain Calamai ordered the *Andrea Doria* to turn hard to port, but the ship responded lazily, and by that time the collision was a certainty. With no pause whatsoever, the *Stockholm* plowed full force into the *Andrea Doria*'s starboard side, ripping an enormous 30-foot-deep wound in her side. She was ripped open from her upper deck down to her double-bottom tanks. Pivoting as the *Andrea Doria* dragged her along, the *Stockholm* did further damage before her captain, H. Gunnar Nordenson, reached the wheelhouse and ordered a reversal of engines and a closing of the watertight doors.

As the *Stockholm* pulled away, thousands of tons of water rushed into the gash in the *Andrea Doria*'s side, and the ship began to list 18 degrees to starboard, which rendered all of her lifeboats on that side of the ship useless. Only eight lifeboats remained to handle the 1,706 people aboard.

The captain made a quick decision and chose not to issue an abandon-ship order. Panic, he reasoned, would be a bigger enemy than the sea at that moment. Instead, he radioed an immediate SOS to all ships in the area, and within minutes the French Liner *Ile de France,* the destroyer escort *Allen,* the freighter *Cape Ann,* the navy transport *Pvt. William H. Thomas* and the *Stockholm* all rushed to the aid of the stricken ship, which continued to list and take on water. The *Ile de France,* with a suitable dramatic flourish, appeared with every light ablaze, and the passengers aboard the *Andrea Doria* gave her a healthy cheer of welcome.

The rescue operations went on all night. By 4:30 A.M. the *Andrea Doria* was completely abandoned. One thousand six hundred sixty-three persons had been rescued. Forty-three were killed in the collision, some as they slept in their starboard cabins. At 10:09 A.M. the proudest ship in the Italian Line slid beneath the surface of the Atlantic and settled to the bottom, some 225 feet below.

For several months the ship lines battled in court, blaming each other for the catastrophe. In January 1957, just before the *Andrea Doria*'s engineering officers were to appear, both shippers agreed to settle out of court. The Italian Line feared that the revelation of the presence of a faulty watertight door would bring up lawsuits by its passengers.

As it turned out, the Italian Line had nothing to fear. Two and a half decades later, filmmaker and adventurer Peter Gimbel dove into the wreck of the *Andrea Doria* and salvaged its safe, which Gimbel opened on television in 1984. In later dives he and his crew surveyed the point of impact from the *Stockholm.* It had hit at precisely the location of the supposedly faulty watertight door. There was nothing left of it. Functioning or not, the door obviously had not sent the *Andrea Doria* to the bottom. A hole in the bottom of the generator room revealed, according to Gimbel, "80 feet of her hull . . . open to the sea." Nothing could have kept the ship afloat after that.

ATLANTIC OCEAN
May 1, 1915

A torpedo from a German U-boat sank the Cunard liner Lusitania *on May 1, 1915, in the Atlantic Ocean off the coast of Ireland. One thousand one hundred ninety-eight drowned or were killed by the explosion.*

The transatlantic steamship companies vied with one another throughout the lush years of transatlantic ship travel to win the "Blue Riband" for having the fastest ship afloat. In the early years of the century, the "riband" floated back and forth between Germany, America and Great Britain. In 1903 the German Lloyd Lines possessed it, and it was then that the British Admiralty helped the Cunard Line to build two of the most luxurious and fastest liners afloat. In return Cunard agreed to include fittings that would allow the ships to be taken over by the Admiralty and used as armed cruisers during wartime.

The first of the two ships was the *Lusitania,* which began its maiden voyage on September 7, 1907, from Liverpool. She was the largest ship afloat at the time and one of the most luxurious. And by the end of her second westbound voyage, on October 5, 1907, she was also the fastest. At an average speed of 23.99 knots, from Queenstown to Ambrose Light, she had clearly won the Blue Riband for England.

In November 1907 the *Lusitania*'s sister ship, the *Mauretania,* was launched, and proved to be the *Lusitania*'s only serious competition. The two passed the ribbon back and forth until 1909, when the *Mauretania* won it and kept it for the next 22 years.

In May 1913, as war drew closer in Europe, the *Lusitania* was secretly refitted. The number one boiler room was converted to a powder magazine, and a second magazine was carved out from part of the mail room. The shelter deck was adapted to accommodate four six-inch guns on either side. When war broke out, in September 1914, the *Lusitania* entered the Admiralty fleet as an armed auxiliary cruiser but continued to make the Liverpool—Queenstown–New York run on a monthly basis.

A contemporary painting depicts survivors of the* Lusitania *disaster floating in the North Atlantic. One thousand one hundred ninety-eight passengers aboard the luxury liner were either drowned or killed by the explosion. *(Illustrated London News)*

The gigantic hole ripped in the hull of the* Lusitania *by a torpedo from a German U-boat on May 1, 1915. *(Illustrated London News)*

During the last few days of April, she was loaded at New York with 1,248 cases of three-inch shrapnel shells, 4,927 boxes of cartridges, 1,639 ingots of copper, 74 barrels of fuel oil and several tons of food supplies. She was obviously not setting out on an exclusively peaceful ocean crossing, and most of the 2,165 passengers were blissfully unaware of the lethal and dangerous cargo upon which they were sitting, sipping their bouillon.

The ship left New York at noon on May 1, 1915, with Captain William Thomas Turner on the bridge. It was her 101st crossing. All went well and serenely until May 6, when Captain Turner received bulletins from the Admiralty advising him of German submarine activity off the Irish coast. The captain ordered all of the lifeboats hanging on davits to be swung out and lowered to the promenade deck, doubled the watch on the bridge, bow and stern and blacked out all of the passenger portholes. The cruiser *Juno* was supposed to escort the *Lusitania* from the vicinity of the Irish coast to home port, but for some unexplained reason, she was never ordered out.

On Friday afternoon, May 7, 1915, Kapitan-Leutnant Walter Schwieger gazed through the periscope of his submarine, *U-20,* and spotted the *Lusitania,* steaming straight ahead at a conservative 18 knots. At 2:10 P.M. the *U-20* fired one torpedo, which struck the *Lusitania* on its starboard side, squarely behind the ammunition-loaded number-one boiler room. Within seconds there was a larger explosion as the ammunition cache ignited.

The *Lusitania* immediately began to list to starboard, rendering her lifeboats on that side useless. Passengers scrambled to the usable lifeboats, but there were far too few to take the passengers off the fast-sinking ship. It only took 18 minutes for the *Lusitania* to go down off Old Head at Kinsale, Ireland. Out of 1,159 passengers and 702 crew members, only 374 passengers and 289 crew members survived. One thousand one hundred ninety-eight were either drowned or killed in the twin explosions.

Survivors sued Cunard for negligence, but on August 23, 1918, a court in New York exonerated the line, stating: "The cause of the *Lusitania* sinking was the illegal act of the Imperial German government, through its instrument, the submarine commander." By then, the *Lusitania* had become a rallying cry of indignation, particularly useful in convincing those in the United States not inclined to join the war in Europe that it was a necessity.

ATLANTIC OCEAN

September 8, 1934

Negligence on the part of the crew and the ship line was responsible for the fire that eventually sank the Morro Castle *in the Atlantic Ocean on September 8, 1934. One hundred thirty-seven died.*

In the pre-Castro 1930s the run between the casinos of Havana, Cuba, and New York was a simple and short one, and the Ward Line provided two luxurious and fast ships to make the trip even more enjoyable: the *Morro Castle* and the *Oriente*. Launched in 1930, they plied the short route often enough for their crews and captains to grow careless.

On the *Morro Castle,* for instance, passengers never remembered a lifeboat drill. Captain Robert Wilmott ordered the removal of several fire hoses from the promenade deck because a woman had slipped on the water from a leaky hydrant and threatened to sue the company. Ironically, the tragic fire that would consume the *Morro Castle* and kill 137 of its passengers would start on that very deck.

The early September 1934 voyage of the *Morro Castle* was an unusual one. On September 7 Captain Wilmott, a rotund and rollicking man who loved the socializing portion of his duties more than the naval ones, climbed into his bathtub to ready himself for dinner. He never climbed out. While bathing he suffered a fatal heart attack.

This left the running of the ship to First Officer William Warms, a company man if there ever was one, as the events of that night would prove. Around midnight of the seventh, John Kempf, a fireman from Long Island, smelled smoke in the writing room, and stewardess Harriet Brown noticed that the linen locker, used to store stationery and winter blankets, was "intolerably hot."

But it would be 2:45 A.M. on the eighth before steward Daniel Campbell would finally pull open the linen closet door, see the flames and collect some crew members to battle the blaze. By the time an alert reached the bridge of the ship, the flames had eaten through thin wooden partitions into a ventilator shaft. From here the fire spread quickly to other ventilating and elevator shafts. Fed by a 20-mile-an-hour wind against which the liner was plowing at a brisk 19.2 knots, the flames began to eat at the insulation of the electrical wiring, creating short circuits that crippled the phone and alarm system. Stewards walked through the corridors banging on pots and pans to waken the passengers.

But on the bridge confusion had set in. Captain Warms reasoned that they were only half an hour from shore (they were off the coast of Asbury Park, New Jersey), and by the time they reached the Ambrose Channel, all would be in the hands of other firefighters. But his optimism was ill timed. The fire was raging out of control now, and the passengers were gathering in the stern. The captain sent out no SOS but did attempt to steer the ship out of the wind to prevent the flames from feeding into the staterooms. No sooner had he given the order to change course when everything aboard the *Morro Castle* went dead. The electric pumps stopped. The foghorn and the whistle became inoperable. The gyropilot and the steering gear refused to respond. Thus, the ship was dark and out of control, drifting in a flaming arc.

It was now 3:31 A.M., and 200 of the 316 passengers were huddled in the stern. No one officially connected with the ship had given them information or directions. Some passengers had brought their life preservers; some had not.

The crew was apparently busy saving itself. Seamen began by lowering boat number 10. Seven crew members and three women jumped into the boat, which had space for 48. The number-one boat was lowered. It loaded 29 crew members and two passengers. Number three contained 16 crew members and one passenger; number 11 hit the water with 16 crewmen and no passengers; number five with four crewmen and no passengers. And so it went. Six boats came ashore in New Jersey between 6:00 and 9:00 A.M. Their available space totaled 408; they carried 85 persons, most of them crew. Back on the ship, six other boats burned on their davits, as did seven balsa rafts.

Meanwhile, Captain Warms argued with his radiomen. Afraid that his company would have to pay salvage fees, he refused to send out an SOS. Finally, at 3:17 he gave permission to send out a "standby" signal. At 3:24, as the batteries in the emergency transmitter were going dead, the radioman took matters into his own hands and sent out the one and only SOS, 28 minutes after chaos had gotten the upper hand.

Receiving the signal, Coast Guard boats set out immediately from several stations in New Jersey, but the boats were dories, and they would take a long time to reach the stricken ship. Three nearby vessels, from seven to 20 miles away, caught the message and attempted to radio back, but by that time the *Morro Castle's* radio was dead.

By 5 A.M. passengers began to leap into the water. Some survived; most drowned. At dawn the first Coast

The Morro Castle *burns out of control off the coast of New Jersey on September 8, 1934. Sweeping safety reforms for ships at sea resulted from this tragedy caused by careless crew members.* (Library of Congress)

Guard dory arrived, and its five oarsmen hauled as many survivors aboard as they could. The *Monarch of Bermuda,* which was 20 miles away at the time of the call, arrived at 7:30 and launched four boats that picked up 71 survivors; the *Andrea S. Luckenbach* fished 21 out of the water; the *City of Savannah,* 65. John Bogan, the owner of the fishing smack *Paramount,* hauled 67 into his small boat and later told reporters, "It was the most horrible sight I ever saw. The water was full of dead."

Captain Warms and 14 men were taken aboard the cutter *Tampa.* In his first words to the captain he insisted that he would not be responsible for tow charges.

The maddening irresponsibility continued on shore. A Ward Line attorney met the crew and warned them against making any public statements. Later, in court, the company claimed the disaster was an act of God. When this defense was rejected by the court, they then tried to characterize it as a communist plot to burn up the ship. No evidence of a plot was discovered, but plenty of evidence of malfeasance aboard the *Morro Castle* was revealed in court: No fire drills, watchmen who were too busy serving passengers to make their rounds and the improper storing of inflammable material were only some of the violations aboard the *Morro Castle.* Backed into a corner, Ward Line attorneys insisted that, despite the fact that 92 of the first 98 evacuees of the ship were crewmen, it was the vessel, and not Ward Line employees, that failed. Finally, on February 26, 1937, the courts decreed that lawsuits amounting to $13,512,261.11 would be settled for $1,200,000.

Some good did come out of the *Morro Castle* disaster. Legislation forcing the United States to accept the International Convention for Safety of Life at Sea was passed by Congress. In addition, sweeping reforms were made in the entire U.S. merchant marine. Sprinkler systems were made mandatory throughout passenger ships, radio laws were modernized and the Federal Marine Inspection Service was enlarged.

ATLANTIC OCEAN
April 14, 1912

Overconfidence in design and a collision with an iceberg caused the sinking of the "unsinkable" Titanic *on April 14, 1912. One thousand five hundred seventeen died.*

The *Titanic,* announced the British White Star Line, would not only be the most luxurious liner afloat, its individual watertight compartments would also make it virtually impossible for the new and giant liner to sink at sea. The company, of course, knew that if more than four of the watertight compartments were breached at once, the *Titanic* would sink. But the odds against that were astronomical, and so the maiden voyage of this truly titanic liner—882.5 feet long, 92.5 feet broad and 104 feet high, capable of a cruising speed of 30 knots, with crystal chandeliers and sweeping staircases, inlaid wood in first class, a special lounge for the servants of the wealthy and the most up-to-date marine machinery available—was perfection at sea. However, there were not enough lifeboats.

That was only one of many human failings that caused the terrible tragedy of the sinking of the *Titanic,* which has been recounted over and over in books and on film. Its last hours were packed with enough foolishness, bravery and cowardice to fill at least half a shelf of adventure novels. And the lessons learned from them would make transatlantic travel considerably safer for future voyagers.

The *Titanic* was sister ship to the *Olympic,* built in the same Irish shipyard and launched shortly after its sibling. The *Olympic* had been in service for almost a year when the *Titanic* made ready for its maiden voyage on April 10, 1912. Its itinerary: Southampton to Cherbourg to Queenstown to New York. Its complement after leaving Queenstown: 322 passengers in first class, 277 in second class, 709 in third class and a crew of 898. Total: 2,206. Its lifeboat capacity: 20 lifeboats with a total capacity of 1,178.

The passenger list was democratic enough: Third class contained immigrants; second class contained the middle class; and the first-class list was packed with the world famous and the wealthy, including John Jacob Astor, Isador Straus and Benjamin Guggenheim.

The beau monde was eager to try new inventions, such as the radio. Thus, on Sunday, April 14, as the *Titanic* was nearing the end of its silken smooth and memorable maiden voyage, the radio room aboard the *Titanic* was flooded with personal messages to be forwarded to Cape Race and on to America. John Phillips, the wireless operator, was inundated with them, so much so that he grew careless about the repeated warnings sent by ships in the area about the presence of icebergs.

It had been a warm winter, and an unusual number of icebergs had broken off the polar cap and were floating southward. Still, the White Star Line had charted a course for the *Titanic* that they felt would carry it safely away from the ice and its dangers. At noon Phillips received a message from the *Baltic.* "Have had moderate variable winds and clear fine weather since leaving," it read. "Greek steamer *Athenai* reports passing icebergs and large quantities of field ice today in Latitude 41.51 degrees north, Longitude 49.52 degrees west."

A contemporary newspaper illustration captures survivors in lifeboats and the* Titanic *poised before its final plunge to the bottom of the North Atlantic on the night of April 14, 1912. *(Daily Sphere)*

In the first in a string of stupidities, the message was passed not to the bridge, but to White Star president Bruce Ismay, who was aboard for this celebratory voyage. He showed it around to the ladies in the first-class lounge but then stuffed it into his pocket and forgot it until 7:15 that evening, when he finally delivered it to the chartroom.

More messages poured into the radio room from the passengers, and these missives took precedence over two receptions from two other ships, either one of which might have saved the *Titanic.* At 7:30 the freighter *Californian* radioed the *Antillian,* reporting three large icebergs. At 9:30 the *Meshaba* contacted the *Titanic* directly, warning that "much heavy pack ice and a great number of large icebergs" lay ahead. Neither of these messages was delivered to the bridge.

At 11:00 P.M. the *Titanic* received one last warning. Just before he shut down his radio for the night, the radioman aboard the *Californian* directly contacted the *Titanic* to announce that the *Californian* was totally hemmed in by ice and had stopped engines. It was close enough to the *Titanic* to see its lights. Radioman John Phillips, his patience worn to a nub by the mountain of transmissions he had made that day, snapped back, "Shut up, shut up, I am busy!"—thus sealing the fate of the *Titanic* and the 2,000 aboard.

Meanwhile, Captain Edward Smith, denied the information he should have been receiving, was working on the instinct that had made him an experienced and respected captain. Sensing the sharp drop in temperature, he posted six lookouts to watch for ice and kept the speed at a steady 22.5 knots. First Officer William Murdock also kept sharp eyes out for ice. At 11:40 P.M. lookout Frederick Fleet yelled out, "Iceberg! Right ahead!"

Murdock snapped out the order to Quartermaster Hitchens, "Turn the wheel hard-a-starboard!" Then he yanked the engine-room telegraph to full speed astern and pushed the button closing all of the watertight doors.

Silent seconds passed. Nothing occurred. And then a telltale shudder ran through the ship, sending it trembling from bow to stern. The ship had missed colliding directly with the iceberg. But an underwater knifelike edge of ice had struck the *Titanic*'s steel plates on her starboard side and sliced a gash beneath the water line long enough to flood the first six compartments, which included the number-five and number-six boiler rooms.

Passengers in third class felt the collision, knew what had happened and panicked. Those in second class were moderately alarmed. Some were amused. Some who had left their portholes open found chunks of ice on their bunks. In first class the passengers still up and about merely watched the iceberg glide by and went back to playing cards.

Captain Smith, conferring with Thomas Andrews, the designer of the *Titanic,* immediately knew the worst had happened. The ship could have stayed afloat if up to four of her watertight compartments flooded. But with *six* flooded and filling, there was no hope. She would definitely sink.

At 11:50 Captain Smith ordered radioman Phillips to send out the CQD international call for help. Second operator Bride suggested to Phillips that he also tap out the new SOS signal. Phillips did, and the *Titanic* became the first vessel in distress to use the new code.

The North German Lloyd steamer *Frankfurt* answered first. Shortly after this the Cunard Liner *Carpathia,* some 58 miles away, received the distress call and immediately changed course, stoking her boilers to the bursting point and disregarding her own safety by steaming full speed toward the ailing *Titanic.* It would be the *Carpathia* that would rescue most of the survivors.

Meanwhile, the icy Atlantic was pouring into the *Titanic* at frightening speed. In the first 10 minutes the water rose 14 feet above the keel.

At 12:10 Captain Smith ordered the lifeboats to be uncovered and women and children to be placed in them first. There had been no lifeboat drill; no instruction in donning life jackets had been given. The passengers were bewildered. But only the top officers and Bruce Ismay knew the worse truth: There were 1,028 fewer spaces in the lifeboats than there were people aboard. The outlook was catastrophic. Before the night was out, more than half the persons climbing toward what they fully expected would be rescue would be dead.

Fifth Officer Harold Lowe was in charge of guiding passengers onto lifeboats, and by all accounts he was unable to maintain the kind of even-handed calmness necessary to bring about an orderly and efficient evacuation. The first lifeboat, number 14, was launched with 55 people aboard. But it would be one of the fuller boats. Number one carried only 12 people—Sir Cosmo and Lady Duff Gordon, her secretary, two Americans, six stokers and Symons, one of the lookout men. And so it went, with an average of 40 people per lifeboat that should have carried 65.

The first-class passengers, first at the lifeboats, were treated with preference. Only four of the women in first class died, three of them by choice when they refused to leave their husbands. Of the 93 women in second class, 15 survived; out of 179 women in third class, only 81 were saved. In fact, at one point, the doors between the third-class section and the upper-class sections were locked. Eventually, the rioting third-class passengers broke through.

There was an almost eerie calmness about the way some passengers met their deaths. Mr. and Mrs. Isador Straus, two millionaires, sat side by side as the ship went down; John Jacob Astor saw his wife safely into a lifeboat and then settled into a chair in the sumptuous first-class lounge to face his fate; Benjamin Guggenheim and his valet went to their cabins, donned evening dress and sat in splendor as they awaited the inevitable.

Only after the last first-class woman was in a boat were the third-class passengers allowed onto the boat deck. Some leaped from escape ladders into the water. Others milled about without direction.

Meanwhile, down below, last, frantic efforts were being made to pump out two of the watertight compartments. They failed, and shortly after midnight the captain ordered the crew to abandon ship. The last lifeboat, a collapsible one, carried four crewmen and 45 passengers, including White Star president Ismay, who leaped into the boat at the last minute, despite the fact that there were still women and children who had not been taken into the lifeboats.

One thousand five hundred seven passengers were still aboard the ship, and some of them now leaped into the frigid water. Captain Smith had ordered distress rockets to be launched into the sky, and the *Californian,* a mere 20 miles away, saw them and did nothing. Its captain thought they were celebratory.

Chaos continued on the ship and in the sea. Most of the half-empty lifeboats did nothing to rescue those who were struggling, drowning and freezing in the water. In lifeboat number five the women refused to allow the officer to search for survivors, despite the fact that they were in the midst of them. In lifeboat number six, just the opposite occurred; women pleaded with crewmen to try to rescue survivors in the water, but the crewmen rowed stoically away from the *Titanic.* Only Fifth Officer Lowe apparently made a concerted effort to rescue the drowning swimmers. He tied his lifeboat to three others and a collapsible craft and circled around to pick up whomever he could. But he was one of the last to leave the ship, and he plucked only four survivors from the icy Atlantic. Only 13 were pulled out of the water by 18 partially loaded lifeboats.

At 2:20 A.M. the *Titanic* began her final dive. The boilers exploded and, loosed from their anchoring supports, rushed forward. With the sound, according to one survivor, of a long freight train leaving the tracks, the huge, unsinkable ship pointed her bow toward the bottom, rose almost perpendicular to the surface of the black water around her and slid beneath the surface. Her decks were full of passengers; her band, who had tried to maintain calm by playing ragtime tunes during the evacuation, struck up the Episcopal hymn "Autumn"; Captain Smith, who had decided to retire from the sea after this voyage, remained on the bridge and went down with his ship.

A whirling vortex was created, drowning some of the swimming survivors. The rest who would live pulled on the oars of their lifeboats, away from the spot where the *Titanic* once sailed.

Within an hour the *Carpathia,* with all of her lights ablaze, arrived on the scene. Between 4:45 and 8:30 she rescued 705 survivors from the lifeboats. There were no survivors left alive in the sea. At 5:40 A.M. the radio operator aboard the *Californian* opened his radio channels and learned, to his horror, what had happened less than 20 miles away.

Bruce Ismay, who had locked himself up in the doctor's cabin as soon as he had been brought aboard the *Carpathia,* was later exonerated by the British board of inquiry, although he was severely chastised by the American one. The blame for the *Titanic*'s sinking was placed on the captain and his senior officers for failure to take notice of the four ice warnings that had been received. None of them could answer the charges; all had gone down with the ship. The captain of the *Californian* was also blamed for not going to the aid of the *Titanic,* even though the radio was, as was the custom in 1912, shut down for the night when the distress calls went out.

The horrific tragedy did produce some positive safety measures for future transatlantic passengers. The required number of lifeboats was revised to accommodate the maximum—rather than the minimum—number of passengers aboard. Boat drill became mandatory. And most important, international regulations requiring radios to remain open and functioning 24 hours a day were instituted, so that cataclysms like that of the *Titanic* could never occur again.

The story of the *Titanic* did not end with the official inquiry. Full of irony, cowardice, bravery and noble self-sacrifice, it became not only the model for seagoing adventure stories for decades to come but also the object of speculation for adventurers. On September 1, 1985, a team of American and French researchers, jointly sponsored by the American Woods Hole Oceanographic Institute and the French Institute for Research and Exploitation of the Sea, finally reached the wreck of the *Titanic,* 73 years after it plunged 13,000 feet to the bottom of the North Atlantic. Murky television and still pictures revealed a ravaged but impressive hulk. Eleven more dives in 1986, in the submersible *Alvin,* revealed that the *Titanic* had split in two on its way to the bottom. Both pieces of the ship were standing upright on the ocean bottom, with the 300-foot bow section embedded in 50 feet of mud some 1,800 feet away from the stern, and both of them in such total darkness that hardly any marine life lived on or near them.

Even now, intrigue and controversy swirl around the *Titanic.* Talk of salvage has arisen on one side of opinion. Relatives of those who perished argue that the wreck is really a gigantic tomb, the final resting place of 1,517 people, and should remain undisturbed.

The oceanographers have refused to reveal the exact location of the wreck, lest fortune hunters try to rob the ship of the staggering wealth that supposedly sank with it. And so the *Titanic* continues to generate its own unique and legendary aura, a mysterious, glamorous source of stories and speculation.

BALTIC SEA

January 30, 1945

The most tragic and underreported maritime disaster in history occurred on January 30, 1945, when an unidentified Soviet submarine torpedoed the Wilhelm Gustloff, *loaded with refugees, in the Baltic Sea. Five thousand three hundred forty-eight people died—some records say 7,200.*

Inexplicably, the worst, most tragic maritime disaster of all time, one that may have killed nearly five times the number of those drowned in the *Titanic* tragedy and more than the sinking of the entire English and Spanish Armadas in the 16th century, has gone virtually unrecorded. Missing from all but a few histories of World War II, it is given a glancing reference every now and then in stories of other, lesser sinkings. It took place during World War II, but even exhaustive studies of naval warfare of that period fail to mention it. It is almost as if the world has drawn a curtain of shame around this, possibly the most tragic of all disasters that ever occurred at sea.

The *Wilhelm Gustloff* was a passenger liner that belonged to Germany's Labor Front. Named after Wilhelm Gustloff, the Swiss official of the Nazi Party who was murdered on February 2, 1936, it was launched on July 25, 1937. An imposing ship weighing 25,484 tons, it was 695 feet long and 78 feet wide, had 10 decks and accommodated 1,465 passengers and a crew of 417.

Before it could be put into service as a passenger liner, the *Wilhelm Gustloff* was absorbed by the navy of the Third Reich and did not sail on its maiden voyage until March 23, 1938. Fitted out as a hospital ship and troop carrier, it was berthed in Gotenhafen, in the north of Germany, on the Baltic Sea.

By 1945, as World War II was drawing to a close, the sea war between Germany and the Allies was largely over. The Soviet fleet had not been particularly effective. Its main force, bottled up in the Gulf of Finland by the German Navy, did manage, nevertheless, to delay the Nazi advance through Poland and the eastern USSR. In the Baltic Sea it launched raids on small surface vessels and aircraft, mosquito fleets of motor-torpedo boats and other light craft and some submarines. By late January 1945 the *Wilhelm Gustloff* had been repeatedly bombed, repaired and pressed back into service.

At 7 P.M. on January 30, 1945, she left the harbor of Gdynia, Poland, a few miles north of Danzig (now Gdansk). Jammed on all 10 of its decks, including the open ones, were German military personnel, technicians, female merchant sailors and an enormous number of civilian refugees trying to escape the advance of the Russian troops toward Danzig.

She must have rested extremely low in the frigid January water. Various reports said that she had up to 10,000 passengers and crew aboard. The official German estimate, reported in *Das Groose Lexicon Des Zweiten Welterkriegs,* was 6,600—which seems more plausible, considering that she was originally designed to carry a total of 1,882, including crew.

It was a bitterly cold night; one can only imagine the monumental discomfort of those crammed on the open deck. Their only comfort was the knowledge that the voyage was to be a short one, to Kiel-Flensburg, on the sheltered peninsula of Germany that juts up and almost joins Denmark.

Two hours from Gdynia, barely into the Baltic, and off Stolpmunde, the *Wilhelm Gustloff* was torpedoed by an unidentified Soviet submarine. At precisely 9:08 she received the full force of the submarine's torpedoes. There was no hope for most of the passengers. One thousand two hundred fifty-two of them did manage to find spaces in lifeboats and rafts (some accounts, broadcast by Finnish radio, said that only 900 did), and these survived.

But the crammed decks and compartments of the hapless ship were jammed with far more people than there were spaces in the lifeboats. The ship sank swiftly. The temperature at the time was –18 degrees centigrade, or slightly less than 0 degrees Fahrenheit.

According to German historians 5,348 (or 7,700 according to other historical sources) passengers were left stranded on its ice-encrusted deck without a means of escape. They perished instantly and passed into a black abyss of almost totally unrecorded history.

BAY OF GIBRALTAR
March 17, 1891

A storm-caused collision with the British battleship Amson *caused the sinking of the steamer* Utopia *on March 17, 1891, in the Bay of Gibraltar. Five hundred seventy-six passengers and crew died.*

The British steamer *Utopia* of the Anchor Line was a sturdy enough steamer. It had made numerous transatlantic crossings before picking up a full complement of some 800 Italian immigrants bound for America on March 16, 1891.

By the next day the ship had stopped for supplies at Gibraltar and was ready to clear the harbor. A raging gale had blown up the night before, and there were various advisories cautioning ships to remain in port. But the *Utopia* ignored them and chose instead to try for the open sea.

It never made it out of the harbor at Gibraltar. Caught in the violent sea and winds, it was slammed into the iron-plated bow of the British battleship *Amson.* It might as well have had a hole blown in its side. Within minutes the enormous swells raised by the storm poured thunderously into the *Utopia,* and she began to sink.

Panicked, some of the immigrants leaped overboard into the stormy sea. They drowned immediately. But even those who remained on board the ship or took to lifeboats were not assured of safety. Although six ships, including the *Rodney,* the *Amson,* the *Immortalite* and the *Freya* rushed to the rescue, 576 people—half crew and half passengers—died, either from drowning or exposure to the stormy waters.

BELGIUM
ZEEBRUGGE
March 6, 1987

One hundred eighty-eight people drowned and 97 were hospitalized when the channel ferry Herald of Free Enterprise *capsized on the night of March 6, 1987, just after leaving Zeebrugge, Belgium, for Dover. Human error and corporate carelessness were the causes of the disaster.*

"From top to bottom, the body corporate was affected with the disease of sloppiness," said Lord Justice Sir Barry Sheen as he concluded the official investigation of the *Herald of*

The steamer Utopia *after its collision with the British battleship* Amson *in the Bay of Gibraltar on March 17, 1891. Five hundred seventy-six passengers and crew died in the tragedy.* (Illustrated London News)

Free Enterprise disaster. The captain of the 7,951-ton Channel ferry was suspended for one year; the first officer for two years. Townsend Car Ferries Ltd. was soundly thrashed in the international press for corrupting the meaning of the name of the ship and bringing on more apparently necessary governmental regulations.

On Friday evening, March 6, 1987, the ship, a car and passenger ferry and one of the new, fast, "roll on, roll off" superships capable of generating increased revenue by moving people and vehicles off and on rapidly, carried 543 passengers on three decks and 84 cars and 36 trucks in its huge, hangerlike bay below. Passengers drove their vehicles onto the boat through huge doors in the deck and stern, then ascended to the passenger levels.

The *Herald of Free Enterprise* normally made the Dover-to-Calais run across the English Channel, but in March of 1987 she had been shifted to the Dover-to-Zeebrugge, Belgium, route because of the refitting of other vessels and a change in the winter schedule.

It was company policy to leave with the bow loading doors open and then close them before the boat cleared the shelter of the port. This was the procedure that was followed as the ferry left Zeebrugge at 7 P.M. on a calm and clear night. But something went woefully wrong. The bow doors failed to close. The crew member normally in charge of securing them had been relieved and was dozing in his cabin. That, too, was routine. If he was not available, others took on the task. Those others were observed by some horror-stricken passengers to be frantically pounding on the giant hydraulic doors with sledge hammers, trying to force them to close before the *Herald of Free Enterprise* hit the open sea.

The captain plowed on, full speed ahead, past the breakwater. And as he did the *Herald of Free Enterprise,* like a thirsty whale, gulped in tons of water. The sea roared into the cavern of the vehicle compartment, flinging cars, trucks and crewmen against the thin bulkheading. Within moments the weight of the constantly increasing intake of water and the displaced cars and trucks tipped the boat to port. It heeled over rapidly and capsized, submerging its passenger decks.

"All the glasses were flying [around the dining room]," said survivor Susan Hames to reporters afterward. "As the ship went over there were people falling, and there was so much glass." Walls became floors; ceilings became walls; people piled up on each other, and careening furniture threatened them. Crewmen pulled distraught passengers up improvised ladders; other passengers swam into the icy waters holding on to anything they could find.

"They came flying down, tumbling on top of me, screaming," said William Cardwell, a port-side passenger. "I thought I was going to die, it was over."

It took slightly over a minute for the *Herald of Free Enterprise* to dig herself into 30 feet of water just outside the breakwater of Zeebrugge. "It was terrible," said a port worker who received the first survivors. "There were women crying because they couldn't find their husbands. There were children clinging to life preservers. Some were in terrible shock."

And some were trapped in the boat and would remain there for days. It had all happened within 20 minutes of leaving the dock, less than a mile from the port, on a starry night in clear, calm seas.

The rescue effort would go on for days: 408 people were saved, 97 were hospitalized, and 188 hapless passengers drowned, all because of a failure to put human safety before business profits and perceived efficiency.

CANADA
NOVA SCOTIA
HALIFAX
April 1, 1873

A huge storm, shortage of fuel and a foundering on reefs converged to cause the sinking of the liner Atlantic *near Halifax, Nova Scotia, on April 1, 1873. Five hundred sixty died.*

"To think that while hundreds of men were saved, every woman should have perished. It's horrible. If I'd been able to save just one I could bear the disaster, but to lose every woman on board, it's too terrible, it's too terrible."

These were the words of Captain John A. Williams, the severely chastened commander of the luxurious steam-and-sail liner *Atlantic,* as he appeared before a Canadian board of inquiry in April 1873. His ship had gone down off Halifax, Nova Scotia, partially because of the weather, mostly because of his lack of judgment and attention. That not one woman survived, and that only one child—pulled by his hair through a porthole by an alert crewman—could not be blamed upon the captain. That was the fault of one of the largest cases of mass cowardice in recorded history.

The *Atlantic* was a modern ship by 1870s standards. Four hundred thirty-five feet long, with a displaceable tonnage of 3,607, she was powered not only by a full complement of sails but by four 150-horsepower engines. She was a heavy ship, with three eight-foot iron decks, a hold filled with tons of coal, luxurious appointments and a capacity of 1,200 passengers and crew. In late March 1873, a mere two years old, she set sail from Queenstown, England, bound for New York with a load of 975 passengers and crew. The passengers were mostly aristocratic; the captain was an experienced sailor who depended more on his instincts than on charts and who, as with many sea captains of the time, had little experience with steam engines.

The first four days out from Queenstown were balmy and tranquil ones. But from then on, the North Atlantic was whipped to a deadly froth by high winds. Turbulent seas forced the captain to reduce speed to a snail's crawl of three knots, and this plus the power needed to stay on course in heavy seas burned an enormous amount of coal. When the ship was 1,100 miles from port in New York, the captain was informed that only 419 tons of coal were left. He remained unruffled. He was used to relying on sail for power, and if they ran out of coal, he assumed that they could always use that.

But the storm stubbornly persisted, and when the coal supply dwindled to an inadequate 100 tons, they were still

A Currier and Ives lithograph captures the Atlantic *foundering on a reef near Halifax, Nova Scotia, on April 1, 1873.* (New York Public Library)

460 miles from New York, and the winds were far too strong and capricious to be employed. Captain Williams decided to divert to Halifax, Nova Scotia, to resupply the ship with enough coal to get it to New York. However, Captain Williams had never been to Nova Scotia, and he failed to consult the charts that would have revealed the extreme hazards in getting there.

At 11:50 P.M. on April 1, 1873, he sited a red light. Thinking it was the Sambro Light, poised at the entrance of Halifax Harbor, he turned his ship over to Third Officer Cornelius Brady and went to bed.

Third Mate Brady knew no more about the dangers of Nova Scotia than did Captain Williams. The red light was not Sambro Light at all. It was Peggy's Point Light, meant to warn ships away from the razor-sharp reefs that surrounded it. So, as the Captain retired, the ship was put on a course directly for the reefs of Peggy's Point—a mistake that would never have been made had either officer consulted the charts that were on the ship's bridge.

At 3:00 A.M. a lookout saw the telltale signs: Waves were breaking on rocks ahead of them—dead ahead. The lookout sang out, the captain rolled out of his bunk and the *Atlantic* crashed headlong into a string of reefs that ripped her open from bow to stern. Three hundred passengers on the lower decks were drowned immediately in their bunks.

Those on the upper levels, awakened by the impact, ran out onto the decks. It was a frigid night, with blasts of cold air from the Arctic freezing even the salt spray on the ship's halyards and decks. The men, in an inexplicably universal display of pusillanimity, abandoned their wives and children and climbed the rigging, hoping to escape the waves that were already washing over the deck, transporting some who were there into the freezing water.

Once awakened, Captain Williams acted quickly. He ordered a line rigged between the foundered ship and the shore, and the crew began to transport passengers by rope to the snow-covered beach.

But then an inexplicable circumstance began to unfold. It could not have been lost on Captain Williams or his crew that the only people they were transporting were men and that by the time they were through every last woman and child had either drowned or frozen to death. One small boy, John Henly, was grabbed by his hair by Richard Reynolds, a crew member, but that would be the only child to survive.

Four hundred fifteen men, including 60 crew members and Captain Williams, made it safely to shore and survived. The captain was relieved of his command by his employer, the British White Star Line, and would pass into obscurity after his ignominious failure in the face of a disaster he himself had at least partially caused.

CANADA
NOVA SCOTIA
SABLE ISLAND
July 4, 1898

Collision in a heavy fog with the British steel bark Cromartyshire *off Sable Island, Nova Scotia, on July 4, 1898, sent the French liner* La Bourgogne *to the bottom of the North Atlantic. Five hundred sixty died.*

In 1896 Guglielmo Marconi invented the wireless. By 1904 it had replaced flags by day, lights by night and horns in fog as the principal means of communication between ships at sea. Thus, those 560 passengers and crew who needlessly died aboard the French liner *La Bourgogne* on Independence Day, 1898 missed, by six years, a safety device that might have prevented the horrendous collision, in a heavy fog, that sent them and their ship to the bottom of the icy waters off Nova Scotia.

On July 3, 1898, a day out of New York on the way to Le Havre, a dense fog settled over the *Bourgogne.* Captain Jean-Paul Deloncle did not order a reduction in speed at all. It was his job to arrive in Le Havre on time, and he was determined to adhere to the schedule. Oswald Kirkner, a passenger, later told London reporters: "Few of the passengers had crossed the Atlantic more than once, but even as amateurs most of us realized we were moving too fast. By midmorning on July 3, visibility had dropped to forty yards. Still, there were no indications that our captain intended to reduce speed."

Not only was the ship moving too fast for safety under the weather conditions, but she was also far off course—some 150 miles north of her assigned route and dangerously near Sable Island, an exposed sandbar one mile wide and 20 miles long. The cause of more than 200 wrecks, it was called "the graveyard of the Atlantic."

Meanwhile, the British steel bark *Cromartyshire,* operating under sail, was groping its way through the fog at a comfortable four knots, sounding its foghorn every minute. It would be the wife of the *Cromartyshire's* captain who would first hear the horn on the *Bourgogne.* She warned her husband, who alerted the first mate. But it was already too late. As Mrs. Henderson, the wife of the *Cromartyshire*'s captain, later told a board of inquiry: "Suddenly, the huge hull of an ocean greyhound loomed up in the mist going at least seventeen knots. Our signal system was hopelessly inadequate. Almost as soon as I caught the first glimpse of the big ship there was a fearful crash. . . ."

The sailing ship was relatively undamaged. But its bowsprit had ripped open a huge hole in the side of the liner, smashing in the starboard boiler hold and engine room. It was 5:10 in the morning, and the *Bourgogne* was sinking and listing to starboard, which made the launching of her port boats impossible.

Still, had the following deplorable events not occurred, most of the passengers might have been saved. The *Cromartyshire* was nearby and ready to take survivors aboard.

Instead, panic and cowardice on the part of the crew of the *Bourgogne* turned the collision into a catastrophe. No attempt was made to effect an orderly evacuation. Instead, crew members commandeered the functioning lifeboats, beating off hysterical passengers, killing some and drowning others.

Charles Duttwellers, a survivor who had been beaten away from a crew-occupied lifeboat, was finally plucked from the water by a small boat from the *Cromartyshire.* He later told reporters, "I saw women shoved away from boats with oars and boathooks. Members of the crew

Lifeboats from the French liner* La Bourgogne *go over the side after its collision, in a heavy fog, with the British steel bark* Cromatyshire *off Sable Island on July 4, 1898. The bark limped home; the liner sank, drowning 560. *(Illustrated London News)*

assaulted many passengers with any implement that came handy. If no instrument was to be had they punched the men and women helpless in the water with their fists."

Dozens of other survivors told similar stories. John Burgi managed to battle his way into a lifeboat and place his aged mother into the boat. "Sailors fighting for their own lives threw my poor old mother into a watery grave," he later told reporters. "They threw me out of the boat five times. Then they beat me with oars and shoved me under the boat. I managed to stay afloat for nine hours and was finally rescued by a party from the vessel that had rammed the liner."

Captain Deloncle and all of his senior officers went down with the ship. But out of 300 women, only one—Mrs. A. D. Lacasse of Plainfield, New Jersey—survived. And out of the 165 survivors of the collision, 100 were crew members. Five hundred and sixty others drowned that awful night.

August Pongi, one of the survivors, summed it up succinctly. "From beginning to end," he told the board of inquiry, "the whole business was a lasting disgrace to the French merchant marine."

CANADA
NOVA SCOTIA
WINGO SOUND
December 24, 1811

Heavy storms off the Baltic station in Wingo Sound caused the sinking of the British warships St. George *and* Defence *on December 24, 1811. Two thousand died.*

The Christmas Eve sinking of the British warships *St. George* and *Defence* was probably inevitable given the weather conditions. Storms had kept them in port, along with a small flotilla of other British warships, at the Baltic station from early November, their assigned sailing date, until December 17, 1811. Even then, the break in the weather was only comparative, and from there until they approached Wingo Sound, they encountered a series of steadily escalating storms.

On December 24 the worst of the storms hit. It was of hurricane intensity, with ice-laden winds of nearly 100 mph. The flotilla was devastated and scattered, and its two lead ships, the *St. George* and *Defence*, devoid of masts and sails, were sent to the bottom with all but 14 of their men. These managed to get off in a small boat to shore, but more than 2,000 others on board and below decks on the two ships went down in a blizzard that made it difficult to distinguish between sky and sea.

When, weeks later, parts of the hulls of the two ships broke loose from either mud or the rest of the ship and floated to the surface, a grisly sight greeted those who flocked to the location of the wreck. There, laid out as if they had been placed there, were 500 bodies, including that of the commanding officer, Admiral Reynolds.

CANADA
QUEBEC
ST. LAWRENCE RIVER
May 29, 1914

The collision in a fog with the Norwegian collier Storstad *sank the* Empress of Ireland *in the St. Lawrence River on May 29, 1914. One thousand twenty-seven died.*

The *Empress of Ireland* and her sister ship *Empress of Britain* were the two fastest ships on the Liverpool-Quebec run when they were commissioned in 1906. Not a particularly luxurious liner, the *Empress of Ireland* was an efficient, comfortable ship that accumulated an unspectacular but safe record of crossings between its commissioning and May 28, 1914, the beginning of its last voyage.

Under the command of Captain Edward Kendall, the *Empress of Ireland* left Quebec at 4:30 P.M. on Thursday, May 28, 1914, with 1,057 passengers and 420 crew members. By approximately 1:30 A.M. on May 29, as its passengers slept, it had reached Father Point on the St. Lawrence River and dropped off its pilot.

It was a calm night, and the *Empress* left the pilot station headed for the open sea. Approximately eight miles away the Norwegian collier *Storstad,* loaded with 11,000 tons of Nova Scotia coal, was making its slow way toward the *Empress*. First Mate Alfred Toftenes of the *Storstad* reckoned that the two ships would pass at a safe distance.

Suddenly a thick fog bank drifted from the land onto the river. Within minutes it had enveloped both ships, making them invisible to each other. Still, there was no cause for alarm. If each kept on her appointed course, nothing would happen.

Captain Kendall decided to take no chances. He ordered his ship to go full speed astern, which stopped it dead in the water. Meanwhile, Captain Thomas Anderson of the *Storstad,* who had been informed by Toftenes that it was "hazy," slowed his ship but did not stop. Instead, he exchanged blasts of the whistle with the *Empress* and pressed on. It was the first of two errors in judgment that would, in a very few minutes, climax in catastrophe.

Both captains strained to see in the darkness and chowder-thick fog. They could see nothing until it was too late. Abruptly, the *Storstad* broke through the fog. Her course was heading her straight for the *Empress,* and within seconds her bow collided with the liner squarely amidships. Coated with ice, the steel bow sliced through the steel plates of the *Empress* as if she were cake. Captain Kendall ordered his engines full speed ahead in an effort to beach his ship, but the engines failed to respond.

And now Captain Anderson, on the *Storstad,* made his second error in judgment. Instead of ordering his engines to reverse, he kept on moving ahead, plowing further into the *Empress,* crushing passengers and crewmen in their bunks and forcing the *Empress* to keel over, which drowned passengers and rendered half of her lifeboats unlaunchable.

Later, in the inquiry, Captain Anderson stated that his mysterious actions were intended to keep the hole he had created watertight by "holding [the *Storstad's*] bow against the side of the *Empress,* and thus preventing the entrance of water into the vessel."

It of course did not. Just the opposite occurred. Tons of water roared into the *Empress,* and she sank in 14 minutes. Miraculously, 217 passengers and 248 crew members managed to clamber to safety. But 1,027 others drowned or were killed by the invading bow of the *Storstad.* It was, at that time, the greatest passenger toll in maritime history during peacetime.

The board of inquiry placed the blame squarely on Captain Anderson's shoulders. But Captain Kendall was not blameless. Canadian Pacific rules stated that in the event of fog or snow on the St. Lawrence, hands were to be stationed at the watertight doors ready to close them. Captain Kendall ordered no such precautions to be taken when he sighted the fog, and the fact that all of the watertight doors were open during and after the collision accounted, at least in part, for the rapid sinking of the *Empress of Ireland* and the staggering loss of life.

CAPE OF GOOD HOPE
1586

A combination of reckless sailing and cowardice combined to sink the Portuguese sailing ship St. James *off the Cape of Good Hope in 1586. Four hundred fifty died.*

There are two important pieces missing from the story of the wreck of the Portuguese sailing vessel *St. James.* Lost in the interstices of time, the name of its captain and the details of the final rescue of 60 survivors of the ship seem available nowhere, thus preventing the tale from being well made. Nevertheless, the story is bizarre enough to be recounted.

The captain of the *St. James* was noted as a heavy wine drinker and a reckless man at the wheel of a sailing ship. His philosophy was to use full sail at all times and leave the rest to fate.

Sometime in the spring of 1586, he was sailing his ship the *St. James,* with his usual reckless abandon, through the Cape of Good Hope. Passengers and crew alike both counseled and begged the captain to use some common sense. The ship was running before a heavy wind; darkness had descended, and they were nearing Madagascar, with its seafront pockmarked with reefs.

The captain apparently turned a deaf ear and near midnight ran his ship and its passengers onto a reef. The ship held stubbornly fast and began to break up. It was obvious that she was going nowhere but to the bottom of the sea. Passengers panicked and scrambled for the lifeboats, which were in woeful disrepair.

The captain knew which lifeboat was seaworthy, and he gathered Admiral Fernando Mendoza, who happened to be aboard, and a few crewmen. Elbowing women and children out of the way, they leaped into the one and only launchable lifeboat. As they sailed away from the shrieking passengers, this abominable group yelled back assurances that they would send help from shore, which was fifty miles away. Neither they nor anyone else would return to the stricken ship with its approximately 500 passengers.

A small group of determined survivors finally managed to repair a damaged lifeboat and launch it. But it was fearfully overcrowded and rode low in the water. Frantic passengers dove over the side of the ship and swam after the foundering boat. Those on board used knives, hatchets and sabers to fend off the drowning and desperate others who tried to save themselves.

The lifeboat was still overloaded, however, and those aboard appointed a Portuguese nobleman as executioner. With highborn calm, he silently pointed at those who seemed too weak to pull their share of the oars. The condemned were tossed overboard by the rest.

After 20 days of this, the now comfortably loaded lifeboat came ashore on the east coast of Africa. Reaching a village near the beach, they came upon the cowardly captain and his party of survivors. The natives of the village were hostile, captured all of the outcasts from the *St. James,* tortured some, killed others and turned the rest free.

The captain and the admiral were among those freed, and they reached Mozambique safely. The captain, absolved of any negligence—possibly because of the friendly testimony of Admiral Mendoza—was given another ship to command, which he managed, several years later, to wreck in much the same way. But this time the captain died in the disaster.

Sixty people survived the wreck of the *St. James,* but the details of which people they were, whether they were taken off the wreck of the ship or whether they were solely the lifeboat sailors who were released by the hostile natives, is unclear. What is known is that 450 perished, and certainly most of them on the *St. James,* which had been first wrecked and then abandoned by its captain.

CARIBBEAN SEA
NEAR VERACRUZ, MEXICO
August 10, 1591

A major hurricane sank the Spanish Grand Fleet in the Caribbean Sea near Veracruz on August 10, 1591. Five hundred died.

Cuba, discovered by Columbus in 1492, was conquered by Spain in 1511, and under the leadership of Diego de Velázquez, it became both a major staging area for Spanish exploration of the Americas and an assembly point for the treasure ships carrying looted plunder back to Spain. Loaded with tons of gold and silver, these galleons were the target of choice of pirates and brigands, who laid in wait just over the local curve of the horizon as the Spanish ships left Havana Harbor.

Thus, by the late 1500s Spain took to combining its treasure ship crossings into one grand, annual transatlantic run. A flotilla of ships, called the Grand Fleet, would manage the yearly trip relatively unscathed by pirates.

There was one hazard that no kind of convoy could avoid, however, and that was the weather. In 1591 four major hurricanes roared through the Caribbean, and one of them coincided precisely with the annual sailing from Havana of the Grand Fleet. Caught in the middle of the storm, the fleet was buffeted and battered cruelly by enormous swells and mast-splitting winds. A dozen vessels with 500 Spanish sailors and soldiers and an untold tonnage of treasure were sent to the bottom of the sea in one afternoon, and it was a sorry-looking remainder of the Grand Fleet that limped back to Havana Harbor for repairs.

The late 1500s had not been kind to the Spanish fleet. A mere three years before, the Spanish Armada had sunk in the English Channel (see p. 275) with an enormous loss of life.

CHINA
SHANTOU HARBOR
March 18, 1921

Fighting among passengers aboard the steamer Hong Koh *in Shantou Harbor sank the ship on March 18, 1921. More than 1,000 died from drowning or fighting.*

Violent animosity between the residents of two Chinese cities caused one of the grisliest maritime disasters of all time on March 18, 1921.

The steamer *Hong Koh,* under British command and captained by Captain Harry Holmes, approached the port of Shantou at low tide that day. Its public rooms and decks were packed with Chinese residents of Xiamen (Amoy) and Shantou. Fistfights and arguments had already broken out between the residents of the two cities, and a general feeling of ill will pervaded the voyage. Captain Holmes welcomed the harbor pilot aboard with a sense of relief.

But his peace was short-lived. The pilot announced in no uncertain terms that a sandbar would not allow the *Hong Koh,* which drew 22 feet, to reach the harbor of Shantou. And he would not, under any circumstances, attempt to take the ship across the bar.

The captain announced to the disgruntled passengers that the ship would have to proceed to Xiamen unload and then return to Shantou. The Shantou residents rioted, smashing furniture and portholes.

The captain ordered his crew to station themselves in the bow of the ship with guns and a hot-water hose aimed at the rioting passengers. "On the count of three, we will fire!" announced Captain Holmes, his entire concentration and that of his crew on quelling the insurgence aboard his ship. What nobody realized was that the *Hong Koh* was drifting toward a razor-edged reef. Within moments it piled up on the rocks with a terrible sound of metal scraping against stone. And within seconds the reef had opened a gash beneath the waterline of the steamer, causing her to list to starboard.

The momentarily chastened passengers, refueled by fear, now began to set upon one another in earnest, each fighting for a place in the lifeboats that were being lowered. Knives, hatchets and axes flashed and fell as hundreds of people scrambled for the lifeboats. Hundreds were murdered on the spot, and the decks were covered with blood.

The demoralized Captain Holmes ordered his crew to fire over the heads of the rioting crowds. It made no difference whatsoever. People were hacked to pieces as they tried to climb into boats; those in the boats were swamped by overwhelming crowds trying to take their places. Lifeboats tangled in their halyards and spilled their human contents into the water. Others were smashed to kindling against the side of the ship.

Captain Holmes was the only British citizen aboard the *Hong Koh.* He ordered his officers to remain with him as the ship began to slip beneath the waves, but none did. They managed to crowd, unscathed, into the last lifeboat. Captain Holmes accompanied his ship to the bottom of Shantou Harbor. Around the sinking ship floated the bodies of over 1,000 people.

CHINA
WUSONG
December 3, 1948

The overloaded and ancient steamer Kiangya, *bearing refugees, collided with an unexploded Japanese mine near Wusong on December 3, 1948. The mine detonated, sinking the ship and killing 2,750 passengers.*

The worst officially recorded marine disaster to that date took place as the Communist armies of Mao Zedong (Mao Tse tung) were sweeping the mainland clean of Jiang Kaishek's (Chiang Kai-shek's) Nationalist forces. It was the end of the Chinese Civil War, when the major cities were falling like November leaves—Nanjing, Hankou, Chongqing and now Shanghai.

On the morning of December 3, 1948, refugees, attempting to escape to Ningbo, were rioting on the Shanghai waterfront, bargaining, pleading, demanding, even killing for the opportunity to leave the city, whose outskirts were already occupied by the invading Communist army.

The aged and rusting coastal steamer SS *Kiangya* had arrived early that day from Nanjing with a passenger load of 2,250 people. The *Kiangya* was designed to carry 1,186, but all rules and regulations were being broken that day to accommodate the screaming, terrified refugees who poured out of various Chinese cities in search of safety.

Thus, the crew of the *Kiangya* paid little heed to the passengers on board the ship throwing their already purchased tickets to the milling, shouting refugees on the pier at Shanghai. Some tickets reached the docks. Others fell into the scum-coated waters of the port. Fights erupted over those that landed on the piers; desperate men dove after those that hit the water.

By 6:30 P.M., when the *Kiangya* lumbered away from its pier, there were more than 3,450 passengers on its decks and in its state- and public rooms. There was hardly an inch of space on any of its decks.

As darkness deepened most of the 3,450 aboard attempted to make themselves comfortable for the night. But scarcely had this happened when a huge flash of light, flames and smoke enveloped the ship. A concussion set off a vibration that rattled the entire length of the vessel, and then a sickening settling to port began as tons of water roared through the gaping hole ripped in the *Kiangya's* hull by a forgotten, abandoned Japanese mine left over from World War II. The *Kiangya,* lying low in the water, had collided directly with it, and the mine had blown a hole in its hull.

The ship began to sink immediately and swiftly. Those who could fought their way upward to the top decks, racing the rising water level that had already drowned hundreds below decks. Again, the shoving and fighting that had taken place on the piers of Shanghai was repeated on the decks of the *Kiangya.* The aged and the weak succumbed first and were unceremoniously flung overboard to make room for the strong.

No radio transmission was possible; the radio room had been decimated by the explosion of the mine. Thus the sinking ship had to hope for river traffic to find it and rescue those who were steadily climbing to the very top decks of the sinking ship. The *Hwafoo* first sighted the ship and sent out the first SOS, which attracted the SS *Mouli.*

Its captain approached with some trepidation. Seeing the 700 hysterical passengers clustered on the *Kiangya,* he feared that they would swamp his own vessel. Sensing this, officers aboard the stricken ship quieted the crowd, and an orderly transfer took place.

The next day the river was dragged for bodies. More than 1,000 were returned to Shanghai and stacked on the piers of the now Communist-controlled city. Two thousand seven hundred fifty died in this, one of the most unexpected and tragic of all maritime catastrophes.

CHINA
ZHOUSHAN ISLANDS
August 29, 1916

Collision in a deep fog off the Zhoushan Islands between the cruiser Hsin-Yu *and the cruiser* Hai-Yung *sent the* Hsin-Yu *to the bottom on August 29, 1916. One thousand Chinese soldiers aboard died.*

The period from 1900 to 1917, when it entered World War I, was a tumultuous one for China. The Boxer Rebellion, the revolution fomented by Sun Yat-sen and the threat from Japan kept the military in constant motion.

On the fog-shrouded evening of August 29, 1916, a major convoy of troops entered the vicinity of the Zhoushan Islands in the Strait of Formosa, bound for Fuzhou. It was an orderly grouping, with the exception of the cruiser *Hai-Yung.* The captain of this particular vessel apparently lost his way in the fog and, instead of remaining on course, began to zigzag frantically through the cautiously proceeding flotilla.

After narrowly missing two other ships, the *Hai-Yung* rammed the cruiser *Hsin-Yu* directly amidships, slicing it almost in two. The 1,000 Chinese soldiers aboard did not have a chance of rescue. They were pitched into the water and drowned or were ground up by the screws of the *Hai-Yung.* Only nine sailors, 20 soldiers and a foreign engineer survived the collision and sinking of the *Hsin-Yu.*

CONGO
BRAZZAVILLE
March 1, 1993

One hundred and forty-seven refugees plunged into the Congo River and drowned while boarding a ferry in Brazzaville, Congo, on March 1, 1993.

In one of the more bizarre maritime disasters of modern times, 147 people drowned before even boarding a hugely overcrowded boat.

In the early 1990s war and turmoil ricocheted through Africa, and with the turmoil huge crowds of refugees fled or were deported from one country to another. In January of 1993 thousands of refugees flooded into Congo during a revolt by Zairian soldiers in which 1,000 people were killed. In February of the same year more than 5,000 refugees came to Congo when Cameroon expelled illegal aliens.

By the end of February the government of Congo had determined to expel all illegal aliens by Friday, March 5, 1993. This produced a stampede of refugees to exit points, one of which was Brazzaville, a 20-minute ferry ride from Kinshasa, Zaire.

Zaire sent a small, dilapidated 200-passenger ferry belonging to the state-owned National Transport Service of Zaire to accommodate the refugees. The captain pulled up to the dock in Brazzaville, lowered his gangplank and 3,000 desperate people rushed forward, some trampling others in a mad dash across the dock and up the gangplank.

Panicking, the captain began to draw away from the dock while the refugees were still loading onto his ship. The gangplank, already bursting with an overload of people, collapsed, flinging those on it into the fast rushing waters of the Congo River. Then the captain, realizing his error, attempted to return to the dock. Thrashing survivors in the water were crushed between the boat and the dock; others were ground up by the ship's propellers. One hundred and forty-seven would-be passengers died in helpless confusion, which precipitated yet more cross accusations and animosity between Zaire and Congo.

EGYPT
SAFAGA
December 14, 1991

Four hundred sixty-two passengers, most of them Muslim pilgrims returning from Mecca, were drowned when the Salem Express, *a Red Sea ferry, rammed coral reefs six miles from its destination in Safaga, Egypt, shortly before midnight on December 14, 1991. One hundred eighty survived.*

The 1,105-ton Red Sea ferry *Salem Express* set sail from Jidda, Saudi Arabia, on December 13, 1991, loaded with nearly 700 passengers and crew. Most of the passengers were Egyptians headed for the Red Sea port of Safaga, 293 miles southeast of Cairo. The vast majority of these were Muslims returning from *unrah,* a pilgrimage to Mecca held outside the regular pilgrimage season. The other Egyptian passengers were workers in Saudi Arabia coming home either for a holiday because their contracts had ended or because they were being deported by the Saudis.

Most of the 36-hour trip was uneventful. But late on Saturday evening, December 14, the ship ran into heavy weather. Forty mph winds whipped up 10-foot waves. The ship, for all its tonnage, bobbed sickeningly in the pounding seas.

Whether the captain was blown off course by the storm or went off course in an effort to make a run for the safety of Safaga will never be known. Shortly before midnight the *Salem Express* struck a coral reef a mere six miles from Safaga and immediately began to break up and sink.

It all happened in a matter of minutes, too short a time for any of the lifeboats to be lowered. Some passengers who survived swore that crew members shoved them away from escape routes and saved themselves. "They wore life preservers and left us and even pushed us aside to escape," said Abdel-Aiti Hassan, a survivor. The facts dispute this. The captain went down with his ship; his body was discovered on the bridge two days later. Only 10 of the 70 crew members survived.

It would be dawn before full-scale rescue efforts could be launched; the raging storm and darkness prevented ships from entering the area. The American and Australian navies sent in rescuers; helicopters brought those who escaped drowning to hastily erected tent shelters in Safaga.

Four hundred sixty-two people drowned; 118 of the survivors were hospitalized. It was one more tragedy that consumed the lives of Muslim pilgrims either in or returning from Mecca.

ENGLISH CHANNEL
August–October, 1588

A succession of hurricane-force winds, faulty decisions and the wounds of war sank the Spanish Armada between August and October 1588 in the English Channel. At least 4,000 died.

"God has seen fit to direct the course of events other than we would have wished," wrote Don Alonzo Pérez de Guzmán, duke of Medina Sidonia, as the once proud Spanish Armada shambled home, half its ships sunk and anywhere from 4,000 to 10,000 of its men dead. That final discrepancy is possibly a testament to the disarray of everything about Philip II's plan to invade England, overthrow the Protestant Elizabeth I and establish himself as ruler of England.

Between August and October 1588, the Spanish Armada was pummeled by hurricane-force winds in the English Channel. By the end of October, it had been totally destroyed, with a loss of life of at least 4,000. (New York Public Library)

Preparations for the invasion began in 1586 under the direction of the Marqués de Santa Cruz, but the course of events to which Medina Sidonia later referred began to take a hand in 1567, with a surprise attack by Sir Francis Drake. This delayed the evolution of the attack plans, and the unexpected death of Santa Cruz delayed them still further.

Still, Philip, goaded by Pope Pius V, was determined. He appointed Medina Sidonia to head up the armada—a peculiar choice, considering that the duke was schooled in neither navigation nor naval warfare, and he confessed to violent seasickness every time he left land. Perhaps it was not God who determined the course of events at all.

In any case, the 130 ships and 30,000 sailors and soldiers set out from Lisbon in May 1588 bound for Flanders, where they were to meet the army of Alessandro Farnese, duke of Parma and, combining forces, overwhelm England. However, a series of skirmishes with the English, combined with the ordinarily foul weather of the English Channel, scattered the armada, which never did make contact with the duke of Parma and his forces.

Disgruntled and down by four ships, Medina Sidonia put his lack of navigational skills to immediate use and determined to sail home via Scotland and the west coast of Ireland. The first fatality of that decision was the *Gran Grifón,* the *capitanas* of the armada. It foundered on the rocks of Fair Isle, taking 1,000 men to the bottom.

The ragtag remnants of the armada sailed on, its ships patched and its supplies running low. Somewhere out of the Hebrides, Medina Sidonia ordered all animals on board any of the ships flung overboard—an odd order, considering that the animals could have been slaughtered for food for his starving men.

The winds blew steadily and relentlessly, tattering the already torn sails of some of the ships of the armada. The *Santa María de la Rosa,* vice-flagship of the fleet and a huge galley powered by sails and galley slaves—mostly captured British seamen—was in particularly bad shape by September, with half her crew down with disease and most of her hull leaking from the cannon holes in her sides.

Her captain, Martín de Villafranca, determined that she would never make it to Spain and endeavored to find shelter and a harbor off Kerry. The winds rose to hurricane force and drove his hapless ship ashore near Dunmore Head, between Great Basket Island and Beginish. Two other Spanish ships, the *San Juan de Portugal* and the *San Juan,* were already at anchor and slowly sinking into the shallow water of the harbor when the *Santa María de la Rosa* skidded into sight, firing its guns for assistance. Her sails were in shreds, her mast splintered. She dropped anchor, but the wind and sea had done their worst. Marcos de Aramburu, the commander of the *San Juan,* wrote in his ship's log, "In an instant we saw she was going to the bottom while trying to hoist the foresail and immediately she went down with the whole crew, not a soul escaping—a most extraordinary and terrible occurrence."

He was not quite right. Three hundred soldiers and sailors did perish as the ship went down, but one Genoese seaman, Giovanni de Monana, rode a plank to shore, where he was captured by the English.

The *San Juan de Portugal* did send raiding parties ashore, captured some provisions and again set sail for Spain. It reached home, but before it did 200 of its crew died on board of disease and starvation.

Meanwhile, at Killybegs, Admiral Alonso Martínez de Leyva, having lost his own ships, the *Santa Ana* and the *La Rata Santa María Encoronada,* in storms off Loughros Bay of Donegal, put his and other men to work patching up the giant galley *Girona.* The harbor was littered with the wrecks of the ships of many nationalities, and the admiral did admirable work. He rebuilt its masts, patched its decks and hull and loaded on cannons, stores and 1,300 men.

On October 26, 1588, the *Girona* sailed out of Killybegs north to Scotland and back down the channel. And the next day, October 27, an enormous hurricane hit her. Her rudder split in two, she foundered side to the wind and went down off what later became known as Port-na-Spagna. Every one of the 1,300 men aboard, including Admiral de Leyva, drowned.

Sixty-three ships of the Spanish Armada sank during the three months of their ill-fated voyage. Sixty-five, in horrendous shape, made it home. At least 4,000 men died at sea. And Elizabeth remained on the throne in England. It was not one of Spain's finest hours.

FINLAND
UTO
September 28, 1994

More than 900 people drowned in the capsizing of the Estonia, *a car and passenger ferry that went down in the Baltic Sea in a storm in the early morning of September 28, 1994, off the island of Uto, Finland.*

Night ferries ply the Baltic Sea carrying passengers and cars between Finland, Estonia and Sweden. They are gigantic vessels that accommodate up to 2,000 passengers and 460 cars and trucks. One of them, the *Estonia,* made the run from Stockholm, Sweden, to Tallinn, Estonia, overnight. Noted for its elegant smorgasbord, its indoor pool and dancing to live music in its bar, the *Estonia* was a popular boat.

At 7:00 P.M. on September 27, 1994, it left Tallinn for Stockholm carrying, along with a large number of Swedish citizens returning from a day of shopping and sightseeing in the Finnish capital, 70 civilian police workers back from a union seminar, 21 teenagers from a Bible school and 56 retirees on a group excursion.

It was a convivial group who ate, drank and danced until shortly after 8:30 P.M., when an icy Baltic storm blew in and roughened the voyage enough to make the band stop playing and pack up. When the passengers retired to their cabins shortly afterward, 20-foot waves were pounding at the boat and 56-mph winds were whipping across its six decks.

As the passengers fell asleep the storm raged on. Shortly before midnight an engineer noticed something alarming. Water was rushing through the front cargo door into the automobile hold. That afternoon Swedish inspectors had inspected the door and found its supposedly watertight seals to be in unsatisfactory condition, particularly the inner door, which doubled as a ramp. But the problem did not seem to them serious enough to affect the *Estonia's* sailing that night.

Now, their conclusions proved to be terribly wrong. "On the TV monitor in the machine room, we could see water rushing in on the car deck," one of the crew members later told investigators. "I think the rough seas somehow broke the entrance to the car deck open. We saw that the ramp was not closed properly. There was something wrong. The outer ramp was closed, but the inner door was not properly attached."

Water was up to the crew's knees as they turned on the pumps in the hold. Fifteen minutes later it was apparent that these pumps were being overwhelmed. The water was rushing in the door now and its level was rising alarmingly.

The bridge was contacted, and at 1:24 A.M. the captain sent out a distress call. For another half hour the hold continued to fill with water that, because of the pounding of the waves of the storm on the ship, was sloshing back and forth, making it increasingly more difficult to control the vessel.

Then, at around 2 A.M. the *Estonia* suddenly capsized. "The boat lurched really severely," related Paul Barney, an Englishman who was sleeping in his cabin when the boat went over. "I was thrown off my bed and things started to slide in the cabin. I tried to make my way up to the exit but that got harder as the ship started to list more and more."

Those on the lower decks had the longest to climb, and panic made the trip longer. Passengers, in their night clothes, tumbled over one another trying to get to the open decks, the life jackets and the yellow life rafts that were falling like leaves from the boat as it listed farther and farther.

Andrus Maidre, a young survivor, later told reporters, "Some old people had given up hope and were just sitting there crying. I also stepped over children who were waiting and holding on to the railing."

It would take only half an hour for the ship to capsize and sink, plunging 242 feet to the bottom of the Baltic, off the Finnish island of Uto.

Mr. Barney reached a top deck but found no life jackets. He made it into a lifeboat with 11 other people. Only six of them survived the seven hours they spent in tossing seas until daybreak, when they were finally rescued. "Hope was beginning to disappear because the weather got really severe," he said later. "There were seven- and eight-feet waves coming over us. Every time we got slightly warmer, we got drenched again."

He was one of only a few who survived the sinking. More than 900 passengers and crew, including the captain, went down to a watery death inside the *Estonia.*

Navy remote controlled submarines later left Turku, Finland, to videotape the wreck of the ship resting on the sea floor. The conclusion from the videotape was that the front cargo door had fully separated from the rest of the vessel. A gap of three feet between it and the hull had allowed the rush of water to enter with enough force to capsize and sink the ferry. It was the same scenario that had befallen the *Herald of Free Enterprise* off Zeebrugge, Belgium, in 1987 (see p. 267). One hundred and ninety-three people died in that disaster.

This, with its toll of more than 900, made it the worst maritime disaster in modern Baltic history and the third-worst ferry accident on record after the *Dona Paz* in the Philippines (see p. 288) and the ferry that sank in the Tsugaru Strait in Japan in 1954 (see p. 284).

FRANCE
LA ROCHELLE
January 12, 1920

A combination of engine trouble and heavy seas sank the French steamer Afrique *near La Rochelle, France, on January 12, 1920. Five hundred fifty-three died.*

The *Afrique,* a small, two-masted, one-funnel steamer owned by the French Compagnie des Chargeurs Renuis and based in Marseilles, sailed the colonial route between France and its West African colonies.

The *Afrique* left Bordeaux on January 11, 1920, with 458 passengers and a crew of 127 bound for Dakar. On the evening of Saturday, January 12, she developed engine trouble in the Bay of Biscay. Her engineers worked feverishly when a heavy sea rose around the ship. The high swells, powered by heavy winds, swept the *Afrique* shoreward, straight for the Roche-Bonnie Reefs, 50 miles from La Rochelle.

The *Afrique*'s radio operator immediately sent out SOS messages, which reached the *Ceylon,* a sister ship of the *Afrique,* the SS *Lapland* and the Belgian liner *Anversville.* But the heavy seas and winds and the shallow water around the reefs kept the three ships from getting close enough to rescue the passengers and crew.

Shortly after midnight of the 13th, two lifeboats were launched from the *Afrique* with great difficulty. They carried a mere 32 people. They were all taken aboard the *Ceylon* to safety, but they would be the only survivors. At 3:00 A.M. the *Afrique* slipped off the reef, was washed into deep water and sank, taking 553 passengers and crew to their deaths. It was the worst French maritime disaster since the loss of the *La Bourgogne* in 1898 with 560 fatalities (see p. 270).

GREAT BRITAIN
ENGLAND
SPITHEAD
August 29, 1792

Gross and fatal misjudgment by workmen caused the sinking of the battle frigate Royal George *while it was in port at Spithead, England, on August 29, 1792. More than nine hundred died.*

One of the worst marine disasters in history took place, ironically enough, in port.

The *Royal George,* England's most celebrated and valuable battle frigate, had a long list of admirals at its helm and an array of 100 guns stationed on either side of its multiple masts. On August 29, 1792, it was in port at Spithead undergoing routine repairs. A small pipe beneath the waterline on the starboard side had ruptured, and while workmen crawled through the ship, sailors lounged about and merchants sold their wares to the admiralty aboard the *Royal George.* While this was happening, the ship was tipping dangerously to port.

A brisk breeze suddenly picked up, a momentary squall that would ordinarily have no effect on an upright *Royal George.* In port, however, she was at the mercy of swells that poured unchecked into her open portside gun ports. Tons of water cascaded into the vessel within minutes, and she sank like a stone, taking the 1,300 men aboard with her. Four hundred managed to swim to safety, but more than 900 drowned, trapped in the giant hull of the *Royal George,* which sank in a lethal whirlpool in shallow water in the port of Spithead.

GREAT BRITAIN
ENGLAND
WOOLWICH, RIVER THAMES
September 3, 1878

The excursion steamer Princess Alice, *loaded with celebrants, collided with the steam collier* Bywell Castle *near Woolwich on the River Thames on September 3, 1878. Six hundred forty-five died.*

Steamer excursions were popular pastimes in the late 19th century, and the regally appointed river steamer *Princess Alice* was one of Britain's most popular excursion boats. Berthed at North Woolwich pier, the steamer regularly made day trips from London Bridge to Gravesend and Sheerness. Those aboard were entertained by music on its deck, games in the parlors or tippling in its oversized saloon, reserved, in the 1870s, for men only.

On the morning of September 3, 1878, the *Princess Alice* left London Bridge with 700 holiday-minded merrymakers bound for Gravesend. At 6:00 P.M. the ship blew several blasts on its whistle and pulled back into the Thames to begin the sail home to North Woolwich pier. Captain William Grinstead decided to hug the south side of the river to avoid a two-knot ebb tide that was working against the ship midstream.

This particular night the maneuver put the *Princess Alice* on a collision course with the steam collier *Bywell Castle.* There were no regulations in 1878 regarding the proper order of passing (port to port is the regulation today), but it was generally agreed that passenger ships had the right of way.

The *Bywell* moved to port, toward the shore, expecting the *Princess Alice* to pass it in midstream. Instead, the excursion steamer continued to hug the shore, heading directly for the *Bywell Castle.* Too late, both pilots spun their wheels, trying to avert a collision. Both ships unleashed huge blasts on their horns; lanterns waved hysterically; shouts and screams were exchanged.

And then, with a horrific shriek, the *Bywell Castle* plowed into the *Princess Alice* directly amidships, crumpling her, splintering her decks and ripping an enormous hole in her hull from top to bottom. The steamer split in two and started to sink immediately.

The tragic collision of the excursion steamer Princess Alice *and the steam collier* Bywell Castle *near Woolwich on the River Thames on September 3, 1878. Six hundred forty-five died in one of the worst steamship wrecks in history. (Illustrated London News)*

The force of the impact catapulted screaming passengers into the water, where they were soon joined by others frantically trying to escape from the sinking vessel and by the bodies of those who had drowned instantly below decks. No one was wearing a life preserver; rescue lines were flung over the side of the *Bywell Castle*, but few survivors from the steamer reached them. Within two minutes the *Princess Alice* had sunk to the bottom of the Thames, carrying 645 persons to a watery death.

Two boards of inquiry were held. One placed the blame on Captain Grinstead for "improper starboarding." The second placed the responsibility on both captains.

GREAT BRITAIN
SCOTLAND
NORTH SEA
July 5, 1988

One hundred sixty-six men died and 65 were pulled from the water after the Piper Alpha *oil rig exploded in the North Sea 120 miles off the coast of Scotland on July 5, 1988. It was the worst disaster ever to strike British oil rigs in the North Sea.*

In the late 1960s Britain began to tap the rich and abundant oil reserves beneath the North Sea. The amount of oil under the floor of the sea seemed enormous, and it would provide a healthy economic lift for the United Kingdom from the 1970s onward. By 1988 there were 123 British oil rigs dotting the stormy waters of the North Sea, prominent among them the huge *Piper Alpha* rig, 120 miles off the shore of Scotland opposite Aberdeen.

Like all oil rigs, the *Piper Alpha* performed a number of duties. It not only serviced oil wells under the sea, carrying the oil by pipeline to a terminal in the Orkneys, it was also a conduit for natural gas in another pipeline that went to Norway's Frigg field. In addition, it served as a transfer point for gas from Texaco's Tartan field and supplied gas to power the nearby Claymore platform.

It was an immense, 34,000-ton structure of steel and wood, built stronger than the oil rigs of the Gulf of Mexico because of the fierce battering it would take in North Sea weather. It was 649 feet high, but most of this height was sunk by six huge steel legs into 440 feet of water. There was a helicopter pad above the water line, and nestled among the rigging with the multiple pipes that contained oil and gas running directly through them were the crew quarters housing the 230 workers who worked in two shifts aboard the rig.

There had been two minor explosions in the North Sea during the week of July 2. On Sunday, the 2nd, the British Petroleum Sullom Voe terminal in the Shetland Islands blew up, causing heavy damage but no casualties. On Tuesday, July 4, the Brent Alpha platform exploded, again with no casualties.

Some of the men aboard the *Piper Alpha* were uneasy. For three days, beginning on Sunday, July 2, workers had complained of a heavy gas smell. On Monday, July 3, worker Craig Barclay phoned his fiancée and told her that he had refused to ignite a welding torch that day because of the strong smell of gas. On Tuesday, July 4, Thomas Stirling phoned his fiancée, Janice Stewart, in Glasgow and complained of a sickening gas smell. He went on to tell her how some workers had donned breathing masks to work that day. It was the last either woman would hear from their fiancés.

At dusk on July 5 half the workers were sleeping in their quarters; half were at work. Beneath the waterline, directly under the crew's quarters, a leak had developed in a compression chamber. Natural gas was forcing itself into the chamber, and it ignited. Two cataclysmic explosions geysered up from the chamber, tore through the quarters and split the rig in two. In seconds an inferno of fire followed, shooting flames between 300 and 400 feet in the air.

Those in the living quarters had no chance of surviving. They were either blown apart, incinerated or tossed into the sea and then buried under the collapsing platform. One hundred sixty-six men died, most of them in their quarters, some who were too slow in escaping.

"It was a case of fry and die or jump and try," Roy Carey, a 45-year-old survivor, told reporters from his hospital bed in Aberdeen. "There was no time to ask—it was over the side or nothing. I just dived—it may have been 60 feet."

It *was* 60 feet, and the survivors either slid down hoses or dove from the platform into the water, where they dodged missiles of flaming debris. No lifeboats were launched; there was no time for that. A rescue boat that rushed to the platform caught fire during the second explosion. Two of the three rescue workers in the boat were killed.

Flames could be seen 70 miles away, and helicopters and planes of the Royal Air Force flew to the tragic scene. By dawn all the survivors had been picked up. The rig still burned, and would for another week.

Investigations were launched, safety improvements were promised and the prime minister and the queen extended their condolences to the families of the victims of the worst oil rig explosion in the short history of British oil exploration in the North Sea.

But that was where the story ended. Because the North Sea has a habit of scrubbing itself, the ecological damage was pronounced minimal. Only the civil damage suits brought by the survivors and the families of the dead were heard, and the payouts made a very small dent in the assets of the oil companies prospecting in the North Sea.

GREAT BRITAIN
SCOTLAND
ROCKALL
June 28, 1904

Faulty judgment by the captain was responsible for the sinking of the liner Norge *after it ran on the rocks of Rockall, Scotland, on June 28, 1904. Five hundred fifty died.*

The Scandinavian-American Line steamship *Norge* was an old but reliable liner, acquired from Thingvalla, another Danish line that cruised the North Atlantic. Launched in 1881 as the *Pieter de Coninck,* it served that line for eight years and then was sold to the Scandinavian-American Line, which renamed it and put it into service between Stettin, Copenhagen, Christina, Christinsand and New York. She was a small ship, capable of carrying 1,100 passengers (50 in first class, 150 in second class and 900 in steerage) at a very conservative top speed of 11 knots.

On June 22, 1904, the *Norge* left Copenhagen with 700 emigrants and a crew of 80. Six days later she had only made the coast of Scotland, near Rockall Island. The weather was foggy, and that night she ran onto the rocks that ring the island. The ship was not extensively damaged at first, but the captain, attempting to free the ship by reversing her engines, ripped several huge holes in both sides of her hull.

The *Norge* immediately sent out distress signals, but before help could arrive she sank, taking 550 persons, most of them emigrants who were trapped below decks, down with her.

GREECE
PAROS
September 26, 2000

The Express Samina, *a ferry carrying passengers through the Greek islands, smashed into a well-marked rock outcropping by the island of Paros and sank on September 26, 2000. The captain and 4 crew members were indicted for manslaughter in the death of 79 people.*

An unused lifesaving device and an empty lifejacket, washed up on the island of Paros are mute evidences of the unnecessary loss of life from the bizarre grounding of the* Express Samina *on a well-marked, rocky outcrop. (AP/Wide World Photo)

Nero fiddled while Rome burned. Aboard the 345-foot, 4,407-ton ferry *Express Samina,* the captain and the crew watched a soccer match on television while their ship smashed into a rocky outcrop marked by a light beacon that could be seen for seven miles. Admittedly, there was a strong wind blowing across the Aegean Sea that night, and crew members, later charged with manslaughter, blamed the crash on the wind. But the Greek court called it criminal neglect, resulting in manslaughter.

The *Express Samina* pulled out of Athens's port, Piraeus, at 5:00 P.M. on Tuesday, September 26, 2000, headed for Paros, which would be the first of six stops as it wound its leisurely way through the Greek islands and eventually ended its voyage in the tiny Lipsis islands near the Turkish coast. There were 447 passengers and 64 crew members aboard—a mixture of a multitude of nationalities and purposes. A large percentage of the passengers, as always, were tourists. Dinner was served and finished, and the population of the ferry settled down to afterdinner pursuits: some watched television; some played cards; some retired to their staterooms. The wind picked up and reached gale status, but the surface of the sea was only slightly disturbed.

Then, suddenly, at 10:30 P.M., with a rending crash, the *Express Samina* slammed into the rocks on the Portes islet, a large rocky outcrop known to all who navigated the area. Plainly marked with a huge light, it warned off all captains and crewmen who saw it. In fact, noted coast guard chief Andreas Sirigos later, "You have to be blind not to see it." But at 10:30 on that particular night, Captain Vassilis Yannakis was not on the bridge of the *Express Samina.* He and several members of the crew were below, in front of a television set, watching the soccer match between Greece's Panathinaikos and Germany's Hamburg.

Even so, the Portes light was visible to passengers on the deck. "It is inexplicable how the ship collided with a well-known rock that carries a light visible from a distance of seven miles," the coast guard chief later marveled. In his testimony, the ferry's number two blamed bad weather, and he described how he grabbed the wheel from the helmsman but failed to steer the ship away. "I did what I could," he added.

"I saw it hit," said survivor Christine Shannon later. "It was well above the top deck. . . . It was like the movie *Titanic.*"

"We were watching television at the time in one of the lounges," said another survivor, Andreas Spanos. "The vessel started listing and there was panic."

All of the lights went out on the ferry, plunging it into eerie darkness. There was no public announcement, according to most of the survivors. "Nobody told us to do anything," said Heidi Hart, an American tourist. "They were just yelling and pushing. We were handed out life vests. People were starting to jump out of the boat without vests. I thought we were going to die the whole time. We got on a boat and they let it down. There was a hole in it."

The ship continued to list. "Lots of people jumped into the sea," continued Mr. Spanos. "I jumped into the sea. I knew the vessel was going to sink once it started listing. There were still lots of people on the ship. It wasn't wavy, but the sea was cold. I could hear people screaming in the distance."

"The ship's left side was touching the sea and it turned into a slide," one survivor remembered. "One after the other, we fell into the sea."

"The ship fell apart as it sank," survivor Zoe Kolida told reporters. "There were people hanging from the railings. Children were crying and old people were screaming. I jumped in and looked back . . . and the ship was gone."

The waters were dotted with terrified passengers, some in life vests, some not, some in boats, some swimming aimlessly, some helping those in distress. A group of four navy commanders and 17 army conscripts, returning for duty on the island of Naxos, were hailed as heroes, saving children and reassuring terrified swimmers. One of them died as he attempted to save one passenger. A huge number of pleasure boats, fishing boats, coast guard cutters and helicopters immediately set out to rescue those who survived.

"I saw a flare after I fell into the sea and a fishing boat from Paros in the distance," said Effi Hiou, who had been in bed and was thrown off the boat as it lunged over and downward. "People, young and old, were calling out. I was swimming. But I was wearing shoes and pajamas. I started to get cold, but thankfully, the water was warm. I had been swimming for one and a quarter hours when I found a plank of wood. I don't know where it came from. A suitcase hit me on the head. Big waves were coming. We kept swallowing water and we said this will be our tomb. We carried on, calling out for help to the fishing boat." Just before midnight the fishing boat pulled alongside Ms. Hiou, and took her aboard along with so many survivors that they had to be piled on top of one another to fit into the boat.

The rescue effort continued all night and into the next day, aided by a British naval ship that joined the search in the middle of the night. More than 400 survivors were plucked from the water. Three hundred fifty-seven were taken to local hospitals on Paros. But 79 drowned in this disaster that never should have happened.

"How can one not be outraged at a shipwreck that has cost so many lives just a mile and a half from the shore, in an area that any captain sailing in the Aegean is familiar with?" asked Merchant Marine Minister Christos Papoutsis during the inquiry.

During that same week, arsonists attempted to burn down the Athens offices of Minoan Flying Dolphins, the operator of the *Express Samina.* They did not succeed, but Minister Papoutsis suspended the operating licenses of 50 aging ferries in a safety crackdown that affected 58 ferries and nine cruise ships. All of them—including, in absentia, the *Express Samina*—fell below European Union safety standards, which require that ships be decommissioned after 27 years of service.

HAITI
PORT-AU-PRINCE
February 17, 1993

Nine hundred passengers drowned when the passenger ferry Neptune *overturned in a rain squall on a trip from Jeremie to Port-au-Prince, Haiti, in the early morning hours of February 17, 1993. The death toll was attributed to panic, overloading and the total absence of lifejackets, lifeboats and government regulation of safety equipment.*

The government of Haiti seems to care little about the fate of most of its population. The city of Port-au-Prince, with an enormous population, has one fire house. At any one time fewer than a handful of traffic lights work. The minuscule public school system educates only the children of the powerful and wealthy, who pay almost no taxes. And so the poor, who make up the majority of the population, must do what they can to survive, and the peasants living on the western end of Haiti's southern peninsula, because of nonexistent roads, must rely on one ferry to take them from Jeremie, that region's only city, to Port-au-Prince, where they can sell their produce and wares.

Normally, two three-tiered ferries make the 100 mile run. But on Tuesday, February 16, 1993, one of the ferries was in drydock for repairs. Only the 163-foot *Neptune* was running that day. It was an old ship, uninspected by the government and bereft of either lifejackets or lifeboats. Hours before sailing time, hundreds of peasants leading livestock, bearing boxes of produce and carrying bags of charcoal paid their fares of between $2.00 and $4.50 and swarmed aboard the vessel. As sailing time neared, more and more peasants pushed their way on deck, often shoving their way aboard without paying. The army was called from the nearby Jeremie garrison to restore order, but it failed. Even after the ferry pulled away from port, canoes carried last minute passengers to the overloaded boat, now bearing nearly 1,500 passengers and an enormous number of cattle. It was chaotic, but the poor people of Haiti were used to chaos.

At 2:30 A.M. on July 17, when three-quarters of the run was completed, the *Neptune* ran into a rain squall. It was not a particularly vicious storm, but it was enough to cause the ancient rusty hulk of the ferry to sway in the swells. Those who were on the open decks shoved their way to sheltered areas. Others, feeling the deepening of the yaws the boat was making from side to side, panicked. The captain shouted over a megaphone to the passengers not to gather on one side of the boat.

But it was too late. Overloaded, unbalanced by a surge of passengers to the rail away from the wind and buffeted by waves, the *Neptune* capsized, spilling its passengers and cargo into the water, instantly drowning the trapped inhabitants of the sleeping cabins on the lower of its three decks.

"The sea was full of people," Madeline Juilen, a survivor, later told reporters. "I kept bumping into drowned people." She and others saved themselves by clinging to floating bags of charcoal. Others survived by holding on to the carcasses of drowned and floating animals. But they were in the minority. "There were lots of children on the boat, more than 100. They all died," said Moise Edward, another survivor. Fewer than 600 passengers were rescued, most of them by U.S. Coast Guard helicopters and boats. More than 900 drowned, making this one of the worst maritime disasters in world history.

HOLLAND
TEXEL RIVER
December 22, 1810

Driven inland by a hurricane, the British frigate Minotaur *broke up on the banks of the Texel River in Holland on December 22, 1810. Five hundred seventy died.*

The British frigate *Minotaur* was carrying a full load of cargo, passengers and 74 cannon when it encountered a raging hurricane off the coast of Holland on the night of December 22, 1810. Controlling the ship was impossible; its captain cut sail and tried to ride the storm out.

But near midnight the enormous sea drove the *Minotaur* up on the sandbanks near the entrance of the Texel River. Battered repeatedly by waves, the ship began to break up, and by 2:00 A.M. on December 23 it was apparent that she was going down, and quickly.

Passengers and crew made for the lifeboats and found, to their horror, that only two were seaworthy. The others were in such disrepair that they would have sunk upon launching. One hundred ten people, the maximum capacity of the two boats, survived. But 570 others, frantically searching the ship for something—anything—that might float and to which they might cling, perished as the ship was beaten to pieces by a relentless and savage sea.

INDIA
MADRAS
April 1902

A cyclone sank the British steamer Camorta *in the Gulf of Martaban near Madras, India, in April 1902. Seven hundred thirty-nine died.*

The British India Steam Navigation Company was formed in 1862 to service passengers and mail between India and England. It proved successful, and this service was later expanded to Australia, East Africa and the Persian Gulf.

In 1880 the iron-hulled steamer *Camorta*, built in Glasgow, joined the fleet. Capable of making 11 knots, she was a reliable, hardworking ship that plied the India-England route from 1880 to 1883, made one trip to Australia in 1883, was transferred in early 1886 to the Netherlands India Steam Navigation Company and then was transferred back again to the British India Company in late 1886, where she remained.

In April 1902 (the precise date is unavailable from all existing records), the *Camorta* departed from Madras with 650 passengers and a crew of 89 bound for England. She did not get far. In the Gulf of Martaban (now called Mannar), between Ceylon (now Sri Lanka) and India, the *Camorta* ran into a giant killer cyclone.

The ship barely had a chance to send out a distress signal before being overwhelmed by the storm-whipped sea. She went down with the entire ship's complement. There were no survivors, and no trace was ever found of her.

INDIA
MANIHARI GHAT
August 6, 1988

An overloaded passenger ferry capsized near Manihari Ghat, India, on August 6, 1988. More than four hundred died.

A passenger ferry overcrowded with more than 565 pilgrims set out from Manihari Ghat, a city in the state of Bihar, 200 miles northwest of Calcutta, to reach a religious site of the Hindu god Shiva on Saturday, August 6, 1988. There had been days of monsoons, and the Ganges River was swollen and turbulent.

Still, the officers and crew of the ferryboat allowed it to be dangerously packed with hundreds more than it could safely accommodate. The ferry was in midstream when the waters seemed to wrest control of the boat away from its crew. Within seconds of reaching the halfway point of its short voyage, it tilted, held a precarious balance for a moment and then capsized, spilling its passengers into the roaring river waters.

It was over in an instant, and all aboard were left to their own will and strength to survive. One hundred fifty people swam to safety or were rescued by frantic witnesses who set out in small boats to try to fish the flailing swimmers out of the monsoon-swollen Ganges.

The following day the Indian government sent divers from New Delhi to recover the bodies of the more than 400 who drowned. Cranes freed the hulk of the ferry, which was broken into several pieces. Scores of the drowned were trapped within the wreckage of the ferry itself. Others washed up at various places on the banks of the Ganges miles from the accident site. Still others were never found.

INDONESIA
MANADO
June 29, 2000

An ancient ferry, overloaded to twice its capacity with refugees from the fighting in Indonesia, sank in a storm on June 29, 2000, drowning nearly 500 people.

From mid-1998 on, violent clashes between Muslims and Christians raged throughout Indonesia. The Moluccas, in the far eastern portion of the Indonesian archipelago, was a particular center of confrontation and retribution. Nearly 3,000 people of both faiths died in one 18-month period. And so flood upon flood of refugees washed from place to place, particularly from the Moluccas.

On Wednesday, June 28, 2000, more than 485 of these refugees, fleeing a massacre in which up to 200 of their fellow villagers had been killed, and desperate to escape the bloodshed behind them, shoved, clawed and crammed themselves aboard a rickety ferry, the *Cahaya Bahari*, meant to carry no more than 250. As in scores of small island ports, these desperate people fought their way up the narrow gangplank, passing their belongings and their children over their heads. Some younger ones shinnied up the *Cahaya Bahari*'s mooring ropes. Their unified object, 200 miles away, was Manado, a small port and the capital of the province of North Sulawesi, where they could wake up in safety the next day.

The captain of the vessel was well aware of the risk of overloading, but these were troubled and unusual times, and he let the refugees fill every available space. A short time later they pulled out of the port, headed west.

On Thursday, June 29, the ferry ran into one of the many storms of the monsoon season. It came up suddenly, thunderously and mercilessly, pounding the ship, which was never meant to withstand a storm of this magnitude. Waves over nine feet high broke over its bow and decks and eventually into its hold. Helpless and eventually rudderless, the *Cahaya Bahari* turned side to the waves, foundered and sank.

Just before its final foundering, the captain radioed that it was taking on water in a storm off the northeast tip of Sulawesi, and it was to there that fishing vessels and maritime patrol aircraft went as soon as the swiftly moving storm allowed them. They found nothing; not a trace, not a floating board, not a lifeboat, not a survivor. Three days later a fishing boat in the Celebes Sea close to the islet of Karakelong, 120 miles northeast of Manado, came upon 10 survivors and one dead body. The survivors were clinging to a piece of debris and each other.

They were all that was left of the passengers and crew of the *Cahaya Bahari*, which had either sunk to the bottom of the sea or had been swept out on strong currents to the mouth of the Pacific Ocean. Nearly 500 died on a fated journey that was meant to rescue them from death.

INDONESIA
SUMATRA
January 21, 1996

Only six miles from its destination in Sabang, in northern Indonesia, the ferry Gurita *sank in a storm on the night of January 21, 1996. Forty-seven passengers were saved; 340 drowned.*

On Friday, January 21, 1996, the 555-ton ferry *Gurita* left Jakarta for Sabang, on the island of Weh, in northern Indonesia. It was a well traveled, expectedly short trip, and a popular one, and the ship was, as usual, overloaded with nearly 400 passengers.

Most of the trip proceeded uneventfully until the ferry neared its destination. Then, one of the frequent and sudden storms that roam the China Sea appeared and began to batter the ship. It lurched, listed and then began to sink with alarming rapidity only six miles from Sabang.

Panic swept through the foundering boat. "I heard noises and screams with many people praying," later related Margaret Crotty, an employee in the Jakarta office of Save the Children. "Many were fighting for life jackets, and I was pushed into one of the cabinets. But I jumped off the boat without a life jacket."

She was one among many. The storm-churned waters were filled with flailing passengers, some with life vests, some without, some in a few of the precious rubber rafts that passed for lifeboats on the *Gurita*.

Ms. Crotty attempted to climb aboard one of the rafts as it bobbed by, but those in it, fearful of its capsizing, shoved her away. She had the presence of mind and the schooling to remove her pants, knot the legs and trap enough air in the trousers to create a balloon that would keep her afloat.

Her leg was bleeding and she was afraid of sharks following her and getting to her before she drowned. It was a grisly experience. More and more dead bodies surfaced and floated in the wake of the *Gurita*, which had now disappeared beneath the waves caused by the storm, which was dissipating.

Ms. Crotty swam for 10 hours, vainly looking for land or some sign of rescue. None came, but at the end of the 10 hours another swimming survivor appeared next to her with a bag of hard candies. "It was like a miracle," she said. "He appeared beside me and then offered me some candies, and that's how I survived for the next five or six hours."

Finally, after 16 hours at sea, she reached a lighthouse and clawed her way up the piles of rocks around it. A police boat, searching for survivors, picked her up in less than an hour.

"It was a good feeling to feel alive," she admitted, but added, "It's a mix of many things. I was shocked from the experience, the dead bodies and the fear of drowning. But then there are my family, relatives and friends to go to."

She was one of the few fortunate ones. Three hundred and forty unwary passengers drowned in this, one of the most devastating of a myriad of fatal ferry sinkings in the waters that surround and separate the islands of Indonesia.

INDONESIA
SUMATRA
February 10, 1999

Three hundred and twenty-five passengers aboard the Harta Rimba, *a timber boat sailing the China Sea from Borneo to Sumatra, drowned as the overloaded vessel sank at midnight on February 10, 1999.*

More than 13,000 islands form the sprawling archipelago of Indonesia, and a huge portion of its population relies upon sea transportation to travel between these islands. The plenitude of ships plying these waters are noted for their lack of maintenance and tendency to overload, and at times the separation between cargo and passenger vessels is conveniently blurred.

This was the situation that led to tragedy aboard the *Harta Rimba* on February 10, 1999. *Harta Rimba* translates into "Heart of the Forest," and the ship was a timber ship designed to carry a crew and timber, and that was all. On February 10 it was also loaded with 325 passengers headed from Kuala Sambas on the west coast of Borneo to Riau province in Sumatra.

"The ship had requested a sailing license to carry timber," a port official in Sintete stated later. "It was not supposed to carry passengers."

It nevertheless did and set sail early on Saturday, the 10th. By nightfall the weather worsened, and large waves began to batter the ship. By midnight a leak had developed below the water line, and the *Harta Rimba* began to sink. Chaos prevailed. There were only 200 lifejackets aboard, and according to Hermanto, the ship's captain, they were all distributed, after which the rest of those aboard were told to cling to blocks of wood and drums to stay afloat.

Later there was some confusion about the state of the weather. Some survivors reported high waves and wind. But "[It] was fine when the accident happened," Suherman, a young laborer who survived, told reporters. "All I know is that there was a leak in the ship, and it started sinking just around midnight on Saturday. I just jumped into the sea like everybody else. Luckily, I found a life vest."

Hundreds of passengers fled into the dark and roiled waters of the China Sea, with no evidence that they would be rescued. Distress calls were radioed from the *Harta Rimba,* but according to navy Lieutenant Hadi Pangestu, "There was a problem with radio communication due to the bad weather."

Nevertheless, rescue vessels did converge on the scene. But all they found were crew members—all of whom were wearing life vests—and very few passengers. The rest, 325 of them, drowned. Commercial vessels, alerted to the scene, continued to comb the waters where the *Harta Rimba* sank and found nothing but floating timber.

An inquiry was established, but two years later, as of this writing, no conclusions had been drawn about responsibility or the reasons that the captain and the crew all survived, while only a handful of passengers did. The license of the *Harta Rimba* was revoked in one of the more futile gestures of recompense in nautical annals.

ITALY
LEGHORN
March 17, 1800

A carelessly thrown match started the fire that caused the sinking of the British frigate Queen Charlotte *near Leghorn, Italy, on March 17, 1800. Seven hundred died.*

The British frigate *Queen Charlotte* was massive, a huge and impressive sailing ship bristling with cannon and tall with square sails. But this impressive leviathan fell victim to a carelessly set fire on March 17, 1800, while she was sailing off the coast of Leghorn, Italy. According to a report in the *London Times,* the fire was set "by some hay which was lying under the half deck, having been set on fire by a match in a tub, which was usually kept there for signal guns."

Within minutes the fire had turned the between-decks area into a raging conflagration. A Lieutenant Dundras took several squads of men into the area to work the pumps, but while he and his men were working, the fire spread above decks to the sails. After a few more minutes it had eaten through the wooden deck. Within moments the enormous iron cannons crashed through the weakened wooden decking, falling through and crushing Dundras and his men below.

The ship sank quickly, taking 700 men—most of her crew—to the bottom. The handful of survivors were rescued by an unidentified American ship.

JAPAN
HAKODATE
September 26, 1954

Bad judgment sent the giant passenger ferry Toyo Maru *out from Hakodate into the teeth of a typhoon in the Sea of Japan on September 26, 1954. The ferry sank, drowning 794.*

Giant passenger ferries ply the Tsugaru Strait between Hokkaido and Honshu, the main island of Japan. The huge northern city of Hakodate, with its 300,000 plus population, feeds these ferries daily with workers and families traveling across a strait known for its strong currents.

September 26, 1954, was not the kind of day for a ferry to have even been out of dock. A raging typhoon was roaring through the Sea of Japan. Nevertheless, the ferry

Toyo Maru, with nearly 1,000 people aboard, was in the strait when the typhoon struck.

It capsized immediately, broke apart and smashed against the rocks near Hakodate. Miraculously, 196 passengers survived by clinging to pieces of the ferry. But 794 others were carried to the bottom with most of the ship.

JAPAN
SOUTH COAST
September 19, 1890

A plot to assassinate a Turkish political figure sent the ramshackle ship Ertogrul *out in adverse weather off the south coast of Japan on September 19, 1890. It sank, drowning five hundred and eighty-seven.*

In late August 1890 Osman Pasha, a high-ranking Turkish official, departed on a political mission to the mikado of Japan. Why such a powerful man as the pasha should agree to board the *Ertogrul,* a notoriously ramshackle ship navigated by a ragtag crew and incompetent officers, is probably the most mysterious part of this mystery story. It is possible that he was spirited aboard. But if this is so, why then did he complete his mission in Japan, after chugging from the Mediterranean and through the Sea of Japan, on a ship whose engines constantly broke down and that leaked and pitched as if she were about to come apart in an instant?

Partially recorded history draws a veil over the whole matter, except to record the end of the story in some detail. On September 19, 1890, the *Ertogrul* left Japan heading for home, her unarmored, wooden-framed hulk powered by two 300-horsepower engines that were barely able to get it up to 10 knots.

A squall came up in the Sea of Japan, and the ship finally began to disintegrate, a situation that had been predicted by the sailors and officers of the fleet of seaworthy ships owned by Turkey before she left. It was also a circumstance that was probably assured by several in Constantinople, since, according to the stories told by some of the 66 survivors of the wreck, Osman Pasha was last heard pounding on the inside of the door of his stateroom. The door had been mysteriously locked from the outside, and no key was available. So the *Ertogrul* became his watery coffin and that of 586 others who also went down with the ship.

JAPAN
TOKYO BAY
July 12, 1918

An unexplained explosion sent the battleship Kawachi *to the bottom of Tokyo Bay on July 12, 1918. Five hundred died.*

The *Kawachi* was a formidable battleship. Built in 1912, this 21,420-ton leviathan was 500 feet long and 84 feet in the beam and sported not only a full complement of 12- and six-inch guns but five 18-inch torpedo launchers as well.

The *Kawachi* only sailed for six years. On July 12, 1918, as she was entering Tokuyama Bay, 150 miles northeast of Nagasaki, she suddenly blew up. What was left sank immediately. There was never a definitive answer to the mystery of the explosion, which killed 500 sailors and officers, but an elaborate investigation concluded that a fire in a below-decks magazine could have set off a series of explosions that eventually ignited the big one. None of the 400 survivors, however, could verify this version of a disaster that still remains unexplained.

JAVA SEA
January 27, 1981

A fire in its hold led to the sinking of the Tamponas II *in the Java Sea on January 27, 1981. Five hundred eighty died.*

The *Tamponas II* was purchased from Japan in 1980 by the Indonesian government–owned Pelni Shipping Corporation. Government-owned passenger ships, stripped of unnecessary equipment to allow as many people as possible to crowd aboard, ply between the islands, and the 6,139-ton, 10-year-old *Tamponas II* was one of the youngest ships in an otherwise aging fleet.

By all accounts the Pelni Shipping Corporation was not a company that made a priority of safety. In fact, after the tragedy of the *Tamponas II,* its master complained bitterly to the Indonesian newspaper *Sinar Harapan* that its vessels, and particularly the *Tamponas II,* were not given nearly enough time between voyages for maintenance. The company, stated the captain, had ignored his complaints that four days between trips were not enough to keep the ships in safe running condition.

On Sunday night, January 25, 1981, the *Tamponas II,* with 1,054 passengers and 82 crew members aboard, was plying its way from Jakarta to the Celebes port city of Ujang Pandang, 1,000 miles to the east, when a fire broke out in a hold that contained 166 automobiles.

For 24 hours the crew battled the fire, but by early Tuesday morning it had gotten out of control, spread to the engine room and caused an explosion. Most of the passengers huddled on the deck as the sea began to swell with an approaching storm.

An SOS went out from the *Tamponas II,* which was by now halfway between Surabaya in east Java and the southern tip of the island of Borneo. The *Sanghi,* a passenger ship nearby, was the first to arrive on the scene, just as the stern of the *Tamponas* went down. There were no lifeboats or rafts aboard the ship, and passengers had to leap into the turbulent waters.

One hundred forty-nine were fished from the water by the *Sanghi.* Three other rescue vessels, including two Indonesian minesweepers, picked up the rest of the 471 survivors, which included 28 crew members and the captain.

As night fell on the evening of the 27th and the storm intensified, 60 rubber rafts were tossed into the water in the hope that survivors would cling to them. None did, and by the 29th the full fury of the monsoon hit and made the search for either survivors or bodies impossible. Six hundred twenty people survived the sinking of the *Tamponas II*. Five hundred eighty died.

LABRADOR
EGG ISLAND
August 22, 1711

Storms off Egg Island, Labrador, were responsible for the sinking of the English Armada on August 22, 1711. Two thousand British sailors died.

The English Armada passed into history ignominiously, the victim of bad judgment, deliberately dangerous navigation and heavy storms off Egg Island, Labrador.

It all began on April 29, 1711, when the enormous British fleet of 61 warships and transports carrying nearly 10,000 sailors, troops and their families set sail from England under the command of Admiral Sir Hovendon Walker. Sealed orders from Queen Anne revealed that the armada was to attack and capture the most heavily protected stronghold in the Western Hemisphere, the fortress at Quebec. It was to be a sneak attack.

Given the size of the fleet and the presence of General John Hill, it seemed a mission destined for success and glory, which would be a total turnaround from the humbling defeats the British had heretofore suffered in previous attacks against Quebec.

The armada stopped first in Boston for provisions and then headed up the Massachusetts coast. Part way there the first misstep occurred. The fleet overtook a French sloop, the *Neptune,* commanded by a Captain Paradis. She was on her way to Quebec, and with almost unbelievable naivete Admiral Walker decided that Paradis would be their ideal guide to Quebec, just the man to guide them through the treacherous waters of St. Lawrence Bay. Counseled against it, Walker persisted and paid Paradis 500 pistoles for his piloting expertise.

So, at breakneck speed the armada, led by a loyal Frenchman, headed toward an encounter with Frenchmen. At approximately 10:00 P.M. on the night of August 22, 1711, a heavy fog descended upon the fleet, scattering its normally tight formation. By the time the fog was blown away by gale-force winds and a biting storm, the eight transports of the armada had been blown onto the razor-sharp reefs that surrounded Egg Island.

Every one of the eight transports was dashed upon the rocks, which split the wooden ships asunder. Within minutes pounding surf smashed them apart still further, and more than 2,000 women, children, sailors and soldiers were swept into the black and freezing waters. Most of the bodies were later discovered washed up on the shore.

The warships and Admiral Walker were luckier. Faster and more maneuverable than the transports, they had missed the reefs and the island. But Paradis, the Frenchman, had fled, and there were no longer enough troops to safely mount an attack on the French garrison at Quebec, which had probably already been alerted. In ignominy and defeat, Walker turned his remaining ships around and headed back to England, where he would fall into disgrace and be forced to spend the rest of his days in the colonies.

There were some survivors of the wrecks, apparently. Along with the skeletons on the shore, French inhabitants a year later discovered other skeletons huddled in hollow tree trunks and in shelters constructed of branches of shrubs. None survived long enough to leave Egg Island alive.

MYANMAR
GYAING RIVER
April 7, 1990

A double-decked ferry capsized in a storm in the Gyaing River on a trip from Moulmein to Kyondo in southern Myanmar (Burma) on the night of April 7, 1990. Twenty-five of the 240 passengers survived; 215 drowned.

On exactly the same night as the disaster aboard the *Scandinavian Star* (see p. 287), a double-decker ferry boat on the Gyaing River between Moulmein and Kyondo, Myanmar (the former Burma), capsized in a fierce storm.

Details of the sinking were and remain sketchy because international journalists were severely restricted in their reporting of domestic news following the suppression of anti-government protests in 1988. However, it was known that enormous gale winds were bending trees and whipping the Gyaing River into a froth as 240 passengers crowded onto the ferry in Moulmein to make the relatively short trip to Kyondo.

The ferry had not gone far when a combination of winds, waves and an uneven distribution of weight on the two-decked ferry caused it to tip, then capsize, flinging passengers and crew alike from the decks. The ship rolled over completely, trapping more passengers within its confines.

Only 25 survivors made it to shore. Two hundred and fifteen people drowned.

NEW ZEALAND
AUCKLAND
November 17, 1874

Faulty navigation worsened a fire aboard the Cospatrick *near Auckland, New Zealand, on November 17, 1874. Four hundred sixty-eight died.*

The 1,200-ton *Cospatrick* was a lithe and agile sailing frigate grandly outfitted with a teak hull and three masts. Used as a luxury passenger liner for its first years, it was then used to lay submarine cable in the Persian Gulf in 1856. By the 1870s the *Cospatrick* had been purchased from its original owner, the Blackwell Company, by Shaw, Savill & Company, which made a business of transporting immigrants in steerage conditions from Great Britain to Australia and New Zealand.

On September 11, 1874, she left England with 429 immigrants. The crossing was long, uneventful and on schedule. But on November 17, off the coast of Auckland, a small fire broke out in the boatswain's cabin, which was also used to store oil and paints. It was not a huge fire, and the crew was able to bring it under control.

And then, for some unexplained reason, either the captain or the helmsman brought the ship squarely into the wind. Fed by this wind, the fire exploded through the roof of the cabin and ignited the headsails. Minutes later the entire bow section of the boat was enveloped in flames.

It took little time for the immigrants below to clamber above. They massed on deck as the fire spread from halyard to halyard and rail to rail. The crew began to lower lifeboats, but this was almost as disastrous as the fire itself. The first boat, overloaded with frantic passengers, some of whom leaped into it as it was being lowered, foundered and sank. The second boat caught fire as it was being lowered. Only two boats containing 81 passengers and crew got off the *Cospatrick*. The remainder burned in their checks.

It took 36 hours for the ship to smolder to its gunwales. Before that the masts burned through and collapsed the deck, crushing and burning scores of immigrants still trapped aboard the ship. The captain, his wife and his small son all drowned.

The two lifeboats, with Second Mate Henry MacDonald in charge, drifted for three days without provisions. On the fourth day a storm drove them apart, and one apparently sank. Neither it nor its passengers were ever found.

By November 26 only five survivors remained alive—MacDonald, three crewmen and one passenger. On that afternoon they were sighted by the British *Sceptre,* which drew alongside and took them aboard.

NORWAY
SKAGERRAK
April 7, 1990

The Scandinavian Star, *a 10,000-ton ferry, caught fire in the early morning hours of April 7, 1990, while making a trip from Oslo, Norway, to Frederikshaven, Denmark. One hundred and ten passengers died and scores were injured.*

Scandinavia is laced with waterways and the countries within it are separated by open sea. Ferries—from hydrofoils to overnight vessels that resemble small cruise ships—ply the North Sea and the Skagerrak, the fjord-fed strait that flows between Norway, Sweden and Denmark.

The *Scandinavian Star,* a Danish-owned ferry, was making an overnight trip from Oslo, Norway, to Frederikshaven, Denmark, on the night of April 6, 1990. It carried 493 passengers, mostly Norwegian, who were traveling to Denmark for the climate and the shopping. In the hold was a full load of cars and trucks.

The ferry had just been purchased by VR-DANO, a Danish company, from SeaEscape, Ltd., of Miami, and this was its first run in Scandinavia. It had had an engine room fire on a trip from Cosumel, Mexico, to Tampa, Florida, in 1988, but no one had been injured.

This night would be different. In the early morning hours of April 7, shortly after the disco, the restaurant and the cafe aboard had quieted and the passengers had retired for the evening, a fire broke out on a lower deck outside of one of the cabins. The crew discovered the fire and extinguished it before it could do any damage. But shortly after this another fire broke out on an upper deck, and, unattended and undiscovered, it ignited into a blaze.

"We woke up in the middle of the night," related Leo Odeland, an Oslo resident traveling with his wife and two daughters. "I heard a noise outside. I went out of the cabin and saw thick, black smoke."

His first instinct was to sound the fire alarm. But: "Nothing happened when I pushed the fire alarm," he said. "I hadn't heard the alarm go off. At another place where there was supposed to be an alarm, nothing was there."

Awakened and panicked passengers rushed from their staterooms looking for guidance and rescue. They found neither. "It was all chaos and no organization," later said Eli Kvale Neilsen. "We had to find the life vests ourselves. It was clear that the crew was not trained for an emergency."

Nor could the crew understand the passengers or the passengers the crew, who were Portuguese and comprehended neither Norwegian, Swedish or Danish. Only the officers were Danish, and they were busy sending out distress signals, which were not received until 2:30 A.M.

With no one to help them, passengers tried to loosen the lifeboats. Some succeeded. Most did not, and meanwhile the suffocating black smoke from the midship fire permeated cabins and decks alike. One hundred and ten passengers died, all from smoke inhalation. No one perished from burns.

Within an hour helicopters and other ships surrounded the listing, 10,000-ton vessel, fishing survivors from the waters of the Skagarrat and dropping firemen by rope from the helicopters. The firefighters picked their way through piles of corpses and miraculously found three passengers alive and uninjured in the midst of the grim carnage.

Tugs towed the listing ship, still smoldering, into Lydokil, above Goteborg on the Swedish coast, and the investigation began. There had been several isolated incidents of arson aboard Scandinavian ferries in the months preceding the *Scandinavian Star* disaster. The cause was determined definitely to be arson, although the arsonist was never caught.

But the extent of the destruction and the large death toll was one that could have been avoided, experts concluded, after inspecting the ship. They found that its smoke detectors had failed, that an automatic fire-extinguishing system did not work and that emergency generators did not work for an hour after the fire. It was one more tragedy that care could have prevented.

PHILIPPINES
TABLAS STRAIT
December 20, 1987

Allowing an apprentice officer to pilot the monumentally overcrowded passenger ferry Dona Paz *through the crowded Tablas Strait in the Philippines on December 20, 1987, caused one of the worst maritime disasters in history, the collision of the ferry with the tanker* Victor. *Both ships were set afire, and at least 3,000 died.*

It was Sunday, December 20, 1987, when the 2,215-ton ferry *Dona Paz,* owned by the Sulpicio Lines, left Tacloban, on Leyte Island, horrendously overloaded with Filipinos anxious to spend Christmas in Manila with their relatives and friends.

As night fell the 3,000 passengers that crowded the cabins and three decks of the *Dona Paz* (she was designed to carry 1,424 passengers and 50 crew members) attempted to make themselves comfortable. It was not an easy task. Up to four people shared individual cots; hundreds sprawled on mats they had laid out in the ship's corridors; hundreds more sat shoulder to shoulder on the decks of the ship.

It was a dark, moonless night, but an uneventful one for 265 miles of the 375-mile trip from Tacloban to Manila. According to the Coast Guard inquiry, by 10:00 P.M. only an apprentice officer was on the bridge, piloting the boat through the busy Tablas Strait, 110 miles south of Manila. The other officers were watching television or drinking beer.

At 10:00 P.M. the ferry was just off Mindoro Island, in the busiest part of the strait. And at precisely that hour, the *Dona Paz* collided head-on with the 629-ton Philippine tanker *Victor,* bound for Masbate Island with 8,300 barrels of oil. Within seconds the heavier, more powerful ferry had ripped into the hull of the *Victor,* peeling open its compartments and smashing into the oil it was carrying.

Barely a minute later the *Victor* exploded with a horrendous roar, sending flaming oil onto the *Dona Paz* and igniting the surface of the water surrounding the two foundering ships.

"I went to a window to see what happened, and I saw the sea in flames," one of the survivors, 42-year-old Paquito Osabel, told a reporter later. "I shouted to my companions to get ready, there is fire. The fire spread rapidly and there were flames everywhere. People were screaming and jumping. The smoke was terrible. We couldn't see each other and it was dark. I could see flames, but I jumped."

Both ships sank almost instantly. Many of those passengers who did not drown were burned. Even those who jumped into the water were soon coated with flaming oil and perished. Pampilio Culalia, who leaped into the water, leaving his 14-year-old daughter, 10-year-old niece and his brother behind, sobbed, "I saw the ship in flames and [as I swam away] I wanted to kill myself. But God shook me and woke me."

Of the more than 3,000 passengers aboard the *Dona Paz,* only 24 survived. Of the 13 crewmen aboard the *Victor,* two were fished from a sea that was littered with the charred corpses of those who had tried to escape. All were suffering from serious burns.

For seven hours the *Don Eusebio,* a passenger ship, circled the area searching in vain for survivors. A search mission of five commercial vessels, two naval patrol craft and three U.S. Air Force helicopters covered a wide area looking for signs of life. They found nothing either.

The next morning decomposing bodies began washing ashore on Mindoro Island, and for the next week, as Christmas came and went, hundreds of charred, bloated bodies began to float to the surface. But thousands more would never be recovered.

"This is a national tragedy of harrowing proportions," said Philippine president Corazon Aquino. "Our sadness is all the more painful because the tragedy struck with the approach of Christmas," she added.

It was the worst sea disaster in Philippine history and one of the worst marine disasters in the history of the world. There were twice the number of casualties of the *Titanic* (see p. 263), and it was only eclipsed in the 20th century by the torpedoing of the *Wilhelm Gustloff* in 1945 (see p. 266).

ROMANIA
GALATI
September 10, 1989

The Mogosoaia, *a Romanian tourist ship, collided with the Bulgarian tug* Peter Karaminchev *on the Danube River near Galati, Romania, on September 10, 1993. Eighteen survived the collision; 161—all passengers on the tourist ship—drowned.*

Pleasure cruises for tourists are one of the few sources of hard currency for Romania. The Danube River runs along the border between Romania and Bulgaria, and these small ships join the extensive commercial traffic that winds from the Black Sea to the former Yugoslavian border.

On September 10, 1989, the *Mogosoaia,* a Romanian vessel packed with tourists making an autumn excursion along the Danube, was beginning its cruise from the port city of Galati, 125 miles northeast of Bucharest on a tributary that connects the port on the Black Sea with the main body of the Danube. The weather was overcast, with a heavy fog, and most of the passengers were inside. There was little to see from the dampness of the deck.

Abruptly, with no warning, the *Mogosoaia* collided with the Bulgarian ship *Peter Karaminchev,* which was tugging a convoy of loaded barges. The two ships loomed upon each other instantaneously, and the *Mogosoaia,* far lighter than the combined tug and its barges, began to sink immediately.

Sailors aboard the Bulgarian tug attempted to rescue the frantic passengers aboard the pleasure cruiser, but it sank out of sight within minutes, leaving only wreckage and 18 survivors on the surface of the river. One hundred and sixty-one passengers on the *Mogosoaia* drowned, trapped in the ship as it sank into the Danube.

TANZANIA
MWANZA
May 21, 1996

More than 500 people drowned in the capsizing of the overloaded ferry Bukoba *on Lake Victoria, near Mwanza, Tanzania, on May 21, 1996.*

Lake Victoria, which borders on Kenya, Uganda and Tanzania, is the world's second-largest freshwater lake (Lake Superior is the largest). Before 1977 East African Railways and Harbors Corporation maintained a joint shipping, air and railway service on the lake, but beginning in 1977 Kenya, Uganda and Tanzania provided separate ferry services on often poorly maintained boats operating with minimal safety and rescue facilities.

It was a far-reaching operation because the area bordering on Lake Victoria is particularly impoverished and most roads are nearly impassable in the rainy season. Thus, a large portion of the population relied on water transport to travel from place to place. With a limited number of ferries available, overcrowding was—and remains—a problem.

On May 21, 1996, a dangerous situation escalated to tragedy. It was the end of the schoolyear for students at a boarding school in Bukoba, Tanzania. Finished with their exams, the students headed home to Mwanza and other parts of Tanzania.

On the night of May 20 at Bukoba, on the western shore of the lake, they and 400 other passengers struggled and battled their way aboard the 500-ton ferry *Bukoba,* an aging vessel scheduled soon to be taken out of service and built to accommodate 425 passengers. A huge gathering tried to crowd aboard the boat, but police battled most of them back from the gangplank, and the ferry pulled away from the dock at 10:00 P.M. to begin the 110-mile overnight trip to Mwanza.

The first scheduled stop was only 15 miles down the coast from Bukoba. Kemondo Bay was not only a boarding place but a shipping location for bananas, and the *Bukoba* was to pick up eight and a half tons of the fruit to deliver to the busy port city of Mwanza.

However, disgruntled would-be passengers who had been prevented from boarding the boat at Bukoba piled into cars and raced ahead of the *Bukoba* to Kemondo Bay. Once again a huge battle raged between the police at this location and determined passengers. "There was no proper procedure for ticketing," one of the passengers later told reporters.

This time the passengers overwhelmed the police and ran up the gangplank and onto the ferry, which put it in a severely overloaded state. More than 600 passengers were aboard, and conditions became nearly unbearable, particularly in the third-class section below decks.

"The ship was overloaded, especially the cargo, the bananas," said Cleophace Kamala, a farmer from Dar es Salaam. "I was in the third-class cabin and there were so many people. There was no space. No space to go and sit or to sleep."

When the ferry left Kemondo Bay it was already listing to starboard, but an hour later it seemed to right itself, and the passengers settled in for the long night ahead.

Then, at 2 A.M. the ferry began to roll from side to side. It continued to do so for another six hours. At 8 A.M. land appeared as they neared Mwanza. Jumanne Rume Mwiru, the captain, turned the ferry south toward the city, and suddenly the boat listed far to starboard again. Several crew members ran along the packed decks telling people to run toward the other side of the vessel. This they did, and as they surged toward the port side either their weight or a rushing wave hit the boat, or perhaps the motion of the captain's turning destabilized the boat. It capsized. The entire port side of the ferry descended into the water, and passengers on the upper decks were flipped, screaming, into the lake.

The *Bukoba* continued to rotate and now began to sink. Businessman Erafto Kinubi, on one of the middle decks, felt the first capsizing moment and tried to find a lifejacket. "I went to where the lifejackets were but there were too many people fighting and there weren't enough jackets anyway," he said later. "Then there was another wave and the ship flipped upside down. I was thrown clear."

Other survivors told of treading water for hours trying to help people who could not swim and helplessly watching as they sank beneath the surface of the sea. Others scrambled atop the overturned hull and listened helplessly to the knocking of people imprisoned beneath them. Ramadan Rashin, a businessman who climbed on the hull, said, "I was praying to God. People were knocking inside, those who were trapped, knocking, knocking, knocking."

Below decks all was terrifying pandemonium. Water rushed into the darkening, crowded space, and people swam frantically against the rising tide. Many of them drowned in the first minutes.

In one compartment Cleophace Kamala, the farmer from Dar es Salaam, a 25-year-old man named John Kamala Luboa and a 13-year-old girl swam up to the inverted floor and clung to pipes, inhaling air from an air-pocket that steadily shrank around them. The water was up to their chests. Corpses nudged their legs in the water beneath them. They banged on the hull with wood blocks trying to attract the attention of rescuers as the air grew thinner. The girl grew tired and sank under the water.

After six hours the two men dove deep down through the corpse-clogged waters toward what had once been the upper decks. They came up against windows and smashed them with their fists. "I was diving down and pushing the bodies out of the way to see the window," Mr. Luboa later told rescuers, who fished him from the water minutes after he surfaced.

Canoes, boats and other ferries converged on the capsized boat. They searched for survivors, pulling those who lived out of the water. All morning they heard the banging from inside the hull, where hundreds were trapped.

Finally, at 2:30 P.M. they chopped a hole in the side of the hull, hopeful of freeing the still living survivors within it. But it was a mistake. A few minutes later the ship sank in 93 feet of water, turning into a watery coffin for all within it and leaving only an oil slick waving among islands of floating hyacinths.

More than 500 people drowned in this, the worst shipping disaster in Lake Victoria's history and one of the worst in Africa's maritime annals. President Benjamin Mkapa termed it a national catastrophe in which entire families were wiped out. "We have lost our citizens," he told a crowd in a memorial service in a soccer stadium. "We have lost the richness of their future."

UNITED STATES
FLORIDA
COAST
July 31, 1715

A hurricane off the coast of Florida sank the twin capitanas, *the two leaders of a Spanish flotilla, on July 31, 1715. More than 1,000 died on the two ships.*

Spanish ships sailing from the New World to the Old were usually carrying plunder from the ancient civilizations Spain conquered in Central and South America. Low in the water from the weight of gold and silver, they were targeted by pirates, who infested the Caribbean like locusts in a plague year. For this reason the large convoys were named armadas; they were armed to the gunwales and then some. As many soldiers as sailors strolled the decks of the galleons, ready to stave off any attack.

The armada that was scheduled to leave Havana Harbor on July 24, 1715, was particularly well armed, since it was heavy with appropriated treasure. No less than 4,000 chests of gold, silver, emeralds, silks, pearls and porcelain nested in the holds of the 11 ships. There were 14 million newly minted gold and silver pesos from Veracruz and emeralds and gold bars from Colombia and Peru. It was more than a king's ransom, and it was all destined for the treasury of King Philip of Spain.

But all the cannons in the kingdom manned by the 2,000 soldiers and sailors aboard the 11 ships were worthless before a storm. And in the Florida Straits this armada ran into a gigantic one.

There had been no warning when they left Havana under benign and balmy skies. But by July 29 the weather had turned murky. A swell began to rock the boats, and the humidity rose. Oppressive, heavy air signaled that a hurricane was on the way.

It struck with full force at 2:00 A.M. on July 31, 1715. Hundred-mph winds slammed into the ships, driving them shoreward onto the reefs off the Florida coast. Every captain knew that the best defense was to sail for the open sea to ride out the storm. The *Grifón,* under the captaincy of Antonio Darie, managed to do this and survived. But it was the only ship that did. Ten ships loaded with a fortune were crazily whirled counterclockwise in the vortex of the storm.

Masts collapsed first, crushing those on the decks beneath them. Then the pounding of the waves and the wind ripped the riggings from the vessels. All 10 ships sank. Three hundred Indians from Florida and Central America were pressed into service as slave divers. Forced to dive until exhaustion, fully half of them drowned in the effort to retrieve 6 million of the 14-million-peso treasure that had sunk beneath the sea in the Florida Straits.

UNITED STATES
HAWAII
HONOLULU
February 9, 2001

Running a test of the emergency surfacing maneuver for 16 civilian observers, the nuclear submarine USS Greeneville *collided with the Japanese shipping boat* Ehime Maru *off Honolulu, Hawaii, on February 9, 2001, sinking the fishing boat and drowning 9 students and crew members.*

The *Ehime Maru,* a 180-foot, 750-ton Japanese fishing boat, was sailing through seas with a medium swell nine miles south of Waikiki Beach, Hawaii, on February 6, 2001. Aboard were 35 people—the ship's crew and captain and teachers and students from the Uwajima Fisheries High School in southwestern Japan, aboard for practical experience on a working fishing boat.

All was progressing uneventfully when suddenly, without warning or a sound, the 360-foot, 6,900 ton USS *Greeneville,* a U.S. Navy nuclear attack submarine, surfaced directly under the *Ehime Maru,* collided with it, shattered it and instantly sank it.

"There was a violent collision, or I should say there was a very loud noise and a jolt that seemed to lift our stern up," Hisao Onishi, the captain of the vessel, told a news conference later. "We heard two cracking noises. I could not see any other ships in the area, and I looked around, thinking we might have hit a floating object."

The power went out on the *Ehime Maru,* preventing the captain from calling for help, but he related how he yelled for everyone to head up to the bridge and into the life rafts. "The ship went down without tilting, almost straight down," he continued. "We couldn't get the life rafts out and were washed into the sea." Akira Kagajyo, a crew member floating near the captain in the sea, told him that the engine room flooded to the ceiling and he was flushed out by the rush of seawater.

As they floundered in the sea trying to find each other and to stay afloat, the USS *Greeneville* suddenly became visible. It was on a public relations voyage for the U.S. Navy with 16 influential civilians aboard. The captain, Commander Scott D. Waddle, was demonstrating the submarine's capabilities, including its ability to surface quickly in an emergency situation. Because of the crowd of civilians, it was discovered later, Commander Waddle had not been able to run a complete check of procedures, including consulting his radar operator, before surfacing, and he rushed the maneuver.

"I could see several people on the [submarine] tower," Onishi told reporters. "They were just looking until the Coast Guard arrived."

It would be 35 minutes before a Coast Guard helicopter and plane reached the scene and a full 50 minutes before patrol boats rescued 26 survivors from the collision. Asked later why the *Greeneville* did not rescue the survivors from the sea, Admiral Thomas Fargo, the commander-in-chief of the U.S. Pacific fleet, said that choppy conditions made it too dangerous for the *Greeneville* crew to open hatches and take survivors aboard. There were waves of three to four feet with a six-foot swell, the admiral said.

But Captain Onishi disputed this, and his dispute nearly caused a severe rupture in Japanese–American relations. "The conditions were calm enough that water did not enter the life rafts," he told reporters.

Nine students and crew members from the *Ehima Maru* were drowned and descended with the ship 2,000 feet below the surface of the Pacific Ocean. Commander Waddle was relieved of his command, and U.S. diplomats and the U.S. Navy had much explaining to do, not only to Japan but to the rest of the world.

UNITED STATES
ILLINOIS
CHICAGO
July 24, 1915

Faulty design, greed and an unequal distribution of passengers combined to capsize the excursion steamer Eastland *at its dock in Chicago on July 24, 1915. Eight hundred fifty-two picnickers died.*

"I thought the damned ship would take the turns on her side like a skipping stone."

So wrote naval architect W. J. Wood after maneuvering the excursion steamer *Eastland* through its "S curve" configuration, designed to reveal any dangerous listing capabilities. Wood found plenty, as his succinct observation attests. So did John Deveraux York, another naval architect, who on August 2, 1913, wrote to the harbormaster of the Port of Chicago, "You are aware of the conditions of the SS *Eastland,* and unless structural defects are remedied to prevent listing, there may be a serious accident."

Neither the authorities nor the St. Joseph–Chicago Steamship Company, the owner of the *Eastland,* apparently paid attention to these warnings from experts. The *Eastland* was too much of a money-maker, ferrying thousands of holiday merrymakers from Chicago on excursions to picnic grounds on the banks of Lake Michigan.

No less than 7,300 people, most of them employees of the Western Electric Company, possessed excursion tickets for the annual picnic of the Hawthorne Club on July 24, 1915. It would take five steamers to transport them from the Port of Chicago to Michigan City, Indiana.

The excursion began early, at 6:00 A.M. Bands played, whistles blew and more than 3,500 picnickers dashed aboard the *Eastland,* which was to lead the flotilla that day. The *Eastland* was licensed to carry only 2,500 passengers, but company officials let it fill over its capacity.

Passengers, eager to be snapped by a professional photographer who was poised on a bridge, crowded to the port rail. The boat began to list—dangerously.

Chief engineer Joseph M. Erickson noted this and opened the ballast tank on the starboard side. But no sooner had he done this than the passengers rushed to the *starboard* railing. He opened the port ballast tanks. But the listing continued and worsened. Sailors were dispatched to the deck to try to redistribute the passengers. It was useless. They were having too much fun, and there were too many of them to heed the instructions of the crew.

Now the list to port increased alarmingly. The 60 crew members, seeing it and knowing what would happen next, jumped ship, literally, landing safely on the dock. Captain Harry Pedersen, on the bridge, shouted to his absent crew to open the inside doors and get the people off the ship.

It was too late. As the formerly festive group of picnickers screamed and began to slide toward the port rail, the *Eastland* tipped and capsized. Passengers, provisions, furniture, everything crashed toward the port side of the hull as the *Eastland* settled onto the river bottom. A mere eight feet of her starboard side jutted above the water, a metal island that was crowded with panic-stricken excursionists who clawed and clung to one another.

Hundreds had already drowned, crushed by the tons of water that roared into the *Eastland* through its open portholes. Rescuers on the shore threw anything that was loose and floatable to the thrashing and terrified people. Workmen using acetylene torches nearby rushed to the scene and immediately went to work cutting a hole in the *Eastland's* hull, where pounding had been heard.

No sooner had J. H. Rista, a torchman, sliced into the hull when Captain Pedersen arrived on the scene, shouting to the workman to stop cutting holes in his ship. The workmen surrounded the captain, who continued to yell irrationally and attempt to pull the torchmen away from their rescue work. Finally, police arrested Pedersen and Dell Fisher, his first mate, who had joined the captain in his

The capsized hulk of the Eastland *lies on its side in the Chicago River while rescuers search for survivors. Eight hundred fifty-two died in this "accident that never should have been allowed to happen."* (Library of Congress)

crazy interference tactics. "After I got rid of Pedersen," Rista later testified before the board of inquiry, "we took out forty people, all alive, out of that hole he had tried to stop me from cutting."

The 40 were fortunate. Eight hundred fifty-two hapless picnickers drowned, and their bodies were hauled off to the Second Regiment Armory for identification.

Hundreds of lawsuits were filed against the company. They would stumble through the Illinois court system for 20 years and finally be thrown out. Astonishingly, the courts eventually ruled that the fault lay with the engineer, who had "neglected to fill the ballast tanks properly."

UNITED STATES
NEW YORK
NEW YORK
June 15, 1904

A wholesale disregard of safety measures coupled with mindlessness caused the fatal burning of the steamer General Slocum *in New York Harbor on June 15, 1904. One thousand thirty-one people died, according to the New York City Police Department.*

President Theodore Roosevelt fired the chief inspector of the U.S. Steamboat Inspection Service because of it. The most sweeping reforms ever instituted in U.S. maritime history were the result of it. Operators of excursion steamers from that point forward would think twice before valuing operating costs over human life.

Those were the positive results of the horrific tragedy of the *General Slocum* on June 15, 1904. The dark side was the long litany of criminal stupidity practiced by its captain, crew and owners, all of which came to light in the exhaustive investigation that followed.

The *General Slocum* was one of a number of excursion steamers that gave pleasure to thousands at the turn of the century. Thirteen years old in 1904, she featured huge, Mississippi-style side wheels that powered her at roughly 18 mph. She was a favorite among church and civic groups to take their members on annual picnics and sails.

On June 15, 1904, a bright and beautiful late spring day, St. Mark's German Lutheran Church, located on Sixth Street in Manhattan, gathered 1,360 children, teachers and chaperones for its annual Sunday school voyage and picnic to Locust Point, just beyond Throg's Neck in the Bronx.

At 9:00 A.M. on the 15th the throng assembled on the pier at Third Street and the East River. Allowing all 1,360 aboard would dangerously overload the boat, but Captain William Van Schaick ignored this, thus committing the first of a string of stupidities.

The worst, however, had happened long before June 15. In violation of all safety regulations, a forward storeroom had been cluttered with cans of oil and barrels filled with leftover excelsior from a previous shipment of crockery. This volatile, flammable combination was ignited shortly after the ship left its pier either by an oiler passing through with a burning torch on his way to the engine room or by a crew member who lit a lamp and dropped the match. (The crew, almost to a man, was composed of former deckhands and truck drivers; experience was at a premium aboard the *General Slocum.*)

Meanwhile, the white and yellow steamer moved serenely upstream through Hell Gate and toward the Bronx. As she neared 83rd Street, Frank Perditsky, a 14-year-old picnicker, noticed smoke coming from one of the holds in the bow. According to sworn testimony, he shoved his way through the crowd and approached a "man with gold braid on his cap"—which was enough to identify Captain Van Schaick. Reporting excitedly that he had seen "smoke coming out of the boiler room," he was dismissed airily by the captain, who told him to "shut up and mind your own business."

The captain later ordered First Mate Edward Flanagan, a former ironworker, to take a seaman and investigate. They found the smoke, all right, but the door to the room from which it was coming was locked. Flanagan got a key and opened the door, and the air he let in caused the smoldering fire to burst into flames.

There was a fire hose nearby, which Flanagan grabbed. But no water emerged from it when he turned the valve. Precious minutes were lost as crewmen went for the donkey engine used to build up water pressure. Even with this working correctly, no water came from the hose, which had neither been used nor inspected since the launching of the ship 13 years before.

Flanagan now had the presence of mind to unscrew the nozzle of the hose. As he did a solid rubber washer fell out. It had been installed to prevent water from dripping on the deck—another mistake which would be repeated 30 years later aboard the *Morro Castle* (see p. 262). Flanagan now rescrewed the nozzle onto the hose. The water pressure forced water into it, and its aged and unused linen casing burst apart, soaking the crewmen who held it.

At this point the fire was absolutely out of control. The sensible thing to do would have been to head sharply for shore, where the smoke from the fire had already been noticed and an alarm turned in to the 138th Street Station of the New York City Fire Department. Firemen had already laid hose out to the end of the longest pier.

But Captain Van Schaick continued on, ordering the pilot to steer for North Brother Island. His rationale, expressed in later testimony, was "a moral responsibility not to risk setting buildings on fire." There was a sand beach on North Brother Island, and it was possible to beach the boat there in shallow water and presumably disembark the now-uneasy passengers who were crowding the decks of the ship.

The General Slocum *after the fire that killed more than 1,000.* (New-York Historical Society)

An artist's rendering of the horrendous burning of the steamer General Slocum *in New York Harbor on June 15, 1904. One thousand thirty-one people died, according to the New York City Police Department.* (New York Public Library)

But Pilot Van Wart fumbled the landing, missed the stretch of sand and steamed the *General Slocum* instead into a rocky cove. He now ran her up on the rocks. Thus, her blazing bow was in the shallows, while her stern, which offered the only safe exit, was in water 30 feet deep.

The fireboat *Zophar Mills* had been alerted and was on its way, but it was a full mile astern of the flaming excursion steamer. The tugboats *John Wade* and *Walter Tracy* were also on their way, but they were three-quarters of a mile away in another direction.

Terror and further malfeasance were now sealing the fate of the horrified passengers. The 10 lifeboats were lashed so thoroughly that the adults among the passengers could not budge them, and no crewmen offered to help. Children fought with one another for life preservers, which were supposed to be of regulation solid cork. Instead, to save money, the Knickerbocker Company, which owned the ship, had substituted granulated cork fitted with seven-inch bars of cast iron to make them come up to the weight standard. They were worse than useless; children leaping into the water with the life preservers on were pulled to the bottom by the iron weights.

But by now children were dying in other ways. Some were trampled underfoot in the panic on deck. Others leaped into the water and were chewed up by the still-turning paddle wheels. And then the flames ate through the decks. They collapsed, sending hundreds more into the burning bowels of the ship.

By 11:30 A.M. it was over. More than 600 bodies were recovered, some of them burned beyond recognition. From then until midnight delivery wagons and coal carts were pressed into service to supplement the ambulances and hearses that carried bodies and the injured to hospitals and morgues in the Bronx.

The next day cannon were brought from the Brooklyn Navy Yard to be fired over the water. The vibrations from the explosions loosened more bodies that were sunk in the mud of the river bottom.

Many of the dead were buried in a mass grave in the Lutheran Cemetery in Queens. To this day, every June 15 a memorial service is held for the dead. The exact figure was never precisely determined. According to the U.S. Steamboat Inspection Service (whose head President Roosevelt would fire), it was "only 938." The New York City Police asserted firmly that the figure was 1,031.

A coroner's jury finally charged Captain Van Schaick, Mate Flanagan and the inspector responsible for checking the ship's fire-fighting equipment with first degree manslaughter. The officers of the Knickerbocker Company were charged as accessories.

Only Captain Van Schaick received a sentence, for criminal negligence, of 10 years at hard labor in the federal penitentiary at Sing Sing. In 1908 President Roosevelt pardoned him because of his advanced age.

NUCLEAR AND INDUSTRIAL ACCIDENTS

THE WORST RECORDED NUCLEAR AND INDUSTRIAL ACCIDENTS

* Detailed in text

Germany
- Oppau
  - * Badische Anilinfabrik (1921)

Great Britain
- England
  - Liverpool
    - * Windscale Plutonium Plant (1957)

India
- Bhopal
  - * Union Carbide Pesticide Plant (1984)

Japan
- Tokaimura
  - * Nuclear processing plant (1999)
- Tsuruga
  - * Nuclear power plant (1981)

Switzerland
- Lucends Vad
  - Underground reactor (1969)

United States
- Alabama
  - Brown's Ferry
    - Nuclear power plant (1975)
- Idaho
  - Idaho Falls
    - * Idaho Nuclear Engineering Laboratory (1961)
- Michigan
  - Detroit
    - Sodium cooling plant (1966)
- Minnesota
  - Monticello
    - Nuclear power plant (1971)
- New York
  - Rochester
    - Ginna Steam Plant (1982)
- Oklahoma
  - Gore
    - * Sequoyah Fuels Corporation Plant (1986)
- Pennsylvania
  - Middletown
    - * Three Mile Island Nuclear Power Plant (1979)
- Tennessee
  - Erwin
    - Nuclear fuel plant (1979)

USSR
- Kasli
  - * Nuclear waste dump (1957)
- Pripyat
  - * Chernobyl nuclear power plant (1986)

CHRONOLOGY

* Detailed in text

1921
- September 20
  - * Oppau, Germany; Badische Anilinfabrik

1957
- * Kasli, USSR; Nuclear waste dump
- October 10
  - * Liverpool, England; Windscale Plutonium Plant

1961
- January 3
  - * Idaho Falls, Idaho; Idaho Nuclear Engineering Laboratory

1966
- October 5
  - Detroit, Michigan; Sodium cooling plant

1969
- January 21
  - Lucends Vad, Switzerland; Underground reactor

1971
- November 19
  - Monticello, Minnesota; Nuclear reactor

1975
- March 22
  - Brown's Ferry, Alabama; Nuclear reactor

1979
- March 28
  - * Middletown, Pennsylvania; Three Mile Island Nuclear Power Plant
- August 7
  - Erwin, Tennessee; Nuclear fuel plant

1981
- March 8
  - * Tsuruga, Japan; Nuclear power plant

1982
- January 25
  - Rochester, New York; Ginna Steam Plant

1984

December 3

* Bhopal, India; Union Carbide Pesticide Plant

1986

January 4

* Gore, Oklahoma; Sequoyah Fuels Corporation Plant

April 16

* Pripyat, USSR; Chernobyl nuclear power plant

1999

September 30

* Tokaimura, Japan; Nuclear processing plant

NUCLEAR AND INDUSTRIAL ACCIDENTS

There is no more timely disaster than a nuclear accident. Fortunately, only one actual meltdown has occurred so far, at Chernobyl (see p. 315), but the nuclear age is young. And because a little is known about both its short-and long-term consequences, no disaster is more frightening.

Ever since Enrico Fermi put one of Albert Einstein's theories to work, the world has experienced a steadily increasing and intensifying series of nuclear accidents. The worst so far is the latest; the very worst is, alas, yet to come.

In a positive sense, nuclear disasters have made us more cautious. Three Mile Island (see p. 312), for instance, put the brakes on the proliferation of nuclear power plants; the Chernobyl disaster brought it to a virtual standstill in America. Without arguing the relative virtue of atomic power, the very fact that huge public outcries have followed each of these disasters forces government and industry to concentrate more on safety practices than ever before and to be more public about these practices or the lack of them.

For if there is a single thread that runs through all the industrial and nuclear accidents of this section, it is a failure to pay proper attention to safety precautions. From Oppau, Germany, in 1921, to Bhopal, India, in 1984, to Chernobyl in 1986, to Tokaimura, Japan, in 1999, safety precautions were either ignored, circumvented or, as in the case of Chernobyl, dismantled.

Monumental carelessness on the part of both management and labor also runs through these accidents. Not one of the industrial plants examined—*not one of them*—had a workable evacuation plan in place for the innocent populace that lived around the plant. In each case evacuations had to be made up as they went along, and in each case these evacuations were delayed enough to increase the number and severity of casualties, often enormously.

In each case human error was the trigger that brought about the cataclysms that ensued. And if the similarity in cause and effect of man-made catastrophes has proved anything, it is that human error is at the root of virtually every man-made disaster. With industrial and nuclear disasters this human error is compounded all too frequently by human greed, by political and governmental expediency, by economic considerations and by deliberate misinformation or withholding of information.

And this is the final dimension that characterizes nuclear disasters and makes them unique: In every case, from the absolutely tight lid that was kept on every detail of the Kasli disaster in Russia's Ural Mountains in 1957 through Windscale, Three Mile Island, Chernobyl, and Tokaimura, official silence or deliberate downplaying of casualty statistics and, what is far worse, the danger posed by radiation is consistent enough to be policy.

Admittedly, some of this has come from ignorance. But an equal amount has also come from governmental design, explained away as a desire not to create panic in the populace. This policy has resulted in hundreds of lawsuits against governments by the families of victims who have died agonizing and puzzling deaths long after the incident that had been passed off as posing little or no danger to the public at large.

With the terrorist tragedy of September 11, 2001 (see p. 147), nervousness in New York City escalated to panic in the populace regarding the safety of nuclear plants not only from accidents but from infiltration or attack by terrorists. Contingents of military personnel were immediately dispatched to various nuclear facilities to secure them. In the process, encouraging facts came to light. In the decade of the 1990s, after 10 years of decreased regulation under the administrations of Ronald Reagan and George Bush, the Clinton administration tightened regulation of nuclear facilities markedly. Physical barriers against the escape of radiation were increased. Armed security guards and electronic surveillance equipment were upgraded and increased around nuclear plants.

The Nuclear Regulatory Commission personally supervised multiple drills of mock intrusion designed to test the efficiency of plant security measures. In addition, new regulations for the screening of workers and random drug and alcohol testing as well as behavioral counseling were mandated and regulated. The enforcement of these regulations led to the FBI classification of nuclear power plants as "hardened" targets. This increase in awareness has been a reflection of a worldwide governmental movement, and since its growth, major nuclear accidents in nuclear power generating plants have not occurred. The Chernobyl catastrophe in 1986 was the last major radiation release from a power plant.

However, there are other nuclear facilities that are still vulnerable to accidents and therefore threats to the environment and the populace near them. Prior to 1980, most of these other nuclear facilities were military ones and therefore tightly guarded. But as more and more of them became commercial enterprises, the possibility of accidents and radioactive contamination increased. The Tokaimura accident (see p. 308), the worst of this kind so far, occurred in a commercial fuel preparation plant. And so, as governments and municipalities continue to use nuclear power, which is cheap and abundant, without spending the proper amount of time on researching methods of preventing disasters and their consequent effects, these tragedies will continue to occur.

The criteria for inclusion in this section were based on the effect the disaster had on the innocent. Certainly, industrial accidents are a frequent occurrence, and that is regrettable and recordable. Certainly, the life of a worker in a factory is not worth more nor less than a person living down the road from the factory. But there are certain compensations involved in injuries or deaths suffered by those employed by an industry.

In the case of each of the included industrial and nuclear accidents in this section, huge numbers of innocent people who had either nothing or very little to do with the industry were victims of the disaster. And that was the reason for that happening's inclusion.

GLOSSARY OF NUCLEAR TERMS

chain reaction The process that occurs when uranium atoms split, emitting neutrons that split other uranium atoms in a continuum, or chain.

China Syndrome A meltdown through the reactor floor, accompanied by a large explosion. The worst possible kind of reactor accident.

cladding The metal that encases the nuclear fuel material.

condenser The heat exchanger in which steam is transformed into water by the removal of heat and the transference of the steam to a cooling pond.

control rods Carbon rods that are dropped between fuel rods, neutralizing the fission, thus slowing or stopping the chain reaction.

coolant Usually water, which removes nuclear-generated heat from the core.

core The place in a reactor in which the nuclear fuel is contained and nuclear reaction occurs.

curie A measure of radioactivity based on disintegration over time.

emergency core cooling system Any sort of system that is present to cool the core in the event that the regular cooling system malfunctions or fails to function.

fission The splitting of a heavy atomic nucleus. This forms two lighter "fission fragments" and smaller particles, such as neutrons. In a nuclear reactor the energy released from the fission provides heat, and part of this heat is converted into electricity.

fuel pellet The basic form of uranium used in reactors.

fuel rod A stainless steel tube holding uranium fuel pellets.

fusion The opposite of fission, and the energy source of stars. In this process atomic nuclei, such as those of hydrogen, are fused, releasing nuclear energy.

meltdown Second to an explosion, this is the most catastrophic occurrence in a nuclear accident. When the cooling systems—primary or emergency—fail to keep the temperature in a reactor within controllable limits, either the metal rods containing the pellets or the pellets themselves melt. Then the core melts into a glowing radioactive mass capable of smashing through the wall or the floor of the reactor's containment walls and releasing a massive surge of radiation.

millirad One thousandth of a rad.

millirem A term used to describe the measuring of the absorption of radiation by human beings. The average person undergoes, as a result of exposure to everything from dental X rays to cosmic rays, 100 to 200 millirems of radiation per year. One chest X ray causes an exposure of from 20 to 30 millirems.

rad The standard unit of an absorbed dose of radiation.

radioactivity The spontaneous disintegration of the nucleus of an atom, which in turn emits radiation. The radiant energy is in the form of particles, or rays, as alpha (positively charged particles), beta (negatively charged electrons) and gamma rays (elements of electromagnetic radiation resembling X rays and considerably more penetrating than other forms of radiation).

reactor The core and its protective shell, or container.

rem An abbreviation of "roentgen equivalent, man." A measure of the quantity of any ionizing radiation that contains the same biological effectiveness as one rad of X rays.

uranium A metal with radioactive properties used as a fuel because it has the ability to undergo continuous fission.

GERMANY
OPPAU
September 20, 1921

An error in mixing chemicals, producing explosive gas, caused the giant explosion in the Badische Anilinfabrik plant in Oppau, Germany, on September 20, 1921. Five hundred died; 1,500 were injured, many seriously.

The cataclysmic explosion of the Badische Anilinfabrik works at 7:30 A.M. on September 20, 1921, in Oppau, Germany, released more than an odd, green, evil-smelling gas into the atmosphere. Mystery surrounded the explosion; wartime activities at the plant were unveiled during the post-explosion investigation; speculation that nuclear experiments were being conducted in the plant persists. The complex mystery of it all has never been solved.

A soup kitchen is set up and utilized by survivors and families of victims of the gigantic explosion at the Badische Anilinfabrik plant in Oppau, Germany, on September 20, 1921. Five hundred died; 1,500 were injured, many seriously. (Library of Congress)

The Badische plant was built in 1913, shortly before the start of World War I. Almost from the very beginning it was engaged in aiding the German war effort. First, it saved Germany from a military collapse when, in the spring of 1915, it supplied artificial nitrates when the British blockade cut off the Chile saltpeter supply. Then, expanded, it manufactured most of the chlorine and phosgene used by the German army when it engaged in poison gas attacks in World War I.

After the war it was converted into a sprawling, imposing complex that covered acres and employed between 10,000 and 15,000 people. It bustled with profitable activity, utilizing chemicals to produce nitrates for dyes and artificial fertilizer. It employed the so-called Haber process, in which nitrogen was extracted from the air and then, in the presence of a catalyst, mixed under high compression with hydrogen. This converted the nitrogen into ammonia, nitric acid, nitrates, fertilizers and ammonium sulfates.

According to Major Theodore W. Sill, a member of the Interallied Mission appointed to study the German chemical industry after the war, the Oppau plant had a long history of accidental explosions. Despite the fact that German engineers developed a new type of steel capable of holding hydrogen under pressure, a compound that could remain solid against 2,000 pounds of pressure per square inch when the hydrogen and nitrogen gases were compressed under a temperature running from 500 to 600 degrees, the tanks *did* erupt. In September 1917 one of the compression tanks blew, killing approximately 100 workers; the concussion from the explosion was powerful enough to kill workmen crossing a bridge half a mile away.

But that was small compared with the cataclysmic eruption that took place in the early morning of September 20, 1921. The shifts changed every morning at 7:30 A.M. at this monster plant surrounded by a small town of 6,500, most of whom either worked in the plant or were family members of workers in the plant.

It was a cold and crisp day, and the departing shift was just concluding its duties. Three trains pulled onto the siding alongside the plant, preparatory to discharging hundreds of workers for the incoming shift.

And then, with a thunderous roar, the main building of the huge plant exploded. According to eyewitnesses the entire structure lifted from its foundations and then descended in 1,000 pieces on those who managed to survive the first blast.

The three trains that had just pulled into the siding were catapulted into the air. They sank back onto the siding and were immediately buried under the descending girders, bricks and tiles of the fragmented factory building.

Immediately after the first explosion, a second, slightly lesser one occurred, leveling more factory buildings and sending voluminous, sickening clouds of chemical smoke into the air.

Outside the factory sites the village of Oppau was leveled, as if it had been bombed from the air. The shock waves spread over a 50-mile radius throughout the Mannheim-Ludwigshafen district. At Eisenheim a train just leaving the station was blown off its tracks and plowed through the wooden sheds where French soldiers of the Army of Occupation were quartered. Twelve soldiers were killed and several were injured by the freak accident caused by the explosion's concussive force.

The nearby villages of Frankenthal and Edigheim were also leveled, and a steady stream of bandaged refugees began to pour from them headed toward the hospitals in Mannheim.

Fires spread immediately from the explosion site and moved on into the surrounding villages. Firemen wearing gas masks made their way to the huge, funnel-shaped crater that was once a factory. Their task was multiple: putting out fires, rescuing those who had been injured, putting mutilated animals out of their misery and hauling the bodies and pieces of bodies out of the debris. As they worked groundwater rapidly filled the crater, making the retrieval of bodies or the survival of the injured within it impossible.

Seventy thousand people attended the mass funeral that took place in Ludwigshafen Cemetery in Mannheim, and expressions of bewildered gratitude came from Germans over the international outpouring of grief, concern and comfort.

But then the questions arose: Did the explosion, like those that occurred during the war, come about because of a tank giving way under the giant pressure of uniting nitrogen and hydrogen? Scientists in Berlin speculated that the cause of the catastrophe was "extreme heat, generated by some hitherto unknown gas explosive, [which] must have led to the decomposition and subsequent explosion of a large quantity of ammonia and sulphate of saltpeter which forms the basis of artificial fertilizer."

True enough, some scientists said, but what was that other "unknown gas explosive"? What sort of experimentation was taking place at Oppau? Why did the second explosion take place? And what accounted for the enormous devastation, the likes of which no one had yet seen?

The events of the following years may have explained it. The beginnings of bombs that would stagger the imagination might very well have been given their first public airing at Oppau in September 1921. All during World War II the nuclear triggering device that would control this awesome power was the subject of a race between the Germans and the Allied powers. The Allies won and won the war. And humankind, for better or worse, entered the atomic age.

GREAT BRITAIN
ENGLAND
LIVERPOOL
October 10, 1957

The overheating of uranium cartridges releasing radioactive iodine caused widespread radioactive contamination surrounding the Windscale Plutonium Plant near Liverpool, England, on October 10, 1957. Thirty-three cancer deaths were attributed to the disaster. There was a temporary suspension of the milk and beef industries of northwestern England.

The Windscale plutonium factory in the Cumberland country of northwest England manufactured plutonium for use in nuclear reactors and atomic bombs and produced certain by-products that were used in medicine. Powered by the nearby Calder Hall atomic power plant, it was thought to be, in 1957, a model of clean and efficient productivity.

But the accident that took place on October 10, 1957, which was England's first nuclear accident—and one of the first in the peacetime world—was the forerunner of hundreds of nuclear accidents that would release radioactivity into the atmosphere. In 1957 the world was naive to the hazards, and little space was given to the accident. But as its aftermath extended and deepened, so did the awareness of its significance.

At 4:15 P.M. on Thursday, October 10, 1957, the number-one pile of uranium at Windscale overheated, and as its temperature rose it released radioactive iodine-131 vapor and some oxidized uranium particles into the air. It would be 15 minutes before the red-hot uranium pile would be discovered; that part of the plant had been shut down for maintenance.

Shortly after it was discovered, workers wearing gas masks and other protective equipment were assigned to use carbon dioxide to extinguish the fire. It was ineffectual.

A sense that this was no ordinary fire began to grow. All of the plant's off-duty safety workers were called back, and all of the roads to the plant were blocked off. By 5:15 safety experts issued conciliatory statements to the press claiming that all danger had departed.

By 9:00 A.M. on the 11th, it was decided to use water to damp down the fire. Two plant officials and a local fire chief hauled a hose to the top of the containment dome and aimed it at the fire. No one knew quite what would happen, and plant workers all over the complex crouched behind steel and concrete barriers.

Fortunately, the water worked, but it also released huge clouds of radioactive steam through the stacks and into the atmosphere. The worst was over, everyone thought; there had been neither an explosion nor a meltdown.

By midday of October 11, nearly all of the 3,000 workers at the plant and the nearby Calder Hall atomic energy plant were sent home. They had been exposed to radiation, and it was obvious that a reevaluation of the situation was needed.

Significant quantities of radioactive iodine-131 had been released into the atmosphere over a 200-mile radius, and at 2 A.M. on Sunday, October 13, police began to knock on the doors of the farmhouses in Cumberland. The milk from their cows, the police warned them, might be radioactive.

By Tuesday the 15th, the milk ban was extended from a 14-square-mile area to 200 square miles, including 600 dairy farms. Approximately 30,000 gallons of milk, worth $11,000, were dumped into the Irish Sea each day until the end of October, and all distribution of milk from the contaminated area was immediately halted.

Beyond that, hundreds of cows, goats and sheep were confiscated, shot and buried. Farmers who slaughtered their animals for meat were told to send the thyroid glands to the Atomic Energy Commission for testing.

Farmers in the area now began to make public the tales they had exchanged among themselves: Even before the accident, sterilization had occurred in their cattle. W. E. Hewitson, a dairy farmer in Yottenfews, stated that he had changed bulls four times in four years, but only a third of his cows either calved or gave milk.

Then it became apparent that the radioactive iodine-131 that safety experts first said had drifted out to sea had not done so at all. There was a marked increase in the radioactivity of the atmosphere after the accident at the Windscale plant.

Several months later British officials conceded to a United Nations conference at Geneva that nearly 700 curies of cesium and strontium had also been released into the air over England and northern Europe, in addition to 20,000 curies of iodine-131. The iodine dose represented more than 1,400 times the quantity American officials later claimed had been released during the 1979 accident at Three Mile Island (see p. 312).

As was so often the case, there were no official follow-up studies regarding the health of residents of the 200-square mile area near the plant. When a local health officer, Frank Madge, used a Geiger counter to confirm abnormal radiation levels in mosses and lichens, representatives of the British Atomic Energy Authority discouraged publication of his findings.

Private studies of health data in downwind European countries later indicated a clear impact of the accident on infant mortality rates. Dr. Ernest Sternglass, interviewed by Harvey Wasserman and Norman Solomon for their study, *Killing Our Own,* remarked, "[It was] as if a small bomb had been detonated in northern Great Britain."

As late as 1981, while the Windscale plant continued to operate without modification, British scientist E. D. Williams stated in the January 1981 issue of *Health Physics Journal* that there were "high cesium levels in people eating fish caught in the path of the Windscale effluent." More than 30 cancer deaths in the vicinity of the Windscale plant have been directly attributed to the 1957 accident.

INDIA
BHOPAL
December 3, 1984

The worst industrial accident in history, the explosion at the Union Carbide Pesticide Plant in Bhopal, India, on December 3, 1984, was caused by a combination of faulty maintenance, laxity in management, outdated equipment, faulty judgment and social factors. At least 2,000 died; 200,000 were injured.

For years India has been a place where a patina of the present is layered over an ancient civilization used to conducting itself in ancient ways. The teachings of Kashmir Shaivism go back 3,000 years, and the lyrics of Brahmin chants are in Sanskrit. Oxen carts still haul produce; electricity is still a distant stranger to parts of the country.

When the government of a new, free India was formed, its character was, at least in part, shaped by the country's spiritual leader, Mohandas Gandhi. Although television now permeates India as it does all over the world, the most popular program—the one that brought the entire country to a virtual standstill—was a dramatization of the great and ancient Hindu spiritual epic, the *Mahabharata.*

Thus, there is always a sense of difference, a feeling of, if not compromise, at least accommodation when Western ideas or industries introduce themselves into India. And sometimes those accommodations carry with them a certain carelessness.

Although Western industries find that locating their plants in Third World countries is profitable because labor is cheap, there are also trade-offs. The labor is usually both cheap and unskilled. And because of the distance from the source, some equipment that exists in these plants goes too long without updating, replacement or even maintenance. Finally, there is the danger that a kind of casualness, a slowing down of the metabolism that is more in tune with the pace of ancient life than of modern life, works against the constant, concentrated vigilance that can prevent industrial disaster by preparing for it.

This was at least a factor in the complex series of events that led up to the worst industrial accident in the history of the world early in the morning of December 3, 1984. The disaster took place in the Union Carbide Pesticide Plant in Bhopal, a small city in the north central region of Madhya Pradesh, in India, midway between New Delhi and Bombay. The plant, a boon to this economically depressed city, was located in its slum section, a community called Jai Prakash Nagar.

At 2:45 P.M. on Sunday, December 2, while children played in the dirt outside the huts crammed together near the plant's entrance, about 100 workers reported for duty for the eight-hour late shift. The plant, which manufactured the pesticide Sevin, had been closed down for some time and had been reactivated only a week before. It was still working at a partial pace, carrying through the process of making the pesticide, which consisted of a mix of carbon tetrachloride, methyl isocyanate and alpha-napthol.

The methyl isocyanate, MIC, was stored in three partially buried tanks, each with a 15,000-gallon capacity. One of the tanks, number 610, was giving the workers trouble. For some reason they could not determine, the chemical could not be forced out of the tank. Nitrogen was pumped into the tank to force the MIC into the Sevin plant, but each time this was done, the nitrogen leaked out.

There was a greater problem with tank number 610, however, and this, plus a leak that had not been repaired in seven days, would set the stage for a major catastrophe. First of all, MIC must, to maintain stability and be nonreactive, be kept at a low temperature. A refrigeration unit designed to keep it at that temperature had, for a still-unexplained reason, been turned off. The chemical was thus warmer than the four degrees Fahrenheit recommended in the plant's operating manual, but just how much warmer was impossible to tell, since the instruments monitoring it were old and unreliable.

In addition, the money-losing plant had undergone further cost-cutting procedures in the past months, and this included the curtailment of maintenance on the noncomputerized, behind-the-times equipment. And finally, new supervisors and operators were in key positions. As a result of this laxity, tank number 610, besides having a faulty valve and not being maintained at the proper temperature, was also overfilled.

Other pieces of the scenario began to come together. At about 9:30 P.M. that night a supervisor ordered a worker to clean a 23-foot section of pipe that filtered crude MIC before it went into the storage tanks. The worker did this by connecting a hose to the pipe, opening a drain and turning on the water. It flowed into the pipe, out the pipe drains and onto the floor, where it entered a floor drain. It flowed continuously for three hours.

All of the workers and presumably the new supervisor knew that water reacts violently with MIC. They also knew that there was a leaky valve not only in tank number 610, but also in the pipe that was being washed. Rahaman Khan, the worker who washed out the pipe, later told the *New York Times,* "I knew that valves leaked. I didn't check to see if that one was leaking. It was not my job." It is generally conceded that it was the water flowing from the hose that triggered the horror that was to follow.

At 10:30 a pressure reading was taken on tank 610. It was two pounds per square inch, which was normal.

At 10:45 the next shift arrived. The water was still running.

At 11:00 P.M. the pressure had climbed to 10 pounds per square inch, five times what it had been a half hour before. Something was obviously wrong. But no one did anything about it, because it was still within acceptable limits. In fact, some workers later testified that that was the usual temperature and pressure of the MIC at the plant.

Then, too, there was the problem of the instruments. Shakil Qureshi, the MIC supervisor on duty, later noted that he thought that one of the readings was probably wrong. "Instruments often didn't work," he said. "They got corroded. Crystals would form on them."

But by 11:30 P.M. the eyes and noses of the workers informed them that something was indeed wrong. Their eyes began to tear. They knew MIC was leaking, but this happened on the average of once a month. They often relied on these symptoms to inform them that a leak had occurred. Suman Dey, a worker, later told reporters, "We were human leak detectors."

V. N. Singh, another worker, discovered the leak at approximately 11:45 P.M. He noticed a drip of liquid about 50 feet off the ground, which was accompanied by some yellowish white gas. Mr. Singh informed his supervisor, Mr. Qureshi, who said that he would look into it after his tea break.

The tea break began for everyone at 12:15. And while this ancient custom went on for 20 minutes, the disaster continued to unfold unchecked.

From 12:40 A.M. on December 3 events began to take place with lightning rapidity. The smell of gas rose alarmingly. Workers choked on it. The temperature gauge on tank number 610 rose above 77 degrees Fahrenheit, the top of the scale. The pressure gauge was visibly inching upward toward 40 pounds per square inch, a point at which the emergency relief valve on the MIC tank was scheduled to burst open.

At 12:45 P.M. the pressure gauge read 55 pounds per square inch, 15 points from the top of the scale. Supervisor Qureshi ordered all the water in the plant turned off, and it was only then that the water in the hose that had been running for three hours was finally found and turned off.

But it was far too late. The water reacted with the MIC, and the leak burst forth. Panicked workers dashed to and fro, blinded and coughing.

An alarm sounded, and within minutes the fire brigade arrived to place a water curtain around the escaping gas. But the curtain reached only 100 feet in the air. The top of the stack through which the gas was now spewing into the atmosphere was 120 feet high, and the gas fountained another 10 feet above that.

A vent gas scrubber, a device designed to neutralize the escaping gas, was turned on. But its gauges showed that no caustic soda was flowing into it. Or, perhaps, the gauge was broken. Who knew at this point? In either case, the gas, instead of being neutralized, was shooting out of the scrubber stack and was being carried on the high winds southward from the plant into the surrounding slums.

There were four buses parked by the road leading out of the plant. Drivers were supposed to man them in an emergency and load and evacuate workers and people who lived near the plant. But no drivers appeared. They, along with the terrified workers, were running from the plant.

At 1 A.M. Mr. Qureshi had run out of ideas. He called S. P. Choudhary, the assistant factory manager, who instructed him to turn on the flare tower, which was designed to burn off escaping gas.

But, explained Mr. Qureshi, with all that gas in the air, turning on the flare would cause a huge explosion. At any rate, a four-foot, elbow-shaped piece of pipe was missing from the flare. It had corroded and was due to be replaced as soon as the part arrived from the United States.

An alternative would have been to dump the MIC into a spare storage tank. There were two spares that were supposed to be empty. But they were not. Both contained MIC.

The workers who remained and tried to control the leak now donned oxygen masks. It was the only way they could breathe. Visibility was down to one foot. The supervisor, unable to find a mask, he said, opted to run away from the plant. He found a clear area, scaled a six-foot fence topped by barbed wire, vaulted over it and fell to the other side, breaking his leg. He was later transported to a hospital, with many, many others.

The gas poured unchecked out of the leak until 2:30 A.M. Jagannathan Mukund, the factory manager, arrived at 3 A.M., and only then, because, he later stated, the telephones were out of order, did he send a man to inform the police about the accident. The company had a policy, he said, of not involving the local authorities in gas leaks.

And to be fair, the sleeping populace *did* hear the emergency sirens going off, but they sounded so often in false alarms that the people in the surrounding slums ignored them and went back to sleep—some of them for the last time.

Outside the factory people were dying by the hundreds, some in their sleep. Others, panicked, choking, blinded, ran into the cloud of gas, inhaling more and more of it until they dropped dead. Thousands of terrified animals perished where they stood.

The outside temperature was only 57 degrees Fahrenheit, which kept the lethal cloud of gas close to the ground, rather than allowing it to rise and dissipate into the atmosphere, as it would have under warmer conditions.

The gas crept into open shacks, killing the weak and the frail immediately. Others woke, vomited and groped blindly to get outdoors, where they filled their lungs with the searing chemical vapor.

"I awoke when I found it difficult to breathe," said Rahis Bano to a reporter afterward. "All around me my neighbors were shouting, and then a wave of gas hit me."

She fell down, vomiting, and her two sons, whom she was carrying, rolled on the floor. She revived herself and grabbed one son. He and she would survive; the son she left behind would die.

Rivers of humanity, tens of thousands of people, stumbled about. Some were trampled. Others simply gave up and sat down. As the cloud spread southeastward it enveloped the Bhopal railroad station. Ticket takers, trainmen and passengers died where they stood.

A hill was located in the center of the city, and thousands rushed toward it thinking they could climb above the gas. "There were cars, bicycles, auto rickshaws, anything that would move on the road trying to get up the hill," said one survivor. "I saw people just collapsing by the side of the road."

New hazards presented themselves; many of the fleeing refugees were run over by cars and buses and emergency vehicles. The police, instead of helping, heightened the panic by roaring through the crowds, their police van loudspeakers shouting, "Run! Run! Poison gas is spreading!"

Hospitals were immediately filled. Doctors and nurses tried to save as many as they could, but Hamida Hospital recorded a death a minute until it finally gave up trying to keep count. Dr. N. H. Trivedi, deputy superintendent of the hospital, told the *Times,* "People picked up helpless strangers, their best friends, their relatives, and brought them in here. They did far more than the police and official organizations."

Most hospitals placed two stricken people in one bed until there was finally no more room, and emergency clinics were set up in stores and on streets.

When dawn finally broke over Bhopal, it lit a scene of cataclysmic destruction. Thousands of bodies—human and animal—littered the streets. No birds sang. The only movement was from trucks sent out to pick up the dead and to search houses for more dead and dying.

Between 2,000 and 2,500 had died, and more than 200,000 would be afflicted for years, possibly for the rest of their lives, with the aftereffects of the Bhopal tragedy. Some were permanently blinded. Others could not sleep, had difficulty breathing or digesting food and had trouble functioning.

For a week the suffering continued in Bhopal's hospitals and clinics. Children between one and six years old seemed to suffer most. The tragedy was made worse by the inability of either medical specialists or parents to do anything. Relatives watched mutely from doorways as doctors placed intravenous feeding tubes in the children's arms and oxygen tubes in their noses and mouths.

For weeks sirens wailed, cremations took place one after another and bodies were buried in mass graves. The worst panic took place 10 days after the accident, on December 13, when Union Carbide announced that it would start the plant up again on Sunday, December 16, to neutralize what remained of the MIC.

Bhopal, normally a city of 900,000, was already depleted. Besides the 2,000 dead and the 200,000 injured, 100,000 others had fled after the disaster. Now 100,000 more took to trains, buses, cars, planes, auto rickshaws, two-wheeled tongas and their own feet to put a distance between themselves and what they perceived to be the site of another possible catastrophe.

Two thousand paramilitary troops and special armed police officers were brought in by the Indian government to supplement the local police force in an effort to prevent the looting of vacated homes and to maintain order in the clinics and refugee camps.

Most of the dead came from Jai Prakash Nagar and Kali Parade, the two slum neighborhoods adjacent to plant, but the brisk night breeze carried the fatal fumes much farther than that.

A year after the accident residents of Bhopal who had been affected by MIC were still suffering. According to authorities in India, an estimated 10 to 20 percent of the 200,000 people injured were still seriously affected. Many were having trouble breathing, sleeping, digesting food and undertaking simple tasks, just as they had right after the leak occurred.

They suffered memory loss, nausea, nerve damage, including tremors, and damage to kidneys, liver, stomach and spleen. A year later, 40 percent of those afflicted were in the same condition, 40 percent had improved and 20 percent had worsened. Medical studies predicted that these would be long-term, perhaps lifelong, afflictions.

The relief effort had become bureaucratic and sometimes contradictory. Cortisone injections were given by one medical team; cough medicine and aspirin by another. One health expert, Rashmi Mayur of the Urban Institute in Bombay, averred that he had come across one victim who had been able to get 250 pills in one day from seven different doctors.

What ultimately became apparent in the tragic unfolding of this disaster was that ignorance was also a culprit. Even as people were dying, Union Carbide factory doctors were telling local physicians that MIC, which is used in 20 to 25 percent of all the world's pesticides, only caused lung and eye irritation. And none of these company doctors had informed the local medical workers ahead of time that a simple antidote for the effects of the chemical was to merely cover the face with a wet cloth. "Had we known this," police superintendent Swaraj Puri later told reporters, "many lives might have been saved."

Six months before the accident, the National Academy of Sciences had said that little or nothing was known about the health effects of most of the 54,000 chemicals used in commercial products, thus making treatment and prevention difficult at the very best.

Afterward, concerted efforts were made to find causes and blame, and there was more than enough to go around. Officials of Union Carbide were arrested when they arrived in India and then freed. They were later charged with criminal negligence, as was the plant supervisor. The government of India filed suit against Union Carbide in the federal district court in Manhattan seeking compensation for the victims of the disaster. The suits are still pending.

It was generally acknowledged that the seeds of the tragedy were planted in 1972, when, under government pressure to reduce imports and loss of foreign exchange, the company proposed to manufacture and store MIC at the plant in Bhopal. Both the local government and the company agreed at that time that the risks would not be high.

Dr. S. R. Kamat, a prominent Bombay expert on industrial health and the hazards of development, probably summed it up most succinctly and accurately: "Western technology came to this country but not the infrastructure for that technology," he told the *New York Times* on February 2, 1985. "A lot of risks have been taken here," he went on. "Machinery is outdated. Spare parts are not included. Maintenance is inadequate. Bhopal is the tip of an iceberg, an example of lapses not only in India but by the United States and many other countries."

JAPAN
TOKAIMURA
September 30, 1999

Forty-nine people were injured and one killed when enormous doses of radiation were released in Japan's worst nuclear accident in a processing plant in Tokaimura on September 30, 1999.

"Inadvertent criticality" is the somewhat tortured phrase used by scientists to describe and identify the bungled process that caused Japan's worst nuclear accident. It occurred on September 30, 1999, in the Tokaimura nuclear processing plant 87 miles northwest of Tokyo, and it was not a grand scale disaster like Three Mile Island in Pennsylvania or Chernobyl in what is now Ukraine (see p. 315), but it was serious enough to kill one worker, injure 49 and force 33,000 residents of the surrounding village to barricade themselves indoors, fearing radiation poisoning, for a number of days. Radioactive iodine-131 would linger in the air over the immediate area around the primary school, kindergarten, shops and homes near the plant for more than a week. And for the entirety of Japan, it would be a terrifying reminder of the havoc, horror and devastation that the dawn of the nuclear age had introduced to them as it snuffed out the lives of 79,000 of their fellow citizens in Hiroshima and Nagasaki on August 6 and 9, 1945.

"A major accident resulting in a radioactive leak has happened," announced Kohi Kitani, president of the JCO Company, the plant's operator. "We apologize from the bottoms of our hearts."

Workers at the plant were mixing a liquid batch of uranium when suddenly, in a flash of blue light, they began a chain reaction—or, in the language of present science, went critical. The atoms began to split, producing fission products that spread rapidly from the scene of the accident in radioactive quantities more powerful than the radiation of the uranium in its original state.

The error that caused the accident was human in an industry that is notoriously unforgiving of human error. The usual and approved amount of powdered uranium that should be poured into the type of tank the workers were using is 5.2 pounds. They attempted to pour 35 pounds of uranium into the tank of nitric acid—more than five times the acceptable amount.

If the workers had been using the proper kind of vessel for the process—a tall, cylindrical container—the reaction might have at least been mitigated. But they used a flatter receptor that was circular, a shape that encourages a chain reaction. And to make matters worse, the container was surrounded with a shell filled with cooling water. The water functioned as a reflector, bouncing neutrons from the uranium back into the tank and improving still further the conditions for a reaction.

All of the workers in the purification chamber received massive doses of radiation. One collapsed on the spot. Others fled. An enormous shower of radiation in the form of gamma rays and neutrons descended over the entire area. An emergency crew immediately went into action, attempting to drain the tank with the uranium and nitric acid mixture by remote control. It did not work. Because of the density of the uranium in the mixture, neutrons that ordinarily would have been absorbed by the other material were not, and the reaction spun out of control.

Hundreds of residents of Tokaimura line up outside the town hall to be tested for radiation exposure after Japan's worst nuclear accident. (AP/World Wide Photo)

The Tokyo Electric Power Company rushed 880 pounds of sodium borate to the plant, designed to absorb the radiation, but there was no way of getting close enough to the processing tank to dump the powder onto it. Finally, the emergency workers smashed the pipes leading to the purification chamber, and the vessel began to cool. It would be 17 hours before chemicals inserted in hoses drenched the area enough to absorb the radiation.

Meanwhile, the East Japan Railway suspended trains to the immediate region of the accident, and all the roads and major highways were blocked off. Residents of the area were told not to drink well water and to avoid contact with the rain that began to fall shortly after the accident.

Yet, with all of these precautions in place, the ventilation system in the building was not turned off, and open ventilators continued to pump radioactive iodine-131 at double the legal limit into the village. It would be a full seven days before the exhaust fan would be turned off and the opening sealed. Eventually, 49 people, including workers in a nearby facility and the doctors and medical personnel who tended to the stricken workers, were exposed to massive doses of radiation.

Now, as the scope, seriousness and implication of the accident began to reach the Japanese public, anger and fear spread. Tokaimura had originally, in the 1950s, welcomed the building of more than a dozen nuclear-type plants in or near the village. (The processing plant in which the accident occurred was forming liquid uranium that would be transformed into a solid for a nearby reactor in the same area.) Whereas Tokaimaura had originally housed only one corporation—Hitachi—it was now home to the nation's nuclear industry, which supplied one third of Japan's electrical power. But the cost had risen. In 1995 there had been a large leak of radioactive coolant at a breeder plant. In 1997 it was the scene of Japan's worst previous accident, in which 37 workers were exposed to radiation after a fire was improperly extinguished and caused a small explosion.

Now it became apparent that the seriousness of the safety precautions that were the rule in the processing plant were either not understood or ignored by the workers. Reports began to circulate that the plant's workers routinely took dangerous shortcuts that violated regulations and were encouraged to do so by plant managers in an effort to increase productivity. U.S. Energy Department officials who toured the plant concluded that managers counted on workers to follow rules but never explained why the rules were important.

That was small comfort to the residents of Tokaimura, who remained barricaded in their homes for weeks following the accident. "I am furious," Tomi Oshiro, who lived close by the plant, told reporters afterward. "It took place right next to people's houses, and still it took a long time before people were warned or any emergency measures were taken."

Equally as perplexing and infuriating to the large population of Japan was the seemingly casual, often contradictory way the government treated this obviously serious happening, at least at first. The chief of the Natural Resources and Energy Agency continued to maintain that an uncontrolled reaction could not occur at a nuclear plant in Japan. He adamantly stated to reporters that he did not intend to instruct power companies to reexamine their safety measures. On the other hand, a top official of the Ministry of International Trade and Industry challenged this and said that the accident was serious enough to cause reconsideration of Japan's entire nuclear power policy.

Prime Minister Keizo Obuchi finally spoke to the public, apologizing for the slow official reaction. "We must obtain decisions from scientists, technicians and experts to assess such accidents," he said. "If they were slow in making decisions, we must make sure that that never happens again."

The most seriously injured worker, Hisashi Ouchi, who had been exposed in a few minutes to 400 times the maximum amount of radiation a nuclear worker is allowed to receive in an entire year, died. He had been unconscious since two weeks after the accident and had been kept alive, it was thought, possibly because of fear of a national backlash against nuclear power. Although in a coma and filled with huge amounts of painkilling drugs, the stricken worker constantly evinced signs of agony to the end of his life.

"He was a victim of a myth perpetrated by the national government and the nuclear power industry that nuclear energy is safe," said Tatsuya Murakami, the mayor of Tokaimura, when told of the worker's death.

"So much has been made of Japan's sophisticated technology that supposedly makes nuclear energy safe," antinuclear activist Chihiro Kamisawa said following the accident. "[This] proves that's absolutely not true."

JAPAN
TSURUGA
March 8, 1981

A leak from a disposal building at the nuclear power plant at Tsuruga, Japan, on March 8, 1981, caused widespread radiation contamination. Fifty-nine workers were exposed to radiation, and Japan's fishing industry was temporarily suspended.

The nuclear power plant at Tsuruga, Japan, a seacoast city of 60,000 located on the far west coast of Japan, opposite Tokyo, was in chronic trouble in the early spring of 1981. And Japan's Atomic Power Commission, like the atomic power commissions of the United States (see Three Mile Island, p. 312) and Great Britain (see Windscale, p. 304), spent as much time misinforming the public as it did in investigating the mishap.

On March 8 a huge leakage of radioactive waste occurred in a disposal building adjacent to the main plant. The first newspaper report did not appear until April 18, more than a month from the day that 16 *tons* of the waste had spilled into the adjoining Wakasa Bay, which flows into the Sea of Japan. The April 18 bulletin merely stated that a crack in a pipe or the storage tanks themselves "might have allowed waste water to seep into general drainage pipes into the Wakasa Bay."

Shortly thereafter the Ministry of International Trade and Industry also announced that it had found radioactivity levels 10 times normal in seaweed near drainage outlets, and the Kyodo News Service accompanied the news release with the discomforting information that the amount of cobalt-60 discovered in the seaweed and the soil surrounding the plant was "5,000 times the previous highest reading . . . the effects on the human body could be serious if the radioactive waste has spread throughout the bay."

It would not be until April 21, six and a half weeks after the mishap, that the Japan Atomic Power Company would make its first statement, in which it acknowledged that some waste-contaminated water had leaked onto the floor of the plant and that 56 men who had been put to work mopping up the water with buckets and rags "had been exposed to radiation at levels considerably below government limits," an assessment the Ministry of Trade and Industry immediately disputed. The announcement went on to speculate that the plant's executives might be indicted on criminal charges. Finally, two days later, the Tsuruga company released more detailed information, which had been withheld, a company spokesman avowed, because of "Japanese emotionalism toward anything nuclear."

The accident, according to the account released by the company, occurred when an operator apparently forgot to shut off a valve, which in turn let water run through a radioactive sludge tank that overflowed and splashed on the floor of the power plant and then seeped into the general sewage system.

Akira Machida, the plant's general manager, attempted to downplay the accident by comparing it with Three Mile Island, "[It was] nowhere near as serious as America's Three Mile Island," he told reporters, but then he acknowledged that the biggest blunder was in failing to report it to the authorities.

Further revelations came swiftly. Forty-five other workers had been exposed to radiation in January, when another pipe had broken in the plant. Thirty-one other accidents had occurred since the plant had opened in 1970.

But the worst was yet to come. Fish and fish products from the immediate area of Tsuruga had been recalled following the March 8 mishap, but no one knew how widespread the contamination of the waters of the Sea of Japan had been. (Several years later, mutant forms of fish continued to be caught in the area, indicating that there was far more contamination than first reports indicated.) Japanese officials could not have forgotten the 1954 furor when 23 fishermen and the tuna catch aboard the Japanese fishing boat *Lucky Dragon* were victims of acute radiation exposure after the U.S. test explosion of a hydrogen bomb near Rongelap Atoll in the Marshall Islands. But their silence in 1981 seemed to indicate that they had.

Finally, in May, the chairman of the board and the president of the Japan Atomic Power Company resigned, accepting the responsibility for the leakages. A government investigation blamed human error, faulty equipment and structural weaknesses.

It would be Japan's first nuclear accident. That human carelessness and cover-up should figure at all in the nuclear industry of the first country in the world to suffer a nuclear holocaust made it enormously significant, and a discouraging comment on the pervasiveness of human carelessness.

UNITED STATES
IDAHO
IDAHO FALLS
January 3, 1961

Sabotage was suspected but never proved in the chemical explosion that blew apart the reactor core at the Idaho Nuclear Engineering Laboratory in Idaho Falls on January 3, 1961. Three died.

The explosion that shattered the core of an atomic reactor at the National Reactor Testing Station, part of the Idaho Nuclear Engineering Laboratory at Idaho Falls, Idaho, is significant in two ways: First, it was the first nuclear accident to occur in the United States. And second, it was symptomatic of conditions that would cause future disasters at other nuclear plants around the world.

The Idaho Nuclear Engineering Laboratory is a huge complex in which research and development projects are conducted for the military, spent nuclear submarine fuel is recycled and military radioactive wastes are stored. Because of the military and therefore largely secret character of the operations of the facility, details remain sketchy.

At 9:02 P.M., Mountain Standard Time, on January 3, 1961, three military technicians were at work, operating a new-style reactor known as Stationary Low Power Reactor Number 1. The reactor was a two-year-old prototype of a small mobile unit that was being developed as a heat and power facility for the armed forces in remote areas.

Suddenly, the core of the reactor blew. A fuel rod shot out of the reactor, piercing the body of one of the technicians and pinning him to the reactor containment, high above the core. The other two men were blown apart and had to be buried in pieces in lead-lined coffins. The radiation level within the container building was so high that it would be weeks before officials dared enter it, even in protective garb.

The cause of the accident was later determined to be "human error," causing an accidental overloading of one chemical against others. In 1981 Stephen Hanauer of the Nuclear Regulatory Commission, in an interview with Harvey Wasserman and Norman Solomon, authors of *Killing Our Own,* a study of atomic radiation in the United States, indicated that "the 'accident' may have been caused deliberately by one of the technicians in a bizarre suicide-murder plot stemming from a love triangle at the plant." More atrocious happenings than this have occurred, but there has, so far as this author knows, been no further substantiation of this steamy analysis of the events of January 3, 1961.

What is known, however, is that the Idaho plant had, as have most nuclear and industrial facilities experiencing accidents, a history of sloppiness. In the late 1960s it was charged with accidentally dumping concentrated uranium on a nearby road. From 1952 to 1970, its management deliberately tossed 16 billion gallons of liquid waste into wells that fed directly into the water table below, causing radioactive contamination seven and one half miles away.

During the 1978 World Series the plant supervisor was consumed by watching the games on a portable TV, which had been sneaked into the plant against regulations and neglected to notice a dangerous buildup of radioactivity in a small nearby uranium-processing column. Or, perhaps even if the game had not gobbled up his attention, he would not have noticed the imbalance in the column, since one recording chart of the plant's monitoring devices had run out of paper two weeks before, and the paper had not been replaced.

At 8:45 P.M. high-radiation alarms were tripped by a bursting force of radiation from the afflicted column. The supervisor and others escaped to uncontaminated areas, and the column was brought under control, but not before 8,000 curies of radioactive iodine, krypton and xenon had been released into the atmosphere—an amount that could easily threaten the health of anyone downwind of the plant. The plant supervisor was later fired, and an investigation of worker alienation and low morale at the plant indicated that these were major factors leading to both the 1961 and 1978 accidents.

UNITED STATES
OKLAHOMA
GORE
January 4, 1986

Faulty judgment was the culprit in the chemical leak at the Sequoyah Fuels Corporation Plant at Gore, Oklahoma, on January 4, 1986. One died.

The bald statistics of the chemical leak released by a storage tank rupture at the Sequoyah nuclear fuel processing plant in Gore, Oklahoma, on January 4, 1986, are unimpressive. But its magnitude goes far beyond its statistics, in a remarkable and disquieting resemblance to the far more cataclysmic disaster at the Union Carbide plant in Bhopal, India, two years before (see p. 305). The very vastness of that catastrophe was supposed to be a lesson in safety, maintenance and preparedness. But the accident in Gore confirmed the axiom that lessons unlearned become errors recommitted.

The facts are these: At about 9:30 A.M. on Saturday, January 4, 1986, workers at the Sequoyah nuclear fuel plant, which is owned by the oil and natural-gas manufacturing company Kerr-McGee, discovered that they had overfilled a shipping container with liquid uranium hexafluoride. The substance is used as a raw material for nuclear fuel, and it was to be shipped in solid form for further processing. The error was caused by a faulty instrument, which caused a container that was rated to hold 27,500 pounds of the substance to be loaded to a weight of 29,700 pounds.

A decision was made to heat the container enough to convert some of the substance to gas, so that the excess could be removed. For some reason never explained, the heating process was delayed until *two hours* after the discovery of the overfill.

Finally, at 1:30 P.M., the workers moved the container into an outdoor steam chest and began to heat it. Several workers remained nearby in the production building. One of them, James Harrison, had just mounted a flight of stairs to the top of the building. Suddenly, the tank exploded. The uranium hexafluoride, now partially a gas, shot skyward, merged with the moisture in the air and fractured into two components: uranyl fluoride, a heavy white powder, and hydrofluoric acid, a highly corrosive gas. Both were initially radioactive, but as soon as the uranyl fluoride powder fell to the ground, only it remained radioactive.

The gaseous hydrofluoric acid, a lethally corrosive substance, was drawn immediately by the ventilation equipment into the production building where the workers were stationed. Thirty-one workers were immediately exposed to the corrosive cloud. They ran, stumbling and gasping, from the building, while the cloud outside, powered by 25- to 30-mile-per-hour winds, began to drift outward over the open countryside, where it would affect 77 unsuspecting people and send them to Sequoyah Memorial Hospital.

James Harrison, the worker who had climbed the staircase inside the production building, was trapped above the slowly rising cloud of gas. His only escape was a route directly through it, and he chose to plunge forward, through the cloud and down the stairs. Before he reached the bottom, he had been blinded and scalded and was choking from the fumes.

He was immediately rushed in an ambulance to the hospital, while the gas, which is widely used in industry to etch glass and make plastics, including Teflon, was working within his body. The gas coagulates proteins on contact and can almost instantly create second- and third-degree burns when it touches the skin. Even in a diluted form, it can burn or irritate the mucous lining of the lungs.

By the time he reached the hospital, Harrison's exposed skin was laced with horrible burns. The corneas of his eyes had been seared by the acid, blinding him. Four hours after arriving at the hospital, Harrison's lungs swelled and hemorrhaged from the effects of the acid gas, and he died.

Others, including motorists and the local sheriff, who had been affected when he responded to the emergency situation at the plant, were treated with oxygen, asthma medication and breathing therapy. They survived, as did nearby residents, a fifth of them children, who were exposed to the gas cloud as it grew more diluted and drifted downwind from the plant. The 30 plant workers who had been exposed to the uranyl fluoride were given alkaline treatments—two Alka Seltzers every four hours—to treat its presence in their kidneys.

Safety-clothed cleanup men were sent out to scrub down highways and dig up topsoil that had been contaminated by radiation, and the investigation began. Within weeks the pattern that characterized the Bhopal disaster began to reassert itself. The plant, it turned out, had been cited for a number of safety violations—15 since 1978. The cylinder rupture was not without precedent. In 1960 17,800 pounds of liquid uranium hexafluoride from a ruptured cylinder that was being heated at a government uranium plant in Peducah, Kentucky, had injured 21 men. In 1966 one worker had been hospitalized when 3,844 pounds of uranium hexafluoride escaped from a cylinder after a valve had been improperly turned at a government uranium processing plant in Femald, Ohio. In 1978, 21,125 pounds escaped from a ruptured cylinder that was being improperly moved at a government plant in Portsmouth, Ohio. And a mere two years earlier, in 1984, the horrendous leak at the Union Carbide installation in Bhopal, India, had devastated a countryside, killing over 2,000 people and injuring 200,000.

One would think that the history of accidents preceding the event at Gore, Oklahoma, would have precluded it. But apparently not. Kerr-McGee, like Union Carbide, was not making much money on the plant and had resorted to cost-cutting shortcuts. As in Bhopal, uncared for and aging instruments did not work properly and thus contributed to the catastrophe.

But that was just the beginning. The similarities multiplied. Both happened with a weekend crew in charge. In both cases the process used was a clear violation of company regulations, and yet a supervisor took part in it. In both cases, despite repeated instructions from regulators, the plants had failed to develop evacuation plans. In both cases innocent people beyond the perimeters of the plant were unexpectedly afflicted with the effects of carelessness, cost cutting and human error. Thus, the parallels were massive and discouraging and presaged an uncomfortable future for populaces living near chemical processing plants.

UNITED STATES
PENNSYLVANIA
MIDDLETOWN
March 28, 1979

The worst nuclear disaster in the history of the United States was blamed on human error, which was in turn caused by design flaws. No deaths or injuries occurred at the plant; there is still contention over infant and fetus mortality after the radiation spread. The chief casualty was the growth of the National Atomic Power Program.

The worst nuclear disaster in U.S. history occurred in one of America's youngest nuclear power plants. The Three Mile Island Unit Two Nuclear Power Generator, owned by the Metropolitan Edison Company and located on an island in the Susquehanna River, approximately 11 miles south of Harrisburg, Pennsylvania, began operation on December 28, 1978. According to a letter sent by consumer advocate Ralph Nader to President Jimmy Carter, the plant was rushed into service in order to obtain a tax break of $40 million, despite the fact that during its initial break-in period the reactor was experiencing mechanical failures and other problems.

Nader was, and still is, opposed to public nuclear power, and that undoubtedly skewed his evaluation of the birth of the plant. Still, there must have been a basic core of truth in his accusations, for just slightly more than three months after it began operating, the Three Mile Island generator showed its flaws in a dramatic and terrible way by leaking radiation over an enormous area and by narrowly missing that most dreaded of nuclear accidents, a reactor meltdown.

At 3:58 A.M. on Wednesday, March 28, 1979, the first of a chain of mishaps occurred at the plant. A pump that provided steam to the electric turbines broke down. This in turn shut down another pump that circulated water through the reactor, which in turn raised the temperature of the reactor, which opened a relief valve designed to bleed off the increased pressure brought about by the rise in temperature. Within the reactor some of the cladding, or sheaths around the fuel rods, melted. The uranium pellets in them apparently did not.

By this time alarms were sounding in the control room, and operators, unschooled in this sort of unprecedented emergency, began to make wrong decisions, while the system itself malfunctioned. The relief valve failed to close, and consequently pressure in the reactor dropped low enough to allow the water to vaporize.

Then, a major error was committed. An operator opened a valve allowing water from this system to enter a waste tank, where it created enough pressure to rupture the plumbing. Sixty thousand gallons of radioactive water flooded the reactor to a depth of eight feet.

A second human error followed rapidly. The emergency core cooling system kicked in, but an operator shut it off.

Now, a pump flooded an auxiliary building with contaminated water, causing a release of steam. Within moments radioactive steam poured up the vent stack and into the atmosphere.

Inexplicably, it would take operators almost three hours to act on these events. It would be 7:00 A.M. before state authorities would be informed and another hour before the authorities would declare a "general emergency."

Even this general emergency was, as in the Windscale disaster (see p. 304), minimized, presumably to prevent panic. Margaret Reilly, of Pennsylvania's Department of Radiation Protection, in one of the most monumental understatements of all time, likened the escape of radiation to "a gnat's eyelash."

However, authorities were aware that a minimum of a million millirems per hour of radiation was present inside the reactor building at Three Mile Island, a lethal dose for anyone directly exposed to it. Monitors 1,000 feet from the vent stacks, where the radioactive steam was spewing into the air, showed levels of 365 millirems of beta and gamma rays per hour.

Three months later Albert Gibson, a Radiation Support section chief who would coauthor the Nuclear Regulatory Commission's final report on Three Mile Island emissions, testified, "All radiation monitors in the vent stack, where as much as 80 percent of the radiation escaped, went off the scale the morning of the accident. The trouble with those monitors is they were never contemplated for use in monitoring accidents like Three Mile Island." Besides the beta and gamma emissions, there were bursts of strontium and iodine-131, which characteristically settles on grass, is eaten by cows and thus enters the milk supply.

On Thursday holding tanks filled to overflowing with radioactive water were opened, pouring 400,000 gallons of water containing xenon-133 and xenon-135 into the Susquehanna River, while federal nuclear officials assured the public that the gases posed "little hazard to persons living downstream of the . . . plant." By the end of Thursday, March 29, detectable levels of increased radiation were measured over a four-county area, and officials at the plant admitted that, contrary to their early assessment, 180 to 300 of the 36,000 fuel rods in the reactor had melted.

At 9:00 A.M. on Friday, March 30, the Pennsylvania Emergency Management Agency reported that there had been a new, "uncontrolled release" of radiation—a puff of contaminated steam. Because of intense radioactivity within the reactor, the temperature had risen high enough in places to break up the water molecules into hydrogen and oxygen, forming a large bubble of hydrogen, which, if large enough, could prevent further reduction, therefore inhibiting the ability of the circulating water to cool down the fuel rods. Thus, a meltdown was possible and becoming more probable.

Now, Governor Richard Thornburgh issued a directive that advised pregnant women and small children to evacuate and stay at least five miles away from the Three Mile Island facility. In 23 schools children were pulled from classes, crammed into cafeterias and ordered not to open windows. From these gathering points, they were transported in sealed school buses to other schools 10 to 15 miles away. ("It was sure hot in that bus with all those windows up," said nine-year-old Kim Hardy from Etters, a community within the five-mile radius.) Parents were then informed of their children's whereabouts.

Fright, but no panic, abounded. An air-raid siren shrieked in Harrisburg shortly before noon, setting off a midday traffic jam of jittery state employees. The alarm was explained away by the governor's office as either a malfunction or the overzealous response of a civil defense official to Governor Thornburgh's directive.

Meanwhile, towns near the plant, such as Goldboro, had been emptying out ever since the beginning of the accident. A small leak of people from the villages had turned into a torrent by Friday, the 30th. Gasoline stations were jammed; telephone switchboards were so over-loaded that callers received nothing but busy signals. Fifteen mass-care centers were established in counties surrounding the Middletown area.

Back at the plant officials were tensely monitoring the bubble of hydrogen, trying to decide whether to allow it to sink to the bottom of the containment vessel by drawing off water—and thus risking a further increase of temperature and the consequent possibility of a meltdown—or starting up the reactor again and trying to saturate the bubble with steam, which would break it up.

A third, venting method was tried, and on Saturday, the 31st, the bubble was reduced enough so that a combination of safety rods and water could hasten the "cold shutdown" of the reactor. The danger of a meltdown passed.

By Monday, April 9, the Nuclear Regulatory Commission (NRC) declared the Three Mile Island crisis at an end and said it was safe for pregnant women and young children to return to their homes, despite the fact that the reactor was still leaking small quantities of radiation into the air and that readings of radiation emissions were still above average. Schools were reopened, government offices returned to business as usual and the civil defense forces were taken off full alert. It would be months before the reactor would be entirely shut down, and further instrument failure would lengthen that process, too.

But the book on Three Mile Island was not closed, by any means. First, there was the business of assigning responsibility for the accident. On May 11, 1979, the NRC issued a report blaming the operators for "inadvertently turn[ing] a minor accident into a major one because they could not tell what was really happening inside the reactor." This juxtaposition of human, instrument and design error ran through the NRC report like a fugue.

The accident began when someone forgot to reopen a set of valves, and operators failed to notice the mistake. Operators apparently paid attention mainly to the pressurizer water indicator, which was misleading them, and failed to watch other instruments that should have informed them that something was wrong. Operators apparently failed to follow the procedure for dealing with a stuck-open pressure relief valve, and so forth.

But in its conclusions, the NRC removed some of the onus from the operators. "Human factors engineering has not been sufficiently emphasized in the design and layout of the control rooms," it admitted in its summary.

In February 1984 Metropolitan Edison Company pleaded guilty to charges that it knowingly used "inaccurate and meaningless" test methods at the Unit Two reactor prior to the accident. The company then disciplined 17 employees—among them a former vice president, shift supervisors, control room operators, shift foremen and managers—for manipulating records of the tests. The penalties ranged from letters of reprimand to the loss of two week's pay.

Another aspect of the continuing story of Three Mile Island was the effect on the surrounding population. As in practically every nuclear accident, there was no evacuation plan in place when the disaster occurred. To compound this, reports released during and after the accident were either deliberately or inadvertently misinformative. Despite the admirable motivation of preventing needless panic, the "gnat's eyelash" analogy seems irresponsible in light of the later findings of other scientists and investigators.

Although the NRC continued to maintain that there had been and was still no significant intensification of radiation as a result of the Three Mile Island accident, Dr. Ernest Sternglass, a University of Pittsburgh Medical School professor of radiology, in a paper presented at the Fifth World Congress of Engineers and Architects at Tel Aviv, Israel, in 1980, stated that figures from Harrisburg and Holy Spirit hospitals showed that infant deaths in the vicinity of Three Mile Island had *doubled,* from six during February through April 1979 to 12 in May through July.

Furthermore, Dr. Sternglass observed, data from the U.S. Bureau of Vital Statistics showed that there were "242 [infant] deaths above the normally expected number in Pennsylvania and a total of 430 in the entire northeastern area of the United States." He based his linkage on the large amounts of iodine-131 released into the atmosphere and the peaking of infant mortality within a matter of months after the release of the I-131.

Dr. Sternglass went on to charge that, as NRC investigator Joseph Hendrie had confirmed on March 30, 1979, individual areas where the steam plume touched the ground were "husky" and in the range of 120 millirems per hour or more, which was easily enough to cause severe damage to fetuses in the womb. In addition, Dr. Sternglass noted that doses of I-131 had impacted people in the path of the plume in Syracuse, Rochester and Albany, New York, and each city had suffered rising infant deaths.

"My daughter got real sick," Becky Mease of Middletown told an NRC panel. "She had diarrhea for three days straight and headaches and she became anemic. I didn't know what to do. My little girl is still getting colds and sinus problems. Now if that's not because of that power plant, you tell me what it is."

Deaths in the Middletown area from thyroid cancer (the thyroid gland is particularly affected by iodine-131) are still monitored by families and organizations. No absolute link has been established, but those who were affected feel that the cause was the accident and the radiation it released into the Pennsylvania countryside. Some cancer victims have sued the Metropolitan Edison Company.

And finally, the third reason for the continued interest in Three Mile Island is the ongoing impact it has had on the nuclear energy industry in America. Prior to Three Mile Island, antinuclear activists were relatively quiescent. On May 6, 1979, after the accident, a crowd of 65,000 demonstrators arrived at the Capitol in Washington, D.C., to demand the cessation of building and the closing of nuclear power plants in the United States.

The Three Mile Island disaster opened the door to an escalation of protest activity that became rocket powered after the Chernobyl disaster (see below). It was responsible for the abandonment of the Shoreham Nuclear Energy plant on Long Island and the virtual halt in construction of nuclear power plants nationwide in the 1980s.

USSR
KASLI
1957

Military and Soviet secrecy has muffled the details of an explosion in a nuclear waste dump near Kasli, in the Ural Mountains of the USSR, sometime in 1957, but a chemical or steam explosion has been theorized as its cause. Hundreds were said to have died; tens of thousands were afflicted.

What was perhaps the most cataclysmic nuclear disaster in the world remains, to this day, unofficially recorded and officially nonexistent, according to the former Soviet government. Even the U.S. Central Intelligence Agency, which apparently knew of the disaster shortly after it occurred, suppressed any mention of it and in fact denied its existence for 20 years, until the Freedom of Information Act forced it to open its files on the subject.

So it is only possible to piece together the fragments of this cataclysm, largely through the diligent research of Dr. Zhores Medvedev, a Soviet émigré scientist who, in 1976, first brought the news of the explosion to the West.

In an article titled "Two Decades of Dissidence" in the British journal *New Scientist,* Dr. Medvedev told of "an enormous explosion, like a violent volcano," in a radioactive-waste dump in the Ural Mountains near the town of Kasli, or Kyshtym, as Dr. Medvedev called it. "The nuclear reactions had led to an over-heating in the underground burial grounds," he continued. "The explosion poured radioactive dust and materials high up into the sky . . . Tens of thousands of people were affected, hundreds dying, though the real figures have never been made public. The large [50-kilometer-square] area, where the accident happened, is still considered dangerous and is closed to the public."

The official response to the article was swift and negative. *Tass* denied it. Sir John Hill, the chairman of the United Kingdom Atomic Energy Authority, wrote a letter to the *Times* of London calling the story "rubbish."

But one month later, Lev Tumerman, another Russian émigré, wrote a letter to the *Jerusalem Post* relating a ride through the same countryside in the Urals in 1960. "On both sides of the road as far as one could see the land was 'dead,'" wrote Tumerman, "no villages, no towns, only the chimneys of destroyed houses, no cultivated fields or pastures, no herds, no people . . . nothing."

Signs warned him to proceed without stopping for the next 30 kilometers and to drive through at maximum speed. "An enormous area, some hundreds of square kilometers, had been laid waste," he concluded.

Finally, in 1979, Dr. Medvedev published *Nuclear Disaster in the Urals.* In it he quoted Soviet scientists who made post-catastrophe studies of plant and animal life and subsequent weather patterns in the area. All confirmed the explosion and its consequent radiation contamination.

Within weeks a special report was released by the Oak Ridge National Laboratory that confirmed the Soviet scientists' reports and stated that a system of 14 lakes had been contaminated by the Kasli blast and that 30 small towns that had been on Soviet maps of the southern Ural region before 1957 were no longer there.

Investigation by reporters of newly released CIA files revealed anecdotal confirmation from survivors who were in the area. One described a huge explosion that shook the ground and spewed a huge cloud of red dust into the air, which settled on the leaves of trees. "Very quickly," the eyewitness said, "all the leaves curled up and fell off the trees," as did vegetables that were covered with the red radioactive dust.

"All stores in Kamensk-Uralskiy which sold milk, meat and other foodstuffs were closed as a precaution against radiation exposure," said another, "and new supplies were brought in two days later by train and truck. The food was sold directly from the vehicles and the resulting queues were reminiscent of those during the worst shortages during World War II."

That was only the beginning. As the effects of the widespread radiation contamination began to be felt, another witness reported, "The people in Kamensk-Uralskiy grew hysterical with fear, and with incidence of unknown 'mysterious' diseases breaking out." Homes were burned to the ground to prevent their owners from reentering them, and these displaced people "were allowed to take with them only the clothes in which they were dressed."

"One of the current topics of conversation at the time," said another survivor, "was whether eating fish or eating crabs from the radioactive rivers of the area was more dangerous . . . Hundreds of people perished and the area became and will remain radioactive for years."

American officials tended to soft-pedal the implications for worldwide storage of nuclear waste. Despite more realistic warnings against risks of nuclear dumping, Richard Corrigan, a spokesperson for the Ford administration in Washington, wrote in the *National Journal* of August 1979, "They [the Russians] don't know what they're doing and we do." It was cheerful news, but disarmingly simplistic when compared with the facts of Windscale (see p. 304) and two catastrophes that were yet to come: Three Mile Island (see p. 312) and Chernobyl (see below).

USSR
PRIPYAT
April 26, 1986

The worst recorded nuclear accident in history, that of the Chernobyl nuclear power plant in Pripyat, USSR, on April

26, 1986, was the result of human error in conducting a test of the system. Thirty-one died in the explosion and fire; more than 100,000 were evacuated from the vicinity of the plant. More than 5 million were exposed to radioactive fallout.

"An accident has occurred at the Chernobyl nuclear power plant as one of the reactors was damaged. Measures are being taken to eliminate the consequences of the accident. Aid is being given to those affected. A government commission has been set up."

Two days after the stupendous disaster at the Chernobyl nuclear power plant, located 70 miles north of Kiev, the capital of the Ukraine, the Soviet government released this terse, businesslike and uninformative announcement.

Some of the facts and some of the effects of this, the worst nuclear disaster in history, had already drifted out of the Soviet Union on winds bringing radioactive waste to Scandinavia, then eastern Europe, then western Europe and finally to the rest of the world, including the United States.

The situation at Chernobyl was frightening. With four 1,000-megawatt reactors in operation, it was one of the largest and oldest of the Soviet Union's 15 or so civilian nuclear stations. There, a cascade of awesome human errors had set in motion, as surely as uranium brought about a chain reaction, a series of events the likes of which the world had yet to experience.

At 1:00 A.M. on Friday, April 25, operators of the number-four reactor, which had gone on line in 1983, began to reduce its power in preparation for an operations test. The test was designed to measure the amount of residual energy produced by the turbine and generator after the nuclear reactor had been shut down. The conclusion of the test would tell these engineers how long the turbine and generator would be able to run if, in some sort of emergency, the reactor were shut down.

It was a routine test. The valves on the main steam line between the reactor and the turbine were to be closed, thus stopping power to the turbine, and the residual energy would then be measured until the turbine stopped.

While this was happening, steam would still be produced by the reactor, which would be slowly reduced to a fraction of its potential power. That steam could either be released into the atmosphere through bypass valves or condensed back to water in a cooling unit. If the operators decided to rerun the test, they could open the valves to the turbine and close the bypass valves.

The difficulty with the process was that if the reactor continued to operate, certain "perturbations," as nuclear experts euphemistically call them, could take place in the reactor, which could in turn increase the pressure and cause the unit to be automatically shut down. Or they could reduce the pressure, causing the automatic flooding of the reactor with emergency cooling water. In other words, no one could tell just what would happen in the reactor under these circumstances, but the operators at Chernobyl, determined to carry through their test without a shutdown of the reactor, *shut off all of the emergency safety systems.*

As astounding as that seems, this is exactly what happened at 2:00 P.M. on Friday, the 25th, as the reactor was reduced to 7 percent capacity. The reactor's emergency cooling system was shut off. Then the power regulating system and the automatic shutdown system were disconnected. It was a little like a fire department responding to a burning fire and then dismantling the fire alarms and fire escapes and going home.

What the operators did was in violation of regulations, but they did it anyway—as countless other operators in other industrial and nuclear accidents had and would—and continued with their routine testing through the afternoon and evening. And all the while, the "perturbations" went on in the reactor.

Some time during this process, a reactor operator received a computer printout that indicated the reactor was in extremely serious danger of overheating unless it was shut down immediately. He ignored it.

Control rods were withdrawn, lowering the power in the reactor below the minimum required by the unit's operating manual. Xenon gases began to build up as the temperature rose in the reactor.

At 1:22 A.M. on Saturday, April 26, these same operators noticed that the power level had risen to the point at which, had the emergency system been engaged, it would have shut down. The operators noted it and kept on testing. If they had stopped at that moment, if they had heeded the warnings the instruments were clearly giving them and reengaged the safety system, the disaster would not have occurred. But they blundered on, ignoring the obvious.

Exactly one minute and 40 seconds later, the reactor blew. There was a loud bang as the control rods began to fall into place. At that instant the operators knew exactly what was about to happen. They desperately tried to drop the rest of the control rods to stop the runaway chain reactions that were taking place in the reactor, the splitting by radiation of superhot water and the reactions caused by the superheating of its graphite shell.

But it was too late. The control rods drop by gravity, and that takes time, and the operators of the reactor had used up all the time there was. Twenty seconds later the fuel atomized. Three explosions tore through the reactor, blowing off its top, sending its 1,000-ton steel cover plate rocketing into the air and ripping off the tops of all 1,661 channels, which were attached to the cover plate and contained the nuclear fuel. The channels became like "1,000 howitzers pointed at the sky," according to Dr. Herbert J. C. Kouts, the chairman of the Department of Nuclear Energy at Brookhaven National Laboratory on Long Island. Powered by these nuclear howitzers, a huge fireball shot up into the sky. The graphite caught fire and burned fiercely and wildly. The reactor was completely out of control and beginning to melt down.

Flames continued to shoot over 1,000 feet into the air. This would continue for two days and nights. The operators within the building were doomed. Emergency alarms went off all through the complex, in which 4,500 workers were employed.

Miles away a startled populace witnessed a gigantic fireworks display of hot radioactive material being flung into the night sky and onto the winds that would eventually carry this material far enough to contaminate a huge nearby area and, eventually, to a much lesser and varying degree, much of the rest of the world.

The reactor continued to burn while emergency teams hauled off the dead and the radiated. Others, their boots sinking in molten bitumen, uselessly attempted to battle the blaze. But, as in other nuclear accidents around the world, evacuation of the populace was delayed while scientists debated the seriousness of the situation.

Finally, at 1:50 P.M. on Sunday, April 27, fully 36 hours after the accident, the local radio station at Pripyat announced that a full-scale evacuation was to begin immediately. The city of 40,000 was to be totally abandoned, and 1,100 buses, some of them commandeered from Kiev, undertook the task. To prevent panic, rallying points were not used. The city was emptied within two hours and 20 minutes.

The countryside around Pripyat, a region of wooded steppes, small villages and moderately productive farms, was less thickly populated. Between Pripyat and Chernobyl lay the Kiev reservoir, fed by the Pripyat River. And at this point radioactive matter was falling like lethal rain onto the thinly settled countryside and into this reservoir that supplied water to the 2,500,000 people of Kiev, Russia's third-largest city.

Meanwhile, at the plant workers were shutting down the other three reactors. The fire continued to burn unchecked. Twenty-five percent of the radiation leaked in the accident was released in the first 24 hours of the fire. The fire would continue for eight days.

On Monday morning, April 28, Swedish monitoring stations detected unusually high levels of xenon and krypton and concluded that, considering the prevailing winds, an atomic accident had occurred in the Soviet Union. Sweden demanded that the Soviets comply with international agreements to notify other nations immediately after a nuclear accident that might threaten those countries with radiation. It was not until 9:00 P.M. that night that the Soviets released the terse statement quoted at the beginning of this entry, a masterpiece of noninformation.

But the truth began to seep out. On Tuesday Soviet diplomats in Europe and Scandinavia approached private nuclear agencies, asking advice on fighting graphite fires. United Press International, frustrated by the silence from official sources, quoted a Kiev woman who communicated with them by telephone. "Eighty people died immediately and some 2,000 people died on the way to hospitals," she told UPI. "The whole October Hospital in Kiev is packed with people who suffer from radiation sickness."

A Dutch radio operator reported a message received from a Soviet ham broadcaster. "We got to know that not one, but two reactors are melted down, destroyed and burning. Many, many hundreds are dead and wounded by radiation, but maybe many, many more," he said, ending with a plea. "Please tell the world to help us," he concluded. This was clearly at odds with the official version of events, which placed the dead at two and the injured at 197.

Now, the heavier products of radiation, the ones that the atmosphere could not dissipate easily and were lethal to human beings, were beginning to fall on Europe. Among a score of elements detected in the fallout were cesium-134 and iodine-131, both easily assimilated by the body and both thought to cause cancer.

By Wednesday, April 30, European countries began to take steps to preserve their own people. In Austria mothers in the province of Carinthia were being advised to keep infants and small children indoors. The Polish government banned the sale of milk from grass-fed cows and issued iodine tablets to infants, children and pregnant mothers in order to protect the thyroid gland against poisoning from iodine-131. In Sweden officials warned people not to drink water from casks that collected rainwater for summer cottages and banned the import of fresh meat, fish and vegetables from the Soviet bloc countries. Evacuation plans were activated for citizens who were traveling or working in the area within 200 miles of Chernobyl. A group of American students studying in Kiev boarded planes for Moscow, then London, then the United States.

By Thursday, May 1, the Soviet bulletins noted that 18 people were in critical condition and that the fire was cooling down. In an effort to control it still further, civil defense forces began to drop bags of wet sand from helicopters hovering over the gaping hole in the top of the reactor. The radioactivity levels within the building were still too high to allow human beings, even in protective gear, to enter.

International help came swiftly. Dr. Robert Gale, the head of the International Bone Marrow Transplant Registery, left Los Angeles for Kiev on May 1. Two days later his associate, Dr. Richard Champlin, and Dr. Paul Terasaki, a tissue-typing expert, joined him. They would have much work to do with the hundreds hospitalized from the accident.

Wind patterns were affecting the radiation levels reported in various European countries. In Sweden it fluctuated between normal and five times the normal amount. Traces of iodine-131 were detected in rainwater samples in the Pacific Northwest region of the United States, but they were not deemed dangerous.

By Monday, May 5, the Soviet government announced that dikes were being built along the Pripyat River to prevent potential contamination and that the leakage of radiation from the plant had virtually stopped. This was not the case, as later studies would indicate. In a report released the following September, a study prepared by the Lawrence Livermore National Laboratory in California asserted: "The nuclear disaster at Chernobyl emitted as much long-term radiation into the world's air, topsoil and water as all the nuclear tests and bombs ever exploded." Cesium, a product associated with health effects such as cancer and genetic disease, does not break down into a harmless form for more than 100 years, and it was sent into the atmosphere in quantities, the study estimated, that were as much as 50 percent more than the total of hundreds of atmospheric tests and the two nuclear bombs dropped on Japan at the end of World War II.

On May 9 the Soviets began the monumental task of encasing the still smoldering wreck of a reactor in concrete. It involved tunneling under the reactor, in order to prevent a "China Syndrome" style of meltdown, which would immediately contaminate the groundwater near the reactor. The massive job was begun by dropping thousands of tons of sand, boron, clay, dolomite and lead from helicopters into the graphite core. Then the huge sarcophagus of concrete was poured and erected.

As May gave way to June, Soviet authorities attempted to protect citizens from the continuing effects of exposure to radiation. On May 15, 25,000 students in the Kiev area received an early vacation when all of the elementary schools and kindergartens were closed early for the summer.

Officials told residents of Kiev to keep their windows closed, mop floors frequently and wash their hands and hair often to reduce the chance of radiation contamination. And for the first time, these authorities acknowledged the dissemination of radiation over the rest of Europe.

The Russian children would be transported by the state to "Pioneer" camps scattered from the Moscow suburbs to the Crimea. More than 60,000 children, in fact, joined the first evacuees from Pripyat, who, like 12-year-old Olya Ryazanova, remembered a fire-blackened nuclear power plant, "a sort of mist, a misty cloud around it," and booted workers washing down the road in front of her home.

On May 15, the day the schools closed, the radioactive cloud had, after first blowing north to Scandinavia and Byelorussia, reversed itself and was hovering over Kiev. Crowds had formed at railroad stations and airports, most of them women and children, and the government had added extra trains and flights out of the city.

As more accurate information began to filter out of the Soviet Union, the scope of the disaster continued to grow. Hans Blix, the head of the International Atomic Energy Agency, confirmed that at least "204 persons, including nuclear power station personnel and firefighters, were affected by radiation from the first degree to the fourth degree." The government newspaper *Izvestia* revealed that more than 94,000 people had been evacuated. Eventually, the official number of dead would be set at 31.

It was learned that a full month before the disaster, a Ukrainian journal had reported management failures and labor dissatisfaction at Chernobyl. Because coal was becoming scarce in the Soviet Union, construction at the plant was speeded up in 1984, and it was suggested that this haste—a fifth nuclear reactor was already under construction at the time of the accident—was partially responsible for the tragedy. But ultimately, the blame was focused on human error, and in June *Pravda* announced that the director and chief engineer of the plant had been dismissed for mishandling the disaster and that other top officials were accused of misconduct ranging from negligence to desertion.

As with any nuclear catastrophe, the story of Chernobyl had no quick ending. In 1991, five and a half years after the accident, Pripyat became a ghost town, a place from which everyone had departed forever, and the sarcophagus that encased the 171 tons of coagulated and resolidified uranium fuel became outdated. Its contents will remain radioactive and dangerous for at least 150 years, and the 20-story-high cube of concrete and steel that was hastily executed to bury that radiation has a life span of only 25 years.

The exploded generator's tortured and twisted mass of nuclear fuel was viewed through special periscopes. According to a reporter for the *New York Times,* it resembled a "nightmarish cave, of great uranium magma oozings solidified into what workers already are nicknaming 'elephant's feet' of deadly radioactive permanence."

In addition to the rebuilding of the gigantic sarcophagus, scientists, workers and officials were faced with the problem of the 800 burial pits of other contaminated material, including trees, topsoil and even entire houses. Some of them were close to Pripyat's water supply, and if the city was ever to live again, these pits would also have to be encapsulated with clay, concrete and steel, or perhaps decontaminated. The independent Soviet republics demanded that three nuclear power generators still operating at the Chernobyl plant be closed by 1995.

Power plant workers were resettled in a town called Slavutich, located 35 miles east of the plant, where even the tree bark has been scrubbed clean of radiation. Still, hot spots have been discovered here and there in the new village, and workers must change their clothes three times en route, twice a day, as they traverse the 18-mile "hotzone" around the plant.

Near Chernobyl 500 tons of dangerously irradiated beef have been stored in 40 refrigerated boxcars ever since the disaster. No one knows what to do with it.

In the now independent country of Ukraine, the long-term effects of exposure to radiation have continued to manifest themselves. Thyroid cancer rates have increased tenfold in Ukraine and as much as 84 times in parts of Belarus.

"As of today," Dr. Valery P. Tereshchenko, of the Institute for Endocrinology and Metabolism in Kiev said in June of 2000, "we have operated on about 1,500 children in Ukraine alone."

"To tell you the truth," Dr. Mykola D. Tronko, director of Ukraine's thyroid cancer institute told a *New York Times* reporter, "even 14 years after the accident, we have more questions than answers."

Finally, under pressure from the United States and Europe to shut down the last working reactor at Chernobyl, Ukraine's president, Leonid Kuchma, agreed to close the plant by December 15, 2000, if the world's seven wealthiest industrial nations would help to finance the major decommissioning and aid in the replacement of Chernobyl's electrical output. The nations agreed, and the long process began in early 2000. But it would not be an easy process. "It's not like closing a big door and putting a big lock in place and then forgetting about it," added Mr. Kuchma. "We have to continue servicing this plant."

He knew the problems from experience. Unit number 2 in Chernobyl was shut down in 1991 after a fire raged through its turbine hall, causing extensive damage to just about everything but the reactor. In 1994 unit number 1 was shuttered after the Group of Seven industrialized nations pledged financing. But now the largest task of all, closing unit number 4 and shuttering Chernobyl's last functioning nuclear dynamo in the ruined unit number 3—and thus the entire plant—loomed large.

First, there was the problem of the deteriorating sarcophagus, which was built hastily and badly. There is still, as of this writing, an enormous amount of radiation within it, and the cost of keeping that radiation locked firmly inside will be substantial. Not only will a corps of engineers be needed to oversee it, but, most important, a new sarcophagus at an estimated cost of $750 million must be built. The original one rests on potentially unstable foundations and has been invaded by rainwater, which has in turn corroded the maze of pipes and steel girders within. It has been in danger of collapse for years.

It is generally also agreed that a heating plant and new electricity supplies to run the support facilities will be needed. The money for this has been pledged by the seven wealthiest nations, but it has, as of this date, not been entirely collected.

In 2001 there were still 6,000 workers and their dependents working in the plant, a far cry from the 28,000 that inhabited the village of Slavutch when Chernobyl was functioning fully, but a sizable number nevertheless. And still, the radiation hovers in the ground. Women who till small gardens in Slavutch wear surgical masks because of the toxins given off every time a shovelful of dirt is overturned. The once-rich farmland is dotted with mounds where radioactive trucks and bulldozers have been buried. The manmade canals that once carried cooling water to and from the reactor swarm with carp and catfish that no one dares eat.

Even before the final shutdown, the so-called red zone—the depopulated area in an 18-mile radius around the plant where 135,000 people once lived—was metamorphosing from a graveyard to a memorial.

Finally, at 1:16 P.M. on December 15, 2000, Ukraine president Leonid Kuchma gave an order over a video hookup to shut down Chernobyl forever. Plant chief Oleksandir Yelchishchev turned a switch that sent containment rods sliding into the core of reactor number 3. Within seconds a dial showed the atomic reaction in the core plummeting to zero. But even as it happened, President Kuchma issued a statement of undeniable reality: "This menacing page of the book of modern history cannot be considered closed," he said.

Nor will it be. Thirty-one workers died in the immediate meltdown at Chernobyl, but in the 14 years following this, 4,000 cleanup workers died of radiation poisoning. An estimated 70,000 have been disabled by radiation in Ukraine alone. Approximately 3.4 million of Ukraine's 50 million people are considered affected by Chernobyl.

Meanwhile, the 66 tons of melted nuclear fuel and 37 tons of radioactive dust remain, for now, in the old sarcophagus. The project of rebuilding it will take until 2007. And even then, the problem of getting rid of the fuel inside will not have been solved.

Chernobyl was to be a triumph of technology. It turned out to be one of the world's greatest challenges to that technology. Progress of this sort comes with a price, the world found, as did the American developers of the first atomic bomb.

"We [were] the great guinea pigs of modern times," said Yevgeny Konoplya, the director of the Radiobiology Institute of the Belarus Academy of Sciences in 1994. "We are getting to prove for the world what radiation can do to humans. We have suffered from the policies of a country that no longer even exists. We have suffered from lies. And we have suffered from other people's belief in technology. We once had a beautiful country. What we have now is pain."

RAILWAY DISASTERS

THE WORST RECORDED RAILWAY DISASTERS

* Detailed in text

Algeria
- Algiers (1982)

Argentina
- * Buenos Aires (1970)

Bangladesh
- * Jessore (1972)
- * Maizdi Khan (1989)

Brazil
- * Aracaju (1946)
- * Mangueira (1958)
- Nova Iguaçu (1951)
- * Pavuna River (1952)
- * Tangua (1950)

Burma
- Toungoo (1965)

Canada
- Ontario
  - * Hamilton (1857)
- Quebec
  - * St. Hilaire (1864)

Chile
- Mapocho River (1899)

China
- Guangzhou (1947)
- Luoyang (1935)
- Yancheng (1938)

Costa Rica
- Virilia River (1926)

Czechoslovakia
- * Pardubice (1960)

France
- * Lagny (1933)
- Les Couronnes (1903)
- * Modane (1917)
- St. Nazaire (1871)
- * Versailles (1842)
- * Vierzy (1972)

Germany
- * Celle (1998)
- * Laangenweddingen (1967)
- * Magdeburg (1939)

Great Britain
- England
  - * Harrow-Wealdstone (1952)
  - * London
    - * (1988)
    - * (1999)
- Scotland
  - * Dundee (1879)
  - * Gretna Green (1915)

India
- Bhosawal (1867)
- * Firozabad (1995)
- * Gaisal (1999)
- * Hyderabad (1954)
- Jasidih (1950)
- Karna (1998)
- Lahore (1891)
- Madras (1902)
- * Mahbubnagar (1956)
- * Mansi (1981)
- Marudaiyar (1956)
- Moradabad (1908)
- * Patna (1937)

Indonesia
- Java
  - * Cirebon (2001)
  - * East Priangan (1959)

Ireland
- * Armagh (1889)

Italy
- * Salerno (1944)
- Voghera (1962)

Jamaica
- * Kendal (1957)

Japan
- Osaka (1940)
- Sakuragicho (1951)
- * Tokyo (1962)
- * Yokohama (1963)

Kenya
- * Darajan (1993)

Mexico
- Behesa (1916)
- * Cazadero (1945)
- * Cuautla (1881)
- Encarnación (1907)
- Guadalajara
  - * (1915)
  - * (1955)
- Los Mochis (1989)
- Mexico City
  - (1895)
  - (1919)
- * Saltillo (1972)
- * Tepic (1982)

Netherlands
- Woerden (1962)

New Zealand
- * Waiouri (1953)

Pakistan
- * Gambar (1957)
- * Sangi (1989)
- Sind Desert (1954)

Poland
- * Nowy Dwor (1949)
- Pzepin (1952)

Portugal
- * Custoias (1964)

Romania
- Bucharest (1918)
- Costesi (1913)

Russia
- Odessa (1876)
- Petrograd (1920)
- Tcherny (1882)

South Africa
- Orlando (1949)

Spain
- Aguadilla (1944)
- * Lebrija (1972)
- San Arsenslo (1903)

Switzerland
- * Basel (1891)

Turkey
- Philippopolis (1879)

United States
- Alabama
  - * Mobile (1993)
- Colorado
  - * Eden (1904)
- Connecticut
  - * South Norwalk (1853)
- Illinois
  - * Bourbonnais (1999)
  - * Chatsworth (1887)
- Massachusetts
  - * Revere (1871)

New Jersey
* Atlantic City (1896)
* Hackettstown (1925)
* Hightstown (1833)
* Woodbridge (1951)
New York
* Angola (1867)
* Brooklyn (1918)
* Queens (1950)
Ohio
* Ashtabula (1876)
Pennsylvania
* Camp Hill (1856)
* Laurel Run (1903)
* Mud Run (1888)
* Shohola (1864)
Tennessee
* Hodges (1904)
* Nashville (1918)
Utah
* Ogden (1944)
Washington
Wellington (1910)

Yugoslavia
* Zagreb (1974)

CHRONOLOGY

* Detailed in text

1833
November 11
Hightstown, New Jersey

1842
May 8
* Versailles, France

1853
May 6
* South Norwalk, Connecticut

1856
July 17
* Camp Hill, Pennsylvania

1857
March 17
Hamilton, Ontario

1864
June 29
* St. Hilaire, Quebec
July 15
* Shohola, Pennsylvania

1867
June 26
Bhosawal, India
December 18
* Angola, New York

1871
February 25
St. Nazaire, France
August 26
* Revere, Massachusetts

1876
January 8
Odessa, Russia
December 29
* Ashtabula, Ohio

1879
January 11
Philippopolis, Turkey
December 28
* Dundee, Scotland

1881
June 24
* Cuautla, Mexico

1882
July 13
Tcherny, Russia

1887
August 10
* Chatsworth, Illinois

1888
October 10
* Mud Run, Pennsylvania

1889
June 12
* Armagh, Ireland

1891
June 14
* Bale, Switzerland
December 8
Lahore, India

1895
February 28
Mexico City, Mexico

1896
July 30
* Atlantic City, New Jersey

1899
August 24
Mapocho River, Chile

1902
September 11
Madras, India

1903
June 27
San Arsenslo, Spain
August 10
Les Couronnes, France
December 23
* Laurel Run, Pennsylvania

1904
August 7
* Eden, Colorado
September 24
* Hodges, Tennessee

1907
September 19
Encarnación, Mexico

1908
May 8
Moradabad, India

1910
March 1
Wellington, Washington

1913
December 6
Costesi, Romania

1915
January 18
* Guadalajara, Mexico
May 22
* Gretna Green, Scotland

1916
November 19
Behesa, Mexico

1917
December 12
* Modane, France

1918
July 9
* Nashville, Tennessee
October 15
Bucharest, Romania
November 2
* Brooklyn, New York

1919
October 5
Mexico City, Mexico

1920
December 22
Petrograd, Russia

1925
June 16
* Hackettstown, New Jersey

1926
March 15
Virilia River, Costa Rica

1933
December 23
* Lagny, France

1935
September 24
Luoyang, China

1937
July 16
* Patna, India

1938
April 5
Yancheng, China

1939
December 22
* Magdeburg, Germany

1940
January 28
Osaka, Japan

1944
March 2
* Salerno, Italy
November 7
Aguadilla, Spain
December 31
* Ogden, Utah

1945
February 1
* Cazadero, Mexico

1946
March 20
* Aracaju, Brazil

1947
July 10
Guangzhou, China

1949
April 28
Orlando, South Africa
October 22
* Nowy Dwor, Poland

1950
April 6
* Tangua, Brazil
May 7
Jasidih, India
November 22
* Queens, New York

1951
February 6
* Woodbridge, New Jersey
April 4
Sakuragicho, Japan
June 8
Nova Iguacu, Brazil

1952
March 4
* Pavuna River, Brazil
July 9
Pzepin, Poland
October 8
* Harrow-Wealdstone, England

1953
December 24
* Waiouri, New Zealand

1954
January 21
Sind Desert, Pakistan
September 24
* Hyderabad, India

1955
April 3
* Guadalajara, Mexico

1956
September 2
* Mahbubnagar, India
November 23
Marudaiyar, India

1957
September 1
* Kendal, Jamaica
September 29
* Gambar, Pakistan

1958
May 8
* Mangueria, Brazil

1959
May 28
* East Priangan, Java

1960
November 14
* Pardubice, Czechoslovakia

1962
May 3
* Tokyo, Japan
May 31
Voghera, Italy

1963
November 9
* Yokohama, Japan

1964
July 26
* Custoias, Portugal

1965
December 9
Toungoo, Burma

1967
June 6
* Laangenweddingen, East Germany

1970
February 1
* Buenos Aires, Argentina

1972
June 4
* Jessore, Bangladesh
June 16
* Vierzy, France
July 21
* Lebrija, Spain
October 9
* Saltillo, Mexico

1974
August 30
* Zagreb, Yugoslavia

1981
June 6
* Mansi, India

1982
January 27
Algiers, Algeria
July 11
* Tepic, Mexico

1988
December 12
* London, England

1989
January 4
* Sangi, Pakistan
January 15
* Maizdi Khan, Bangaldesh

1989
August 10
Los Mochis, Mexico

1993
January 30
* Darajan, Kenya
September 22
* Mobile, Alabama

1995
August 20
* Firozabad, India

1998
January 5
Karna, India
June 3
* Celle, Germany

1999
August 2
* Gaisal, India
March 15
* Bourbonnais, Illinois
October 5
* London, England

2001
September 2
* Cirebon, Indonesia

RAILWAY DISASTERS

As the first means of mechanical transportation, the railroad enjoyed a long period of unchallenged supremacy and comparative safety. Slow speeds, a conservative amount of track mileage, and a small number of passengers combined to keep the accidents and the casualty numbers low.

The very first days of rail travel were, in fact, restricted entirely to traffic in and out of mines. The first recorded railway's existence is an illustration of a narrow-gauge mine railway at Leberthal in Alsace, included in *Cosmographae Universalis,* by Sebastian Munster, published in 1550. The first railway of any consequence seems to have been a line made of balks of timber from coal pits at Wollaton and Strelley near Nottingham, England, in the late 16th century. True to tradition, it was designed to take the coal from the pits to the River Trent, outside Nottingham.

It would be another three centuries before railways were used to transport people. The first railway in the world to carry fare-paying passengers was the Ostermouth Railway (also known as the Swansea & Mumbles Railway), which opened for business in April 1806 in Ostermouth, England. Horses and sails were used to move the cars at speeds slow enough for the railway to eventually, once the novelty wore thin, lose customers to horse buses skimming along a turnpike road next to the railway. It was the last and only time that a railroad lost out to buses.

The first railroad to be built in America came about in 1795, when a short length of wooden track was laid on Beacon Hill in Boston to carry building material for the State House. By 1825 John Stevens tested the first steam locomotive in America on a circular track at his home at Hoboken, New Jersey. Stevens would, four years later, begin service on America's first steam-powered railway, the Pennsylvania Railroad. Within a couple of years, the B & O would best Stevens with a verticle boiler that developed 1.43 horsepower. Small wonder that the early days of railroading were relatively accident free.

Still, by 1833 the wrecks began. On November 9 of that year, a carriage overturned on the Camden & Amboy main line between Spotswood and Hightstown, New Jersey, and 12 of 24 passengers, including Commodore Vanderbilt, the later head of the New York Central, were seriously injured (see p. 358).

Other European, South American and Asian countries developed rail lines at the same time, with the same primitive signal systems, laminated iron rails, brittle cast iron wheels and link and pin couplings between cars. Still, until midcentury most accidents were caused by derailments caused in turn by separating rails, slow-motion collisions and cows wandering onto the tracks.

But by 1853 railways, particularly in America, were catastrophes waiting to happen. With its land grant policy, in which railroads were given land and loans only as track mileage was completed, the U.S. government actually encouraged flimsy and hasty railroad construction. As a result practically every government-subsidized road was poorly and dangerously built. Whereas the English were constructing their railroads carefully and safely in the middle of the 19th century, with an eye for permanence, laying double rows of tracks, erecting substantial bridges, viaducts and tunnels and eliminating curves and grades, the American way was to put it all up quickly and expediently.

Thus, the foundation for the major categories of train wrecks was laid simultaneously with the hasty laying of track: Head-on and rear-end collisions were the result of single tracks and primitive signal systems; derailments came about through sloppy track laying and brittle wheels and axles; bridge disasters came about from poorly designed and hastily built bridges; telescopes resulted from a lethal link and pin coupling method between cars; crossing accidents resulted from lack of communication on the lines; fires resulted from a combination of superheated steam boilers in the engines and wooden cars heated by coal stoves and illuminated by oil lamps.

Many of these failings that would cause grisly accidents and terrible loss of life before the turn of the century would be corrected. Cheap steel would bring sound rails, bridges, wheels and axles. Automatic electric signals, double track and the Westinghouse brake would reduce the danger of collisions. The Miller platform and buffer method of coupling cars through tension and compression would reduce telescoping of passenger cars. The adoption of electric lighting and steam heating at the end of the century would reduce the fire hazard greatly, and the adoption of all-steel passenger cars in 1907 would virtually eliminate it in all but the developing world.

Even so, the most up-to-date technology cannot cancel out human failure. As more sophisticated signal devices were developed, signal men, conductors and engineers sometimes ignored them, resulting in catastrophe. As the possibility of higher speeds was introduced through more advanced equipment, engineers misjudged track conditions and drove their high-speed trains off the rails into abutments or off bridges.

More recently societal conditions have been responsible for train wrecks. The overcrowding of trains in Bangladesh and India, where poverty dictates that train and bus travel are the only means by which enormous segments of the populace can travel, has resulted in calamitous tragedies with enormous loss of life. Inebriated train operators have also become a hazard. And drug use has necessitated mandatory random drug testing for American railroads.

Despite the hazards, railroading remains a relatively safe mode of transportation. As the technology of railroading has developed mightily in Europe and Japan, safety has increased along with the rate of speed. The demands of the superfast trains of France and Japan have dictated that new and safer roadbeds and configurations be built. Furthermore, by the early 1960s, railroads and the governments that subsidized them realized that if they were to compete with airlines for traveling passengers, they would have to increase their speed and offer unique services that airlines could not offer.

The convenience of being transported from the center of one city to the center of another has remained constant. But the upgrading of systems to accommodate new high-speed travel by rail has varied country by country, from nonexistent to superb. Japan was the first to seriously undertake high-speed rail travel. In 1964, its first bullet train—or, as it is known in Japan, Shinkansen—debuted on the Tokaida line, between Tokyo and Shin-Osaka, with a sleek white-and-black profile and a speed of 131 miles per hour. It was a wakeup call for other countries, who were swift to develop their own high-speed trains. Still, Japan has continued to outpace the rest of the world in this sort of travel.

As of 2002, Japan has five separate bullet services, on tracks built specifically for these trains. On the Tokaida line series, 300 trains operate at a top speed of 168 mph. On the Sanyo line, there are 500 trains traveling at a top speed of 186 mph. On the Tohoku line, there are 200 trains operating at speeds of between 150 and 170 mph. Along the Joetsu line, 200 trains operate at a top speed of 170 mph. And along the Honkuriku line, trains operate at an average speed of 162 mph. Thus, it is possible to travel between Hiroshima and Kokura—a distance of 120 miles—in only 44 minutes at an average speed of 164 mph, the fastest scheduled train speed in the world.

Ranking second to Japan, France has developed the Train à Grande Vitesse, better known by its initials, TGV. The first TGV was an experimental gas-turbine model tested in the 1960s but never put into service. Practical use would have to wait until 1981, when the Paris Sud-Est, the first electric TGV—the trademark orange train—was launched for service between Paris and Lyons. It established a world speed record of 236 mph in 1981, but since then, it glides along at a consistent average speed of 135 mph, which makes it the slowest of present-day TGVs.

The gray TGV Atlantique, introduced in 1989, bettered the first TGV world record, reaching a speed of 320 mph. Traveling along the Atlantic coast of France from Paris, the present TGV Atlantique travels at an average speed of 186 mph.

The third generation of TGV trains, the TGV Reseau, introduced in 1993, took high-speed rail design an important step forward by pressure-sealing its passenger cars. Until that time, trains entering tunnels at more than 100 mph experienced sudden pressure changes that were difficult on passengers' ears. The TGV Reseau was designed to be used on all of France's high-speed lines.

A further step forward, brought on by necessity, is the TGV Duplex, a distinctive gray train designed to meet the overwhelming demands on the Paris-Lyons run. By 1996, this popular connection between the two major cities was running with trains only three minutes apart, and so the duplex, capable of carrying 45 percent more passengers than the ordinary TGV, with only 4 percent more drag, has temporarily solved the problem. Currently, the Paris-Lyons run is being extended to Marseille. Dubbed the TGV Mediterranée line, it will also accommodate, in duplex silver-and-blue cars, TGV Reseau and TV Sud-Est trains.

France has also expanded its design beyond its borders. The AVE was exported to Spain in 1991 for the run between Madrid and Seville. The Thalys, introduced in 1996, connects France with Belgium, Germany and the Netherlands. And the Eurostar is in service now in the Chunnel, the English Channel Tunnel between London and Paris and Brussels. Unhappily, the track beds in England have severely curtailed its speed; it only reaches 186 mph on the French side of the Chunnel. However, by 2003, the first major new rail line in more than 100 years in the United Kingdom will be completed, allowing high speed from London to Brussels, which will translate into a two hour, 20 minute trip from London to Paris, or two hours from London to Brussels.

TGVs are extraordinarily comfortable. On their high-speed tracks, the ride is as smooth as an aircraft traveling in clear skies; there are no vibrations. The insulation is so complete that passengers can whisper to each other and still be heard. In addition, these trains are environmentally friendly. Because of their high-tech low-friction design, they need very little energy to keep them going once they have reached their cruising speed. At 165 mph, the engine can be switched off, and the deceleration is hardly noticeable.

Germany, which began to develop its ICE, or InterCity Express, high-speed trains in 1981 is only inhibited by the lack of high-speed roadbeds. The only existing one currently in use is the Berlin-Hannover line, which is nearing completion as this is being written. On existing lines, the ICE reaches speeds of 125 mph; on the completed sections of the Berlin-Hannover line, it tops out at 186 mph. What the ICE lacks in speed, it adds in luxury. Standard class surpasses first-class accommodations of other high-speed trains; the decor is stunning, and the comfort and space are remarkable.

In the United Kingdom, studies have gone forward to build high-speed lines north of London, since increased passenger traffic will make the current lines obsolete and incapable of carrying the passenger load by 2010.

The United States has lagged considerably behind these other countries in upgrading its rail system. In fact, because of the gigantic growth of airlines after World War II, it has allowed its rail system to deteriorate. Gone are the luxury trains of the past. The Twentieth Century Limited and the Super Chief are just warm memories for a generation that once experienced them.

Although one-third of the freight in America is carried on rails, the cataclysmic desertion of the railroads for the airlines by passengers has forced practically all of the individual railroads to lose gigantic amounts of money. In 1971, the government took over passenger service from bankrupt railroads and formed the National Railroad Passenger Corporation, which later became Amtrak, serving more than 500 communities in 45 states and employing about 21,000 workers.

Amtrak has yet to make money. By 1997, it had gone through $22 billion in government subsidies, and Congress formed the Amtrak Reform Council to monitor the railroad's finances. Amtrak made a commitment to become self-sufficient before its reauthorization expired in December 2001.

Taking its cues from Europe and Japan, who turned their railroad problems around with high-speed trains, Amtrak developed the Metroliner, a semi-high-speed train between New York and Washington, D.C., and New York and Boston. In 2000, it introduced the Acela, a true high-speed train on the same line.

Amtrak did not, however, become self-sufficient in December 2001, despite the precipitous drop in airline ridership following the crashing of two commercial airliners by terrorists into the towers of the World Trade Center in New York City on September 11, 2001 (see p. 147). Although Amtrak reported a 6 percent surge in its passenger traffic following the tragedy, it only accounted for slightly more than a 1 percent decrease in its loss of ridership for the year. Its northeast corridor trains, which include the Metroliner and Acela, have consistently made money, but the rest of the system has just as consistently lost it.

Obviously, the future of passenger rail traffic is in high-speed trains. France, Germany and Japan have proved this. However, in the United States, the extremely effective airline lobby has succeeded in garnering the lion's share of government transportation subsidies, thus starving Amtrak.

Another advantage to high-speed travel, in addition to its environmental friendliness, convenience and comfort, is its relative freedom—so far—from accidents. In more than two decades there has been only one major railway crash involving a high-speed train, while freight and ordinary speed trains, running, in too many cases, on roadbeds that were installed generations ago, continue to add to the casualty total. It appears that social change will dictate not only the development of rail travel but also its safety.

Like the other disaster categories in the book, events in this section were chosen in terms of human, not material, loss. An arbitrary cutoff figure of 50 deaths was used, and the only deviations from this figure occurred when historic firsts were deemed important enough to be included.

ARGENTINA
BUENOS AIRES
February 1, 1970

Human error caused the collision of an express train and a commuter train near Buenos Aires, Argentina, on February 1, 1970. One hundred forty-two died; hundreds were injured.

On Sunday night, February 1, 1970, a cross-country express train of Argentina's Bartolomé Mitre railroad, composed of two diesel locomotives and 21 passenger coaches containing 500 people, was nearing the end of its 1,000-mile journey from the northern city of San Miguel de Tucuman to Buenos Aires. It was traveling at 65 mph and was 50 minutes behind schedule.

Ahead of it a commuter train packed with 700 weekenders returning to Buenos Aires from the fashionable northern suburb of Zárate had experienced mechanical difficulty and was stalled five miles outside the Pacheco station, which was 18 miles from Buenos Aires.

A signalman, Máximo Bianco, on duty near the crash site should have warned the approaching express of the stalled commuter train. But for some reason he did not, and the express plowed into the back of the stalled local train at 65 mph.

"We were going very, very fast," said survivor María Isabel Algodén, "when all of a sudden everything exploded and people went everywhere."

The last five cars of the commuter train were crushed beyond recognition. Other cars telescoped into one another, mangling their occupants and upending some of the coaches. Wreckage was strewn over a wide area on either side of the tracks outside Pacheco.

Rescue crews were dispatched immediately. Firemen installed mobile power stations to illuminate the wreckage, and air force helicopters flew blood plasma, medical kits and surgical instruments to the macabre, floodlit scene.

Ambulances, trucks, commercial buses and private cars were pressed into service to transport the injured to an emergency hospital that was set up in the Pacheco railroad station. There, doctors from Buenos Aires labored to save whomever they could. The less seriously injured were taken into Buenos Aires and its hospitals.

The dead were first lined up along the tracks and then taken to the Pacheco and Benavidez stations, where they were displayed for identification purposes. One hundred forty-two persons died, most of them in the rear cars of the commuter train. Hundreds more were injured.

At first authorities blamed terrorists. Only 90 minutes before the crash occurred, terrorists had attacked a railroad station three miles from the site of the collision and had made off with $400. Some link between the two incidents was thought to exist.

But further investigation led to the arrest of signalman Máximo Bianco and two of his fellow workers for failing to engage the warning signals that would have informed the engineer of the express that the commuter train was stalled in front of him. Once more, human error had caused a tragic rail accident.

BANGLADESH
JESSORE
June 4, 1972

Human error caused the collision of an express train with a standing train in the station at Jessore, Bangladesh, on June 4, 1972. Seventy-six were killed on the train and the station platform; 500 were hospitalized.

Trains in Bangladesh and India are frequently loaded beyond capacity. It is not unusual to see passengers on the roofs of cars, between cars or hanging outside the cars' compartments. And this was the condition of an express train that left Khulna, a port town in the southern part of Bangladesh, headed north to the city of Jessore on June 4, 1972.

Despite its overload, the express made good time and entered the city limits of Jessore at nearly full speed. But the Jessore stationmaster threw a wrong switch, and the express was channeled onto the same track as that of a waiting train in the station. The moving train plowed into the standing one at maximum speed. The impact of the collision flung the locomotive off the track, and 10 coaches behind it telescoped into one another, hurtling ahead into the splintered remnants of the stationary train.

Passengers on the moving train were flung to the tracks and the station platform. Those in the waiting train were trapped or crushed. Seventy-six died immediately. More than 500 others received injuries serious enough to hospitalize them.

BANGLADESH
MAIZDI KHAN
January 15, 1989

Bangladesh's worst train disaster, the collision of an overloaded express train with a mail train near Maizdi Khan on January 15, 1989, was caused by human error—the switching of both trains onto the same track. One hundred thirty-six died; more than 1,000 were injured.

On January 15 in Tongi, a city in central Bangladesh, hundreds of thousands of pilgrims gathered for a Muslim religious festival. The devout came from all over Bangladesh to take part in the ceremonies, and a train, filled to overflowing with more than 2,000 pilgrims, made its way north from Dhaka to Tongi and then southeast to Chittagong. Passengers were everywhere—in seats, in the aisles, on the platforms, on the roofs of the cars. It was not an unusual occurrence, but in this case it was a terribly dangerous one.

Several days before, a new signal system had been initiated on this line, and later official explanations blamed what was about to occur on railroad personnel being confused by the new system. The express bearing the more than 2,000 pilgrims had almost reached Tongi. It was speeding at 50 mph near the village of Maizdi Khan when suddenly a mail train appeared, headed in the opposite direction on the same track. There was no time for either train to brake, although the mail train had almost stopped. They crashed head-on.

The impact flung the diesel locomotives of both trains off the tracks, and the first two coaches of each train were carried with the locomotives. Other cars telescoped into one another, picking off the passengers on the roofs as if they were billiard balls. "I saw coaches flying up to 15 feet as the collision occurred," said a soldier who, with some 250 other military men, was involved in winter exercises near the tracks. "It was a terrible scene with hundreds of passengers—men, women and children—shouting for help."

The carnage was appalling. Bodies, some of them without limbs or heads, were strewn over the countryside. Screams and moans filled the air. The soldiers were the first on the scene, and they pulled more than 100 dead passengers from the mangled cars and laid them side by side along the tracks.

Medical teams and ambulances arrived, but the job was overwhelming. More than 1,000 were injured, 100 of them critically enough to require hospitalization. Hundreds were taken to hospitals in Tongi, five miles north of Maizdi Khan; hundreds more were taken to Chaka, the capital, 22 miles to the south.

Police were brought in to control the crowd of thousands who thronged the fields around the wreckage trying to find loved ones among the rows of bodies. Twenty-six would die in the hospitals. A total of 136 would perish, and more than 1,000 would be injured in this, Bangladesh's worst railway accident.

BRAZIL
ARACAJU
March 20, 1946

The worst train wreck in Brazil's history occurred near Aracaju on March 20, 1946, when an overcrowded commuter train derailed. One hundred eighty-five died; several hundred were injured.

Brazil's worst train wreck also caused a bizarre aftermath. An overcrowded suburban commuter train carrying 1,000 passengers was apparently too heavy to negotiate a steep incline near Aracaju, the capital of the Brazilian coastal state of Sergipe. It derailed; its cars uncoupled, telescoped and piled up at the bottom of the incline. Hundreds were trapped in the crushed cars; 185 died and several hundred were injured.

Grief-stricken relatives descended on the scene and discovered the surviving passengers. Enraged at the accident, they turned on the engineer. The thoroughly terrified man fled on foot and finally surrendered himself to authorities in the nearby town of Laranjetras. Several of the survivors, blaming him for the crash, tried to lynch him, the engineer told the local police, who took him into protective custody.

BRAZIL
MANGUEIRA
May 8, 1958

Human error—a wrong switch thrown—caused the head-on collision of two commuter trains near Mangueira, Brazil, on May 8, 1958. One hundred twenty-eight died; more than 300 were injured.

It was pouring on May 8, 1958, in the suburbs north of Rio de Janeiro. A driving, torrential downpour limited the visibility of anyone caught in it. Thus, the terrible headon collision of two commuter trains late that afternoon could have been partially blamed on the weather. But a post-accident inquiry laid the vast majority of the blame on officials of the state-operated Central do Brasil line, and Brazilian president Juscelino Kubitschek immediately dismissed three top officials of the line.

The accident occurred as a crowded commuter train pulled out of the station at Mangueira. Picking up speed, it raced onward, its packed cars warm and light in comparison with the driving rain that was mercilessly pelting the train. It was an ordinary run, made countless times by both the passengers and the engineer. But shortly after he pushed his train to full speed, the engineer was suddenly blinded by an oncoming headlight. Somewhere along the chain of command a wrong switch had been thrown, and the outgoing commuter train had been routed onto the same track. The two trains crashed head-on at high speed.

Both leaped skyward. Cars were telescoped or flung on their sides by the enormous impact. Rescue workers, nurses and physicians from Rio de Janeiro, 10 miles to the south, toiled in the driving rain to pry open cars, following the moans and screams of survivors beneath the mass of tangled, smoking steel.

It would be daylight of the next day before some sort of accurate casualty figures could be reached. One hundred twenty-eight people died, and more than 300 were injured, some of them critically.

In a bizarre postscript hundreds of hysterically indignant residents of the northern suburbs of Rio de Janeiro rioted on the night following the crash, venting their frustration and fury on railroad stations along the route of the two trains. They completely wrecked the station at Engenho de Dentro, the most populous town, and attempted to tear down three other stations but were finally repelled by police in riot gear.

BRAZIL
PAVUNA RIVER
March 4, 1952

Overloading, outdated equipment and unrepaired track caused the collision of two suburban trains on the Pavuna River bridge in Brazil on March 4, 1952. One hundred nineteen died; hundreds were injured.

The rebuilt Pavuna River bridge, the scene of an appalling wreck in 1950, was the site of a second train crash on the morning of March 4, 1952.

The suburban trains of the Central Railroad of Brazil were habitually overloaded in the 1950s, and a general overhaul of conditions on the line was being discussed in 1952 by the joint Brazil–United States Commission for Economic Development. But the Brazilians on the commission gave priority to improving the long-distance distribution of food and raw materials over the state of suburban rail traffic. It was nevertheless a frustrating problem for Brazilian workers and industry chiefs who saw thousands of man-hours lost while workers waited for trains on which they often could not find room.

Thus, a dangerously overloaded train composed mostly of old and decrepit wooden coaches left Rio de Janeiro at 8:30 A.M. on March 4, 1952, bound for Juiz de Fora in the state of Minas Geras, 100 miles north of Rio. Not only was every seat taken and the aisles full, but clusters of passengers were clinging to the outsides of the wooden cars and riding the bumpers between them. Fatal accidents occurred daily on the line because of this practice, but a shortage of equipment, authorities complained, prevented any improvement.

The rolling stock was not the only part of the Central Railroad in need of repair and replacement. Some of its rails were in dire need of repair. The day before the coach train left for Juiz de Fora, a freight train had derailed in the middle of the Pavuna River bridge, but there had been no casualties and no inspection of the tracks at the site of the derailment.

At 8:40 A.M., just 10 minutes after the coach train left the station at Rio, several of its wooden cars crossed that same piece of defective track. Like the cars of the freight train, they derailed, swinging across the parallel track of the bridge reserved for trains traveling in the opposite direction.

At that precise moment a modern electric suburban train loaded with commuters thundered onto the bridge headed into Rio. It smashed directly into the wooden coaches with their hapless passengers packed inside and clinging to the outside. The cars splintered into thousands of fragments.

Bodies were sent spinning through the air, onto the tracks, over the side of the bridge and into the river.

One hundred nineteen would die in the three cars, and hundreds more would be cruelly injured. A squadron of ambulances was sent out from Rio de Janeiro, together with rescue squads from four fire stations. Local buses and even trucks were pressed into service. A special train was sent to bring the bodies back to the capital, and special police squads, mindful of the riots associated with major crashes (see p. 330), were stationed in a perimeter around Rio's Dom Pedro Segundo Station, where the ill-fated train had originated.

Investigations were initially hampered. The engineer, it seemed, uncoupled the locomotive and fled in it, abandoning it some miles down the track. A peculiarity in Brazilian law stated that an engineer, if arrested at the scene of an accident, could be held indefinitely without bail, but if he succeeded in escaping arrest for 48 hours, he could remain free unless his responsibility was formally established by the court. He was set free but later charged and imprisoned.

BRAZIL
TANGUA
April 6, 1950

A lack of warning signals and torrential rains were responsible for a passenger train plunging through a weakened bridge over the Indios River near Tangua, Brazil, on April 6, 1950. One hundred ten died; 40 were injured.

Torrential rains soaked the ground and filled the rivers to overflowing near Rio de Janeiro in early April 1950. A 22-car midnight train left Rio on April 5 jammed with Holy Week vacationers bound for Victoria in Espirito Santo State. The train on the Leopoldina Railway was a modern one, with the comforts of some sleepers. By 1:30 A.M. it had reached the vicinity of Tangua, some 55 miles from Rio, and most of the passengers were asleep.

It was at this point that the tracks crossed the Indios River, which days of downpours had swollen until it overflowed its banks on either side. If a warning system had existed, the engineer would not have been permitted to take his train across the railway bridge at Tangua. But no such system existed, and the train eased its way across the bridge.

The bridge's foundations, however, had been hopelessly undermined by the raging floodwaters of the Indios. The train only reached the halfway point when the bridge gave way. The locomotive, two baggage cars and three coaches uncoupled themselves and plunged into the river, wrapped in the girders of the bridge. Miraculously, the other passenger cars remained on the tracks.

There were more than 200 passengers in the first three cars, and all of them sank immediately beneath the white water of the Indios. Some were able to escape by swimming through broken windows and open doors. These survivors together with trainmen and passengers from the remainder of the train worked feverishly in the dark, rainy night to free some of the injured. Forty were saved, but 110 died, trapped in the sunken cars or swept away in the turbulent water.

The bodies were taken by rail to Rio de Janeiro and Nietheroy for identification. Highway access to the scene of the wreck was blocked by landslides. The following day a railway crane was dispatched to grapple the three sunken passenger cars, baggage cars and engine from the still-raging river. It would take three days to recover the train.

CANADA
ONTARIO
HAMILTON
March 17, 1857

Failure to maintain a railroad bridge caused a train to break through the bridge and fall into the Des Jardines Canal near Hamilton, Ontario, on March 17, 1857. Sixty died; 20 were injured.

One of the major rail disasters of the 19th century occurred on the night of March 17, 1857, when the engine, tender and two coaches of a train burst through the rotted timbers of the Des Jardines Canal bridge, plummeted 18 feet through the frigid winter air and smashed through the ice of the canal.

The bridge was a rickety affair, and engineers of the Great Western of Canada line that traversed it were grimly aware that there was no guardrail to prevent derailed trains from plunging off its sides. Had there been one on the bridge, the 1857 disaster might have been prevented, for the train was a local and traveling at a slow speed.

However, the engine appeared to have struck something at the entrance to the bridge and derailed itself onto the timbers. These gave way immediately, and with a huge sigh of steam and an enormous clatter of splitting wood, the entire train stove through the bottom of the bridge, hurtling toward the frozen canal.

The engine was the first to break through the ice. It was lost in the water immediately. The baggage car somehow uncoupled itself and slid harmlessly across the ice to the far shore, injuring its three occupants only slightly.

The two coaches followed the engine; one flipped over on its roof, which splintered when it smashed into the steel of the locomotive's boiler. The other drove like a battering ram into the other car, killing every one of its occupants. The only survivors, aside from the fortunate three in the baggage car, were rescued from the second car. They were dug from the wreckage that night by scores of rescuers from Hamilton who worked by the light of locomotive lamps and torchlights. They improvised stretchers made from ladders to raise the injured from the icy canal to the remains of the bridge. Some survivors were found floating on pieces of ice; others wandered dazed on the intact surface; still others were hauled, half drowned and in shock, from the interior of the second coach. Altogether, 60 persons died and 20 were injured in the wreck.

A train of the Great Western line of Canada plunged through a rickety bridge and into the Des Jardines Canal, near Hamilton, Ontario, on March 17, 1857. Sixty died. *(Illustrated London News)*

CANADA
QUEBEC
ST. HILAIRE
June 29, 1864

The worst bridge disaster in North America occurred at St. Hilaire, Canada, on June 29, 1864, as a result of human negligence; the engineer of a passenger train failed to stop before crossing the open St. Hilaire drawbridge. Ninety were killed; more than a hundred were injured.

The number of railway disasters that have occurred because of disregard of rules is legion and chilling, and the worst bridge disaster on the North American continent occurred for just that reason. An 11-car train of Canada's Grand Trunk Railway left Quebec on June 29, 1864, loaded with 354 passengers, most of them recent German and Norwegian immigrants.

The St. Hilaire drawbridge, over which the train was to pass, hovered 45 feet above a heavily traveled canal. Because of the constant use of the drawbridge, the company had a strict regulation stating that all trains were to come to a full stop before crossing.

The engineer of this particular train apparently ignored both his own eyes and company regulations. Approaching the drawbridge—which had been opened to accommodate a string of six barges—at full speed, the engineer had no chance of preventing the entire train from plunging 45 feet into the canal below.

The engine smashed a hole through one barge, sinking it and dragging it partway to the bottom of the canal. The following cars splayed crazily, some diving straight into the water, others landing on the barges. Ninety people, trapped in the cars of the train, were killed in the crash, but the engineer responsible for the tragedy was miraculously thrown clear of the wreck and survived.

CZECHOSLOVAKIA
PARDUBICE
November 14, 1960

Both engineers of two passenger trains ignored speed and right-of-way regulations and collided near Pardubice, Czechoslovakia, on November 14, 1960. One hundred ten died; 106 were injured.

The village of Pardubice, located 68 miles east of Prague, Czechoslovakia, was the scene of a horrendous head-on collision of two speeding passenger trains on the night of November 14, 1960. Tight restrictions on news gathering in this communist bloc country prevented details from reaching the rest of the world.

What *was* revealed was that both engineers ignored speed and right-of-way regulations, and the two trains collided at top speed shortly after 6:00 P.M. just outside Pardubice. Both trains derailed, and 110 people were killed. One hundred six others, seriously injured, were taken to hospitals.

FRANCE
LAGNY
December 23, 1933

Heavy fog and excessive speed combined to cause the collision of a Strasbourg-bound express with the stationary Nancy express and a commuter train near Lagny, France, on December 23, 1933. One hundred ninety-one died; 280 were injured.

It was the night before Christmas Eve of 1933. A heavy fog fed by the frost-covered ground made it slow going for motorists on the roads surrounding Paris. Even within the city, the haloed streetlights were barely effective.

A commuter train making its tortuous way through the fog stalled near the village of Lagny, 15 miles east of Paris, and the Nancy express, brimming over with students on holiday and other merrymakers leaving Paris early for Christmas, pulled up behind it. The revelers aboard the train—among them two members of the Chamber of Deputies, Henri Rollin and Gaston Poitevin—were unfazed. It was warm in the wooden cars; it was safe. And to assure this safety, trainmen set up red flares along the track, supplementing them with torpedoes.

But the fog was heavier than anyone could have imagined. The Strasbourg express, traveling on the same main line as both the commuter train and the Nancy express and already an hour late, was making 65 mph—an excessive speed on this part of the line under good conditions and an insane speed in the heavy fog of the night of December 23.

The engineer of the Strasbourg-bound train saw no flares and heard no torpedoes. At 65 mph, his train slammed into the rear of the stationary Nancy express, crushing the wooden cars to splinters and flinging bodies, luggage and chunks of cars onto the frosted ground on either side of the tracks. By the time the Strasbourg train stopped, it had plowed through every car on the Nancy train, telescoping whatever it left of each coach.

The devastation was indescribable. The dead and dying strewed the tracks. Hardly anyone was left alive in the Nancy train, and not a single person was injured on the Strasbourg express.

The inhabitants of Lagny rushed to the site of the accident. Local doctors did what they could. Rescuers built fires along the track to illuminate the grisly scene and began to haul the dead from the pile of charred and blood-soaked wood that had once been a passenger train.

Partway through their efforts, they froze in their activity as the sound of another train approaching at a high rate of speed pierced the fog. Dozens of flares had been set up to warn off other trains on the main line. The engineer of the Meaux express claimed later that, in the fog, the lights looked green, but a strange presentiment caused him to look again. The fog cleared for a moment, and he saw the red flares. He slammed on the brakes so suddenly that passengers on the train were thrown from their seats. With a huge shrieking of brakes, it stopped a mere 350 yards short of the wreckage of the Strasbourg and Nancy trains.

Had it not been for the reflexes of the engineer of the fourth train, the carnage would have been worse. As it was, 191 died, 200 were taken to Paris hospitals and 80 others were transported to hospitals and private homes in Lagny, Pomponna and Torigny by rescue trains, fleets of ambulances and private cars. It was, and would remain, one of the gravest accidents in the history of French railroads.

FRANCE
MODANE
December 12, 1917

A military order to move a severely overloaded troop train near Modane, France, on December 12, 1917, resulted in the derailment and plunge of the train into a gorge. More than 1,000 soldiers were killed; hundreds were injured.

It would be 15 years before there would be a full accounting of one of the worst railway disasters in history. The reason: It occurred in wartime, the train was loaded with servicemen and the responsibility for the wreck lay squarely on the shoulders of the military officers in charge.

In early December 1917 a troop train was packed far beyond its safe capacity with more than 1,200 war-weary French soliders on their way home for Christmas. The war had been going badly for the French, and this respite was seen as a necessary one for morale.

Alarmed at the weight and balance problems resulting from the overloading, the engineer refused to leave the station until the load was decreased. A French staff officer confronted the engineer, warning him that he would be court-martialed for refusing to follow orders if he did not pull his train out immediately. The engineer remained adamant, citing safety regulations. The officer produced a pistol and warned that the engineer's offense was a capital one and that he would therefore have him executed for refusing to obey wartime orders. Faced with this ultimatum, the engineer agreed to begin what would be a fatal journey.

The overweight train, loaded with celebrating soldiers, navigated the flat portion of its journey uneventfully. But part of its return trip would take it through the Alps in the southeastern corner of France. It entered the steeply graded Mount Cenis Tunnel near the village of Mondane. Partway through the tunnel the overloaded train wrested itself away from the engineer's control and accelerated far beyond safe limits.

At the bottom of the grade at the end of the tunnel was a wooden bridge and a sharp curve. Below it was an immense and deep gorge. The train blasted the bridge to tinder, shot off the rails and fell sickeningly into the gorge below. When it struck bottom it burst into flames, which consumed the entire train in a matter of minutes. More than 1,000 soldiers died in the crash and the flames. Hundreds were injured. Only a handful of survivors emerged from this monumental and avoidable disaster. Ironically, one of these survivors was the engineer.

FRANCE
VERSAILLES
May 8, 1842

A rear-end collision of two excursion trains in Versailles, France, on May 8, 1842, killed 54 passengers. The cause was a combination of overloading and a broken axle on the lead train.

The first recorded major railway disaster in the world occurred in the late afternoon of May 8, 1842, following the celebration of King Louis Philippe's birthday. Thousands of Parisians had traveled by rail and spent the day on the grounds of Versailles, wining and dining and witnessing a regal ceremony held in the midst of the royal palace's fountains.

When the ceremonies ended there was a rush to board the spartan wooden coaches of the return train to Paris. Extra engines had been added to accommodate the anticipated crowds, and the first train to leave needed at least two locomotives to pull the enormous load of passengers that had been jammed into its cars and then, in the practice of the day, locked in.

Shortly after leaving Versailles the lead engine on the first train broke an axle. The engine nosedived to the tracks, stopping instantly. The second engine then plowed into the first, bursting its boiler and setting both engines afire. Traveling at full speed, the wooden coaches careened into the two engines and split asunder. Moments later the fire from the boilers had spread to them, sending them up in huge pyres of oily flame.

Locked in or crushed beneath the splintered cars, scores of helpless passengers were incinerated by the furious fire that roared through the entire train. The official death toll was listed at 54, but according to unofficial sources the number exceeded 100.

FRANCE
VIERZY
June 16, 1972

Two trains crashed into the collapsed Vierzy tunnel in France on June 16, 1972. One hundred seven died; 90 were injured.

The Vierzy tunnel, located 60 miles northeast of Paris, has stood for nearly two centuries, not an unusual age for structures in Europe. But this one was apparently not con-

A lithograph captures the first recorded major railway disaster in the world: the rear-end collision of two excursion trains leaving the celebration of King Louis Philippe's birthday on May 8, 1842; 54 were killed. (New York Public Library)

structed with the usual care of its contemporaries, and on June 16, 1972, it collapsed with a roar.

Moments later two trains traveling in opposite directions entered the heavily used tunnel. Both were moving at top speed; both crashed into an immense pile of fallen rock and shattered ceiling timbers almost simultaneously. The locomotives of both trains climbed the rock pile and shot skyward through the jagged hole in the ceiling of the tunnel. The following cars remained on the tracks but telescoped into one another.

Help arrived almost immediately, but fear of igniting the diesel fumes that permeated the tunnel forced rescuers to move slowly and carefully. The slightest spark could have set off an explosion that would have dwarfed the train wreck in catastrophic consequences. Thus, it would be three days before 107 bodies and 90 injured passengers would be pried from the tangled wrecks that jutted half in and half out of the remains of the Vierzy railway tunnel.

GERMANY
CELLE
June 3, 1998

A high speed Inter City Express train traveling from Munich to Hamburg on June 3, 1998, struck a concrete pillar of an automobile overpass near Celle, Germany, collapsing the overpass on the train. Ninety-eight died and more than 100 were injured.

Like France's TGV, Japan's Bullet Train, and, to a lesser extent, Amtrak's Metroliner, Germany's Inter City Expresses (ICE) are a sleek, modern, safe and rapid way to travel among that country's major cities. Aerodynamically designed, painted silver and red, possessing airline type seats, restaurant cars and telephones, they continue to be the transportation of choice for businessmen. Their top speed is 175 mph on special tracks, and from 1991, when they were introduced, until June 3, 1998, none of the 104 trains in the fleet had had a fatal accident.

On that morning Inter City Express 884 left Munich bound for Hamburg, 500 miles away. At 10:59 A.M. it was exactly on time as it neared the village of Celle. There was a little-used automobile overpass bridge just outside the village, and one of the railroad's employees, who was working along the tracks, had parked his car on it.

The Inter City Express sped under the overpass at 125 mph. And then something went very wrong. One of the 13 cars immediately following the engine sideswiped a concrete pillar in the overpass, and the bridge collapsed on the train.

"I saw nothing on the tracks as I went under the bridge," the unharmed engineer told German television later. "Then there was a vibration. Then the automatic brakes came on."

The engine had severed itself from the cars of the train. The first two cars behind the locomotive remained on the tracks, but the others behind them splayed themselves like shuffled cards, careening over an embankment as they took leave of their wheels. Following this, the bridge crashed down on the next two cars, burying them, and the following six cars folded up and accordioned into the mass of twisted steel and concrete.

"There was a loud noise, like a helicopter, a crashing, and then there were railway carriages in front of my door," a woman who lived near the tracks told reporters.

This was followed by a momentary silence and then an eruption of screams as people clawed their way through broken windows. Seats and suitcases were thrown clear of the train. Splinters of metal and chunks of concrete from the bridge flew. The automobile that had been parked on the overpass was a mass of twisted metal intertwined with that of the passenger cars of the train.

"There were legs, arms, from adults and children, men and women," one of the 1,200 rescue workers who converged upon the scene told reporters. The dead were covered with sheets and arranged on stretchers in a grassy field next to the tracks. Medical workers set up a field hospital to perform emergency surgery. Those survivors who could make the trip safely were taken by helicopter to hospitals in Hamburg, Hanover, Bremen and elsewhere. The Red Cross chartered six buses to transport other injured passengers to hospitals. Ninety-eight people died and more than 100 were injured in this catastrophic accident, the worst to take place on Germany's premiere ICE line since its 1991 inauguration.

GERMANY
LAANGENWEDDINGEN
June 6, 1967

The failure to report a defective grade crossing in Laangenweddingen, East Germany, caused the collision of a commuter train and a gasoline truck at the crossing on June 6, 1967. Eighty-two were killed; 51 were injured.

The grade crossing alongside the railroad station in the little town of Laangenweddingen, located near Magdeburg, 80 miles southwest of Berlin and 20 miles from the West German border, was defective in July 1967. The barrier, which should have lowered and prevented vehicles from crossing the tracks, snagged repeatedly on an overhead telephone line and failed to close. There was a crossing guard, and it was his duty to report this sort of danger to his supervisor, who would order the proper repairs to be made. But from the time of the first discovery of the defective crossing gate, no such warning had been issued. At 8:00 P.M. on the evening of July 6, 1967, the lack of a report resulted in tragedy.

That night a gasoline truck owned by the East German state-operated oil corporation and loaded with 4,000 gallons of gasoline approached the grade crossing. Seeing no barrier, the driver assumed that it was safe to cross the tracks and so drove ahead.

At that instant a double-decker commuter train roared into sight. There was no time for the train to stop or the truck to dislodge itself. The train slammed full speed into the truck; it exploded in an incendiary spectacle that shot flames to enormous heights. The cars of the train were set afire; the explosion blew apart the engine and the first four coaches, which were heavily loaded with children.

Flaming gasoline poured into the cars, onto the tracks and onto the roof of the railroad station, which caught fire and burned to the ground. Inside the trains children were burned to cinders instantly. Seventy-nine people died on the spot; three more died of their injuries later in the local hospitals at Magdeburg and Bahrendorf. Fifty-one were injured, and Red Cross volunteers, doctors, firemen and policemen labored for hours cutting away wreckage and pulling the injured and the dead from the twisted mass of fused and still-hot metal.

An investigation was immediately initiated, and five days later the crossing operator and his supervisor were arrested and charged with manslaughter. The results of their trial were never released by the East German authorities.

GERMANY
MAGDEBURG
December 22, 1939

The ignoring of a signal by the engineer of the Berlin express caused him to crash at full speed into the rear of the Cologne express near Magdeburg, Germany, on December 22, 1939. One hundred thirty-two died; 109 were injured.

The collision of two express trains on the morning of December 22, 1939, near Magdeburg, Germany, revealed as much about the price the German people were paying for war, even in its early stages, as it did about human error's responsibility for railway accidents.

From 1920 until 1936 German railways were models of efficiency to the world. But with the ascension of Hitler and his Third Reich this would take a dramatic turn. Although military efficiency was consistently high in the early days of World War II, it would be at an elevated and, in this case, tragic cost to German civilians.

By 1939, three years into the Nazi regime, there was tremendous congestion on German railways, which had never before been subject to quite so much strain with quite so many untrained personnel. Not only was the rail system overloaded with troop transports, military supply and personnel trains and freight traffic, it was also deprived, because of military conscription, of its best trained men. Their places were taken by reemployed pensioners, promotions among the present personnel and hastily trained workers pressed into service from other industries.

During the Christmas season of 1939, this situation was exacerbated by extra holiday, troop transport and troop leave trains, which had to be switched into the normal schedule. In addition, the normally crowded passenger trains were filled to capacity on December 22.

On that morning, because of holiday demands, the Cologne express made an unscheduled stop at the Genthin station, just outside Magdeburg. There was a thick early morning mist that considerably reduced visibility.

At Genthin an automatic stop signal lit, announcing the presence and the location of the Berlin-Cologne train in the station. But the engineer of the Berlin-Neunkirchen express either ignored or did not see the signal and ran through it at high speed. The express rammed the rear of the stopped Cologne train, slamming three of its third-class coaches to the roadbed and demolishing its baggage car. The locomotive of the Berlin-Neunkirchen train, deflected by the impact, left the rails and overturned, carrying five telescoping cars with it. The engineer was killed instantly, and a fireman was gravely injured.

The Third Reich immediately clamped a lid of silence on the details of the wreck; because America was not yet at war with Germany, however, some American journalists managed to file stories at least giving the outlines of it. The brief communiqué from official sources admitted to 132 dead and 109 injured, but witnesses at the scene estimated a far larger casualty figure.

GREAT BRITAIN
ENGLAND
HARROW-WEALDSTONE
October 8, 1952

Ignoring signals by the engineer of one train caused the multiple collision in the Harrow-Wealdstone station, near London, on October 8, 1952. The worst crash in the history of British Rail to that date, it killed 112 and injured 165.

The worst crash in British Rail history took place at the worst possible time—at the height of the morning commuter rush hour. An estimated 1,000 persons jammed the platform of the Harrow-Wealdstone station, some 11 miles west of London, on the morning of October 8, 1952. Most were awaiting the arrival of the local from Tring and West Hertfordshire to London, but a few had arrived early to board the many commuter shuttles that used this customarily crowded station.

The local loaded up at 8:19 and was just pulling out of the station when the Night Scot, an express from Perth to London, late and trying to make up time, roared into the station on the same track as the local. It had ignored the clear signals set by signalman A. G Armitage at Harrow number one box and, oblivious, hurtled into the rear of the local train, flinging coaches everywhere and ripping an enormous hole in a footbridge that ran beneath the tracks.

According to observers, bodies seemed to fly through the air, and some dropped through the hole in the footbridge. It was a catastrophic scene, but not the end of the carnage. Seconds later the Manchester express, also late and trying to make up time, sped into the station on an adjacent track that was now blocked with some of the cars of the demolished commuter train. The express, powered by two enormous engines, careened into the cars. Both engines left the tracks, soared vertically into the air and came down directly on the platform, which was crowded with waiting commuters.

Scalding gouts of steam erupted everywhere. Severed electric lines rained sparks and swung lethally over the scene. Within minutes thousands of rescuers rushed to the station, freeing those who were reachable, bandaging horribly mangled bodies, administering morphine to those who were trapped or maimed or dying.

It would be two days and nights before the two rear coaches of the local would be reached by cutting with acetylene torches, and there the rescue operation reached its most hopeless nadir. There were few survivors and scores of dead in these two cars alone. One hundred twelve died; 165 were seriously injured in that morning of horror at the normally placid commuter stop of Harrow-Wealdstone.

GREAT BRITAIN
ENGLAND
LONDON
December 12, 1988

A three-train collision at Clapham Junction, just south of London, on December 12, 1988, killed 33 and injured 115.

"It is sheer, bloody hell," said James McMillan, the assistant chief fire brigade officer viewing the carnage at Clapham Junction on the morning of December 12, 1988. Clapham Junction is a nerve center for 2,200 daily commuter trains to the south coast and southwest of England. More than 220,000 commuters a day pass through it from London's Waterloo station.

For several weeks in November and early December, British Railways, in an ongoing modernization program, had been replacing signals, which warned the multiplicity of trains on a variety of tracks when to proceed or stop. As a result, some signals behaved erratically during the changeover.

At 8:13 A.M. on December 12, at the very height of the morning rush hour just south of Clapham Junction, an inbound long-distance train from Bournemouth, traveling between 60 and 70 mph, plowed full force into a stopped commuter train, which in turn slammed into an empty train headed in the opposite direction.

"We were doing 60–70 miles an hour when there was this almighty explosion," said Chris Reeves, a survivor who had been in the buffet car of the train that had rammed the stationary train in front of it. "We were sitting next to the aisle," he continued, "and when the crash happened we were thrown down the train. There was furniture and enormous lumps of metal flying everywhere. The roof split open like a ripe tomato, and that's how we got out."

Ronald Arlette, who was in the stalled train, recalled the same sort of detonation. "There was an almighty bang," he said, "like an explosion. The carriage went up and we flew over and over. We ended up halfway up the bank."

There was a steep embankment a few hundred yards south of a roadway bridge over the tracks, and it was here that the wreckage of both trains was flung, the torn remains of the buffet car wedged between the empty train and the commuter train.

"There were people underneath me and under the metal and there were people with metal in their bodies," recalled Keith Larner, another passenger on the inbound Bournemouth train. "It was the most horrible sight I have ever seen."

More than 150 rescuers converged on the scene immediately. "There were pieces of bodies lying all over the place in the wreckage," one of 30 ambulance drivers said afterward. Doctors entered the mass of twisted steel, amputating limbs and performing other surgery on the spot to free those who were trapped in the debris. At the same time bloodied survivors staggered out of the wreckage and across four electrified tracks to the embankment, up which they struggled before being given first aid and shelter in a boys' school and a pub.

A total of 33, including the engineer of the long-distance train, died in the accident. One hundred and fifteen were taken to hospitals with serious injuries. The cause of the accident was a technical disconnect in the signaling system that came about as a result of the upgrading work.

GREAT BRITAIN
ENGLAND
LONDON
October 5, 1999

The Great Western Express slammed into a Thames turbo commuter train that had run a red light on October 5, 1999. Forty were killed and many injured.

When British Railways turned from government to private ownership, a marked deterioration in service and safety was noted by regulators. Safety budgets were trimmed to assure profits—which were healthy—said an investigative report issued after the devastating collision of the Great Western Express and a commuter train near Ladbroke Grove, in western London, on October 5, 1999.

It all occurred at 8:11 A.M., at the height of the morning rush hour. The commuter train had left Paddington station headed away from London. The Great Western Express was heading toward London on an intersecting track. It had the green light, but for some reason never explained, the engineer of the commuter train ran the red light on his track, putting his train directly in the path of the hurtling express.

The Great Western slammed into the commuter train broadside, lifting and splitting its cars asunder. The two diesel engines on both trains erupted in a huge fireball that ignited the cars nearest them, turning them into white hot infernos.

Brendon Bentley, a survivor, told reporters later, "I remember just a big surge and everyone being knocked down and then bits of seat and everything coming from over my shoulders and then just darkness. . . . [Then] I could see the flames, and there were some people trapped underneath the seats. We had to try and get over to them and try to give them a hand, but we couldn't. So then the first priority was just to get off the carriage."

Of the cars involved in the fire, Tony Thompson, an officer of the British Transport Police, noted that, "All the seats are gone. Everything is gone. They're just shells."

At first, 70 fatalities were announced, but further exploration disclosed a death toll of 40. Both trains had fortunately been sparsely populated.

The engineer of the commuter train, a man who had only been in his job for two months after 11 months of training, had run a red light and was at fault. But two other factors, investigators believed, could have prevented the crash if the companies involved—Railtrack, First Great Western and Thames Trains—had done proper maintenance and upgrading.

There was a notorious signal—signal 109—near the site of the crash. It had been cited in eight incidents over six years when train drivers complained that they could not see the lights because of power cables and gantries. Nothing had been done to increase the visibility of signal 109 by October 5, 1999.

An unidentified woman cries during the open-air service of remembrance for the victims of a train crash in Ladbroke Grove, west London. The Great Western Express and a Thames turbo commuter train collided killing more than 40 people. (AP/Wide World Photo)

The wreckage of two railway trains lies across the tracks, following a crash near Paddington Station in west London Tuesday, October 5, 1999. The two trains collided near the station at the height of the morning rush hour, killing at least eight people, leaving many more injured or trapped in the wreckage. (AP/Wide World Photo)

The Great Western Express possessed the Automatic Train Protection system, a signal of danger and red lights ignored that was widely used on continental Europe. But it was switched off on October 5, 1999, because it was, as the company phrased it, "not operational."

"We all know if the railways were in public ownership, A.T.P. would have been installed in trains by 1994 or 1995," said Mick Rix, a spokesman for the rail drivers' union at post-accident hearings. "I think the public, rightly, is now making a clear demand that safety must be put before cost."

GREAT BRITAIN
SCOTLAND
DUNDEE
December 28, 1879

Storms caused the buckling of the Tay River Bridge near Dundee, Scotland, on December 28, 1879. Seventy-five people were killed when a train plunged through it; there were no survivors.

The Tay River Bridge at Dundee, Scotland, was completed in 1877, and it was hailed worldwide as a triumph of engineering skill. The longest bridge in the world, it was also the most modern, the strongest, the most graceful. But a mere two years after its completion, it would be the scene of one of the saddest disasters in the history of rail travel, one in which there was not a single survivor.

The train that left Edinburgh at 4:15 P.M. on December 28, 1879, was traveling in hard conditions. A raging storm with hurricane-force winds had lashed the countryside all day, battering at houses and seawalls and swelling streams and rivers. One of the rivers that was at least three times its normal depth was the Tay, and its boiling waters had been ramming the pilings of the Tay River railbridge since the storm began.

The train started to cross the multiple-span bridge at 6:15 and traversed almost half of its length before disaster struck. Midway through the central span, the train lurched as girders began to buckle beneath it. Thirteen girders gave

A view from the storm-damaged Tay River Bridge near Dundee, Scotland, following the plunge into the river of the Edinburgh express on December 28, 1879. (London Illustrated News)

way, pulling the structure out from beneath the wheels of the train and flinging it 88 feet into the river below.

It was only a matter of minutes before the waters closed over the train, leaving nothing in evidence but stray pieces of it, floating bits and pieces of luggage and boards from the cars that shattered on the rocks or were crushed by the steel girders that hurtled downward once the train had shaken them loose. Seventy-five people—all the passengers and crew—died in this wreck.

GREAT BRITAIN
SCOTLAND
GRETNA GREEN
May 22, 1915

Human error on the part of a signalman caused the crash of a troop train and 2 passenger trains near Gretna Green, Scotland, on May 22, 1915. Two hundred twenty-seven were killed; 223 were injured.

The year 1915 was one of grim records. In January a troop train plunged into a ravine in Guadalajara, Mexico, resulting in Mexico's worst train wreck (see p. 349). On

The burned-out hulk of one of the cars of a troop train that collided with a passenger train near Gretna Green, Scotland, on May 22, 1915. Two hundred twenty-seven were killed and 223 injured in the wreck, caused by a signalman's faulty judgment. (London Illustrated News)

May 22 Great Britain's worst railway disaster took place at Gretna Green, in Scotland. Whereas the Mexican tragedy was caused by forces of nature, the wreck in Scotland was reportedly the fault of one man, a signalman who threw the wrong switch.

Gretna Green is a small town in which several tracks of the Caledonian Railway intersect. At 6:00 A.M. on May 22, 1915, a troop train loaded with 500 soldiers bound for the Western front roared into view. The signalman waved it on and into a direct collision course with a local train traveling in the opposite direction. Within minutes both trains hit head-on, turning the engines into mangled, hissing interweavings of steel and telescoping a series of passenger cars into one another.

The signalman had no sooner left his post to try to assist in rescue efforts when the London-Glasgow express appeared. With no signal to slow it, the express rammed both trains, igniting heavy ammunition and gas cylinders that were being carried in a baggage car of the troop train. Within an instant, the wreckage was turned into an exploding inferno in which human beings were incinerated by the score.

A horrendous consequence of the soldiers' battle dress made rescue a dangerous proposition. Each of the soldiers was wearing an ammunition belt, and the heat of the fire began to set off the belts, causing them to explode, thus killing not only the wearer of the belt but rescuers in the vicinity.

Fire brigades, swarms of rescuers, nurses and doctors toiled all day and night to try to save whomever they could. Arms and legs were set, transfusions were given, battle-field operating rooms were set up and amputations were performed on the spot. Long before they got to the war, most of this company was decimated. Out of 500 soldiers and hundreds of civilians, 227 died and 223 were wounded. It was a catastrophe of immense proportions, one of the worst of the war, and it occurred, ironically, hundreds of miles and a channel away from the battlefields.

INDIA
FIROZABAD
August 20, 1995

Three hundred fifty-eight people were killed and more than 400 injured in a cataclysmic collision of two express trains on August 20, 1995, at the village of Firozabad in northern India.

India's train system has long been known not only for its hugely overcrowded, sometimes ancient equipment, but its ubiquitousness and affordability. The state-owned Indian Railways carries millions of passengers every day and about 377 million tons of goods every year over an estimated 37,000 miles of track for very little money. A second-class, air-condtioned coach ticket from Delhi to Bombay—a distance of about 600 miles covered in 17 hours—costs approximately $45. But that is a luxury millions of Indians necessarily forego. Instead, thousands pack into the smelly, hellishly hot unreserved coaches or sometimes bully their way into reserved coaches, especially in northern India.

Another unique hazard of train travel in India is the presence of cows and other livestock on the tracks. Cows are held sacred by the Hindu faith, and every effort is made not to disturb, much less hit, them. However, sometimes they are hit and killed, without much damage to the train.

On August 20, 1995, a cow was one third of the cause of one of India's worst train wrecks. Another third of the cause was the failure of a signalman to raise his signal. The final third was the rail system itself, which scheduled two fast express trains only nine minutes apart and which is only partially electrified and depends upon manually controlled signals from signalmen instead of automatic switching.

Sometime after 2:00 A.M. on the morning of August 20, The New Delhi–bound Kalindi Express hit a cow that was on the tracks in the town of Firozabad, in northern India, 185 miles southeast of New Delhi. The body of the cow became entangled in the engine and damaged the train's brake system, forcing it to stop. Lai Sharma, the signalman, was either unaware of the stopped train or neglected to check the tracks, and he gave the go-ahead signal to another express, the Pureshotham Express from the eastern Hindu pilgrim city of Puri, to proceed at full speed.

At 2:55 A.M. the Pureshotham Express plowed into the rear of the stalled Kalindi Express. The engine of the Pureshotham Express surged ahead, crushing and ripping apart the last six cars of the other train and scattering the cars and six cars of its own train on either side of the tracks.

Most of the passengers were asleep when the crash occurred. One of the survivors, Manas Papnak, told police later that he heard a loud thud, and, as he stepped out of his coach into the darkness, he saw mutilated bodies and pieces of human beings flung everywhere. "The area reverberated with the cries of the injured and dying," he said.

There were 1,300 passengers aboard the train from Puri and more than 900 on the Kalindi Express. It would take cranes and blowtorches to open up the wreckage of the cars to retrieve bodies of the dead, who were then piled 20 deep on trailers that were pulled by tractors to a local hospital and a playground. Three hundred and fifty-eight people were killed and more than 400 were injured in the worst train tragedy in the history of India. Ironically, another of India's serious railway accidents, in 1981, was also caused by a cow on the tracks (see p. 344).

The signalman and his family disappeared after the accident, and angry opposition lawmakers in the Indian Parliament spent a day denouncing and demanding that the current government resign for its failure to assure safety on the state-run railroad. The signalman was never found, nor did the government fall, nor were many improvements made in the railroad system.

INDIA
GAISAL
August 2, 1999

Three hundred three people were killed and more than 400 were injured when two trains, headed in opposite directions on the same track, collided on August 2, 1999, at Gaisal station in West Bengal state, India.

The worst sort of accident imaginable on a railroad is the head-on collision of two high-speed trains on the same track. Preventing it is the first order of business in safety precautions, and therefore this sort of cataclysmic wreck is very rare indeed.

But on August 2, 1999, the notoriously slipshod and overcrowded, uncomfortable and essential rail system of India once more defied the statistics. And once again, human error exacerbated by outdated equipment caused a catastrophe of horrific dimensions.

Gaisal station is a small dot on the map of the behemoth railway system in India. Located approximately 300 miles north of Calcutta in eastern India near the border of West Bengal and Assam, it stands beside a well-traveled main line of intercity trains.

In the early morning of August 2, the Awadh-Assam Express, carrying a full complement of passengers and bound for Guwahati, Assam's commercial capital, traversed its assigned route, which took it past Gaisal. Unbeknownst to the engineer of this express, however, the Brahmaputra Mail, headed in the opposite direction for New Delhi and carrying a large number of soldiers from the Border Security Force and the Central Reserve Police Force, had been switched onto the same track.

For between four and nine miles in either direction, these two trains, carrying between them a total of 2,500 passengers, hurtled at 60 mph toward each other, neither aware of the other's presence. At exactly 1:55 A.M. they met, head on, at Gaisal station with an explosive force that sounded like the detonation of a heavy bomb. The engine of the Awadhi-Assam Express was blasted into the air and landed across the next track. It dragged seven coaches with it, and they in turn crashed into five cars of the other train. The force of the collision ripped the cars apart and set the majority of them ablaze in a grim, 50-foot-high pyramid of twisted steel, glass, seats and people.

"It was like a nightmare," said one soldier who survived. "It was completely dark and it sounded like a very loud explosion." His car had flown through the air and crushed the car of the other train as it descended. "I don't remember how I came down," he said.

It was raining and was stiflingly hot, which made the escape through openings in the mangled steel or through broken windows even more difficult for the hysterical survivors, many of whom were cruelly injured.

As the sun rose and the temperature climbed to 110 degrees, rescue and recovery teams descended upon the gruesome scene. Most of the bodies they were able to pull from the still smoldering cars were burnt or mangled beyond identification. Heavy cranes arrived along with thousands of onlookers. "Oh Lord," Brig. A. Parmar, the soldiers' commander, said when he came upon the scene. "Instead of death at the war front, they met their ghastly end here."

It would be days before the wreckage and bodies could be cleared from the tracks as wave after wave of workers with blowtorches whittled down the deadly pyramid of charred and mangled cars. The toll was terrible. Three hundred and three died, and over 400 were injured, many of them seriously enough to lose limbs.

India's railway minister, Nitish Kumar, announced his resignation over the crash, citing "criminal negligence" and stating that he was resigning to "punish himself" for his "moral responsibility." His resignation was refused at first by the Indian government, which later accepted it. But commentators in the Indian press and on television generally shrugged off Mr. Kumar's move, pointing to Indian Railways' management for its fault in a safety record that included approximately 300 accidents a year, two-thirds of which had been blamed on staff negligence.

Four signalmen responsible for keeping the trains on separate tracks disappeared, but five senior officers in the division where the tragedy occurred were dismissed.

Investigation later revealed that the trains involved in the collision had not been fitted with automatic warning systems, which would have forced the trains to brake in the event of a danger signal—which was not given. Although the equipment was available, it had been installed only on suburban trains. "It will be three to four years before they come into common usage," Railway Ministry Secretary Shanthi Narain said at the investigation.

In addition to this, it was also revealed that the Gaisal station master did not have a tracking panel, which would have informed him that two trains were speeding toward each other on the same track. Its presence might have prevented this, India's second worst-railway catastrophe following the Firozabad collision in 1995 (see p. 341).

INDIA
HYDERABAD
September 24, 1954

Monsoon rains weakened a bridge over the Vasanti River near Hyderabad, India, causing it to give way beneath a train on September 24, 1954. One hundred thirty-seven were killed; 100 were injured.

The eight-car Hyderabad to Kazipet express had not traveled far from Hyderabad in the minutes just after midnight on September 24, 1954. There were 319 persons aboard,

The overcrowding of trains in India contributes greatly to the abnormally high death tolls in that country's train wrecks. United Nations

many of them dozing in the fetid heat. It was the end of the monsoon season; rains had produced floods everywhere, and one of the rivers that was at flood stage was the Vasanti, outside Hyderabad.

The tracks crossed the Vasanti via a bridge 50 miles east of Hyderabad. The bridge had been weakened by the floodwaters, but other trains had crossed it safely that day. The express was not so lucky. It had almost reached the other side of the bridge when the span swayed, groaned and gave way beneath the weight of the train.

The engine and seven of its eight cars plunged into the river with the bridge. The far end of the span gave way first, thus forcing the cars to telescope into one another on their way down. Some people were crushed before they reached the river; others were trapped and drowned.

Salvage crews worked all night and through the following day. More than 100 injured people managed to swim to safety or were dragged from the swollen river by rescuers. One hundred thirty-seven died either in the train or in the waters of the river. Bodies washed up miles from the scene of this, the worst railway wreck to date in Indian history.

INDIA
MAHBUBNAGAR
September 2, 1956

Monsoon rains weakened a bridge over a gorge near Mahbubnagar, India, and a train was flung into the gorge as the bridge collapsed on September 2, 1956. One hundred twelve were killed; hundreds were injured.

Monsoon season in India turns placid ponds into seas and barely visible streams into raging rivers. This was the case in the province of Hyderabad in September 1956, at the beginning of the yearly monsoons.

The Central Rail line of India traverses several of these normally inconsequential streams via bridges. In the southwest portion of India, near the large city of Hyderabad, these bridges are often extended nearly 100 feet in the air, traversing sickeningly deep gorges.

Such was the case between the cities of Secunderbad and Dhone. The Central Line passed 85 feet in the air over a normally placid stream. But on September 2, 1956, this stream had become a torrent, and its incessant battering had weakened the support struts of the towering trestle bridge that spanned the gorge near Mahbubnagar. A guard on duty that night would have stopped the passenger train that reached the bridge at midnight had he been there. But unfortunately, when the train arrived, he was a mile away, checking on another bridge weakened by the monsoons.

The engine traversed the span without incident. But the resulting vibrations traveled down the struts of the bridge, and within seconds after the locomotive reached the other side, the entire steel structure gave way. The tender and first two cars plummeted like stones to the bottom of the gorge, followed by girders from the bridge, which landed on the train, burying and crushing it. One hundred twelve persons were believed to have perished, either from the fall or from drowning. Hundreds more were injured.

INDIA
MANSI
June 6, 1981

The worst train wreck in India's history to that date was caused by an engineer's decision to brake for a cow on the tracks on the rainy night of June 6, 1981, near Mansi. Two hundred sixty-eight bodies were recovered; more than 300 were missing.

Cows are sacred in India. To harm one is forbidden. And the engineer of a nine-car passenger train passing near Mansi, in the northeastern state of Bihar, 250 miles northwest of Calcutta, on the monsoon-whipped Saturday of June 6, 1981, was certainly aware of this. A devout Hindu, he would not, under any circumstances, add further bad karma to the cycle of his lives by harming a cow. And yet on that rainswept night, as his train, loaded with more than 1,000 passengers, approached a bridge over the Baghmati River, there it was: a cow standing on the track.

The engineer applied the brakes too suddenly, too fast and at just the wrong place. The momentum of the train carried it forward; the following cars, sliding on rain-slick rails on a roadbed that had been made, by the heavy rains, soft and insecure, derailed and whipped forward. Seven of the nine coaches and the locomotive plunged over the embankment, off the bridge and into the whirling waters below. It was over in an instant; the cars catapulted downward one after the other, careening and crashing into one another before they sank swiftly into the monsoon-swollen river.

Survivors in the last two cars which remained on the tracks moaned as they faced the fate of the others. There was nothing to be done but run for help to take out the bodies. There was no sign of life in the river below.

Within hours help arrived. Fifty-nine divers, 110 soldiers and scores of villagers from Mansi and other nearby settlements searched the wreckage and the riverbanks for miles and for days. Two hundred sixty-eight bodies were found, but more than 300 passengers from the tragically fated seven cars were missing and assumed dead. It was the worst railway wreck to that date in the history of India.

INDIA
PATNA
July 16, 1937

A monsoon-softened track bed accounted for the derailment of a passenger train near Patna, India, on July 16, 1937. One hundred seven were killed; 65 were injured.

One of India's worst train wrecks occurred in daylight in clear weather on July 16, 1937. The Delhi-Calcutta express had reached the vicinity of Patna, some 275 miles northwest of Calcutta in the Vindhya Mountain range, without incident. The monsoon season had begun, and there were reports of flooding, but none of the rivers in the vicinity had reached the extreme flood stage.

Still, the monsoons had apparently softened part of the roadbed on a stretch of track that ran adjacent to a steep and precipitous embankment. The engine lost traction and leaped from the tracks, carrying seven of the train's nine cars down the embankment. Rolling and pitching down the steep slope, the first two cars became uncoupled from the last three and landed at the bottom of the ravine. The force of the impact telescoped one car completely into the other. Seconds later the remaining five cars landed on top of the engine and the first two coaches, crushing them completely.

"Bodies were strewn around as if it were a battlefield," was the description one observer gave reporters. The passengers in the seven cars involved in the accident were all Indians; the last two cars were occupied by Europeans, and for some unexplained or fortuitous reason, these two became uncoupled and remained safely on the tracks. The occupants of these cars immediately set about rescuing as many as they could from the cataclysmic pileup at the bottom of the ravine. One hundred seven passengers were dead, and 65 were injured.

INDONESIA
JAVA
CIREBON
September 2, 2001

Forty passengers were killed and 60 injured when a passenger train collided with a parked locomotive in Cirebon, Java, on September 2, 2001.

As in much of Asia, the railway system on the thickly populated island of Java in Indonesia is old, in need of great repair and heavily overcrowded. Passengers hang from the windows and gather on the roofs of packed cars. Thus, accidents occur fairly frequently.

At 4 A.M. on September 2, 2001, the train from Jakarta at the western end of the island was on its way to the popular Indian Ocean resort of Yogyakarta, midway across Java. It had traveled 125 miles east of Jakarta and was approaching the small railroad station of the port city of Cirebon at that hour. An empty locomotive was in the station momentarily, waiting to be cleared to move on.

But the engineer of the passenger train had fallen asleep, and his train collided full force with the other engine, which capsized and rolled over on top of one of the passenger cars. Cars piled upon cars and accordioned into others as the force of the forward motion of the passenger train turned expected order into instant chaos.

Local police and rescuers rushed to the scene of devastation. Passengers, some dead, some severely injured, sat motionless, slumped in their seats. Others sat next to the tracks, screaming. The rescuers set to the wreckage with hacksaws to cut away the metal sheeting that trapped survivors inside the mangled cars; others used ropes to pull uninjured passengers through smashed windows. There was no heavy lifting equipment immediately available, and rescuers could hear the screams of trapped survivors held prisoner in the wreckage.

As of this writing, forty people were reported killed and more than 60 seriously injured in this, one of Indonesia's worst train accidents among many.

INDONESIA
JAVA
EAST PRIANGAN
May 28, 1959

The uncoupling of several cars of a passenger train in an act of sabotage sent the cars careening into a ravine in East Priangan on the island of Java on May 28, 1959. One hundred forty were killed; 125 were injured.

Sabotage was responsible for the worst train wreck in the history of Indonesia, in which a trainload of innocent passengers plunged into a ravine in the East Priangan regency of West Java province on the island of Java.

Indonesia in 1959 was a country in ferment. Under pressure from the United Nations, the Dutch had given way to nationalist pressure in 1949, and an independent republic of Indonesia was formed. Sukarno was elected president of a parliamentary form of government, but his administration was marked by inefficiency, injustice, corruption and chaos. Inflation soared in the 1950s, and economic depravation spawned a popular revolt. It began in Sumatra in 1958, and widespread disorders caused Sukarno to become more and more authoritarian. In May 1959 there were numerous incidents that killed both the innocent and the involved. Sukarno would dissolve parliament in 1960.

One of these incidents was the train wreck of May 28 in West Java. The Bandjar-Bandung express was loaded with 500 passengers and, early that morning, was traveling through mountainous terrain slashed by ravines hundreds of feet deep. The train was not moving fast; there were numerous curves and steep inclines, and it was on just such an incline that someone uncoupled the engine from the entire complement of passenger cars.

Set free, the cars rolled backward, gaining speed alarmingly. There were apparently no independent braking devices on them, and in moments the cars left the tracks, careened crazily on the edge of a precipice and then plunged hundreds of feet into a ravine. Miraculously, some passengers escaped uninjured, but 140 were killed and 125 were seriously injured in the wreck.

At first the conductor was charged with negligence, but further investigation determined that an unidentified person (he or she would never be captured and brought to trial) had uncoupled the cars from the engine. Speculation grew that Sukarno might have invented the charges to cover up still another example of inefficiency or corruption in his government, but the dissolution of parliament in 1960 forestalled any possibility of a public inquiry.

IRELAND
ARMAGH
June 12, 1889

A faulty decision by a conductor led to the collision of two excursion trains near Armagh, Ireland, on June 12, 1889. Three hundred were killed; hundreds were injured.

Two excursion trains left Armagh in Northern Ireland on the morning of June 12, 1889, loaded with more than 1,200 holiday-spirited youngsters from the local Methodist Church Sunday school. They were headed for the resort village of Warrenpoint on Carlingford Bay. The distribution of the seven- to 16-year-old picnickers between the two trains was decidedly uneven: The first train was jammed with 970 children and teachers crammed into 13 cars and two vans. The remaining 300 or so climbed into the relative comfort of the wooden cars of the second train.

It all began peacefully enough. The first train spun its wheels, spouted steam, strained and departed on schedule. Shortly after this the half-empty second section eased out of the Armagh station.

But soon afterward the first train encountered Kilooney Ridge, a normally navigable but steep slope. The over-loaded train made it halfway up the ridge, but the sheer weight of its cargo finally ground it to a halt. The engineer, Thomas Magrath, increased the steam. The train groaned forward a foot and then stopped again. The second train eased along behind the first, and its engineer, seeing that the lead train was stalled, slowed and waited at the Annaclare bridge.

Meanwhile, the conductor of the first train was in the midst of making a strange decision. Reasoning that the only way to move his train forward was to lighten the load, he ordered the last seven coaches uncoupled. Both the engineer and the assistant conductor argued vehemently against it. It was obvious that those cars would continue to roll back into the waiting second train.

But for some reason, the conductor's aberrant reasoning won, and the assistant conductor uncoupled the coaches. They began to slip back, first slowly, then swiftly, as the conductor, now apparently having second thoughts, commanded men to run alongside, shoving rocks in the wheels to slow the cars.

It was a futile effort. The cars began to careen downhill at breakneck speed, heading straight for the engine of the second train. Within minutes the last car smashed into the engine, splitting asunder with a horrendous tearing of wood against metal. The other cars telescoped into it, flinging pieces of carriages and passengers down a 70-foot embankment.

With the exception of two men and two girls, every person in the last two cars was killed. Hundreds lay on the slope, some terribly injured. It would be hours before rescuers from the village would be able to clear the bodies from the wreckage and transport the injured to hospitals. The conductor who caused all of this was brought before a Board of Trade inquiry and given a reprimand.

ITALY
SALERNO
March 2, 1944

The stalling of a train in a tunnel near Salerno, Italy, on March 2, 1944, caused the death by asphyxiation of more than 400 passengers and crew.

Ordinarily, railroad wrecks are the result of head-on and rear-end collisions, derailments, collapsed bridges or switching errors. These are the most common hazards, and there are safety regulations and precautions to prevent them. But no one could predict the circumstances that would lead to one of the most tragic train wrecks in the history of European rail travel.

On March 2, 1944, at the height of World War II, a train loaded to capacity with military and civilian passengers approached Salerno, Italy. A long tunnel marks the last few meters into the city, and for a reason never explained this train stalled midway through the tunnel. The fumes from the engine accumulated and spread through the cars, killing more than 400 passengers. Trapped and asphyxiated in their seats, they had no chance of escaping than the lethal contamination.

JAMAICA
KENDAL
September 1, 1957

Mechanical failure caused the derailing of a train near Kendal, Jamaica, on September 1, 1957. One hundred seventy-five died; more than 750 were injured.

An S-curve in a track outside Kendal, Jamaica, is an anathema to train engineers. It appears suddenly and unexpectedly and has been the scene of numerous accidents caused by errors in judgment on the part of trainmen. In 1938 a train failed to negotiate it, leaped from the tracks and crashed, claiming 85 lives.

The tragic train wreck of September 1, 1957, would occur in precisely the same spot. This accident involved a 12-car train loaded with 1,500 passengers returning from a holiday excursion to Montego Bay. The vacationers were all from Kingston and were part of a trip organized by Catholic agencies in that city.

The train approached the curve at 11:15 P.M. on a balmy, absolutely clear evening. It was traveling at a high but not excessive rate of speed and should have been able to negotiate the curve. But two factors apparently conspired to cause the tragedy. First, the train's brakes failed to hold—or so the engineer, Garnish Lurch, told investigators later. Second, the coupling between the third and fourth cars was faulty. It had been loosening at intervals on the way to Montego Bay, and yard workers there had tightened it, but apparently not enough. At the middle of the S-curve, the coupling let go, and the momentum whipped both the engine and all of the cars from the tracks.

The locomotive and first three cars landed on level ground, and their occupants received only injuries. But the last nine cars dove headlong into a ravine. The force smashed cars into other cars, ripping tops and sides from the carriages and flinging occupants through smashed windows or holes in the cars' sides. One car's momentum actually sent it climbing up the far side of the ravine, where it hung precariously while rescuers climbed gingerly through it searching for survivors or hauling off bodies. More than 750 were injured and 175 died in the worst train wreck in Jamaica's history.

JAPAN
TOKYO
May 3, 1962

Human error—the running of a signal—caused the collision of a freight train and a commuter train near Tokyo on May 3, 1962. One hundred sixty died; more than 300 were injured.

Three miles north of Tokyo and just outside the Mikawashima station, at 9:30 P.M. on Constitution Day, May 3, 1962, a freight train ran through a blocking signal designed to keep it from entering an occupied track, jumped the tracks and sideswiped a six-car electric commuter train. The commuter train, packed with celebrants, was outbound from Tokyo.

"Blue electric sparks filled the air and then everything went dark," Shoji Iwasaki, a factory worker and survivor, told the *New York Times*. "People stumbled about, wailing and screaming. I broke a window glass and jumped out and started to climb down the embankment."

The embankment was a 30-foot incline next to the tracks. It was the only means of escape, and most of the stunned passengers who, like the young man, had managed to crawl through the automatically opened emergency doors or through broken windows either stumbled onto the freight tracks or began to crawl down the embankment.

And then tragedy struck.

A second nine-car commuter train rammed into the wreckage of the first commuter train and the freight engine. Cars were hurled in several directions. The freight cars were destroyed and the boiler of the steam locomotive exploded with a dreadful roar, spewing scalding steam in all directions. It spread through the leading cars of the commuter train, scalding the survivors of the first wreck.

"The other train came crashing into our wreckage," continued Iwasaki. "The leading car toppled down and pulled four others after it. It rolled down the embankment, pinning and crushing many people who were fleeing for safety. It was horrible."

Cars continued to tumble down the embankment, grinding fleeing passengers beneath them and piling up, finally, at the bottom of the incline. "It rained bodies," one observer recalled.

Rescuers arrived under a thick, superheated fog of spent steam. The grisly business of removing the dead and the injured went on through the night and into the next day. One hundred sixty people died in the catastrophic wreck, and more than 300 were injured. Three weeks later, following an investigation, nine railwaymen were indicted on charges of criminal negligence.

JAPAN
YOKOHAMA
November 9, 1963

A cracked rail was responsible for a three-train collision near Yokohama, Japan, on November 9, 1963. One hundred sixty-two died; 72 were injured.

Two horrendous tragedies occurred on the same crisp autumn day in Japan in 1963. Early on Saturday, November 9, an explosion rocked through a coal mine in Omuta, killing 446 men. Six hours later three trains were involved in a colossal collision outside Yokohama. One hundred and sixty-two people perished in this wreck.

At 10 P.M. that night a fast-moving freight train traveling on Japan's main north-south Tokaido trunk line derailed when the train passed over a section of track that contained a cracked section of rails. Three of the freight train's cars flung themselves across two adjacent passenger train tracks.

At almost the same instant two passenger trains, traveling in opposite directions, roared into the site. One was a commuter train headed south from Yokohama to its southern suburbs. It plowed into the three derailed cars, smashing them to timber, and shot off the southbound track directly into the path of a northbound train headed for Tokyo. With a horrible screeching of steel, this train was sent sideways off its rails.

Cars telescoped on both commuter trains, and overhead electric lines, ripped from their fastenings by the crash, sent firework displays of sparks fanning into the night sky. Small fires started and were quickly extinguished by a rain that had just begun to fall.

Rescue workers with searchlights rushed to the scene within minutes. Firemen and medical workers cut away twisted and fused pieces of the passenger train, trying to separate the cars and open up escape routes for survivors. A steady stream of ambulances sped the injured to the 13 Yokohama hospitals in the area, while Japanese television broadcast lists of the dead and injured throughout the night. One hundred sixty-two died and 72 were injured, 48 of them critically. A 28-year-old American, William Scott of Colorado Springs, Colorado, was among the dead.

November was an election month in Japan, and the two disasters became political fodder, particularly for the foes of Premier Hayato Ikeda's conservative government. Until the November 21 election, investigations, charges and countercharges of negligence and irresponsibility flew back and forth while the grieving relatives buried their dead.

KENYA
DARAJAN
January 30, 1993

More than 200 passengers on the Mombasa-Nairobi Express in Kenya were killed when the train was derailed on a bridge weakened by floodwaters, and plunged into the Ndethia Geitha River on January 30, 1993.

The opening of the Mombasa-Nairobi railway at the turn of the century also opened up the Kenyan highlands to European settlement. A main line that runs from the Indian Ocean resort of Mombasa to Kisumu on Lake Victoria was and still is not only a vital link between the two points, but a scenic one, too. At one point it passes a mere 40 miles from Mt. Kilimanjaro, rising majestically out of the highlands.

But the railroad is also subject to the ravages of the wild nature through which it travels. Century-old tracks and bridges have received little attention and less repair, and the bridges particularly are at the mercy of the raging floods that result from each rainy season. January of 1993 was a month beset by unusually heavy rains and fiercely angry floods.

On January 30 the Mombasa-Nairobi Express left Mombasa on schedule, headed for Kisuma. Halfway there, in a region of uninhabitable shrubland near Mount Kilimanjaro, the train, carrying 600 passengers, approached a bridge that crossed the Ndethia Geitha River, which was swollen beyond flood stage. The floods had inundated some tracks and swept away Kenya Railway's communication system.

The bridge, built in 1898, had had few repairs, and the ceaseless battering of the flood waters had, unbeknownst to the engineer and crew of the approaching train, washed away one of the bridge's supports. The train hit the bridge; the bridge buckled and collapsed, and the front of the train was flung into the raging floodwaters beneath. The engine plummeted into the riverbank and shattered. Five cars of the train followed it, twisting and turning as they plunged into the froth of the river. A quick-witted conductor, sensing what was happening, pulled an emergency cord that released the third-class coaches from the rest of the train, and these cars remained upright on the tracks leading to the collapsed bridge.

Some screaming passengers were swept in the fallen cars away from the site. Those who could not break windows and escape drowned in the cars. Others swam free and wandered off into the brush.

It would be some time until police and rescuers reached the scene. Army divers were used to pull bodies from the entirely sunken cars; police and other rescue workers used power saws and axes to cut through the coaches all or partially visible on the river's banks.

The toll was terrible. More than 200 of the 600 passengers and crew on the train died in this, the worst railway crash since Kenya had gained independence from Britain in 1963.

MEXICO
CAZADERO
February 1, 1945

Human error—a faulty switch setting—caused the collision of an excursion train and a freight train in Cazadero, Mexico, on February 1, 1945. One hundred died; 70 were injured.

At 11:00 P.M. on January 31, 1945, a trainload of religious pilgrims left Mexico City bound for San Juan de Los Lagos, south of Aguascalientes and about 300 miles northwest of Mexico City. A religious festival for the Virgin Mary was to be held there the next day, and the pilgrims were confident that they would arrive by sunrise in order to participate in the festivities and the devotional services.

The special excursion train provided by the National Railways was not a comfortable one. It was old and contained wooden coaches that had seen years of service. The seats were hard and spartan and not necessarily designed for a good night's sleep. But the pilgrims were content. It was, for some of them, the high point of their lives.

By 12:47 on the morning of February 1, the excursion train had reached the tiny hamlet of Cazadero, 100 miles north of Mexico City. It was plainly marked. Lanterns hung from its rear cars. But a fast-moving freight train, misdirected onto the same track by a faulty switch setting, overtook the excursion train just outside Cazadero and smashed into it. The nine coaches were swept off the track like debris as the steam locomotive of the freight careened forward and then left the rails itself.

Three cars burst into flames instantly, burning alive most of their occupants. Other passengers were flung to the tracks or the roadbed. Still others were trapped in the other cars.

Rescuers from Cazadero and neighboring towns rushed to the scene, extinguishing the fires and attempting to drag survivors and bodies from the charred and tangled wreckage. News of the wreck was telegraphed to Mexico City, and a relief train loaded with doctors and supplies set out immediately. By the time it arrived most of the injured had been taken to San Juan del Río, 20 miles from Cazadero, where a small regional hospital was overwhelmed by the rush of casualties. One hundred pilgrims were killed; 70 were seriously injured in still another railway wreck caused by human error.

MEXICO
CUAUTLA
June 24, 1881

Faulty judgment on the part of an engineer caused the plunge of a train through a weakened bridge near Cuautla, Mexico, on June 24, 1881. A fire compounded the tragedy. Two hundred sixteen died; 40 were injured.

The most tragic train wreck in Mexican history was also one of the grisliest of all railroad accidents.

On June 24, 1881, a troop train loaded with more than 300 soldiers and 60 civilians approached a wooden bridge that spanned the rain-swollen San Antonio River near Cuautla. According to reports in the *New York Times,* there was some consternation among some of the passengers as they noted that the swirling floodwaters had almost reached the tracks along the embankment.

The bridge was not inundated, but the river was only a few feet beneath the trestle's tracks and moving swiftly. Nevertheless, the engineer chose to proceed.

Within minutes the bridge began to sway, made topheavy by the train's weight and weakened considerably by the force of the floodwaters. Moments later it gave way, toppling the train and its contents toward the river. But the engine and cars did not plunge directly into the roiling waters. They careened into the embankment, thus saving the train's occupants from drowning.

But their safety was only momentary. A freight car full of brandy requisitioned for military personnel aboard the train split open and poured its contents onto the engine and the passenger cars below. Sparks from the engine ignited the brandy, which turned the first few wooden cars into horrific infernos. Burning liquid poured into the open cars, setting fire to the military men trapped inside.

Most of the 60 civilians, all of whom were in the last cars of the train, escaped either unscathed or with varying degrees of injuries. But the soldiers in the first cars died horribly, "wrapped," according to the *Times* reporter, "in a sheet of flame." The engineer and conductor were both scalded to death by the steam from the engine's boiler.

Forty of the civilians were seriously injured; 216 passengers, most of them soldiers, died in this, one of the grimmest of all transportation disasters.

MEXICO
GUADALAJARA
January 18, 1915

Loss of control by the engineer on a steep grade accounted for the derailment of a train near Guadalajara, Mexico, on January 18, 1915. More than 600 died; scores were injured.

Even in the best of times, the military tends to obfuscate the precise details of incidents that result in great loss of life to that military's personnel. In 1915 Mexico was awash with civil unrest and violence. President Victoriano Huerta had resigned, partly because of U.S. military intervention under President Woodrow Wilson, and one of the revolutionaries, Venustiano Carranza, was the ostensible head of the country. But bands of revolutionaries, led by Francisco "Pancho" Villa and Emiliano Zapata, continued to disrupt the countryside.

Still, Carranza hoped to present a face of relative order in a country that had been split asunder by civil war, and so, when a railroad wreck of awesome proportions and horrendous fatalities occurred near Guadalajara on January 18, 1915, he and his military aides made certain that the world received no official news of it. The only report of this tragedy came from a letter written in February 1915 by American missionary Mrs. John Howland to the American Board of Commissions for Foreign Missions.

Guadalajara province was secured by Carranza's troops on the 18th of January, and as a reward to his troops, Carranza ordered that their families be sent by train from Colima to join them. The train carried 20 cars, but this was inadequate to accommodate the number of people. "The roofs [were] covered with men and women and many slung under the cars in a most perilous position even for ordinary travel," wrote Mrs. Howland. "At the top of the steepest grade, coming down," she continued, "the engineer lost control, the cars rushed down the long incline, throwing off human freight on both sides and finally plunging into an abyss.

"Nine hundred people were on the train and only six were unhurt. More than six hundred were killed outright," the letter concludes, thus making this one of the most lethal train wrecks of the world.

MEXICO
GUADALAJARA
April 3, 1955

No reason was given for the derailment of a passenger train over a gorge near Guadalajara, Mexico, on April 3, 1955. Three hundred were killed; hundreds were injured.

An astounding death toll of 300 persons made the April 3, 1955, calamity near Guadalajara one of Mexico's worst train wrecks. Certainly the horror of it ranks it as one of the world's worst railway catastrophes.

The popular night express from Guadalajara to the popular Pacific coast resort of Manzanillo, due west of Mexico City, was packed on the first weekend in April 1955. It was Holy Week, and thousands of vacationers were vacating the cities. A large number of Mexicans headed for the Pacific beaches, including Manzanillo. All weekend long, trains ran with passengers standing in the aisles and vestibules.

The night express was no exception. On April 3 it was packed, and there was a holiday mood aboard as it rounded a curve headed for a bridge that spanned a 600-foot-deep canyon near the little town of Alsaba.

Suddenly, celebration turned to tragedy as nine cars derailed and plunged into the canyon, tumbling over one another, telescoping at the bottom, crushing 300 passengers and injuring hundreds more. It would be a day before rescuers could climb down the steep sides of the ravine and extricate survivors from the tangled mass of twisted steel at the foot of the canyon.

MEXICO
SALTILLO
October 9, 1972

Drunkenness in the cab and excessive speed were the causes of the derailment of a passenger train near Saltillo, Mexico, on October 9, 1972. Two hundred eight were killed; hundreds were injured.

More than 1,600 Mexicans made the 60-mile pilgrimage by train from Saltillo and Monterrey to the shrine of St. Francis at Catorce, in the central part of Mexico, on Wednesday, October 9, 1972. It was the saint's day, and the pilgrims, composed of entire families from grandparents to grandchildren, made requests and promises at the shrine and then joined the festival celebrating the Day of St. Francis.

When the festivities and the ceremonies were over, they boarded the train for the short night journey back to Saltillo and Monterrey. In the locomotive of the train, far more secular activities were happening. The engineer and four crewmen had procured tequila and some women in a small whistlestop. A party continued through most of the journey.

By the time the train neared Saltillo, the engineer and his companions were thoroughly drunk and, unfortunately, reckless. At 10 P.M. the train rumbled down a moderately steep grade that led to a bridge two and a half miles south of Saltillo. The track curved at a six-degree angle before it reached the bridge, and because of this, the posted speed limit was 35 mph.

The train was doing 75 mph as it entered the curve. The locomotive fairly flew off the track, derailing 13 of the 22 passenger cars behind it. Most of them overturned, and four of them burst immediately into flames. "There were cars cut in half," said a policeman who arrived on the scene shortly after the catastrophe occurred, "bodies everywhere, and women and children crying and screaming, 'Get me out of here!'"

Dazed and injured survivors staggered about in the darkness, stunned and in pain. The screams of the dying and injured were like a grim chorus. Some survivors found the engineer and his companions, who were still drunk. Enraged, they grabbed the engineer and, ripping loose a cord from within the crushed and flaming coaches, attempted to lynch him and the train's conductor. A special inspector, Arnulfo Ochoa, intervened and promised to have the authorities deal with the dazed and shaken engineer and conductor, whom he then shepherded off to an arriving ambulance. Blood tests at the hospital to which they were taken confirmed their extreme drunkenness.

The long and ghastly job of cutting through the twisted and tangled wreckage began immediately and would continue for two days. Policemen, firemen, Red Cross workers, soldiers and civilian volunteers from a wide area worked through the night, the next day and the following night to extricate some who were trapped and would live. Cars were piled upon cars, and rubble was everywhere. Two hundred eight persons died and hundreds were injured in this, one of the worst rail crashes in Mexican history.

MEXICO
TEPIC
July 11, 1982

An eroded roadbed accounted for the derailment of a train near Tepic, Mexico, on July 11, 1982. One hundred twenty died; hundreds were injured.

The weather was bad on the night of July 11, 1982, in the mountainous regions of western Mexico. A train carrying 1,560 passengers from Nogales, on the Arizona border, to Guadalajara had safely negotiated much of the treacherous track that hugged the sides of mountains and sometimes ascended as much as 1,000 feet above the gorges below.

José Luis Velasco, the engineer, was a careful man, and there was probably nothing he could have done to avert what happened that night. A section of roadbed at the top of an 800-foot ravine had eroded. Fatefully, the weight of this particular train caused it to give way. Ten cars tipped, balanced for a moment and then plunged into the ravine. Coaches piled upon coaches, crushing those and their occupants under them. Others telescoped. Amazingly, some passengers were flung from the wreckage and so survived.

Rescue crews arrived on the scene, but the steepness of the mountain, the imminent danger of mudslides and the blinding rain hampered their efforts. Rescue helicopters were finally called in and airlifted the most gravely injured to hospitals in Guadalajara. Once this had been accomplished, the grim job of cutting the corpses out of the mangled wreckage began. One hundred twenty people, including one American, died in the crash. Hundreds more were injured.

NEW ZEALAND
WAIOURI
December 24, 1953

A swollen river caused by a volcanic eruption caused the bridge accident of a train near Waiouri, New Zealand, on December 24, 1953. One hundred fifty five died; there is no record of survivors.

The eruption of the 9,000-foot-high volcano Mount Ruapehu, near Waiouri, New Zealand, in December 1953 was described by seismologists as "minor." And on the scale of volcanic activities it undoubtedly was.

The eruption was, however, intense enough to send millions of gallons of water barreling down the River Wangaehu, which in turn weakened the Tangiwai Railroad Bridge. On Christmas Eve it was about to be crossed by the Wellington-Auckland express, a nine-car passenger train loaded with hundreds of well-wishers on their way to welcome Queen Elizabeth II, who was making a rare visit to New Zealand.

Partway across the span, the train broke through the sagging and swaying trestle and plunged into the roaring river. Some cars floated, momentarily, but most sank like boulders, drowning 155 hapless passengers.

PAKISTAN
GAMBAR
September 29, 1957

Inadequate signals caused the collision of a passenger train and an oil train in Gambar, Pakistan, on September 29, 1957. Three hundred died; 150 were injured.

Nearly 300 people were burned to death or died from the collision shortly before midnight of September 29, 1957, of a Karachi-bound passenger train and a stationary oil train at Gambar, near Montgomery, West Pakistan. More than 150 were injured as the overcrowded, speeding passenger train crashed into the oil train, which was stopped in the yards of the Gambar railroad station.

Apparently, there were inadequate signals, and the passenger train had been routed onto the wrong track. By the time the engineer of the passenger train realized his situation, he had driven his locomotive full force into the oil train, which exploded with a thunderous roar, collapsing the railroad station and literally blowing apart the passenger train's engine and forward coaches.

The search for bodies and survivors went on through the entire night of September 29, and by daylight it was apparent that it would go on for hours more. The charred, smoking wreckage contained the bodies of hundreds; others had been blown into the countryside. It would be one of the worst rail wrecks in Pakistan's history.

PAKISTAN
SANGI
January 4, 1989

In Pakistan's worst railway disaster in history, a passenger train was incorrectly switched to a siding in Sangi village, where it crashed into a parked freight train on January 4, 1989. Three hundred and seven were killed and 700 injured.

Trains in Pakistan are often named after Muslim holy men, and the *Zakaria Bahauddin* was one such train, traveling the 500-mile overnight route south from Multan to Karachi. On the night of January 3, 1989, it left Multan severely overloaded—another customary condition of trains in both Pakistan and India. Capable of carrying 1,408 passengers, it was weighed down with nearly 2,000 that night.

Early in the morning of January 4, the *Zakaria Bahauddin* approached the village of Sangi in Sind province, a little more than halfway to Karachi. The main line bypassed the village, and the train was traveling at approximately 35 mph as it neared the village.

Then, abruptly, instead of continuing on its intended path, the train veered off onto a siding. An empty 67-car freight train was parked in the siding, and the passenger train roared into it, sundering boxcars, ripping passenger cars open, spilling passengers onto the siding and under the cars of the still plummeting train. The locomotive overturned, taking the first three of the 16 cars of the *Zakaria Bahauddin* with it. It was in these cars that most of the deaths and injuries occurred.

Medical and rescue teams rushed to the scene of the accident, and ambulances took some of the injured to the seven local hospitals, whose 300 beds filled rapidly. Helicopters airlifted other survivors to Karachi and other cities.

The engineer of the passenger train miraculously survived, but 307 people died and 700 were injured in this, the worst railway disaster in Pakistan's history, caused by a signalman's inattention. The unidentified signalman who switched the passenger train onto a collision course with the freight was arrested and jailed for manslaughter.

POLAND
NOWY DWOR
October 22, 1949

Excessive speed was blamed for the derailment of a passenger train near Nowy Dwor, Poland, on October 22, 1949. Two hundred were killed; 400 were injured.

Not much news seeped through the government censorship agencies of the Eastern bloc during the Cold War. Soviet authorities were particularly loathe to admit tragedies that occurred. So for years and years, as far as the rest of the world knew, no train wrecks, air crashes, ship sinkings or other disasters that are the unfortunate consequence of modern-day living occurred in the countries of the Eastern bloc.

Occasionally, however, news did leak out from "unofficial sources." In late October 1949 news came from Poland of its worst rail wreck in 30 years, and one of the worst of the century. Two hundred people were killed and nearly 400 were injured when the Danzig-Warsaw express, traveling at excessive speed, left the rails on a curve near the town of Nowy Dwor, northwest of Warsaw. No further details were ever provided to the world outside Poland.

PORTUGAL
CUSTOIAS
July 26, 1964

The overloading of a passenger car caused its derailment in Custoias, Portugal, on July 26, 1964. Ninety-four died; 92 were injured.

The worst train accident in the history of Portugal occurred because of the overloading of one passenger car. Designed to carry a maximum of 70 passengers, one of the cars of the Automara express was packed with 161 holiday revelers on their way from the seaside resort of Povoa de Varzim to Oporto, Portugal, on July 26, 1964.

A short six miles from the end of their journey, near the village of Custoias, the overloaded car came uncoupled, jumped the track and raced down an embankment. It would take seven hours to excavate the smashed debris of the car and unearth survivors and the dead. Sixty-nine people died instantly in the accident; 92 were seriously injured, and of those another 25 died in the hospital, bringing the total of dead to 94.

SPAIN
LEBRIJA
July 21, 1972

Human error—the ignoring of a signal—caused the collision of two trains near Lebrija, Spain, on July 21, 1972. Seventy-six died; scores were injured.

Servicemen seem to have suffered the most casualties in the world's worst train disasters. The reasons are logical.

Until very recently, troops have been transported primarily by train. Off-duty servicemen on leave generally return home from camp in large numbers, frequently on overcrowded trains.

The latter was the case on July 21, 1972, when the Madrid-Cadiz express, carrying more than 500 passengers in its 14 cars, slammed into a similarly packed local train making its way out of the station at Lebrija.

The post-crash investigation revealed that the engineer of the local, which was loaded with Spanish sailors on leave, had ignored a signal and moved directly into a collision course with the express, which had the right of way. The local sustained most of the damage and all of the fatalities, but there were 103 people injured on both trains, some of them seriously. The 76 dead—all sailors on the local—were pulled from the wreckage by other servicemen either on the trains or from the American Polaris submarine base, which was located very near the site of the wreck.

SWITZERLAND
BASEL
June 14, 1891

A collapsed bridge caused the accident of June 14, 1891, in Basel, Switzerland. One hundred twenty were killed; scores were injured.

The train, consisting of two engines and 12 wooden carriages, departed from Basel en route to Delsburg and began to move across the newly constructed Monchenstein bridge at a normal and respectable rate of speed. The Birs River, which the bridge crossed, was placid; there was no indication that a strong current or a storm-swollen cataract had weakened the bridge.

Yet it collapsed immediately after the two engines had cleared the span. The three carriages directly behind the engines uncoupled themselves from the remaining cars and plunged to the river, twisting as they went and flipping the engines over on their sides. The engineers and stokers were crushed to death.

Behind them 120 passengers were killed instantly, trapped and drowned in the wreckage in the river. The occupants of the third car were seriously injured but at least survived.

The rescue work proceeded very slowly. Some bodies were disentangled from the wreckage and laid out for identification. But many, many more were left in the wreckage, and for weeks they would drift one by one to the surface and come ashore in the small riverfront villages downstream from the crash site.

UNITED STATES
ALABAMA
MOBILE
September 22, 1993

In the deadliest crash in the history of Amtrak, the Sunset Limited left a bridge over an Alabama bayou near Mobile on September 22, 1993, and plunged into the water below after the bridge had been hit and damaged by a towboat pushing six barges. Forty-seven died.

Amtrak's *Sunset Limited* wends its way through the Southwest from Los Angeles to Yuma and Tucson to El Paso, then to the border town of Del Rio and on to San Antonio. After this it turns toward the Gulf coast, where it stops at Houston, New Orleans, Gulfport, Biloxi, Pascagoula, Mobile and then on to Miami.

It was just about 2:00 A.M. when it pulled into Mobile, Alabama, in the early morning of September 22, 1993, and when it left there a half hour later, it consisted of three locomotives, a baggage car and crew dorm, three coaches, a lounge, a dining car and a sleeper. The trip east and south of Mobile was through swamp and bayou country. Unruly rows of magnolia and cypress defined the water ways, along which a constant stream of barge traffic inched along the Mobile River towards Birmingham and Tuscaloosa.

At the same time that early morning, the towboat *MV Mauville* was making its way north through a heavy fog pushing six barges loaded with coal and wood chips. The towboat's captain, Andrew Stabler, had gone below at 11:30 P.M. and was sleeping soundly, and the wheel of the boat was manned by Willie Odeon, a 13-year employee of the Warrior and Gulf Navigation Company, which owned the towboat and barges.

Odeon was unfamiliar with the area and so was unaware that he had left the Mobile River and entered the Big Bayou Canot, a channel over which an Amtrak railroad bridge passed. Barges were forbidden from the channel because the trestle only cleared the surface of the water by seven to 12 feet, depending upon the tide.

The *Mauville* had neither a compass nor nautical charts of the area, nor was it required to carry them by law. It possessed radar, but Mr. Odeon did not know how to use it, nor was he required to by law.

At a little before 2:45 A.M. Mr. Odeon saw a shadow in the fog that looked to him like another towboat and barges, and he headed toward it, expecting to tie up alongside and wait out the fog. One of his six barges slammed into the object at 2:45.

And now the *Sunset Limited,* traveling at 72 mph, approached the trestle. Eight minutes later, at 2:53, it started across the bridge. But the *Mauville* had struck not another barge, but the bridge, and it had forced the tracks out of alignment 41 inches. It was enough to force the edge of the bridge's center span directly into the path of the oncoming *Sunset Limited.*

With a grinding of metal and squealing of steel wheels against tracks, all three engines and the first four cars of the train became momentarily airborne, then plunged into the water of the bayou. One of the engines exploded in flames, and the cars following it dove into the muck and the water, some floating for a while, one descending 30 feet to the bottom of the bayou.

"My watch stopped when I hit the water at 2:55 a.m." 72-year-old survivor James Altosimo told reporters later. "I was asleep, and a jolt woke me up, knocked me from my seat, and by the time I awoke, water was coming into the car."

Mr. Altosimo and other passengers climbed through windows and swam for what remained of the trestle, where they hung on for an hour watching the engines burn 40 to 50 feet away.

"The only light was from the fire," Mr. Altosimo continued. "and it was so hot that I had to splash water on my back to cool off. The stench of the diesel fuel was terrible. We were scared of the water catching on fire."

"It was like a combination of the worst roller coaster and the worst horror show," said another survivor, ". . . with all those fires and electrical wires snapping and we all thought that we were goners."

From the top of the 50 feet that was left of the 498-foot-long wood and steel trestle, the passengers in the three cars remaining on the track made their way furtively to the rear of the train, where they exited, just short of the bridge.

"When it first started, there was a grinding-like noise, like maybe the track was bad or something," said one of these survivors. "Then it got worse, and the train began to jerk and twitch and all of a sudden it came to a dead stop. I got up and opened the door and looked down, and all the train ahead of me was down there and my car had stayed on top of the track."

Survivors were holding on to logs and pieces of wreckage. Those who could swim tried to save those who could not. And the towboat pulled seven people from the warm, black, fuel-stained water.

For the surviving passengers, it seemed like an interminable wait in the midst of chaos for help to come. Crew members from the train never appeared. The feeling of isolation was heightened by the thick fog and the glow thrown off by the burning engine.

"We were throwing sheets and blankets out of the cars that were still on the track to the people who were wet from being in the water. Some people were hysterical. There were a lot of old people on the train," said Edward Mouton, a passenger in one of the three rear cars. "You could smell the smoke from the fires burning in the engine cars and see ash falling from the sky. But it was so foggy you really couldn't see the water."

And still no help arrived. The reason became important in the later investigation. A full eight minutes elapsed from the time that the towboat hit the bridge until a distress call was sent to the Coast Guard. "Mayday. Mayday," it said. "I've lost my tow. There's too much fog. Don't know exact location."

If information had gone instantaneously to the Coast Guard, the investigators concluded, there would have been time to warn the train to stop. But eight precious minutes remained unused. And those on the towboat knew what had happened. The captain had been awakened by the collision with the bridge. He was in the pilot house when all of them heard a swoosh and saw an explosion of fire.

It was not until 3:08 that the Coast Guard received a frantic message from the towboat: "It's real bad here. There's a train that ran off the track into the water and there's lots of people that need help, and there's a fire. Hurry, and get out here, Coast Guard. I'm going to try to help some of them. I'll get back to you."

It would be 4:25 before the first Coast Guard vessel entered the dead end bayou, and the first helicopter arrived at 5:20. By 9:00 A.M. there were seven Coast Guard aircraft there or nearby.

As for a warning to the train before it reached the bridge: The bridge had been built in 1906 and had no warning lights. There was a sensor system that warned of breaks in the track. But the misplacement was in the bridge, not the tracks. More than a decade before the crash, the Federal Railroad Administration, following a 1979 freight train derailment on a bridge in Devils Slide, Utah, had recommended a multimillion dollar installation of detection devices on all railroad bridges. The federal government had determined that it was too expensive to install.

And so the deadliest crash in the history of Amtrak claimed 47 lives in Mobile in 1993. Forty-five died from drowning; the other two were burned to death.

"I've always felt safe on Amtrak," Esther Lucius told reporters after the calamity. "I don't like flying." Asked if her feelings had changed, she answered, "I'll have to think about that."

UNITED STATES
COLORADO
EDEN
August 7, 1904

Swollen rivers caused the wreckage of one bridge to collide with another, collapsing it just as a train passed over it near Eden, Colorado, on August 7, 1904. Ninety-six died; scores were injured.

Collisions of trains are not uncommon. But collisions of bridges are quite another phenomenon, destined for the record books. And such an occurrence resulted in one of the United States' more terrible train disasters.

Much of the land of the American West is scarred with arroyos—gullies that once contained a river or small stream. Pacific train number eight, which because of the St. Louis World's Fair had earned itself the nickname World's Fair Flyer, crossed a number of these arroyos on its trip from Denver to St. Louis and back again. Arroyos were usually just part of the scenery, and the bridges that forded them were maintained in a desultory manner, as were the wagon bridges, some of which were never maintained but left to age in the relentlessly extreme weather.

The beginning of August 1904 brought heavy rains to Colorado, and the normally dry arroyos were no longer unobtrusive. They overflowed with boiling mountain streams tumbling toward valleys miles away. On the night of August 7, the World's Fair Express paused momentarily at the small mountain town of Eden and then pushed on toward St. Louis.

One of the oddest of all railway wrecks took place as a result of the collision of two bridges—one wrecked by floodwaters, one an innocent bystander over an arroyo near Eden, Colorado. Trestle and train were both flung into the floodwaters on August 7, 1904. (Frank Leslie's Illustrated Newspaper)

If its only task had been to cross Steele's Hollow Bridge, its trip would have been as serene as it usually was. But on this particular night coincidence conspired with nature and brought about a catastrophe. Just as the seven-coach express began its traverse of the railroad trestle, an ancient wagon bridge a short distance upstream collapsed with a roar.

Propelled by the floodwaters in the arroyo, the timbers of the wagon bridge acted like battering rams and slammed into Steele's Hollow Bridge just as the train was in midspan. Trestle and train collapsed simultaneously and were flung into the water. Nothing was left of the three main spans of the bridge. The train's locomotive, baggage car and chair and smoking cars dove into the stream together. A quick-thinking porter, seeing the disaster occurring ahead of him, grabbed the air brakes on the two sleepers and dining car, saving himself and his fellow passengers from the horrible fate of those in the rest of the train.

Frank Mayfield, the fireman of the train, was thrown clear. He landed on the embankment and was knocked unconscious. Later, his eyewitness account describing the wreck and early rescue attempts formed the basis of most news accounts of the disaster.

What he saw were pullman cars sitting serenely on the tracks, no sign of the engine and passenger cars with their roof burst open by the impact. Passengers and wrecked cars alike were driven miles downstream by the raging floodwaters, and the army of more than 500 rescuers would spend days and nights digging debris and bodies out of the muddy river banks. Ninety-six died and scores were injured in this freak accident.

UNITED STATES
CONNECTICUT
SOUTH NORWALK
May 6, 1853

Disregard of both a signal and a speed limit caused the bridge accident over the Norwalk River near South Norwalk, Connecticut, on May 6, 1853. Forty-six were killed; 25 were injured.

The year 1853 was a bad year for drawbridge accidents in the United States. Two occurred that year; one was minor, with neither an injury nor the loss of life. But the other was horrendous and is generally considered to be the worst U.S. railroad accident until that time.

The second wreck took place on a drawbridge spanning the Norwalk River near South Norwalk, Connecticut, at approximately 10:30 on the morning of May 6, 1853. The bridge was a well-marked one, with a warning tower consisting of a 40-foot pole from which a red ball the size of a basketball was suspended. If the ball was lowered and visible, it was safe for the trains of the New York and New

Haven railroad to proceed. If it could not be seen, that meant that the drawbrige was open.

At 10:15 that morning the steamboat *Pacific* whistled to pass the closed bridge. The bridge tender, a man named Harford, raised the warning ball for trains and opened the bridge, allowing the *Pacific* to steam through.

Meanwhile, an express passenger train bound for Boston from New York shot through the Norwalk station trying to make up lost time. Its engineer was one Edward Tucker, and his safety record was anything but spotless. It was he, in fact, who had been the engineer in the New York and New Haven's first wreck, in 1849, its first year of existence. He had been severely injured in the head-on collision in Greenwich, Connecticut, although an investigative team had concluded that he had put his train on the wrong track on the advice of his conductor. Still, that accident was caused by inattention to signals, and perhaps that had something to do with the tragedy Tucker was about to cause as he sped his express toward the Norwalk River Bridge.

It was a clear morning. Visibility was unlimited and unimpeded. Once a curve had been negotiated, there were 3,000 feet of straight track to the bridge, during which the warning tower was visible at all times. But for some reason, no one—engineer, conductor or fireman—saw it or, apparently, looked for it. "It's not my duty to look for it," fireman George Elmer testified later.

The train roared on at full throttle, despite the posted 10-mile-per-hour speed limit. Approximately 370 feet from the edge of the 60-foot gap between the tracks' end and the stone tower of the wooden drawbridge, Tucker saw the river—and no bridge. He blew two blasts of his whistle, the signal to the brakeman in the passenger cars to apply the hand brakes. But the brakemen had seen the coming catastrophe at the same time as Tucker had, and instead of applying the brakes, they jumped for their lives.

Showing a similar disregard of responsibility, the conductor, fireman and engineer all leaped from the train without a word of warning to the passengers. The railroad employees rolled free just as the locomotive shot into space, leaped across the 60-foot gap and smashed into the bridge's concrete pier. The tender, two baggage cars and two passenger cars followed. The remaining passenger cars would have also made the leap had it not been for passengers in the last two coaches who grabbed the idle hand brakes and brought their cars to a squealing stop. The third passenger car, caught at the edge of the precipice, cracked open like a rectangular egg, spewing seats, luggage, floor and ceiling timbers and terrified passengers into the Norwalk River. One eyewitness told the *New York Times*, "Many of the seats and the dislodged window sashes, with a crowd of timber fragments, were propelled, some of them, fully across the gulf, and two of the passengers, who were seated just at the spot where the car snapped asunder, were thrown a full twenty feet forward and pitched with frightful force upon the ruins of the second and first cars."

The steamboat *Pacific* stopped immediately and set about trying to rescue survivors, who were frantically swimming to the surface. Sailors and passengers alike on the *Pacific* dove into the waters to assist survivors as the wrecked passenger cars shifted and began to sink.

Meanwhile, on shore a group of South Norwalk citizens had witnessed not only the accident but the bailing out of the train's crew. The group surrounded engineer Edward Tucker, who had broken his leg in his leap for life. One mob member had already produced a lynch rope, but he was challenged by other men who felt that hanging was too good and that shooting would be the proper fate for Tucker. The argument lasted just long enough to allow the police to arrive and extricate Tucker from the mob.

The same afternoon a jury was convened, and although Tucker vowed that the warning ball was not in place, other witnesses stated that it was, and both he and the directors of the railroad were found guilty of extreme negligence.

No criminal charges were pressed, but public indigination forced New York and New Haven officials to require all trains henceforth to come to a full stop before every drawbridge. In addition, Connecticut set up the state's first Board of Railroad Commissioners to investigate all railroad accidents and impose stricter safety rules for railroads.

Forty-six people died in this tragedy, and 25 were injured. There would be more bridge accidents in the future, but this dramatic one would initiate a growing movement toward stricter railroad regulation and safety practices.

UNITED STATES
ILLINOIS
BOURBONNAIS
March 15, 1999

The Amtrak City of New Orleans *collided with a tractor-trailer truck carrying steel at a crossing at Bourbonnais, Illinois, on March 15, 1999. Eleven were killed and 50 injured.*

The *City of New Orleans* is one of Amtrak's crack trains, leaving Chicago in the early evening and arriving in New Orleans on the late afternoon of the next day. On the night of March 15, 1999, the *City of New Orleans* carried 196 passengers, 19 Amtrak employees and two employees of the Illinois Central Railroad through the dark of the Illinois countryside.

By 9:45 most of the inhabitants of the double-decker first-class sleeping car had crawled into their bunks for the night. The train was traveling at 79 mph, its top speed for the area. It approached the gate crossing at the small town of Bourbonnais, where no express trains stopped. The lights were all green; the gates were down. At 27 seconds before the train reached the crossing, the engineer sounded his whistle.

And then, without warning, at 9:47 P.M. the night exploded as the engine struck the back of a trailer truck carrying heavy steel reinforcing bars from the Birmingham Steel Company. The impact split the engine in two and derailed most of the cars of the train. They splayed themselves over the countryside, several catching fire, including the dining car and the double-decker first-class sleeper, which had been slammed up onto one of the train's two locomotives. Split in two and ravaged by fire, it became an incinerator of some of the passengers trapped within it. Other cars were flung into two freight cars on the side tracks, one full of steel and the other full of furnace residue from the Birmingham Steel Company.

Thick smoke pervaded the area, and the injured stumbled from the overturned cars into the arms of the members of the Bourbonnais volunteer fire department. "I don't think I've ever seen anything like that," said fire chief Michael Harsbarger later. Flames could be seen for miles.

Eleven died and more than 50 were injured. And now the contention began. Both the engineer and the driver of the tractor-trailer survived, although both were injured enough to be taken to a local hospital. The engineer maintained that the truck had tried to circumvent the gates that had all been working properly. And tire tracks in the mud around the gates plus an eyewitness who was behind the truck seemed to confirm this.

The truck driver vowed, however, that the gates came down *after* he had entered the crossing and that he was trapped between them. Another eyewitness confirmed this story.

Inspection by the National Transportation Safety Board found that all the signals and gates were working properly. A public board of inquiry was convened, and no criminal charges were placed against the truck driver. The fine points of what occurred that March night were never unraveled.

UNITED STATES
ILLINOIS
CHATSWORTH
August 10, 1887

Human negligence led to the weakening of a railroad bridge, which caused an excursion train to fall through it at Chatsworth, Illinois, on August 10, 1887. Eighty-two died. There is no record of injuries.

In the same way that the terrible South Norwalk bridge disaster of 1853 (see p. 354) brought about regional safety precautions, the even more disastrous bridge accident of August 10, 1887, at Chatsworth, Illinois, raised the responsibility for these regulations to the federal level with the Interstate Commerce Commission and later the U.S. Department of Transportation.

In this event, as in the earlier disaster, human negligence was the prime culprit. This time, however, it was not the engineer of the train but the head of a railroad track-work gang who was responsible for the needless loss of 82 lives.

Timothy Coughlin was the man in charge of a railroad gang cleaning dry brush from the tracks near a small 15-foot wooden trestle that spanned a shallow, usually dry culvert near Chatsworth, Illinois, on August 10, 1887. Their method was to collect the weeds in piles and burn them.

At quitting time Coughlin cautioned his men to make sure the fire was out; but once he gave the order, he failed to follow it up by checking each of the smoldering piles. That night an excursion train of the Toledo, Peoria & Western Railroad left Peoria bound for Niagara Falls. It was packed with residents of the farmlands of central Illinois on their way to a holiday at the falls. As the train sped through Illinois shortly after midnight, its engineer spotted what seemed to be a brushfire on the tracks ahead of him. Too late, he realized that it was a wooden bridge burning. The smoldering remnants of the burned weeds that Coughlin's crew had left that afternoon had set the trestle on fire.

The engineer reversed the engine and whistled for brakes. But he was going too fast, and it was too late to prevent the engine from plunging straight into the flames. The lead locomotive made it across the gulf safely, but the weight of the second engine snapped the fire-weakened timber supporting the tracks, and the entire wooden bridge folded in upon itself and the train, toppling both into the creek. Nine passenger cars crashed, one on top of the other, splintering apart and crushing some of their occupants. The 10th car, a sleeper, managed to brake to a stop at the edge of the trestle, but the speed of the train drove the second sleeper into it, telescoping both cars and killing most of the occupants.

It was an inferno of a wreck, needless in its cause, horrendous in its consequences and so impressive to the people of the Midwest at that time that it became part of its regional folklore. "The Bridge Was Burned at Chatsworth," attributed to T. P. Westendorf, contained the following vivid description of that flaming night:

The mighty crash of timbers
A sound of hissing steam
The groans and cries of anguish
A woman's stifled scream.
The dead and dying mingled,
With broken beams and bars
An awful human carnage
A dreadful wreck of cars.

UNITED STATES
MASSACHUSETTS
REVERE
August 26, 1871

Outdated equipment was responsible for the collision of two excursion trains in Revere, Massachusetts, on August 26, 1871. Thirty-two died; more than 100 were injured.

One of the best known of all American railroad accidents was noted not for its immense casualty figures (32 were killed), but for the impact it had on the public. It occupied the newspapers and minds of mid-Victorian America for months afterward.

August 26, 1871, was an unusually busy Saturday for the Eastern Railroad out of Boston, mainly because of three weekend events: two major religious revival meetings and, in an interesting juxtaposition, a military muster. Because of the crush of people, 192 trains left the Eastern's Boston depot each day of the weekend.

The main line ran from Boston north to Salem via Lynn, with several branches along the north shore. At 8:30 on Saturday evening, August 26, a slow-moving local switched onto this main line from a branch near Revere, Massachusetts. It was a particularly dark night, and according to later reports in the *New York Times*, the engineer of an express, moving at 30 mph, did not see the rear of the slow-moving local until his train was almost upon it. He immediately hit the outdated hand brakes. Although other railroads had already switched to the new Neihouse air brake, the Eastern had not, and the old hand brakes were useless on rails that were slick from rain.

With a roar the express slammed into the rear of the local, forcing the engine two-thirds of the way through the rear car. Steam pipes on the engine's boiler erupted, blasting live steam into the car and searing its occupants. Simultaneously, hot coals from the firebox mixed with the smashed kerosene lamps in the two last cars, setting the wood of these carriages on fire. Flames engulfed both coaches. Thirty-two passengers were either crushed to death or incinerated by the consuming fire, and more than 100 were injured, some of them critically.

American newspapers fed the insatiable appetite of the public for months following the collision of two excursion trains in Revere, Massachusetts, on August 26, 1871. *(Frank Leslie's Illustrated Newspaper)*

Boston erupted with cries of "deliberate murder" directed against the railroad's management. Resultant lawsuits nearly bankrupted the company, forcing it finally to modernize and monitor its equipment.

UNITED STATES
NEW JERSEY
ATLANTIC CITY
July 30, 1896

Human error was blamed for the broadside collision of the Philadelphia express with a West Jersey excursion train at "Death Trap," a track intersection in Atlantic City, New Jersey, on July 30, 1896. Sixty were killed; hundreds were injured.

> When I was at the crossing, I saw the train coming with unslackened speed, and I shouted to [my fireman] Newell, "My God, Morris, he's not going to stop!" Then I followed the first impulse, and leaped from my seat to the floor, and then to the step. I hesitated about jumping after I was on the step, and then, through some unaccountable impulse, I sprang back into the cab. Had I leaped I would have been buried beneath the wreck. . . . When the crash came, my engine broke loose and ran down the track. When I ran back, the sights and sounds I witnessed unnerved me, and I have been in a tremble ever since. I shall never forget the sight of that Reading engine as she rushed toward us.

Thus engineer John Greiner described to reporters the moment of impact of a broadside collision that took place early in the evening of July 30, 1896, at "Death Trap," an insidious and potentially lethal intersection of tracks located at Atlantic City, New Jersey. For years trains of the West Jersey and Pennsylvania lines had had close calls at the crossing, saved only by the alertness of signalmen who gave the white light for safety and the red light for danger to trains approaching Death Trap.

Around 7:00 P.M. on July 30, 1896, the West Jersey excursion train, a slow mover out of Atlantic City, approached Death Trap and was given the white light to proceed. Engineer Greiner of the West Jersey train was aware that the Philadelphia express of the Reading line was approaching the same intersection from another direction and at high speed, but he followed his signals and proceeded to enter the intersection. He had done everything right. Upon leaving the drawbridge just before the crossing, he had whistled for signal instructions and had received them, and then had eased himself into the notorious crossing.

What was going through the mind of Edward Farr, the engineer of the Philadelphia express, will never be known, but it must have had nothing to do with heeding signals. Despite the clear red light that should have prevented him from entering the crossing, he barrelled in at 50 mph, seemingly unaware that there was another train—the Jersey excursion train—directly in front of him. He smashed full force into the side of the second car of the Jersey train, lifting it from the tracks and hurtling it down an embankment and then into a marsh. The second, third and fourth cars telescoped into one another and spun off the tracks and down the embankment.

The express train's engine ricocheted off the cars it had devastated and, carrying the first car of the express with it, hurtled into a ditch, where it immediately set fire to the wooden cars into which it smashed. Its boiler exploded, sending scalding steam into the wreckage and killing some of those who had survived the initial impact.

It was a grisly sight. All of the wooden cars caught fire. The speed of the express flung bodies like discarded dolls through windows and from one end of smashed cars to the other. Engineer Farr was dismembered, and parts of his body were found the length of the car behind the engine. Individual acts of heroism abounded as passengers and rescuers alike risked the fire that consumed the scene.

By dawn the boardwalk was littered with bodies and bandaged survivors. Sixty people were killed; hundreds were injured, and many were mutilated for life.

UNITED STATES
NEW JERSEY
HACKETTSTOWN
June 16, 1925

A collision with a mudslide caused the wreck of an excursion train in Hackettstown, New Jersey, on June 16, 1925. Thirty-eight died; 38 were injured.

The wreck that occurred when an excursion train from Chicago to Hoboken, New Jersey, ran into a mudslide at 3:30 A.M. on June 16, 1925, would have been a minor accident had the engine been a diesel. But 1925 was the age of steam, and the violent explosion of the train's boiler turned a mishap into a disaster.

The train, consisting of two coaches, four pullmans and a diner, was crammed with 182 German-Americans on their way to the port of Hoboken. From there the celebrators would board the SS *Republic* for a summer holiday in Germany.

By 3:30 on June 16, all of the passengers, thoroughly relaxed by an afternoon and evening of celebration, were serenely asleep. The engineer had encountered a huge cloudburst around 3:00 A.M., but by 3:30 it had passed, and he had opened the throttle in order to utilize the downslope of Rockport Sag to pick up momentum. It was necessary. The upgrade in Hackettstown lost both speed and time for trains on rigid schedules.

Under normal conditions this was an easy, expected routine. But this night the cloudburst had caused an immense mudslide partway along Rockport Sag. Mud covered both the embankment and the tracks, and the engine's headlight failed to pick this up until it was already on top of the mud. For 160 feet the engine careened crazily, barely remaining on the track. A siding switch was hidden beneath the mud.

The engine struck the switch and derailed. It instantly churned down a 12-foot embankment, carrying the first four cars with it. Reaching the bottom, the engine toppled over on its side, killing the engineer, the two firemen and the head trainman instantly. The first and second cars slammed crossways, one on the engine, the other across the tracks, and the first two pullmans derailed on either side of the mud pile.

Aside from the railroad employees, one other known passenger, a decapitated woman, was the only fatality at that moment. But then the engine's boiler exploded, shooting steam through the cars and horribly scalding the awakening passengers within.

Their screams brought rescuers from the surrounding farms to the scene, which had now turned nightmarish. Scores of charred bodies and survivors so badly burned they pleaded for either morphine or death were dragged from the destroyed cars. One man, his face completely burned away, was reported by a local paper to have withdrawn a roll of bills from his pocket and roamed the accident site offering money to anyone who would put a bullet through his head.

Thirty-eight passengers died; 38 more were taken to hospitals with serious injuries. The remaining 100 or so passengers were put aboard another train to continue their trip toward their now-ruined holiday.

UNITED STATES
NEW JERSEY
HIGHTSTOWN
November 11, 1833

A broken axle caused the first derailment in U.S. history in Hightstown, New Jersey, on November 11, 1833. One person was killed; several were injured.

The first passenger-car wreck in U.S. history took place on November 11, 1833, in Hightstown, New Jersey, on the Camden & Amboy line. Only one passenger was killed, but the effects of this accident were curious and far reaching.

The accident occurred when the train, roaring along at 25 mph—a high speed for its time—snapped an axle on one of the passenger cars. Instantly, all of the cars on the train derailed, and the wooden coaches, careening off the track, broke apart. The passengers were tossed around, bruised and cut, and some were flung from the cars. James C. Stedman, a jeweler from Raleigh, North Carolina, became the first passenger fatality in U.S. railroad history when he died, several hours later, from injuries received in the accident.

Two prominent men happened to be aboard the ill-fated train that day. One, ex-President John Quincy Adams, escaped uninjured. The other, Commodore Cornelius Vanderbilt, was flung from his carriage and rolled down a 30-foot embankment. One of his lungs was punctured and several of his ribs were broken. He was transported back to his home in New York, where he was reported to hover between life and death for a month.

When the commodore recovered, he vowed never to invest a penny of his enormous fortune in railroads. He was convinced that they were a bad and dangerous investment.

He later changed his mind greatly. In 1862, at the age of 68, when his fortune from steamboats had climbed to $11 million, Commodore Vanderbilt decided to forget his accident and invest in railroads. By the time he died, 15 years later, he controlled the entire New York Central system.

UNITED STATES
NEW JERSEY
WOODBRIDGE
February 6, 1951

Speed was the culprit in the derailing of an express train at Woodbridge, New Jersey, on February 6, 1951. Eighty-four died; more than 100 were injured.

While a new railway bridge was being built over the New Jersey Turnpike, traffic on the Pennsylvania Railroad's Jersey Shore line was diverted to a winding bypass. In the shape of an S, the diversion carried the caveat: "Trains and engines must not exceed a speed of 25 miles per hour." So much for warnings, especially when they are offset by a company schedule that requires commuter trains to make the 35-mile run from Newark to Red Bank in 44 minutes, including stops at various points on the Jersey Shore.

On February 6, 1951, *The Broker,* one of the Pennsylvania's most posh and popular commuter trains, set out on its usual trip between Newark and Red Bank. It was one of the last trains on the Pennsylvania to be pulled by a steam locomotive, and its engineer was Joseph Fitzsimmons, a 47-year-old veteran with a superior record of safety.

What caused Fitzsimmons to favor his timetable over a clearly stated speed limit will forever remain a mystery. Not even the inquiry after the accident could reveal that information from the befuddled engineer.

At 5:43 P.M. on February 6, pulling seven cars packed with commuters, he plunged into the beginning of the S-curved diversion at 50 mph. The violent lurching caused

Rescue workers struggle to pull survivors from the Pennsylvania Railroad wreck in Woodbridge, New Jersey, in February 1951. Wheels and shattered cars are strewn over the embankment from which the train hurtled when a temporary wooden trestle buckled. (American Red Cross)

by the navigation of the first part of the curve galvanized conductor "Honest John" Bishop into immediate action. He tried to shove his way through the crowd of commuters to the emergency stop lever. But by then the locomotive had plunged on into the tight conclusion of the S-curve, and within seconds it had shot off the track, become momentarily airborne as it cleared an embankment and then crashed on its side, hauling all seven cars with it.

Fitzsimmons was thrown clear. His fireman and the passengers in the twisted steel cars fared considerably worse. Pandemonium reigned in the cars, where some commuters had already died and others were dying. Passengers kicked, pummeled and clawed their way through broken windows and mangled doors. Eighty-four died in the crash, most of them in the first four cars. More than 100 were injured.

UNITED STATES
NEW YORK
ANGOLA
December 18, 1867

Defective equipment caused the derailing of the Lake Shore express near Angola, New York on December 18, 1867. Forty-three were killed; hundreds were injured.

For months after it occurred, the "Angola Horror," as the newspapers in Buffalo and elsewhere called the derailment wreck of December 18, 1867, was given a great deal of ghoulish treatment in the national tabloid press and magazines. The city of Buffalo did nothing to dissuade the tabloids and in fact added to the grisly carnival. It held a mass funeral for the accident victims in the Exchange Street depot a mere three days before Christmas of that year.

It was truly a horror, an avoidable tragedy caused by a defective axle on the rear car of the eastbound Lake Shore express on the afternoon of December 18, 1867. Speeding to make up time, the train approached the bridge spanning Three Sisters Creek near Angola, New York, without incident. But just before it reached the bridge, a wheel on the bent axle hit the "frog," or upthrusting part of a switch. The jolt was enough to derail the car, which in turn derailed the car ahead of it.

Swinging crazily on and off the track, the last car finally slammed into a bridge abutment. The impact burst the car's coal stove apart, scattering red-hot coals around the car and on some of the passengers. Within minutes the wooden coach was ablaze, and all of its 42 passengers were burned alive.

Meanwhile, the detached second car was dragged half on and half off the track for 300 feet before it broke loose from the rest of the train and plummeted down an embankment. Amazingly, though the car was smashed apart, only one passenger was killed.

UNITED STATES
NEW YORK
BROOKLYN
November 2, 1918

Excessive speed, outdated equipment, defective brakes and an overworked motorman combined to cause the derailment of a Brooklyn Rapid Transit train outside Malbone Tunnel in Brooklyn, New York, on November 2, 1918. Ninety-seven died; 95 were injured.

Three conditions conspired to both cause and worsen the crash of a Brooklyn Rapid Transit train that derailed just outside the Malbone Street Tunnel in Brooklyn during rush hour on November 2, 1918.

First, most of the cars on the train were ancient wooden ones that had not been replaced by the steel coaches the Public Service Commission had ordered for the line. After the accident, in which all of the fatalities and injuries occurred in the wooden cars and none in the steel ones of the train, Public Service commissioner Travis H. Whitney accused New York mayor John F. Hylan of dragging his feet and not signing an agreement that would have mandated the steel cars.

Second, the motorman, 25-year-old Edward Anthony Lewis, was exhausted. He had just recovered from influenza. His baby had just died. That day he had already put in a full shift as a train dispatcher and was working overtime as a motorman during rush hour to earn extra money.

And finally, and most crucial of all, the train's brakes were defective and failed to hold when Lewis applied them.

The train, loaded with office workers—most of them women—on their way home, approached the Malbone Street Tunnel at about 30 mph. The posted speed limit was 6 mph, but Lewis, in his testimony at the inquiry, said that the brakes failed, and instead of slowing the train picked up speed.

On a curve just before the tunnel's entrance, the train left the track. The lights simultaneously went out in the cars. The decrepit wooden coaches shattered as they capsized and piled into one another, crushing and impaling their passengers on jagged pieces of walls, flooring and ceilings. In all, 97 people, most of them women, were killed, and 95 more were injured, many of them critically.

UNITED STATES
NEW YORK
QUEENS
November 22, 1950

Defective equipment was responsible for the rear-end collision of two Long Island Railroad trains in Queens, New York, on November 22, 1950. Seventy-nine were killed; 363 were seriously hurt.

The Long Island Railroad, a subsidiary commuter line of the Pennsylvania Railroad, has never been known for its reliability or modernity. It is therefore both gratifying and amazing to note that there have been few major accidents on the LIRR.

The worst wreck to occur on the line occurred during rush hour on November 22, 1950, on a stretch of track just outside the Richmond Hill section of Queens. A local 12-car train bound for Hempstead was ordered, by overhead signals, to slow before entering the station. The motorman, William Murphy, heeded the signals and slowed to 15 mph. According to Murphy, the brakes began to grab at this speed, eventually stopping the train dead.

Just before the stretch of track on which the Hempstead train was stalled was a blind curve, and so Bertram Biggam, the rear car flagman, debarked from the back of the train with a red lamp to warn off any approaching trains.

Eventually, the Hempstead train was given the go-ahead signal, and Murphy revved its electric motors. Biggam abandoned his task and climbed back aboard the train. But the brakes appeared to be locked. The train stayed in place.

Just then, the Babylon express, traveling at 40 mph, rounded the curve. Its motorman, Benjamin Pokorney, had received the same go-ahead signal that Murphy had received and had no reason to believe that the track was not clear. The express smashed into the rear of the Hempstead train, burrowing under and into the last car, flinging it into the air and killing almost everyone in that car and then plowing ahead through subsequent passenger coaches.

Mayhem resulted. Pokorney was killed, but it would be hours before his body could be dug from beneath the mountain of twisted debris. All in all, 79 people died, most of them in the last car of the Hempstead train, and an astonishing 363 were seriously injured.

The Long Island Railroad would be branded a "disgraceful common carrier" by New York mayor Vincent Impellitari and forced to pay out $11 million in damage suits. It then began a decades-long improvement program that has yet to be completed.

UNITED STATES
OHIO
ASHTABULA
December 29, 1876

A collapsed bridge caused a train to plunge into a gorge near Ashtabula, Ohio, on December 29, 1876. Eighty were killed; 68 were injured.

No charges of wrongdoing were ever filed against the Lake Shore and Michigan Southern Railway following the collapse of a 152-foot iron trestle bridge outside Chicago on the night of December 28, 1876. And yet, two days after the inquiry, Charles Collins, its chief engineer and one of the two men responsible for the safety of the bridge, committed suicide. Five years later Amasa Stone, the designer of the bridge and president of the line, also committed suicide. The natural conclusion was that both facts and blame were missing from the inquiry, the first and the model for subsequent boards of inquiry into any railroad accidents involving the deaths of passengers.

The famous Ashtabula Bridge Disaster, as it was destined to be called, was the most deadly railroad accident up to that time, and it took place on the same line (Commodore Vanderbilt's Lake Shore Road) and in the same season as the Angola Horror (see p. 360). There was a blinding snowstorm on the night of December 29, 1876, which slowed the Pacific Express from New York to Chicago to a 10-mph crawl. It was an 11-car train composed of two engines, three sleepers, a smoker, a parlor car, two coaches and four baggage cars. The first engine, nicknamed *Socrates,* was piloted by 17-year veteran Daniel McGuire, and he was having great difficulty seeing several feet ahead of him. Drifting snow and sheets of wind-whipped snow had all but eliminated his visibility.

At 7:30 P.M. he reached the iron truss bridge that reached from one embankment of the Ashtabula Creek to the other. He inched across the span and reached the other side with the front wheels of his engine when he felt a sickening, sinking sensation. The struts and underpinnings of the bridge had gone soft and were collapsing under the weight of the train.

McGuire raced his engine and managed to drive *Socrates* on to safe terrain. But the sudden momentum uncoupled the engine behind him, and it and the rest of the train began the long plummet to the bottom of the creek. The engine plunged into the deep ravine first, followed by car after car, telescoping and splintering apart as they smashed into one another. Almost instantly the burning coal stoves that heated each of the cars ignited the sperm oil from their illuminating lamps, sending up huge exclamations of flames and incinerating the passengers in their seats and berths.

McGuire slid down through the snowdrifts of the embankment and encountered the injured engineer of the second engine, "Dad" Folsom. "It's another Angola Horror," said Folsom, and it was, only worse. Out of almost 200 passengers, 80 died, 19 of them so badly burned that they would never be identified. Only 52 escaped uninjured. The magnitude of this Christmas season tragedy prompted the Ohio state legislature, and eventually the federal legislature, to establish permanent boards to investigate fatal accidents.

UNITED STATES
PENNSYLVANIA
CAMP HILL
July 17, 1856

Missed schedules, inadequate signals and human error were responsible for the head-on collision of two trains near Camp Hill, Pennsylvania, on July 17, 1856. Sixty-six children on an excursion train died; 60 were injured.

Human error, inadequate signals and missed schedules conspired to cause this monstrous head-on collision of an excursion and a passenger train near Camp Hill Station, Pennsylvania, on July 17, 1856. Sixty-six children on the excursion train were killed. (Frank Leslie's Illustrated Newspaper)

In the 1850s and 1860s it was not uncommon for passengers to take seats in the middle of a train because they believed that this midway location was the least vulnerable to either a head-on or rear-end collision. They had reasons to be cautious, and not the least of these had to do with the most serious and violent train wreck in the United States in terms of human casualties: a head-on collision that took place near dawn on July 17, 1856, in Camp Hill, Pennsylvania.

Before sunrise on that particular day, some 1,500 children of Philadelphia's St. Michael's Church gathered at the Master Street Station preparatory to boarding two trains for a massive picnic-outing at Fort Washington. One train would leave at 5:00 A.M., the other at 8:00 A.M.

Naturally, there was a huge rush to board the first train; none of the 1,500 children of Irish-American immigrants wanted to wait more than three hours in a train station before beginning the trip to the picnic grounds. The priests succeeded in regaining order and loading 600 young people on the 5:00 A.M. train, pulled by the locomotive *Shackamaxon,* with 21-year-old engineer Henry Harris at the throttle.

It took longer than anticipated to load the train; once the doors had been closed, they had to be opened again to accommodate more passengers who adamantly refused to wait, and it would be 5:30 before the train would ultimately leave the Master Street Station. The delay would be a fatal one.

During the chaos in the Master Street Station, the regular local train from Gwynedd to Philadelphia was making its slow way along the single track of the North Pennsylvania (now the Reading) Railroad. It was the practice then to pull trains that did not have the right of way onto sidings. This right of way was determined by schedules. If a train was on schedule, it had the right of way. If it was more than 15 minutes behind, it was required to pull to the siding and let the train going in the opposite direction through—a dangerous practice that depended on accurate watches and the discretion of conductors. The conductor aboard the local was William Vanstavoren, a young man who had trained under conductor Alfred Hoppel, who, ironically, was the conductor on the excursion train headed in the opposite direction.

Both conductors had reason to be nervous. The local, scheduled to depart from Fort Washington at 6:00 A.M., left at 6:14. The excursion train was a half hour late. Told that the special had not passed through, Vanstavoren, according to the subsequent inquiry, ordered his engineer, William Lee, to proceed anyway. Lee reportedly challenged this order, and Vanstavoren assured him, "It's all right. Just sound your whistle like hell and go slow."

Lee did just this, inching his three-car train (one engine, a baggage car and a passenger car carrying 12 passengers) at 10 miles per hour toward Philadelphia. The passengers, nervous at the sound of the whistle and the slow speed, wisely moved to the back of the car.

At Camp Hill, at the entrance to an S-curve sandwiched between two 20-foot embankments, Lee spotted the excursion train coming directly toward them. It was clipping along at 35 miles per hour, confident that it had the right of way. Lee slammed on his brakes and applied reverse throttle and then jumped, as did his fireman.

Henry Harris, at the controls of the excursion train, saw the almost stationary local at the same time, but it was far too late to avoid a collision. He and his fireman also jumped, but Harris neglected to even try to apply either brakes or reverse throttle. He miscalculated his jump, landed between the tender and the engine and the first car and was crushed to death in the impact that occurred seconds later.

With a roar the engine of the excursion train slammed into the engine of the local, bursting both boilers and catapulting both engines first into the air and then backward onto the wooden coaches of their trains. The passengers aboard the local had long since abandoned the train and escaped uninjured. But the children aboard the excursion train were trapped in the wooden cars, which instantly splintered and caught fire. Children were flung through broken windows and to either end of the collapsing, fiery infernos.

Hoppel, the conductor of the excursion train, was thrown clear, and he oversaw the fire fighting efforts of volunteer fire departments from Camp Hill and Chestnut Hill, who were able to save five of the cars of the special. The others burned to the wheels. Sixty-six children were killed, and 60 more were critically injured.

Conductor Hoppel would be arrested and tried for murder but would be acquitted. Conductor Vanstavoren, distraught beyond sanity, repeated over and over, "All of it, all of it is my fault." Along with some trainman companions, he ran to Edge Hill, commandeered a handcar and then borrowed a buggy that he drove into Philadelphia. Once there, he found a pharmacy, bought arsenic and morphine, went to the offices of the railroad and swallowed the poison. An hour later he died.

"Railroad Butchery," was the description of the wreck in the *New York Times*. The reason for it all was an inadequate warning system that broke down as a result of the faulty judgment of two good men.

UNITED STATES
PENNSYLVANIA
LAUREL RUN
December 23, 1903

Debris on the tracks was the initial cause of the huge train wreck of December 23, 1903, near Laurel Run, Pennsylvania. Sixty-four died; nine were injured.

One of the more bizarre railroad wrecks of the 20th century took place in a driving snowstorm on December 23, 1903, just outside the tiny hamlet of Laurel Run, Pennsylvania. The Duquesne Limited from Pittsburgh to Connellsville, which enjoyed the reputation of being the fastest train on the B & O line, had gained its stature by averaging speeds of 60 mph. It was traveling at just this speed on the night of December 23, despite a steady snow that had reduced visibility to nearly zero.

But snow or not, the Duquesne Limited would not have been able to avoid the catastrophe it was hurtling toward that night. A slow-moving freight of the Nickel Plate line, moving in the opposite direction on a parallel track, had lost practically an entire gondola load of heavy railroad ties. The engineer of the train was unaware of this as he passed the Limited going in the other direction.

Minutes later the speeding passenger train rounded a curve and plowed full speed into the mountain of wooden ties. The engine became airborne, climbing the ties and twisting off the tracks, carrying the baggage car with it. A smoker, running just behind the baggage car, leaped over the baggage car and rammed the engine, slicing off its steam dome and sending scalding steam into the smoker.

The sleeper behind the smoker roared on, ripping up track as it went, and came to rest, precariously poised, on the edge of an embankment. The dining car followed this, toppling over and strewing its occupants around like jackstraws. Dining steward Benjamin Nicholas, dazed by the crash, pulled himself together instantly and became one of the two heroes on the scene.

After seeing to the inhabitants of the dining car, Nicholas crawled through the smoker and its 40 dead and dying occupants and reached the engine, which was still erupting steam. Tearing his coat to shreds, he stuffed it into the ruptured pipe and then climbed down and turned off the engine's boiler valve.

Thomas J. Baum, the train's baggage master, was trapped in the baggage car, which, after colliding with the engine, had ricocheted into the Youghiogheny River. Severely injured, he freed himself and climbed the embankment just in time to hear conductor Louis Hilgot, whose face had been burned into an unidentifiable mass, shout to anyone who could hear that number 49 was due behind them and would have to be flagged down.

Despite his injuries, Baum dragged himself to the tracks behind the wreck. There he could plainly hear the approaching engine of number 49 charging through the snowstorm.

Baum ripped off his coat and, with a package of pocket matches, set it afire and waved it at the approaching train. Number 49 saw the signal and ground to a stop just three feet from where Baum, faint from loss of blood and exhaustion, had collapsed on the tracks.

Rescuers from number 49, including two detectives who arrested several men who were looting the dead, rushed to the scene. Sixty-four persons, all of them men and most of them passengers in the smoker, were killed. Nine were injured. And 500 yards of track on both roadbeds were reduced to twisted, jagged steel.

The shattered, twisted, burned-out remains of the last car of the number six train of the Lehigh Valley line excursion of the celebrating Total Abstinence League in Mud Run, Pennsylvania, on October 10, 1888. Most of the 64 victims of the crash were crushed in this last car when train number seven rear-ended it. (Library of Congress)

UNITED STATES
PENNSYLVANIA
MUD RUN
October 10, 1888

Outdated equipment accounted for the rear-end collision of two trains at Mud Run, Pennsylvania, on October 10, 1888. Sixty-four died; 100 were injured.

"Telescoping," or the passing of cars into and sometimes through one another like the joints of a telescope, was one of the most feared and calamitous consequences of head-on on or rear-end collisions in 19th-century America. The culprit was the American system of link and pin coupling, which allowed car platforms to bounce at varying levels. In England powerful compression coupled the cars, and telescoping was virtually unknown. But it would be 1869 before the Miller platform and buffer—a locking together of the car sills by a strong tension-compression coupling—would be invented, and decades before it would be universally employed in America.

The eight excursion trains that hauled 5,000 members of the Total Abstinence Union of Wilkes-Barre, Pennsylvania, from Wilkes-Barre over the Lehigh Valley line to the mountains near Hazelton on October 10, 1888, were not top-of-the-line equipment. All of the passenger cars were wooden and rickety; each of them was outfitted with outmoded link and pin couplings. Even the first engine on the seventh excursion train contained only steam brakes, and the engineer had to signal by a whistle to the engineer of the second engine to apply them. Each of these factors would add up to a catastrophe.

The eight special excursion trains took the 5,000 merrymakers to the Laurentian campgrounds in the morning and back in the evening at 10-minute intervals, a separation that was twice the usual space between trains in 1888 and thus thought to be safe. As further precautionary measures, two brakemen were stationed as lookouts in each engine cabin, and extra brakemen were assigned to the rear of each train.

Train number six of the eight-train contingent contained, in its last car, a group of particularly joyous celebrants who had apparently been less than abstemious during the day. Shortly beyond the station at Mud Run, this train was halted. Hannigan, its brakeman, debarked and walked the 400 feet from the rear of the last car of loud celebrants to the edge of the station platform and hung out a red warning light.

Train number seven, meanwhile, was approaching Mud Run, tooling along at a conservative 20 miles per hour. The track curved into the station at Mud Run, and the signals read all clear ahead. But train number six was still halted, and as the seventh train rounded the curve, brakeman Hannigan's pathetically small warning light suddenly appeared.

Henry Cook, the engineer of the seventh train's first engine, spotted the warning when he was almost abreast of it. Too late, he blasted his brake signal, simultaneously applying his steam brakes. But the train scarcely slowed. With a horrendous screeching of splintering wood, it rammed into the rear of the sixth train, telescoping the last car halfway through the next one. Two hundred passengers were trapped in these cars, and 64, most of whom were in the last coach, were killed instantly, crushed by the impact and folding together of the two cars. One hundred other passengers were badly injured, some of them dismembered.

"Oh! What a tongue can tell or pen picture this most dreadful calamity," wrote Matt J. Meredith, in a commemorative book titled *First Anniversary of the Mud Run Disaster.* "The roasting, scalding engine under which were crushed those poor young children," he went on, in antique style, and with less than punctilious accuracy (most of the dead were adults), "and the car ahead being telescoped and the lives crushed out of those who but a few moments since were full of life. Oh, God, why visit upon your unhappy children such a death."

UNITED STATES
PENNSYLVANIA
SHOHOLA
July 15, 1864

A 13-1/2-hour departure delay caused the head-on collision of a troop train and a passenger train at Shohola, Pennsylvania, on July 15, 1864. Seventy-four were killed; there is no record of the injured.

A Union Army troop train carrying hundreds of Confederate prisoners of war was scheduled to leave Jersey City, New Jersey, at 4:30 A.M. on July 15, 1864, bound for a Union prisoner of war camp in Elmira, New York. But when the departure time arrived, it was discovered that three Confederate prisoners were missing. A full-scale search was launched through the trains, through the station yards of Jersey City and through the ship that had transported the prisoners up the coast to Jersey City.

The search would take all day, and it would be after 5 P.M. before the three would be found hiding out in the hold of the ship that had brought them there. The 18-car military train would then depart from Jersey City 13-1/2 hours late, well off schedule.

The Erie track near Shohola, Pennsylvania, is a single one, and on that single track, heading in the opposite direction that night, was an enormous 50-car coal train. It had been telegraphed a go-ahead from the control operator at the Lackawaxen station and was proceeding at full throttle.

The two trains met head-on at a curve in the tracks near Shohola. Both engines were catapulted off the tracks, ending up in a steamy tangle on an embankment. The cars of both trains telescoped into one another, smashing asunder. The coal cars, traveling at a greater speed and containing greater mass and weight than the passenger cars, crushed both cars and occupants instantly upon impact. Seventy-four passengers (51 prisoners, 19 guards and four engine crew members) were killed, thus making this one of the most mortally catastrophic head-on collisions in the history of railroading in the United States.

UNITED STATES
TENNESSEE
HODGES
September 24, 1904

Human error was blamed for the collision of two trains at Hodges, Tennessee, on September 24, 1904. Sixty-three died; there is no record of the injured.

Train number 12 from Bristol to Chattanooga and train number 15 from Chattanooga to Bristol traveled the same single track on the Southern line. The schedule always allowed smooth passage. At New Market, a station away from Hodges, Tennessee, depending on the signals and the time, one of the two trains would take to a siding, allowing the other to pass.

On the afternoon of September 24, 1904, the signals and the schedule clearly indicated that number 12 had the right of way. But for some reason that died with them, the engineer and stoker of number 15 ignored both signals and their orders and steamed straight ahead at full throttle through New Market.

A few minutes later, at Hodges, number 15 slammed head-on into number 12, flinging engines and cars off the tracks on either side. Sixty-three passengers and crew members would die in the conflagration that resulted, and the blame would be fixed on the dead engineer of number 15.

UNITED STATES
TENNESSEE
NASHVILLE
July 9, 1918

Human error was responsible for the worst rail crash in number of fatalities in U.S. history, a head-on collision of an express and a workers' train near Nashville, Tennessee, on July 9, 1918. One hundred one were killed and 100 were injured.

There is an ironic circumstance of history surrounding the worst railroad crash in number of fatalities in the United States. Practically all of the victims of the crash were black. It took place in Tennessee. And it took place in July 1918, exactly 12 months before the Chicago race riots that would kill 31 and injure more than 500 in a tense atmosphere of racial tension in the nation.

Thus, the terrible wreck of July 9, 1918, in Nashville went virtually unnoticed, while the subway disaster in Brooklyn in November of that same year (see p. 360), which claimed four fewer lives, received enough attention for history books to incorrectly indicate that it was the nation's worst train wreck in terms of fatalities.

The disaster occurred shortly after 7 A.M. on the morning of July 9 between the Shops and Harding stations on the fringes of Nashville. Train number one of the Nashville, Chattanooga & St. Louis line, loaded with black workers in Tennessee's munitions plants, was a low-priority train that had orders to remain on the double tracks at Shops and allow train number four to rocket by on the single express track. Following this, it was to pull onto the track for its short trip to Harding, where the munitions plant was located.

Engineer Kennedy waited while a freight sped by, and speculation afterward was that he mistook it for the express—a shaky assumption, considering the experience of Kennedy. For whatever reason, Kennedy, after the freight disappeared, pulled out onto the single track and pushed his engine on pulling several ancient wooden passenger cars at 50 mph toward Harding.

He traveled a very short distance before slamming full force into express number four, traveling at 50 mph in the opposite direction. The roaring head-on collision reduced the two engines to steaming scrap metal in an instant. The first two cars of number one telescoped and shot into the engines, shattering into jagged splinters, killing every occupant of each car. One hundred one people were killed and 100 more were injured, and the papers of the nation would hardly mention it.

UNITED STATES
UTAH
OGDEN
December 31, 1944

An engineer's heart attack was the cause of a rear-end collision near Ogden, Utah, on December 31, 1944. Fifty died and more than 80 were injured.

Human error of the most uncontrollable sort was apparently responsible for an appalling rear-end crash that took place in the early morning hours of December 31, 1944, on the salt marshes near Great Salt Lake in Ogden, Utah. The engineer of the second section of the Pacific Limited of the Southern Pacific line suffered a heart attack just before plowing into the rear of the first section.

The two trains were loaded with soldiers on holiday furlough and civilians intent on celebrating New Year's Eve in San Francisco. The first section was a comfortable distance ahead of the second. This was 1944; signals and safety precautions were sophisticated.

Then, shortly after dawn, an express ahead of the first section signaled that it was having difficulty. It slowed and then stopped; its brakeman debarked from the last car and set a series of warning flares along the tracks behind the halted train.

Morning mists were drifting off Great Salt Lake, across the tracks and into the marshes. The flares were shrouded in them, removing their sharp definition and thus their shock capability. Still, they were visible to an alert engineer.

However, as surviving crewmen described later, James McDonald, the engineer of the second section, seemed to be moving and reacting sluggishly as he drove his train at 65 mph straight for the rear of the stationary first section of the Limited. When it came into view and the men around him began to shout, he applied the brakes. But by then it was too late.

The engine slammed into the last car of the parked train, demolishing it and shoving it off the track. The next car, a Pullman, miraculously escaped destruction, but the following two cars were catapulted into the marshes.

Fifty people were killed, 29 of them in the last car of the first section. More than 80 were injured. More might have died from their injuries if it had not been for two army hospital cars attached to the first section, which were far enough forward in the 18-car train to remain undamaged and serviceable. McDonald, examined by company doctors, was found to have had a heart attack.

YUGOSLAVIA
ZAGREB
August 30, 1974

Drinking in the cab of a train caused the crash of a train at high speed into the station at Zagreb, Yugoslavia, on August 30, 1974. One hundred seventy-five were killed; there was no official report of injuries.

The worst rail accident in the history of Yugoslavia took place in the station at Zagreb on the night of August 30, 1974.

A solitary passenger train, the Belgrade to Dortmund express, traveling at nearly 55 mph, roared into the station at Zagreb that night, its engine at full throttle. A subsequent board of inquiry would accuse the train's two engineers of drunkenness, and to the horrified witnesses of this cataclysm, it was the only plausible explanation for the bizarre scene that unfolded before them.

Out of control, the train crashed into the station platform and derailed, splaying its cars onto the platform and adjacent tracks. Electric power cables used for commuter lines exploded into sparks and fell on the metal cars, electrocuting some of the occupants as they tried to escape.

It would be hours before 50 passengers, trapped by the power lines and the overturned cars, could be rescued. They were the lucky ones. A staggering 150 passengers were killed by the crash and 25 others were electrocuted. The two engineers were acquitted of drunkenness but received reprimands for speeding.

SPACE DISASTERS

THE WORST RECORDED SPACE DISASTERS

* Detailed in text

United States

- Florida
  - * Cape Canaveral (1967)
  - * Atlantic Ocean (1986)

USSR

- * (1967)
- * (1971)

CHRONOLOGY

* Detailed in text

1967

- January 27
  - * Cape Canaveral, Florida; Fire in space capsule
- April 23
  - * USSR; *Soyuz I* crash

1971

- June 30
  - * USSR; *Soyuz II* tragedy

1986

- January 28
  - * Atlantic Ocean; Explosion of *Challenger*

SPACE DISASTERS

Thankfully, this section is a brief one. In the more than three decades of its existence, modern space exploration has claimed only 15 lives—21 if you count the six American astronauts who died in airplane accidents and other ancillary activities. And perhaps there are other casualties never reported by the Soviet space program.

None of these lost lives are those of innocent victims. Even schoolteacher Christa McAuliffe volunteered and knew full well the hazards she faced when she entered the *Challenger.* That, of course, is little comfort to the surviving families of those who died, but when examined against the entire landscape of human disaster, it does, at least, circumscribe catastrophes in space.

Space exploration has been largely dominated by the two world superpowers—the United States and the USSR. From that monumental moment in 1957 when *Sputnik I* was launched, the exploration of space moved from the pages of science fiction stories to the front pages of newspapers and from there to the imaginations of the world. To those who had trouble flying in an airliner and who now watched astronauts on the moon, the leap in achievement seemed almost beyond comprehension. But the public loved it. Reservations for the first passenger flight to the moon sold at a brisk clip in New York City in the early 1960s.

That the race between the two superpowers to put a man in space, then men on the Moon, then people on space stations, then people on Mars and other planets would eventually extract its toll in human tragedy was scarcely believable in the early days of the space race. Even the high economic cost failed to dampen the exciting, adventurous enthusiasm of scientists and nonscientists alike. Up went the satellites, space stations and space shuttles, until outer space began to resemble an ill-tended backyard full of floating garbage left by those who played in it.

It would take a tragedy the size of the 1986 *Challenger* explosion to sober up the world, it seemed. The adolescent love affair with outer space ended for many that January day. Experiment, the world learned, carries with it danger and responsibility.

In our eagerness to explore outer space, we had perhaps neglected some of the inner space of thoughtful preparation, both in technology and our own emotional and mental development. We had, in short, not paid enough attention to the natural priorities that regard the preservation of human life as the highest goal.

It was a sad, sobering and perhaps overdue moment. Looking back at the terrible deaths of those who had been killed before in the space race, it was easier to see what was ahead. More exploration, certainly, more monumental achievements, but caution on the way to them.

Ironically enough, one of the lessons that has been learned from the four space tragedies that have occurred so far is that, as in civilian airline travel, the most dangerous moments in any flight occur during takeoff and landing. With the exception of the fire on the launchpad at Cape Canaveral that killed astronauts Grissom, White and Chaffee, all took place in the first few seconds of blastoff or the last few seconds of landing.

Thus, as this is written, more and more launchings are being aborted on the launchpad, as we learn from our disasters. Space exploration has passed through its infancy, into its adolescence, and is now approaching its maturity. More and more space flights are being conducted every year. Project Mercury, initiated in 1958, dims into the past realm of initial experiments. Now, the mainstay of the National Aeronautics and Space Administration (NASA) is the space shuttle, whose flights seem to have become nearly as frequent and expected as those of a commercial airline.

Since 1981, more than 100 shuttle flights have been made, and along the way, history was also made. In 1983, during its first Spacelab mission, a shuttle delivered the first astronaut to represent the European Space Agency (ESA) to a space lab.

Numerous communication satellites were deployed by space shuttles throughout the 1990s. On April 24, 1990, a shuttle deployed the *Hubble Space Telescope.* In February, 1994, Sergei K. Krikalev was the first cosmonaut to fly aboard a shuttle. In June 1995, the first named shuttle, *Atlantis,* became the 100th United States space launch, and conducted the first docking with the Russian space station *Mir.* It carried the largest crew to date aboard a shuttle—eight men—and facilitated the changing of crew members aboard the *Mir.*

October 29 was a historic day in the development of the shuttle program when Senator John Glenn, the first man to travel into space for the United States, made a second voyage aboard the shuttle *Discovery.* President Clinton attended the launching, making him the first U.S. president to do so. Less than a year later, on May 27, 1999, the beginnings of the International Space Station were delivered by shuttle *Discovery,* which made other trips to the *Hubble* telescope to make necessary adjustments and conduct routine servicings. The 107th shuttle flight of *Endeavour* was to once again dock and deliver materials and supplies to the International Space Station.

Each of these flights has contained its own delays and glitches, usually in its computers. And most have experienced temporarily aborted takeoffs of varying lengths. Safety, born of disaster, has averted other disasters for more than 25 years. It is an enviable record, and one that will hopefully be extended for years to come.

UNITED STATES
FLORIDA
CAPE CANAVERAL
January 27, 1967

Three Apollo I *astronauts perished in a simulation exercise at Cape Canaveral, Florida, on January 27, 1967, when a fire caused by a spark from a faulty wire ignited the pure oxygen in their space capsule.*

The *Apollo I* astronauts, expected to be America's first men on the moon, were selected carefully: Virgil Grissom was 40 years old, an Air Force lieutenant colonel and one of the seven original Mercury astronauts. Edward I. White II was 36, a lieutenant colonel in the Air Force and the first American to "walk" in space. Roger B. Chaffee was 31 and a Navy lieutenant commander. This was to be his first spaceflight.

The Apollo space program, first introduced by President John F. Kennedy in May 1961 and designed to place a man on the moon by 1970, had operated since its inception on the razor-thin edge of uncertainty. An expensive, extensive program, it was constantly being scrutinized by Congress and the administration when budget-tightening time came.

The Apollo program itself had undergone a series of technical glitches that had also delayed the program and increased its budget demands. In fact, the moon shot for which Grissom, White and Chaffee were preparing in January 1967 had been postponed twice, in February 1966 and November 1966.

In January it seemed as if they were headed toward a "go," and a series of simulations that precede any space launching began. The space capsule had been positioned atop its booster rocket for several days of trial blastoffs.

Early in the morning of January 27, 1967, flight controllers arrived at the Mission Control Center in Houston and took their seats at the communication consoles. Simultaneously, at Cape Canaveral's Kennedy Space Center, workers in the blockhouse and launchpad prepared the capsule and rocket for a takeoff simulation.

At 1:00 P.M. the three astronauts, in full space regalia, climbed into the 12 × 13-foot spacecraft poised 218 feet in the air atop its *Saturn 1* rocket. The orange gantry surrounded the linked space vessels and the connections that linked them. All that was missing was the fuel in the spacecraft and the rocket. For two hours, until approximately 3:00 P.M., the astronauts checked the instruments in the cockpit and then signaled that they were ready for the hatch to be closed and locked over their heads.

For the next few hours they would be undergoing what was known as a "plugs out" simulation of the countdown, blastoff and first three hours of spaceflight. "Plugs out" meant that, as in a real blastoff, all the electrical and life-support connections between the spacecraft and the gantry would be severed, and the spacecraft would then depend completely on its own inner power.

Two days before, the astronauts had undergone a similar "plugs out" exercise, except that in that one the pure oxygen with which the space capsule would ultimately be filled was not pumped in. Instead, the hatch was left open, and the mixture of gases—predominantly nitrogen and oxygen—that characterizes Earth's atmosphere was allowed to drift in.

However, this one would be closer to the real thing. NASA (the National Aeronautics and Space Administration) had decided to use pure oxygen in the space capsule rather than a mixture of gases because the equipment would be simpler to install and would weigh less. Some scientists in the organization and outside were disquieted by this choice. Pure oxygen is highly combustible, particularly at the pressure volume of 18 pounds per square inch that was contemplated for the Apollo. Although the Gemini craft flew with pressure at five pounds per square inch, the Apollo scientists decided to increase this to approximate the 15-pound-per-square-inch pressure volume of Earth's atmosphere.

At approximately 6:00 P.M., with the hatch closed and the oxygen level at 16, there was a loss of communication between Houston and the astronauts. It was worked on and corrected, and the simulated blastoff was scheduled for 6:41 P.M. The astronauts were in their seats—Colonel Grissom was in the command pilot's seat on the left, Colonel White was in the middle and Commander Chaffee occupied the right seat.

At 6:31 a casual voice, never identified, came over the communication system to the blockhouse at Cape Canaveral and the control center in Houston. "Fire—I smell fire" the voice said.

Two seconds passed.

"Fire in the cockpit!" shouted the unmistakable voice of Colonel White. Three more seconds of silence elapsed. A hysterical cry came over the intercom: "There's a bad fire in the spacecraft!" There was the sickening sound of scuffling, frantic movement and unintelligible shouting. The craft was filling up with black smoke, and it was starting to glow from the heat within—all within 11 seconds. Finally, one last communication blasted from the intercom. It was the strangled voice of Commander Chaffee. "We're on fire!" he pleaded. "Get us out of here!"

Apollo astronauts (left to right) Gus Grissom, Ed White and Roger Chaffee shortly before the tests on the launchpad that ended in their deaths by fire on January 27, 1967. (NASA)

And those were the last words from the astronauts, who were burning to death, while a camera dispassionately recorded, in minute detail, their death throes. They didn't have a chance. A faulty wire near Grissom's couch had ignited a spark, which had ignited the pure oxygen, setting a fire that burned at 2,500 degrees Fahrenheit. They were roasted alive and asphyxiated at the same time in 21 seconds. The only way they could have opened the hatch would have been to unscrew it with a ratchet tool, an operation that would have taken at least 90 seconds.

Rescue workers with gas masks tried to get to the whitehot capsule and open the hatch, but they were beaten back by the heat and incredibly dense smoke. Five minutes later rescuers clad in asbestos were finally able to enter the capsule, where they encountered a grisly sight. The astronauts had struggled out of their restraining harnesses. What was left of Colonel Grissom and Colonel White—little more than their bones—was found lying at the hatch. Commander Chaffee lay below. Pieces of skin containing intact fingerprints were grafted to the hatch. Grissom and White had been tearing at it with their bare hands. There was nothing left of the astronauts' space suits. They had been burned completely through.

"[There may have been a small fire] at first largely absorbed by the spacecraft structure," said Dr. Robert C. Seamans, Jr., deputy administrator of NASA, before a congressional panel investigating the tragedy. "It may have continued for as long as 10 seconds," he continued. This would encompass the first cries of alarm from the astronauts. "A more intense fire may have then developed," he went on, "causing the rapid increase in cabin pressure. This fire was probably extinguished by the depletion of oxygen."

The astronauts died, then, in a space of time somewhere between 12 and 20 seconds, not without suffering terribly. Lieutenant Colonel White, Lieutenant Colonel Grissom and Lieutenant Commander Chaffee were all given heroes' burials in Arlington National Cemetery, a fitting honor for the first American casualties of the space age.

UNITED STATES
FLORIDA
ATLANTIC OCEAN
January 28, 1986

While millions watched, the Challenger *space shuttle exploded on blastoff over the Atlantic Ocean near Cape Canaveral on January 28, 1986, killing all seven astronauts aboard. A combination of low temperatures, O-rings that malfunctioned, and NASA's determination to launch the shuttle caused the disaster.*

In the age of modern telecommunication, we have become accustomed to seeing history as it happens. Millions watched while Jack Ruby killed Lee Harvey Oswald on television live. The evening news carried the Vietnam War into our living rooms, and, as violence escalated in American society, the theory was postulated that, since that time when war appeared on our TV screens as information, we had become a nation inured to violence, numbed by actuality-as-television-drama.

But nobody who was sitting before a television set on the morning of January 28, 1986, was immune to the escalating horror they witnessed as the space shuttle *Challenger* and its seven occupants exploded in a million fragments 74 short seconds after they had taken off from Cape Canaveral, Florida. The personal shock and loss ran like an electric current through practically everyone in the nation, for this was the famous first flight that would take an ordinary citizen into space. Schoolteacher Christa McAuliffe was the Everywoman of the 20th century, and she was also the victim of one of its worst tragedies. Not since the assassination of John F. Kennedy 23 years before had such a shared sense of loss and sadness and bewilderment united ordinary people on the street.

The *Challenger* flight was a storybook mission. The space shuttle was a known quantity. It had succeeded before; it would succeed again. America felt good about its space program. The fact that there was a predominantly civilian crew aboard *Challenger* was, of itself, a positive sign. We had turned the corner toward space travel for the common person.

Scheduled to fly the *Challenger* that January day were mission commander Francis R. (Dick) Scobee; the pilot, Commander Michael J. Smith of the Navy; Lieutenant Colonel Ellison S. Onizuka of the Air Force; Dr. Ronald E. McNair, a physicist; Dr. Judith Resnik, an electrical engineer; Gregory B. Jarvis, another electrical engineer; and science teacher Christa McAuliffe.

Behind the scenes, the scenario was a little less sanguine. Bad weather at the Kennedy Space Center in Florida had already caused two postponements in three days of the *Challenger* launching.

It had originally been scheduled to leave on Sunday morning, January 26. But the weather reports on Saturday had been ominous, predicting heavy rain, and so the Sunday launch was scrubbed. As it turned out, the weather forecasters were wrong, and it was a balmy, blissful day, ideal for a space launch. But the cancellation, once put in motion, had to stand.

On Monday, January 27, the skies clouded up, but meteorologists predicted a clearing trend, so the *Challenger* astronauts suited up and boarded the ship. But a handle on the shuttle latch malfunctioned. Mindful of the *Apollo* disaster (see p. 371), those in command dispatched workers to free or replace the recalcitrant handle. It would take the workers two hours to complete the job, and by then stiff winds had blown up and the launch was again postponed, this time to 9:38 Tuesday morning, January 28, 1986.

But even that launch date and time were the subject of long and worried conferences. The winds that had postponed the January 27 launch preceded a cold front that was about to visit Florida. The forecast issued in midafternoon of the 27th called for freezing temperatures throughout the night.

"When we saw those predicted temperatures, I just knew we had to talk about it," recalled Allan J. McDonald, an engineer who represented Morton Thiakol Inc., shuttle contractors and the manufacturer of the booster rockets. Mr. McDonald argued vehemently that evening against proceeding with the countdown. He and other Morton Thiakol engineers were concerned about the effect of the cold weather on the solid-fuel booster rockets. Their concern was focused on the O-rings, the rubber seals that contained the booster's hot gases.

These seals were obviously critical, and they had had a history of problems. A double set of synthetic rubber Os shaped like giant washers, they fit around the circumference of the rocket casing and were intended to fill the tiny gap that remained after two steel rocket segments were bolted together. In the past the enormous pressures during blastoff had sometimes dislodged these seals. Besides, they had never been tested at temperatures lower than 53 degrees Fahrenheit.

So when McDonald got on the phone to Utah to confer with other engineers at Morton Thiakol, he returned with a unanimous recommendation for delay. The engineers suspected that cold weather could only heighten the chances of a failure of the O-ring seals. The rubber, they argued, would harden and shrink at these low temperatures, increasing the likelihood that the seals would open up. The consequences, they concluded, could be catastrophic.

NASA took this under advisement, but NASA was under pressure to proceed with this launch. They were scheduled to conduct 15 shuttle flights in 1986, which was six more than in any previous year. This was to be the symbolic 25th flight, which would demonstrate that it was safe and that space vehicles could be reused. And it was to be the first flight with an ordinary citizen aboard.

According to later testimony, NASA brought extreme pressure on Morton Thiakol to agree to a go-ahead. Finally, Morton Thiakol reluctantly agreed.

As the weather forecast predicted, the temperature fell below freezing. At 6:00 A.M. it was 27 degrees on the launchpad, and the shuttle's external tank was coated with frost and ice. A special ice team checked the shuttle and boosters three times—at 1:30 A.M., 7:00 A.M. and 11:00 A.M. They noted readings of seven and nine degrees in the righthand booster rocket, much lower than in the left booster—an indication that liquid hydrogen might have been leaking from the rocket. In fact, there were indications that temperatures on the strut connecting the rocket to the external fuel tank were as low as eight degrees below zero. All of this could have signaled danger, but the ice team's mandate was merely to check for excess ice, and so their infrared temperature readings were not reported to high-level officials who made the decision about lifting off.

At 9:07 A.M. the seven astronauts boarded the shuttle. The photograph that will remain as one of the public's permanently stored images of that day shows them smiling, cheery and waving. It was a historic occasion, and the excitement was written on each of their faces.

They climbed into the shuttle, arranged themselves at their stations and drew on extra gloves. It was cold in the

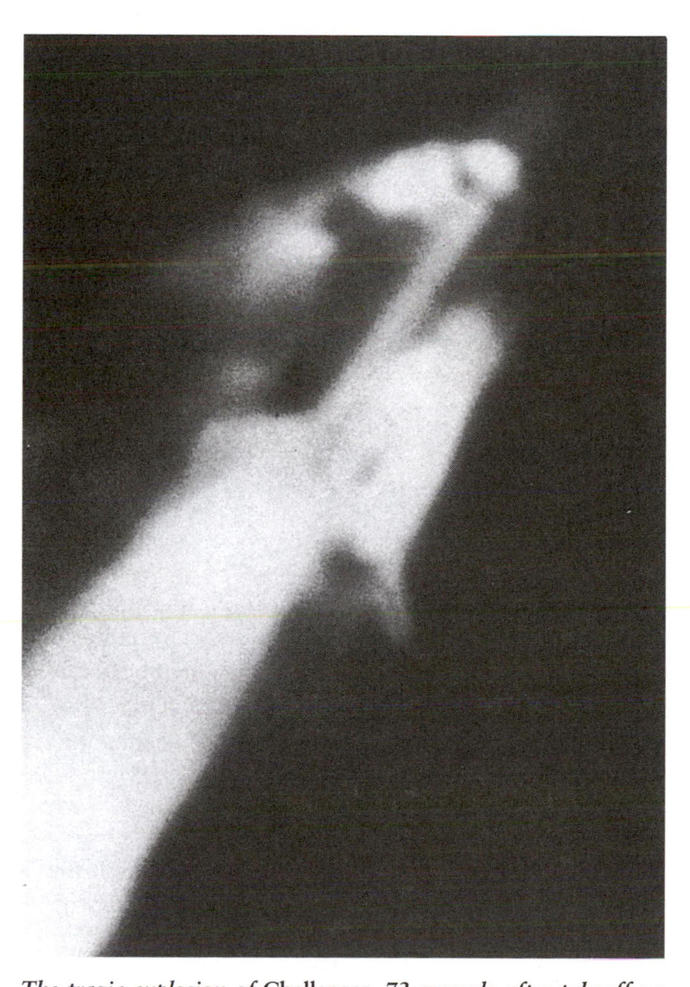

The tragic explosion of Challenger, *73 seconds after takeoff on January 28, 1986. The death of all seven crew members halted the U.S. space program for two and a half years.* (NASA)

spaceship. Shortly after they settled in place the astronauts were told that the liftoff had again been put on hold to wait for the sun to warm up and melt some of the ice on the capsule and rockets. The delay would last for two hours. The crew waited patiently—Francis R. (Dick) Scobee, the commander, and Commander Michael J. Smith at the controls, Dr. Judith Resnik in the center of the shuttle at the flight engineer position, Lieutenant Colonel Ellison Onizuka to her right, and in the mid-deck, Christa McAuliffe, Dr. Donald McNair and Gregory Jarvis.

At T minus nine minutes the countdown was resumed. It was 38 degrees on the launchpad, 15 degrees colder than it had been for any previous launching at the cape or anywhere else. At 6.6 seconds before liftoff, the *Challenger*'s three main engines roared to life. At zero, the two 149-foot booster rockets ignited.

"Liftoff," announced the commentator in Houston. "Liftoff of the 25th space shuttle mission, and it has cleared the tower."

Several thousand spectators, including family members of some of the crew, shivered and cheered as the boosters and the shuttle began their slow ascent into the heavens. It all looked so easy and smooth and impressive.

But less than a second into the flight, there was trouble. A puff of black smoke shot out of the lower part of the right booster at a location covered by a seal. And this was probably the cause of the coming tragedy. The puff of smoke went undetected at the time and was found only when photos were examined the next day.

Twelve to 13 seconds after liftoff, the smoke, which had spread and blackened, disappeared. The computers monitoring the mission registered no warnings or problems.

At 40 seconds into the flight, when the main engines had been throttled down to 65% thrust, the shuttle encountered heavy, shifting winds. It responded by automatically pivoting the booster and main engine nozzles to maintain the correct trajectory.

The ill-fated Challenger *crew (left to right, front row) Michael J. Smith, Francis (Dick) Scobee and Ronald McNair; (rear) Ellison Onizuka, Christa McAuliffe, Gregory Jarvis and Judith Resnik.* (NASA)

At 52 seconds the three engines began their steady throttling up to full power. "Challenger, go with throttle up," Mission Control radioed.

"Roger, go with throttle up," responded Scobee calmly.

At 59 seconds the *Challenger* reached its maximum dynamic pressure, when the vibrations of thrusting rockets, the momentum of the ascent and the force of wind resistance combined to exert incredible stresses on the shuttle structure. At this moment the O-rings would be tested to the extreme.

A new plume of smoke now appeared on the lower side of the right booster. The pressures, which should have been equal between the two boosters, started to diverge. The right booster's pressure dropped alarmingly, indicating a leak. The fire in the O-ring was being fed by escaping fuel. Both the primary and backup seals had ruptured, and at 73.175 seconds into the mission, there were flashes of light and a series of explosions. At 73.621 seconds there was a sudden surge of pressure in the main engines. Intense heat shut down one of them.

Now the superheated propulsive gases set off a chain of events that led to the explosion of propellants in the huge primary fuel tank. Flames from the leak severed the struts that held the rocket's base to the fuel tank. As the booster pivoted outward, its nose swung in and ruptured the tank, releasing its hydrogen, and a fierce explosion occurred. The shuttle and its rockets were consumed in an immense fireball.

At this moment the *Challenger* was about nine miles above Earth and seven miles out over the Atlantic Ocean from the Florida coast. Spectators looked on in horror as the shuttle, soaring so serenely against the crystalline blue of the morning sky, suddenly blew apart in a huge orange flash and then, trailing a white plume, arched over and fell back to Earth. On thousands of TV screens, including the ones in Christa McAuliffe's school, the identical, horrible scene was unfolding.

From Houston at Mission Control, there was a long, terrible pause. "Obviously a major malfunction," stated Stephen Nesbitt, the public relations officer describing the liftoff. "We have no downlink," he added, meaning that all communication with the *Challenger* had ceased. There was a long pause, and then Nesbitt came back on the line. "We have a report," he said, "from the flight dynamics officer that the vehicle has exploded."

Debris would rain down on the Atlantic for hours, making immediate salvage operations impossible. It would be March 10 before Navy divers would find the crew compartment, with its crew inside, in 100 feet of water, 15 miles northeast of Cape Canaveral. The cabin, it was learned, had remained intact until it hit the ocean, where it broke apart on impact. It is believed that the seven were alive and perhaps conscious during that long, nine-mile plunge to the surface of the ocean. It was theorized that the seven had met their deaths either through the shock of the initial blast, the sudden depressurization of the cabin or by the force of the tumbling, nine-mile descent.

In whatever way, they were gone, and the U.S. space program would be thrown into confusion and reassessment. It would be a long, long time, experts predicted, and take many, many safety precautions, before another *Challenger* would be launched from the pad at Cape Canaveral into another blue sky of another clear Florida morning.

USSR
April 23, 1967

A defective parachute caused the crash of Soyuz I *on April 23, 1967, in the USSR. Vladimir Komarov, its sole astronaut, was killed in the crash.*

Vladimir M. Komarov was a Muscovite who had served in the Soviet Air Force from the age of 15 but was almost dropped from the Soviet space program because of a heart murmur. The tall, dark-haired astronaut was a jet fighter pilot with a scholarly nature that had, in 1954, won him admission to the Zhukovsky Air Force Engineering Academy in Moscow.

His training as an aeronautical engineer made him an ideal candidate for space, and in the fall of 1964, he and two companions, scientist Konstantin P. Feoktistov and doctor Boris B. Yegrow, tested the first of the eight-ton Vosknod spaceships. The three made 16 orbits of Earth that fall, and Komarov proved that his physical disability had been either exaggerated or cured.

In 1967 Komarov was selected to test the *Soyuz I,* the first Soviet-launched spaceship in two years. It would make him the first Soviet astronaut to make two trips in space.

There had been tragedies and near tragedies in the space programs of both the United States and the USSR before this trip. Early in 1967 three U.S. astronauts had been killed in a fire during ground tests of the *Apollo* capsule (see p. 371). In March 1965 Colonel Pavel I. Belyayev, guiding the *Vosknod 2* to Earth, was suddenly faced with the failure of the spaceship's equipment for automatically controlled reentry into Earth's atmosphere. Fortunately, manual controls were available, and he was able to land it safely in a dense forest several hundred miles from the intended landing site.

As was the practice in the Soviet Union, Komarov's wife, Valentina, was not informed of her husband's mission to fly the *Soyuz I* until after it had been launched. "When Velodya goes away on an assignment, he doesn't say where he is going or why," she later related. "And this time as well, he just flew off. I felt, of course, that something was about to happen."

The launchpad from which the colonel and his *Soyuz I* craft were launched was only a few yards from an obelisk marking the site of the launching of *Sputnik I,* on October 4, 1957, at Baikonur, the Soviet Union's space center in Kazakhstan, 1,200 miles southeast of Moscow.

It was the practice of Soviet astronauts not to wear space suits in flight, and Colonel Komarov arrived on the morning of April 23 two hours before blastoff wearing a bright blue nylon pullover, sports trousers and light shoes.

The launching was a textbook one, and the colonel entered orbit easily. After the first circle of Earth, he reportedly radioed to ground control a message that sounds anything but spontaneous: "This ship is a major creative achievement of our designers, scientists, engineers and workers," he said. "I am proud that I was given the right to be the first to test it in flight."

U.S. tracking stations were monitoring the experiment closely, and partway into the flight they were aware that all was not well aboard the *Soyuz I.* It was not responding as easily as the colonel's sanguine statement indicated, and it was clear, according to U.S. space officials, that the flight would not extend beyond 24 hours.

The experiment had actually included the launching of a second space vehicle to dock with the colonel's ship, but that was apparently scrubbed early in the colonel's flight.

Real problems began to develop when Colonel Komatov began reentry procedures. According to U.S. observers, he tried to bring his ship in on the 16th orbit, after 24 hours of flight, but he was unable to do so because he could not maneuver it properly to fire the braking rockets.

The colonel and his craft circled the world twice more while he fought to control the tumbling spaceship. Finally, on the 18th orbit, the retrorockets fired, and he appeared to be coming in successfully. "Well done!" was the cry from ground control.

"Everything is working fine," replied Komarov, who was over Africa at this moment.

When the spaceship reached 23,000 feet, its landing parachute was to be deployed. But the frantic tumbling in outer space had taken its toll. The parachute lines had become hopelessly tangled, and the parachute did not open. And the *Soyuz I* streaked toward Earth at a frightening rate of speed. Colonel Komarov struggled to regain control of the ship, but there was no backup for an unopened parachute. He died on impact.

Moscow mourned; thousands passed his bier, which was placed on public display. Two days later, in a ceremony attended by every top dignitary in the Soviet government, his ashes were placed in an urn, and he was buried in the Kremlin wall. The Russian space program would be set back another year by this.

USSR
June 30, 1971

A faulty valve or hatch seal caused a sudden decompression during the landing of the Soyuz II *spacecraft in the USSR on June 30, 1971. All three astronauts aboard were killed.*

On June 5, 1971, the *Soyuz II* spacecraft with Lieutenant Colonel Georgi T. Dobrovolsky, Vladislav N. Volkov and Viktor I. Patsayev aboard was launched into orbit. Its mission although heavily censored by the Soviets, was thought to be to attempt a linkup with the unmanned *Salyut* space station and to conduct some secret, in-space experiments—which included a study of Earth from space and an examination of the long-term effects of weightlessness on human beings.

The blastoff was uneventful, and the astronauts docked at the *Salyut* station smoothly. First, the docking mechanism was locked, and pressure was equalized between the two vehicles. Then, floating through the hatch of the docking tunnel, Viktor I. Patsayev, the 37-year-old civilian test engineer who had been trained specifically for work on orbital stations, entered the *Salyut* laboratory.

While Mr. Patsayev began to connect electrical and hydraulic lines between the craft, the other engineer, Vladislav Volkov, followed him through the crew transfer tunnel. The *Soyuz* commander, Lieutenant Colonel Dobrovolsky, at first stayed in the ferry craft to handle communications with ground control while the systems on the *Salyut* were connected. Finally, he, too, swam through the air of the crew tunnel.

Once abroad, all of the men reported that they were in good spirits and then began their task of setting up an outer space experimental laboratory, the first of its kind in or out of this world.

There was great optimism on the part of Soviet leaders. Premier Aleksei Kosygin said, to them and the world, "We express confidence that you will cope well with this responsible and complex assignment, whose fulfillment will be a major contribution to implementation of plans for developing space for the good of the Soviet people and the whole of mankind." The laboratory was soon in place, in a 55,000-pound space station orbiting around Earth every 88.2 minutes. The astronauts' habitat was not exactly grand; there was a pressurized module of the station, including its crew quarters of 3,500 cubic feet, or the size of a 40-foot house trailer. In addition, there was a service module with a propulsion system for orbital corrections linked by hydraulic lines.

Once establishing themselves, the three space explorers settled down to a daily routine. Although, as usual, very few details were conveyed to the rest of the world, it seemed that this was going to be a sustained mission. This was confirmed when the crew was ordered to place the station into a higher orbit, which would reduce the pull of the rarefied atmosphere and prolong the lifetime of the station. The station was amply stocked with food, treated for preservation over a long period of time. On June 9, four days into the mission, Lieutenant Colonel Dobrovolsky conducted an in-space fashion show, floating upside down and pedaling the air and demonstrating for TV a new tension suit designed to keep muscles in condition on long spaceflights.

Once more, the orbit was raised, and it was obvious that the Soviets were attempting to set an in-space endurance record. For the next few days the astronauts followed a strict regimen: Following breakfast and their usual morning exercises, they took samples of their own blood, ran a set of cardiovascular tests and checked the calcium content in their bones—an important experiment, since weightlessness causes a washing out of calcium from the bones. The afternoons were devoted to experiments with a gamma telescope to study cosmic rays and to make spectrographic studies of Earth's natural formations.

The experiment lengthened enough to be relegated to the back pages of the world's newspapers and disappear entirely from TV news broadcasts. By June 27 the three had lived in space for 23 days, five days beyond the previously known capacity of a person to endure weightlessness in spaceflight and record normal pulse, blood pressure and respiration rate.

Soviet doctors continued to monitor the men. "Each day is now a step into the unknown," said one doctor.

And then it was time to come home. Tass, the Soviet news agency, reported that on June 29 the order was given to return, and the three astronauts "transferred the materials of scientific research and the logs" to the *Soyuz II* for return to Earth.

"After completing the transition operation, the astronauts took their seats in the *Soyuz II* ship, checked the onboard systems and prepared the ship for unlinking from the *Salyut* station," the communiqué continued.

At 9:28 P.M. Soviet time the two vehicles separated. "The crew of the *Soyuz II*," said Tass, "reported to Earth the unlinking operation passed without a hitch and all the systems were functioning normally."

At 1:35 A.M., June 30, the *Soyuz II* reentered Earth's atmosphere. "Its braking engine was fired and functioned throughout the estimate time," Tass went on. "Communication with the crew ceased according to the set program."

The spacecraft braked, the parachute system was put into motion and the soft-landing engines were fired. "The flight of the descending apparatus ended in a smooth landing in the pre-set areas," Tass said.

And now, the tragic ending began to unfold. "Landing simultaneously with the ship, a helicopter-borne recovery group, upon opening the hatch," said Tass dispassionately, "found the crew of the *Soyuz II* spaceship . . . in their seats, without any signs of life." They had died, apparently on reentry, without a word to ground control about its happening or its causes.

A board of inquiry looked into the disaster, and the mystery was soon solved. At the moment they landed, a faulty valve or a hatch seal fault in the crew module had caused sudden decompression, and all three had died instantly from the abrupt change in pressure.

The tragedy would stall the Soviet space program for a full two years. Coming as it did after the death of Colonel Vladimir Komarov (see p. 375), it was a stunning, staggering blow to world public confidence in the possibility of probing outer space safely.

BIBLIOGRAPHY

Bonillo, Denise, ed. *School Violence,* New York: H. W. Wilson Co., 2000.

Brown, Walter R. and Norman D. Anderson. *Fires.* Reading, Mass.: Addison-Wesley, 1976.

Brown, Walter R., Billye W. Cutchen and Norman D. Anderson. *Catastrophes.* Reading, Mass.: Addison-Wesley, 1979.

Bultel, Hal. *Inferno!* Chicago: Henry Regnery Co., 1975.

Butler, Joyce. *Wildfire Loose.* Kennebunkport, Me.: Durrell Publications, 1987.

"Captain X." *Safety Last: The Dangers of Commercial Aviation.* New York: Dial Press, 1972.

Carlson, Kurt. *One American Must Die.* New York Congdon and Weed, 1986.

Clarke, James W. *American Assassins.* Princeton, N.J.: Princeton University Press, 1982.

Clive, Irving, ed. *In Their Name—The Oklahoma City Official Commemorative Volume.* Oklahoma City: Oklahoma Publishing Company, 1995.

Collins, Cindy. *Humanitarian Challenges and Intervention: World Politics and the Dilemmas of Help.* Chicago: Westview Press, 2000.

Davie, Michael. *Titanic.* New York: Alfred Knopf, 1987.

Demaris, Ovid. *Brothers in Blood: The International Terrorist Network.* New York: Charles Scribner's Sons, 1977.

Dobkin, Marjorie Housepian. *Smyrna 1922.* Kent, Ohio: Kent State University Press, 1988.

Dobson, Christopher and Ronald Payne. *The Never Ending War: Terrorism in the '80s.* New York: Facts On File, 1987.

———. *The Terrorists.* New York: Facts On File, 1982.

Dunbar, Seymour. *A History of Travel in America.* Indianapolis: Bobbs-Merrill, 1946.

Dwyer, Jim, David Kocieniewsky, Deidre Murphy and Peg Tyre. *Two Seconds under the World.* New York: Crown Publishers, 1999.

Eddy, Paul, Elaine Potter and Bruce Page. *Destination Disaster: From the Tri-Motor to the DC-10: The Risk of Flying.* New York: Quadrangle, 1976.

Edwardes, Michael. *British India.* New York: Taplinger Publishing Co., 1967.

Emerson, Steven and Brian Duffy. *The Fall of Pan Am 103.* New York: G. P. Putnam's Sons, 1990.

Farrington, S. Kip, Jr. *Railroading around the World.* New York: Castle Books, 1955.

Gadney, Reg. *Cry Hungary! Uprising 1956.* New York: Atheneum, 1986.

Garrison, Webb. *Disasters That Made History.* New York: Abingdon Press, 1973.

Godson, John. *Unsafe at Any Height.* New York: Simon and Schuster, 1970.

Grayland, Eugene C. *There Was Danger on the Line.* Auckland, New Zealand: Belvedere, 1954.

Groenewald, Gulia, ed. *World in Crisis: The Politics of Survival at the End of the 20th Century.* New York: Routledge, 1996.

Gunston, Bill, ed. *Aviation Year by Year.* New York: DK Publishing, 2001.

Hamilton, James A. B. *British Railway Accidents of the Twentieth Century.* London: Unwin, 1967.

Hamlyn, Paul. *Railways.* London: Hamlyn Publishing Group Ltd., 1970.

Hartunian, Abraham H. *Neither to Laugh nor to Weep: A Memoir of the Armenian Genocide.* Boston: Beacon Press, 1968.

Hooper, Finley. *Roman Realities.* Detroit: Wayne State University Press, 1979.

Howard, Roger. *Great Escapes and Rescues: An Encyclopedia.* New York: ABC-CLIO, 1996.

Howland, S. A. *Steamboat Disasters and Railroad Accidents in the United States.* Worcester, Mass.: Dorr, 1846.

Hyams, Edward. *Terrorists and Terrorism.* New York: St. Martin's Press, 1974.

Hyde, George E. *A Life of George Brent.* Norman, Okla.: University of Oklahoma Press, 1968.

International Federation of Red Cross and Red Crescent Societies. *World Disasters Report 1997.* New York: Oxford University Press, 1997.

Jerrome, Edward G. *Tales of Railroads.* Belmont, Calif.: Fearon Pittman Publishers, 1959.

Johnson, Thomas P. *When Nature Runs Wild.* Mankato, Minn.: Creative Education Press, 1968.

Kelner, Joseph, with James Munves. *The Kent State Coverup.* New York: Harper and Row, 1970.

Kennett, Frances. *The Great Disasters of the 20th Century.* London: Marshall Cavendish Books Ltd., 1981.

Larimer, J. McCormick. *The Railroad Wrecker.* Muskogee, Okla.: Muskogee Press, 1909.

Lattimer, John H. *Kennedy and Lincoln.* New York: Harcourt Brace Jovanovich, 1980.

Lenz, Harry M. III. *Assassinations and Executions, An Encyclopedia of Violence, 1865–1986.* New York: McFarland, 1988.
Longstreet, Stephen. *City on Two Rivers; Profiles of New York—Yesterday and Today.* New York: Hawthorn Publishers, 1975.
Marshall, John. *Rail Facts and Feats.* New York: Two Continents Publishing Group, 1974.
Marx, Joseph Laurence. *Crisis in the Skies.* New York: David McKay, 1970.
Matthews, Rupert. *The Fire of London.* New York: The Bookwright Press, 1989.
McClement, Fred. *Anvil of the Gods.* New York: J. B. Lippincott, 1964.
———. *It Doesn't Matter Where You Sit.* New York: Holt, Rinehart & Winston, 1969.
McKee, Alexander. *Dresden, 1945: The Devil's Tinderbox.* New York: E. P. Dutton, 1984.
Medvedev, Zhores. *Nuclear Disaster in the Urals.* New York: Vintage, 1980.
Meltzer, Milton. *The Terrorists.* New York: Harper and Row, 1983.
Michener, James. *Kent State; What Happened and Why.* New York: Random House, 1971.
Miller, Ducas, ed. *Rescue: Stories of Survival from Land and Sea.* San Francisco: Thunder's Mouth Press, 1997.
Morris, John V. *Fires and Firefighters.* New York: Bramhall House, 1955.
Mould, Richard. *Chernobyl Record: The Definitive History of the Chernobyl Catastrophe.* Boston: Institute of Physics Pub, *2000.*
Nash, Jay Robert. *Darkest Hours.* New York: Nelson-Hall, 1976.
Negroni, Christina. *Deadly Departure: Why the Experts Failed to Prevent the TWA Flight 800 Disaster, and How It Could Happen Again.* New York: HarperCollins, 2000.
Nock, Oswald, S. *Historic Railway Disasters.* London: Allan, 1966.
Obenzinger, Hilton. *New York on Fire.* Seattle: Real Comet Press, 1989.
Pomerantz, Gary. *Nine Minutes, Twenty Seconds: The Tragedy and Triumph of ASA Flight 529.* New York: Crown Publishers, 2001.
Pryce-Jones, David. *The Hungarian Revolution.* New York: Horizon Press, 1970.
Reed, Robert C. *Train Wrecks.* New York: Bonanza Books, 1968.
Rolt, Lionel T. *Red for Danger.* London: Bodley Head, 1955.
Rothberg, Robert, ed. *From Massacres to Genocide: The Media, Public Policy and Humanitarian Crises.* Washington, D. C.: Brookings Institute, 1998.
Sanders, James, *The Downing of TWA Flight 800.* New York: Zebra Books, 1997.
Sayre, Nora. *Sixties Going on Seventies.* New York: Arbor House, 1973.
Sobel, Lester A., ed. *Political Terrorism.* New York: Facts On File, 1975.
Soboul, Albert. *The French Revolution, 1787–1799.* New York: Vintage, 1975.
Sterling, Claire. *The Terrorism Network.* New York: Holt, Rinehart and Winston, 1981.
Stover, John F. *American Railroads.* Chicago: Chicago University Press, 1961.
Wasserman, Harvey, et al. *Killing Our Own.* New York: Delacorte Press, 1982.
With, Emile. *Railroad Accidents.* Boston: Little Brown, 1856.

ELECTRONIC SOURCES

"Air France Concorde Begins Training" 2 Sept 2001 <http//dailynews.yahoo.com>

"Amid Tragedy, She Stayed Alive" 10 Feb 1999, 25 Sept 2001 <http://library.newsday.com>

"At Least 80 Killed in Indonesian Train Crash" 2 Sept 2001, 30 Sept 2001 <http://www.cbc.ca/cp/world>

"At Least 80 Die in German Train Crash" 4 June 1998, 28 Sept 2001 <http://nytimes.com/qpass-archives>

"At Least 40 Die in Collision on Railroad in Indonesia" 2 Sept 2001, 28 Sept 2001 <http://nytimes.com/qpass-archives>

"Authorities Focus on Trucker" 18 Mar 1999, 30 Sept 2001 <http://archive.abcnews.go.com/sections/us/DailyNews>

"British Train Crash Toll 70" 7 Oct 1999 28 Sept 2001 <http://nytimes.com/qpass-archives>

"China Theatre Fire Kills 300" 8 Dec 1994, 22 Sept 2001 <http://library.newsday.com/cgi-bin/display>

"China Jails 23 over Disco Fire" 22 Aug 2001, 21 Sept 2001 <http://news.bbc.co.uk/hi/english/world/asia-pacific/newsid>

"Christmas Night Fire Kills 311 in Central China" 28 Dec 2000, 20 Sept 2001 <http://www.wsws.org/articles/2000/dec.2000/chin-d28.shtml>

"Chronology of LAPD Officers' Trials" 9 Sept 2001 <http://www.law.umkc.edu/faculty/projects/ftrials/lapd/kingchronology.html>

"Clues to Gulf Air Crash" 25 Aug 2000, 29 Aug 2001 <http://news.bbc.co.uk/hi/english/world/middle_east/newsid.stm>

"Co-Pilot Chanted 'I Rely on God' . . ." 12 Aug 2000, 1 Sept 2001 <http://www.washingtgtonpost.com/ac2/wp-dyn/A12845-2000Aug11>

"Cold Comfort of Hot-Headed Enemies, The" 3 Aug 1997, 5 Sept 2001 <http://nytimes.com/qpass-archives>

"Columbine Tragedy, The" 27 April 1999, 12 Sept 2001 <http://denver.rockymountainnews.com/drmn/columbine>

"Concorde Accident Report" 25 July 2000, 20 Aug 2001 <http://dnausers.d-n-a.net/dnetGOjg/250700.htm>

"Concorde Crash Fiery End" 26 July 2000, 2 Sept 2001 <http://www.crashpages.com/fc/af-concorde/fieryend/downinflames.shtml>

"Couple's Escape, A" 29 Sept 2000, 26 Sept 2001 <http://www.library.newsday.com>

"Death Toll in Amtrak Collision Reaches 13" 18 Mar 1999, 30 Sept 2001 <http://www.disasterrelief.org/Disasters/990316Amtrak2/>

"EgyptAir Crash Report Puts Spotlight on Pilot" 19 April 2001, 31 Aug 2001 <http://www.cnn.com/2001/US/04/19/egypt.crash/index.html?s=2>

"Eighty Killed as Two Blasts Rip Through Coal Mine" 13 Mar 2000, 19 Sept 2001 <http://www.smh.com.au/news.0003/13/world/world08.html>

"Evidence Disputed" 17 Dec 2000, 31 Aug 2001 <http://abcnews.go.com/sections/is/DailyNews/EgyptAir000811.html>

"Explosions Rip Mexican City . . ." 22 April 1992, 19 Sept 2001 <http://www.emergency.com/mxcoblst.htm>

"Familiar Agony: Carnage in a Market Lane" 31 July 1997, 5 Sept 2001 <http://nytimes.com/qpass-archives>

"Fatalities in H.S. Attack" 20 April 1999, 12 Sept 2001 http://abcnews.go.com/sections/us/DailyNews/denver990420.html>

"Ferry Crew Charged with Manslaughter" 28 Sept 2000, 27 Sept 2001 <http://europe.cnn.com/2000/WORLD>

"Ferry Survivors Say Tanzanian Vessel Was Overloaded" 23 May 1996, 26 Sept 2001 <http://nytimes.com/qpass-archives>

"Fire, Stampede Claim 45 Lives at Bangladesh Garment Factory" 27 Nov 2000, 22 Sept 2001 <http://www5.cnn.com/2000/ASIANOW/south>

"Fortier Resumes Talks with Prosecution" 20 June 1995, 11 Sept 2001 <http://www.cnn.com/US/OLC/faces/Suspects/McVeigh/fortier-talk6-20index.html>

"Forty Killed in Indonesian Train Collision" 2 Sept 2001, 30 Sept 2001 <http://www.usatoday.com/news/world>

"Greek Ferry Crew Speak of Grief" 1 Oct 2000, 27 Sept 2001 <http://www.cnn.com/2000.WORLD>

"Gulf Air Begins Compensation . . ." 22 Aug 2001, 29 Aug 2001 <http://cnn.worldnews.printthis>

"Gunmen Planned Bigger Attack" 26 April 1999, 12 Sept 2001 <http://abcnews.go.com/sections/us/DailyNews/littleton_main990426.html>

"Inadequate Safety Measures Behind Rail Disasters in India" 16 Aug 1999, 30 Sept 2001 <http://www.wsws.org/articles/1999>

"India Resumes Search for Rail Victims" 4 Aug 1999, 30 Sept 2001 <http://www.cnn.com/ASIANOW/south>

"Indian Train Crash Kills 51" 7 Jan 1998, 28 Sept 2001 <http://nytimes.com/qpass-archives>

"Indian Train on Wrong Track" 3 Aug 1999, 30 Sept 2001 <http://www7.cnn.com/ASIANOW/south>

"Kenya Airways Crash" 30 Jan 2000, 20 Aug 2001 <http://dnausers.d-n-a.net/dnetGOjg/300100.htm>

"Killers' Co-Worker Turns Himself In" 2 May 1999, 12 Sept 2001 <http://denver.rockymountainnews.com/shooting/0618arst1.shtml>

"Klebold's Date Bought 2 Weapons" 1 May 1999, 12 Sept 2001 <http://denver.rockymountainnews.com/shooting/0437scho0.shtml>

"Locked Gates Prevented Workers Escaping Death" 30 Nov 2000, 22 Sept 2001 <http://www.wsws.org/articles/2000/nov2000/fire-n30.shtml>

"Massive Probe Begins Into Deadly Japan Blaze" 1 Sept 2001, 4 Sept 2001 <http://www.nytimes.com/2001/09/01/international/01WIRE-JAPA.html>

"Mastermind in Trade Center Bombing to Be Sentenced" 8 Jan 1998, 7 Sept 2001 <http://europe.cnn.com/US/9801/08/yousef/index.html>

"Mine Explosion in China Claims 32 Lives" 14 Dec 1998, 19 Sept 2001 <http://www10.cnn.com/WORLD/asiapcf/9812/14/china.mineblast/index.html>

"Missing Money, Missing Lives: . . . Zhili Fire Incident" Jan 1999, 20 Sept 2001 <http://www.hrichina.org/crf/english/00winter/00W14_Zhili%Fire.html>

"Nine Officials Charged in Sewer-Line Explosions Case" 22 April 1992, 20 Sept 2001 <http://www-tech.mit.edu/V112/N23/sewer,23w.html>

"No Survivors from Gulf Air Crash" 24 Aug 2000, 29 Aug 2001 <http://news.bbc.co.uk/hi/english/word/middle_east/newsid_893000/893877.stm>

"NTSB: Last Seconds of Data Pulled from EgyptAir Flight Recorder" 18 Nov 1999, 24 Aug 2001 <http://www.cnn.com/US/9911/17/egyptair.05/>

"Oklahoma City Bombing" 21 April 1995, 20 Aug 2001 <http://www.fema.gov/OKC95/okc026.txt>

"Oklahoma City Bombing Trial" 1995 11 Sept 2001 <http://www.cnn.com/US/9703.okc.trial/>

"Oklahoma Prosecutor to Seek Death for Bombing" 6 Sept 2001, 12 Sept 2001 <http://www/nytimes.com/2001/09/06/national>

"One Hundred Forty-three are Reported Dead in Crash of Russian Airplane in Siberia" 4 July 2001, 2 Sept 2001 <http://nytimes.com/qpass-archives>

"Philippines Air Crash" 19 April 2000, 20 Aug 2001 <http://dnausers.d-n-a.netGOjg/190400.htm>

"Report of Accountability Review Boards, Kenya-Nairobi Bombings" 7 Aug 1998 <http://www.state.gov/www/regions/africa/board/nairobi.html>

"Rescuers Search Indonesian Air Crash Site" 28 Sept 1997, 31 Aug 2001 <http://nytimes.com/qpass-archives>

"Runway Mistake Suspected in Taiwan Jet Crash" 3 Nov 2000, 29 Aug 2001 <http://nytimes.com/qpass-archives>

"Singapore Airlines Crash" 31 Oct 2000, 2 Sept 2001 <http://www.aviation-safety.net/database/2000/001031-0.htm>

"Swissair Plane Crashes Off Coast of Nova Scotia" 3 Sept 1998, 24 Aug 2001 <http://www.cnn.com/WORLD/americas/9809/03/swissair.crash.04/>

"Terrifying Time on the Open Sea, A" 12 March 2001, 25 Sept 2001 <http:// www.library.newsday.com>

"The Colorado Tragedy" 22 April 1999, 15 Aug 2001 <http://dialog.carl.org>

"The Crash of ValuJet Flight 592" 26 May 1996, 24 Aug 2001 <http://www.cnn.com/US/valujet.592/index.html>

"The Concord Crash . . ." 26 Jul 2000, 29 Aug 2001 <http://nytimes.com/qpass-archives>

"Three Days of Hell in Los Angeles" 29 April 1992, 9 Sept 2001 <http:/www.emergency.com/la-riots.htm>

"Toll Rises to 35 in Mount Blanc Tunnel Fire" 27 Mar 1999, 22 Sept 2001 <http://nytimes.com/qpass-archives>

"Trash May Have Sparked Tokyo Blaze" 4 Sept 2001, 4 Sept 2001 <http://www.nytimes.com/apolnline/international/AP-Japan-Fire.html>

"Trial of the Sheik" 20 Jan 1995, 8 Sept 2001 <http://inic.utexas.edu/menic/utaustin/course/oilcourse/mail/usa/0008.html>

"Two Hundred Thirty-three Die in China Dance Hall Fire" 30 Nov 1994, 22 Sept 2001 <http://library.newsday.com/cgi-bin/display>

"Ukraine's Troubled Mines" 12 Mar 2000, 19 Sept 2001 <http://news.bbc.co.uk/low/english/world/europe/newsid_674000/674542.stm>

"World Trade Center Bombing" 3 March 1993, 17 Sept 2001 <http://www.milnet.com/milnet/wtc.htm>

"Worst Terrorist Attack on U.S. Soil . . ." 30 Dec 1995, 11 Sept 2001 <http://www.cnn.com/US?OKC/deaily/9512/12-30/index.html>

"Zhili Fire—Wounds Are Never Healed" Dec 1999, 21 Sept 2001 <http://www.cic.org.hk/ce_99dec.htm>

INDEX

Italic page numbers indicate illustrations. **Boldface** numbers indicate airline flight numbers.

B

D

E

F

G

H

I

J

K

L

M

Q

R

S

T

U

V

Z